WATER DEFICITS AND PLANT GROWTH

VOLUME VI

Woody Plant Communities

CONTRIBUTORS TO THIS VOLUME

H. E. BARRS

B. CALLANDER

W. R. CHANEY

D. DOLEY

F. DUHME

T. M. HINCKLEY

P. R. JARVIS

H. G. JONES

P. E. KRIEDEMANN

J. J. LANDSBERG

S. G. PALLARDY

H. RICHTER

G. R. SQUIRE

R. O. TESKEY

D. WHITEHEAD

WATER DEFICITS
AND PLANT GROWTH

EDITED BY

T. T. KOZLOWSKI

DEPARTMENT OF FORESTRY
UNIVERSITY OF WISCONSIN
MADISON, WISCONSIN

VOLUME VI

Woody Plant Communities

1981

ACADEMIC PRESS
A Subsidiary of Harcourt Brace Jovanovich, Publishers

New York   London   Toronto   Sydney   San Francisco

ACADEMIC PRESS, INC.
111 Fifth Avenue, New York, New York 10003

United Kingdom Edition published by
ACADEMIC PRESS, INC. (LONDON) LTD.
24/28 Oval Road, London NW1   7DX

Library of Congress Cataloging in Publication Data
Main entry under title:

Water deficits and plant growth.

  Includes bibliographies and indexes.
  Contents: v. 1. Development, control, and
measurement--v. 2. Plant water consumption and
response--[etc.]--v. 6. Woody plant communties.
  1. Plants--Water requirements. 2. Growth
(Plants). 3. Plant-water relationships. 4. Plant
diseases. I. Kozlowski, T. T. (Theodore Thomas),
Date.
QK870.W38   582'.013   68-14658
ISBN 0-12-424156-5 (v. 6)   AACR2

CONTENTS

1. SOURCES OF WATER

WILLIAM R. CHANEY

2. CONIFEROUS FORESTS AND PLANTATIONS

D. WHITEHEAD AND P. G. JARVIS

3. TEMPERATE HARDWOOD FORESTS

T. M. HINCKLEY, R. O. TESKEY, F. DUHME, AND H. RICHTER

8. CLOSELY RELATED WOODY PLANTS

STEPHEN G. PALLARDY

LIST OF CONTRIBUTORS

Numbers in parentheses indicate the pages on which the authors' contributions begin.

H. D. BARRS (325), Division of Irrigation Research, C.S.I.R.O., Griffith, New South Wales, Australia

B. CALLANDER (471), Department of Physiology and Environmental Studies, School of Agriculture, University of Nottingham, Sutton Bonington, Loughborough, Leicestershire, England

WILLIAM R. CHANEY (1), Department of Forestry and Natural Resources, Purdue University, West Lafayette, Indiana 47907

D. DOLEY (209), Botany Department, University of Queensland, St. Lucia, Queensland, Australia

F. DUHME (153), Lehrstuhl für Landschaftsökologie, Technische Universität München, D-8050 Freising-Weihenstephan, Federal Republic of Germany

T. M. HINCKLEY* (153), School of Forestry, Fisheries and Wildlife, University of Missouri, Columbia, Missouri 65211

P. G. JARVIS (49), Department of Forestry and Natural Resources, University of Edinburgh, Edinburgh EH9 3JU, Scotland

H. G. JONES (419), East Malling Research Station, East Malling, Maidstone, Kent, England

P. E. KRIEDEMANN (325), Division of Irrigation Research, C.S.I.R.O., Griffith, New South Wales, Australia

J. J. LANDSBERG (419), Long Ashton Research Station, Long Ashton, Bristol, England

* Present Address: College of Forest Resources, University of Washington, Seattle, Washington 98195

STEPHEN G. PALLARDY (511), School of Forestry, Fisheries and Wildlife, University of Missouri, Columbia, Missouri 65211

H. RICHTER (153). Botanisches Institut, Universität für Bodenkultur, Vienna, Austria

G. R. SQUIRE (471), Department of Physiology and Environmental Studies, School of Agriculture, University of Nottingham, Sutton Bonington, Loughborough, Leicestershire, England

R. O. TESKEY* (153), School of Forestry, Fisheries and Wildlife, University of Missouri, Columbia, Missouri 65211

D. WHITEHEAD (49), Forest Research Institute, New Zealand Forest Service, Private Bag, Rotorua, New Zealand

* Present Address: College of Forest Resources, University of Washington, Seattle, Washington 98195

PREFACE

The emphasis in this volume is on the water relations of woody plants in a community context. The ecosystems discussed include forest trees as well as trees grown for fruit crops and leaf crops.

The opening chapter presents a quantitative overview of sources of water available to woody plants. Separate chapters follow on water relations of coniferous forests, temperate hardwood forests, tropical and subtropical forests and woodlands, closely related woody plants, apple orchards, citrus orchards, and tea plantations.

For each of the plant communities discussed, emphasis is given to hydrological cycles; water use and transpiration; absorption of water; effects of environmental factors on soil and plant water balance; effects of water deficits on physiological processes; vegetative and reproductive growth; yield of harvested products; drought resistance; and cultural practices affecting plant water balance and yield. The subject matter is sufficiently varied so as to make this volume useful to both researchers and those involved in the practice of growing woody plants for wood and fruit crops and for their esthetic values. In this volume water deficits are characterized in bars or megapascals (Mpa) depending on preferences of individual authors (1 bar = 0.1 Mpa).

The contributors to this volume were chosen for their research productivity and extensive experience with the particular plant communities discussed. I wish to express my sincere thanks to each author for his scholarly contribution as well as cooperation during the production phases.

T. T. KOZLOWSKI

CONTENTS OF OTHER VOLUMES

Volume V. Water and Plant Disease

CHAPTER 1

SOURCES OF WATER

William R. Chaney

DEPARTMENT OF FORESTRY AND NATURAL RESOURCES, PURDUE UNIVERSITY,
WEST LAFAYETTE, INDIANA

I. INTRODUCTION

Of the myriad of environmental factors influencing distribution and growth of woody plants, water is of paramount importance and is usually the most limiting throughout the world. This situation exists even though global supplies of water are immense. However, the interaction of air circulation patterns, topography, temperature, and edaphic factors result in uneven distribution and availability of this moisture. Consequently, the forms of moisture utilized by trees, and the adaptations necessary for its utilization, are quite diverse.

This chapter will consider the use and importance to woody plants of moisture that occurs in the atmosphere, the soil, and internal tissues. The physical, anatomical, and physiological factors that affect the availability and absorption of the various forms of moisture will be emphasized.

WORLD WATER RESOURCES

Of the world's total water supply, only an extremely small part is available to woody plants during any one year. Approximately 93% of the earth's 165 trillion acre-feet of water is in the oceans and seas. The re-

1

Water Deficits and Plant Growth, Vol. VI
Copyright © 1981 by Academic Press, Inc.
ISBN 0-12-424156-5

TABLE I
ESTIMATED QUANTITIES OF WATER IN THE EARTH'S HYDROSPHERE[a]

Source	Quantity (billion acre-feet)
Total water	165,000
Total fresh water	11,000
Groundwater to 12,500 feet	8,200
Lakes and streams	118
Atmosphere	12
Soil moisture	6.5
Plants and animals	0.9
Annual precipitation	89
Annual runoff	17

[a] Adapted from Russell and Hurlbut (1959).

maining 7% is fresh water and consists primarily of groundwater with lesser portions in polar ice, glaciers, lakes, rivers, the atmosphere, and soil. A summary of estimated quantities of water in various segments of the earth's hydrosphere is shown in Table I.

The part of the hydrologic cycle of greatest importance to woody plants is the annual precipitation cycle. Approximately 89 billion acre-feet of water fall on the land surface of the world each year (Russell and Hurlbut, 1959). The common precipitation forms are rain, drizzle, snow, sleet, and hail. Although fog, dew, and frost may contribute significantly to the transfer of moisture from the atmosphere to the earth's surface, they are not considered to be precipitation (Critchfield, 1966). Except for the relatively small amounts of water extracted from the groundwater reserves by deep-rooted species, practically all water used by woody plants is identified with the annual precipitation cycle and involves the use of relatively short-term, low-capacity storage media like the atmosphere, the soil, and plant tissues. There is almost twice as much water in the atmosphere as in the soil, and a significant amount of water exists in tissues of plants (Table I).

The worldwide distribution of mean annual precipitation is shown in Fig. 1. On land the greatest amounts of precipitation occur in equatorial latitudes, decreasing toward the poles with a great deal of irregularity imposed by land forms and bodies of water. Although annual precipitation provides insight into the volume of water moving in the hydrologic cycle, it does not directly relate to the moisture available to trees because trees do not utilize precipitation per se; they use water found in the soil and atmosphere. This quantity is a function not only of precipitation, but of condensation, evapotranspiration, runoff, and percolation into the ground water system.

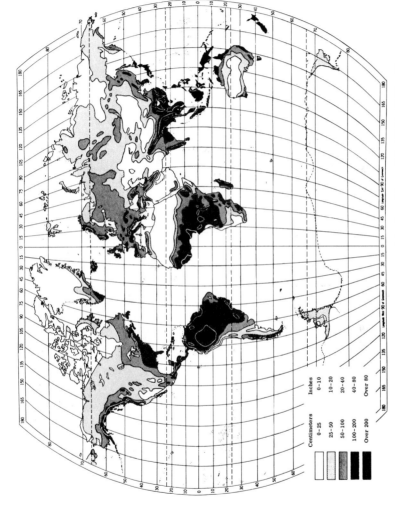

Fig. 1. World distribution of mean annual precipitation. (From Critchfield, 1966, with permission of A. J. Nystrom and Company.)

As important as the precipitation totals accumulated each year, are the characteristics of seasonal distribution and frequency that can significantly influence the availability of moisture to woody plants. Moisture in the atmospheric phase of the hydrologic cycle that condenses as fog or dew also can be very important to trees. Figure 2 shows the mean annual number of days with fog in various parts of the world. Atmospheric moisture is important to some woody plants only seasonally when precipitation is limited, but in some areas it constitutes the primary source of moisture.

II. SOURCES OF MOISTURE FOR WOODY PLANTS

Amounts of water available to woody plants are determined by climatic conditions, soil characteristics, depth to which plant roots extend, proliferation and density of root systems, morphology of shoot systems, spatial relationships among plants, and internal storage and transport. Soil moisture provides the bulk of water utilized by woody plants, but critically important amounts of moisture in the form of dew and fog often are absorbed through the foliage. Internal redistribution of water among tissues and organs also can be important when soil and atmospheric moisture is limited.

A. ATMOSPHERIC MOISTURE

1. Fog

In regions of the world where fog occurs regularly, tree or shrub vegetation may exist even when precipitation records suggest that moisture would be a limiting factor. Fogs and mists are clouds in contact with the ground and contain small water droplets that do not settle on horizontal surfaces and thus do not register in a rain gauge. However, when fog moves horizontally and contacts branches and foliage, the small droplets are deposited and in time combine to form larger drops that run off and fall to the ground. This phenomenon has been referred to as "fog drip," "cloud drip," "occult precipitation," and "horizontal precipitation" (Kittredge, 1948).

In certain coastal regions, a warm, moist air mass frequently moves onto the land surface. Fog results from the cooling effect of ocean currents or as a result of topographic changes that force the moisture-laden air mass to rise. One of the best known examples of this occurrence is the northern coast of California with its narrow belt of redwoods (Went, 1955). That fog drip is an important means of supplying moisture to coast redwoods probably originated with Cannon (1901), who implied that red-

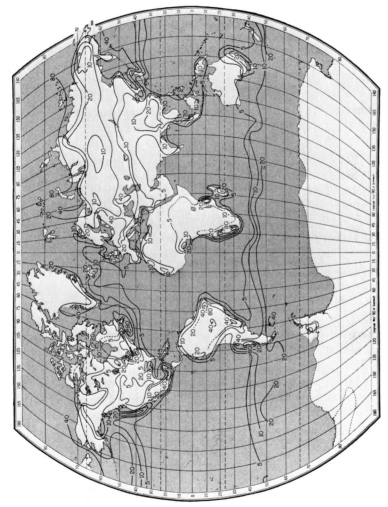

Fig. 2. World distribution of mean annual number of days with fog. (From Berry et al., 1945, with permission of McGraw-Hill Co.)

woods were restricted to the fog belt. Cooper (1917) took a more modest approach, suggesting that fog was only an added factor but still essential. He stated that for full development of redwood forests, heavy winter precipitation was necessary, but alone was not sufficient. Abundant summer fog also was essential. Byers (1953) noted that the best redwood stands were in coastal valleys or river flats. Around Monterey Bay, fog often reached such trees as Monterey cypress (*Cupressus macrocarpa*) and Monterey pine (*Pinus radiata*) on the mountain slopes but seldom settled on the major redwood groves below. Byers (1953) suggested that the effect of the fog belt was to be found in factors other than fog drip. Reduction of evapotranspiration through reduction of the number of hours of sunshine and the daytime temperature were considered most important because the fog was actually a stratus cloud layer with a base above 300 feet.

That fog drip does occur, however, was demonstrated by Oberlander (1956), who placed rain gauges under trees around the San Francisco peninsula in Central California. Dripping was most evident under trees that were exposed and unprotected by any hill to interrupt the wind. Growth of seedlings appeared to be stimulated beneath parent trees that collected and dripped moisture.

Just north of San Diego, California, is a small area where Torrey pine (*Pinus torreyana*) occurs. This pine has the distinction of having the most restricted range of any American pine (Harlow and Harrar, 1969). It is limited to the foggy Santa Rosa Island and the upper slopes facing the ocean where the fog comes in closest contact with the ground. The trees disappear from the lower slopes and only a few hundred feet inland, an apparent demonstration of the importance of condensation of fog droplets (Went, 1955).

Went (1955) described a similar phenomenon along the Mediterranean coast. The annual rainfall of 380 mm on the plateau behind Oran in Algeria is enough to support growth of some pines, but is not adequate for holly oak (*Quercus ilex*). This oak is found, however, along a strip 10 feet wide that follows the ridge of the plateau. Here, coastal fog hits most often.

Other fog forests have been reported in the Sierra Madre Oriental Mountains of eastern Mexico that are scattered from northern Mexico to the State of Oaxaca in the south. The prevailing northeasterly winds gather moisture over the Gulf of Mexico. After passing over the arid coastal plain the moisture-laden air flows up the eastern slopes of the mountains and produces heavy rains and frequent fogs. This occurs especially between 1300 and 2400 m elevation where dense, luxuriant forests dominated by oaks and other tree species exist. Rain gauges modified with screen wire to catch fog precipitation showed pronounced percentage increases in fog interception during the dry season (Vogelmann, 1973). As

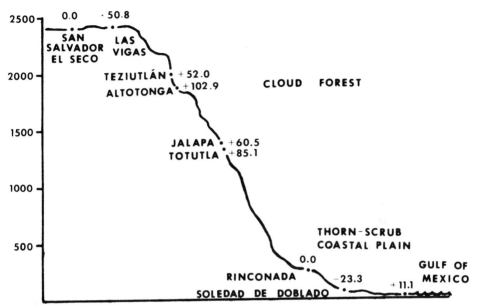

Fig. 3. Relationship of elevation to percentages of water collected by combination fog and rain gauges over plain rain gauges during the dry season in eastern Mexico. (From Vogelmann, 1973.)

shown in Fig. 3, for example, Altotonga at an elevation of 1898 m recorded a 102.9% increase in water collected due to fog during the dry season, whereas San Salvador el Seco on the high plateau did not record any fog. The effect of fog interception was dramatically shown by an isolated group of pines in a dry area near Las Vigas (Fig. 4). The soil under the canopy was saturated to a depth of 8 to 10 cm by the drip of intercepted fog. Beyond the canopy the soil was powder dry (Vogelmann, 1973).

Miranda and Sharp (1950) also reported fog forests in eastern Mexico that were limited to certain steeper slopes (1830 m) near the upper edge of the mountain escarpments. These forests were dominated by *Weinmannia pinnata* and contained species of *Quercus* and members of the Lauraceae.

Experiments on Mt. Wellington in Tasmania (Twomey, 1957) demonstrated that the pines growing there could intercept significant amounts of water from the frequent fog (Fig. 5).

Fog deserts, areas with very little precipitation but with frequent fogs, occur in dry regions bordering a coast. The deserts of Peru and Chile and the Namib desert along the southwestern coast of Africa are typical.

Fig. 4. Fog interception by pines near Las Vigas, Mexico. The dark shadow below the trees is soil wetted by fog moisture dripping from the needles. (From Vogelmann, 1973.)

As the wind blows from the warmer part of the oceans across the cool Humboldt and Benguela currents near these shores, thick coastal fog or low stratus clouds form and are blown over the land. In addition, mountains rise immediately behind the coasts and cause more fog as the air rises against them. Where fog does not occur on the lowlands near the coast, vegetation is practically lacking, but at elevations where fog hangs most of the year, shrub vegetation of unexpected density is encountered (Meigs, 1966). The fog and continual cloudiness that enshroud the lower Andes' slopes in Peru is called the garua. In many parts of the Atacama in Chile, no rain has ever been recorded. Here low heath and brushlike

Fig. 5. Mt Wellington in Tasmania covered in clouds. (From Twomey, 1957.)

forms of plants are dependent on coastal fog for moisture. Further inland the land is devoid of vegetation (Billings, 1970).

That significant amounts of fog can be intercepted by trees has been recognized for a long time and picturesquely described in several stories. One of the first was in a collection of letters by Gilbert White (1788) on the natural history of his parish in England. He wrote,

In heavy fogs on elevated situations especially, trees are perfect alembics, and no one that has not attended to such matters can imagine how much water one tree will distill in a nights time, by condensing the vapor which trickles down the twigs and boughs, so as to make the ground below quite in a float. In Newton Lane, in October 1775, on a misty day, a particular oak in leaf dripped so fast that the cartway stood in puddles and the ruts ran with water though the ground in general was dusty.

Another was recounted by Cannon (1901).

On the hogback of the Santa Moreno mountains lived a woodchopper, in a place once heavily covered by a redwood forest, but where there is left only an occasional large tree. Like other mountaineers, he must use water for culinary purposes at least, and in lieu of a convenient spring or well he has devised a unique tree-well. The chopper has fashioned the ground beneath a large redwood into the form of a trough at the lower end of which he has placed a barrel from which he obtains sufficient water for his needs.

Quantitative measurements of fog drip have been made in only a few locations in the world using a variety of measuring devices. These are usually rain gauges or other containers that have suspended over them wooden reeds, metal rods, screen wire, or even excised tree branches that intercept and drip fog. Considerable controversy exists about these fog gauges which have been criticized for overestimating fog precipitation and for attributing to vegetation a greater capacity for moisture capture than may exist in reality (Kerfoot, 1968). Nevertheless, it is accepted that fog and mist are widespread and are deposited on vegetation.

Classical measurements of fog deposition have been made on Table Mountain in South Africa by many investigators (DeForest, 1923; Kerfoot, 1968; Marloth, 1903, 1905). Here a fairly constant rate of precipitation from fog (approximately 250 mm/month) that often exceeds the rainfall occurs from month to month. Fog precipitation at the windward edge of forests (2 to 3 mm/hr) is often 6 to 10 times greater than in the open, but this is essentially an edge effect and does not extend far into the forest.

Ekern (1964) measured "cloud drip" under Norfolk Island pine (*Araucaria excelsa*) on the relatively dry island of Lanai that lies in the trade wind flow on the leeward side of West Maui. Approximately 3.8 mm/day of cloud moisture were collected beneath a tree. The actual amount intercepted was probably even more because much of the water did not reach the ground.

In the Green Mountains of Vermont clouds envelop slopes above 760 m in spring and fall more than in midsummer. During an 8-week period in summer, 895 ml more moisture were collected by fog gauges than by standard rain gauges. This amounted to a 66.8% increase of available moisture at 1100 m elevation (Vogelmann *et al.*, 1968). Since the measurements were made at the time of year when the frequency of fogs was low, it appeared that fog precipitation was more significant during some seasons than others.

Another example of the amount of water that fog drip may add to the soil was unpublished work from the Pacific Northwest Forest Experiment Station cited by Kittredge (1948). In an 85-yr-old stand of Sitka spruce (*Picea sitchensis*) and western hemlock (*Tsuga heteophylla*) near the Oregon Coast, fog drip was measured with rain gauges and compared to the amount of moisture collected in the open. An excess of 285 mm of water that represented a 44.6% increase was collected under the trees during an 18-week period from May 1940 to December 1941. In a single week, when gauges in the open showed only 0.25 mm, those under the trees caught 20 mm.

In the Namib desert in Africa, an annual deposit of 35 to 40 mm of fog moisture near the coast has been reported, and 20 to 30 mm were deposited on horizontal plates on the ground 25 miles inland. Vertical objects were considered to intercept even more moisture (Meigs, 1966). Several other quantitative measurements of fog drip are reviewed by Kerfoot (1968).

The amount of fog drip was indirectly measured by Means (1927), who determined the percentage of moisture content beneath trees and in the open on the hills behind Berkeley, California. Soil moisture percentage under a 25-yr-old Monterey pine (*Pinus radiata*) that was 15 feet high was 24.4% compared to 7.8% soil moisture 10 feet from the tree. For a *Eucalyptus* tree the soil moisture values were 22.9% under the tree and 9.4% 10 feet from it. These differences in percentage of moisture content were equivalent to 73 and 59 mm of rain under the pine and *Eucalyptus*, respectively.

Numerous factors affect the amount of fog that is intercepted by foliage and drips to the ground. Canopy height and arrangement, wind velocity and turbulence, and water content of fog determine the amount of water passing through a canopy. Total leaf surface area and the shape, arrangement, surface characteristics, and spatial distribution of the leaves influence the efficiency of interception. Went (1955) stated that conifers were more efficient fog catchers than broad-leaved trees. Merriam (1973) agreed that this was probably so. However, the shapes and sizes of leaves that are best for intercepting fog are still questionable (Kerfoot, 1968).

Based on experiments with a fog wind tunnel and artificial leaves of aluminum and plastic, Merriam (1973) concluded that an exponential equation provides a good approximation of the fog intercepted by leaves. The equation $S = S_c[1 - e^{-(F/S_c)}]$, in which S = storage, S_c = maximum storage, e = base of natural logarithms, and F = amount of fog that will impinge on a unit area, provides a means of estimating storage and fog drip when appropriate climatic and vegetative characteristics are known. In the above equation F can be determined from $F = wvtE$, where w is the water content of the fog in mass per unit volume, v is the velocity in length per unit time, t is time, and E is a complex efficiency factor. Fog drip expected from light and heavy fogs over time has been calculated using the above equations for a forest situation that had an assumed storage capacity of 1.25 mm (Fig. 6). An impinging fog of 0.1 mm/hr could result from a fog water content of 0.35 gm/m^3, flowing at a 2 m/sec through a forest canopy with an efficiency factor of 4%. Drip is calculated as the difference between impinging fog and storage. After 5 hr, drip would hardly be noticeable in light fog but would amount to 0.6 mm in heavy fog.

Both theoretical and observed situations indicate that significant amounts of fog moisture can become available to trees and shrubs and that in some localities this moisture is essential for growth and development.

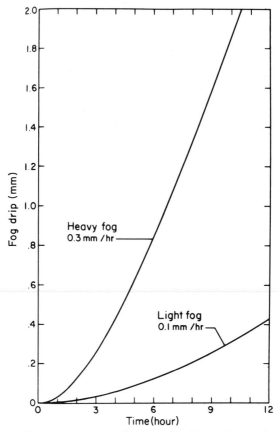

Fig. 6. Expected moisture drip from leaves during light and heavy fogs of given duration. (From Merriam, 1973.)

2. Dew

Although the value of dew for the life of arid zone plants has been
investigated for over two centuries, it is still unresolved and controversial.
On many days in many parts of the world, plants are covered with dew in
the early morning. However, because dew evaporates soon after sunrise,
its importance as a source of water often has been ignored. Nevertheless,
evidence exists that suggests that in some localities, dew improves the
water balance of plants or is criticial for their survival (Gindel, 1965;
Went, 1955).

The deposition of dew or, at temperatures below freezing, frost is a
similar phenomenon resulting from condensation of water vapor from the
atmosphere when in contact with foliage or surfaces cooled below the dew

point by nocturnal radiation. Two types of dew are recognized: (1) dewfall in which water vapor is transferred from the atmosphere to a surface; and (2) distillation that involves movement of water vapor from the soil with subsequent condensation in upper soil levels or on the surface of vegetation (Milthorpe, 1960). The distinction between dewfall and distillation is important in consideration of plant water balance. Dewfall on plant leaves is an accretion of water, whereas distillation is a redistribution of water within the soil/plant/air system. Distillation could in fact be detrimental by increasing soil moisture depletion by providing a short-cut for the transfer of water from the soil to the atmosphere. The validity of many studies to quantify amounts of dew deposited is compromised because no attempt was made to separate dewfall from distillation (Monteith, 1963).

Dew gauges are usually a plate of metal, glass, or wood placed horizontally at varying heights above the ground. The one used most extensively was designed by Duvdevani (1947). It consists of a wooden block (5 × 32 cm) with a standardized painted surface on which the occurrence and quantity of dew can be determined gravimetrically or read optically according to the pattern of dew drops. Table II shows the maximum amount of dew deposited nightly on various types of dew gauges in numerous places around the world. Duvdevani (1964) found a minimum of 115 and a maximum of 232 nights with dew in Israel. The amount of dew was higher in the dry summer than in the winter. Dew was a common phenomenon in the hot Jordan valley on nearly half the nights. In the southern Negev Desert it was even more frequent. Approximately 4.7 mm of dew was deposited annually in the Jordan Valley and 31 mm was deposited along the coastal plain.

As emphasized by Monteith (1963), a great deal of uncertainty exists concerning the amount of dew deposited on gauges and its relation to the amount deposited on vegetation. Since dew formation depends partly on

TABLE II

MAXIMUM DEW DEPOSITION ON ARTIFICIAL SURFACES[a]

Location	mm/night
Israel	0.45
Jamaica	0.43
England, south	0.43
Germany, Munich	0.43
Germany, Baltic Coast	0.37
Moravia	0.25
France, Montpellier	0.22
U.S.S.R., Moscow	0.22
Rumania	0.17

[a] From Monteith (1963).

the physical properties of the surface on which it is formed, it is impossible to ascertain precisely from the dew measured on different models of dew gauges how much dew is deposited on the surface of leaves and shoots.

Masson (1952) set 0.9 mm as the extreme limit of condensation attributable to dew with the average deposition in heavy dew of the order of 0.2 mm/night. Maximum values of 0.1 mm/night were found in Sweden and Egypt (Arvidsson, 1958). These amounts represent about 2 to 5% of the evaporation rate on the following day.

An often-cited study by Harrold and Dreibelbis (1951) reported exceedingly heavy amounts of dewfall measured with lysimeters in Ohio. The maximum values were 2 mm/night and 250 mm/year. These values are theoretically impossible and suggest that intensive fog precipitation may also have been measured (Milthorpe, 1960; Monteith, 1963). Fritschen and Doraiswamy (1973) also employed a weighing lysimeter to quantify dew accumulation for 2 days on a 28-m-high *Pseudotsuga menziesii* tree in the state of Washington. Moisture accumulation of 6.4 and 10.9 liters, which represented 15 and 19%, respectively, of evaporation during the following days, were measured. These unusually high amounts of deposition, purported to be dew, likely represent fog interception also.

Water vapor flux at the earth's surface is governed by radiation, temperature, humidity, and wind speed. Hofmann (1958) combined all four variables and an analysis of surface heat budgets to derive an equation for the rate of dew deposition. Monteith (1963) used this equation to determine condensation rates at 15°C for various wind speeds and relative humidities (Fig. 7). In a study of dew deposition on the surface of short grass in southern England, Monteith (1957) found that the optimum conditions for dewfall were a clear sky for rapid cooling, a relative humidity at sunset of at least 75%, and a wind speed of 1 to 3 m sec^{-1} throughout the night. With wind speeds less than 0.5 m sec^{-1}, mixing of the air was negligible, and distillation accounted for all dew produced. With wind speeds greater than 5 m sec^{-1} evaporation always occurred. A greater quantity of dewfall than that found in short grass was predicted to be deposited on shrubs and trees that project above the region of laminar flow near the ground. Furthermore, sparse distribution of shrubs and trees would probably favor a concentration of the total dewfall on these plants. The total deposit on each tree was considered likely to be greater than if the whole surface was uniformly covered with vegetation. A similar idea was expressed by Deacon *et al.* (1958) who suggested that the natural habit of desert species to grow in isolated clumps may represent an optimum spacing for maximum dew collection. These suggestions contrast with that of Brooks (1928), who thought that dew deposition on trees would not exceed that on grass.

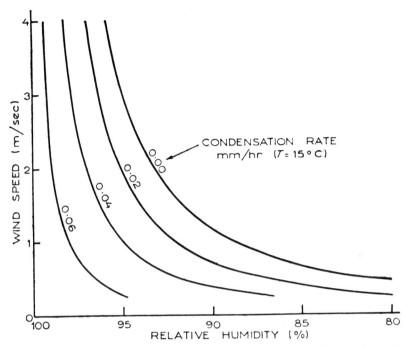

Fig. 7. Computed condensation of dew on a horizontal leaf for given wind speeds and relative humidities. (From Monteith, 1963.)

Dew deposited on leaves could be useful in two ways. It could reduce transpirational losses from leaves while the dew evaporates from leaf surfaces, or it could be absorbed by leaves to reduce internal water deficits. Monteith (1963) argued that it was fallacious to consider net water loss from isolated and heavily dewed plants to be less than that from plants of the same species with a closed canopy. Because potential evaporation normally is greater than potential condensation, net water loss from isolated plants would be greater than the loss from a closed canopy covering the same ground area. Waggoner *et al.* (1969) investigated the relative rate of evaporation from wet leaves and the potential rate of transpiration from dry leaves. They concluded that the rate of evaporation from wetted crops was twice that from unwetted ones. Thus, while transpiration losses were reduced when dew was evaporating, dewfall did not save an equivalent amount of water loss from a plant canopy. The possibility still exists and the question remains whether or not in arid regions the seemingly negligible transpiration reduction is sufficient to tip the balance in favor of plant survival.

One of the early studies on water absorption by aerial parts of plants was that of Lloyd (1905) in Arizona. He showed that the shrub *Fouquieria splendens* (ocotillo) could utilize water absorbed through the stem or bud scales for leaf growth. Spalding (1906a,b) came to the same conclusion for ocotillo as well as for several other desert shrubs. Species included in the genera *Amelanchier, Carya, Corylus, Juglans, Malus, Morus, Prunus, Pyrus, Sambucus, Viburnum,* and *Vitis* were all shown by Brierley (1934) to absorb water through the leaves under laboratory conditions. Water balance studies on several evergreen shrubs and trees in Michigan during winter showed that on frosty nights branches absorbed three to four times as much moisture as was lost by transpiration on a cold winter day (Gates, 1914). Absorption of moisture deposited on leaves was suggested as the mechanism by which evergreen shrubs and trees could maintain a satisfactory water balance when the soil was frozen. Michaelis and Michaelis (1934) made the same suggestion based on results that showed water absorption by spruce branches in the Swiss Alps. Numerous other examples of absorption of water by leaves were reported in the excellent review by Stone (1957).

Controversy about the value of dew exists because not all studies indicate that moisture deposited on leaves is readily absorbed. Walter (1955) concluded that absorption of dew in the Namib of southwestern Africa was unimportant and attributed its value solely to reduced transpiration. Waisel (1960) found that absorption of water by *Tamarix aphylla* in Israel occurred only when saturation deficits were high and during very long periods of continuous spraying. Under natural conditions, however, very little water was absorbed. Nevertheless, it was suggested that the negligible amount of moisture absorbed might be of value to the trees in recovery from sublethal water deficits.

Negative transpiration, or the redistribution of water through the plant into the soil as a consequence of absorption of moisture by the leaves has been reported. Breazeale *et al.* (1950) and Breazeale and McGeorge (1953) reported that tomato and corn plants could move water back into the soil and that water could be pumped back out through the roots even when the roots were sealed into empty flasks. Haines (1952, 1953) attempted to repeat these experiments but could find no evidence that water moved from leaves to roots. Breazeale's results were attributed to diurnal changes in temperature of the flasks that resulted in condensation. Stone *et al.* (1956) also found that water moved through ponderosa pine (*Pinus ponderosa*) and out the roots into empty flasks only when temperature of the flasks was allowed to fluctuate. Seedlings rooted in soil with water potential below the permanent wilting point failed to show any backward movement of water into the soil when dew was added to the shoots. However, seedlings that received dew survived for 20 days longer than seedlings not exposed to dew. It was concluded that the prolonged survival of seedlings receiving dew resulted from resaturation of the needle tissue with a concomitant reduction in the amount of water removed from the roots.

The factors associated with the entry of water into leaves and its redistribution within the plant appear to be the same as those associated with transpiration and water transport. These include surface characteristics of leaves, resistance to movement within the plant, and the water potential gradient across the atmosphere–plant–soil continuum. To the same extent that the cuticle prevents loss of water from leaves it also prevents uptake (Larcher, 1975). Wettability of the cuticle is an important prerequisite for absorption and varies with differences in surface conformation, pubescence, and composition of the cuticle. Consequently, the ability of water to wet leaf surfaces varies widely among different species (Schönherr, 1976; Slatyer, 1960). Permeability of the cuticle to water also varies among species. Plant cuticles consist of a cutin polymer matrix that contains varying proportions of other components such as cellulose, polyuronic acids, proteins, and phenolic compounds. Extractable waxes are embedded in the cuticular matrix. Permeability of cuticles to water is a function of the types and quantities of waxes in the matrix (Haynes and Goh, 1977).

Since stomata are cuticular invaginations, they do not offer a ready port of entry. Surface tension and air in the stomatal cavity also are barriers. The cuticle often is thin at the basal region of trichomes and thus is the site of the most rapid rate of water entry. In epiphytic Bromeliaceae, a significant part of the water supply enters through scales specialized for imbibition (Larcher, 1975). In woody perennials, reports of specialized absorbing cells are rare. Meidner (1954) observed rapid disappearance of water from leaves of *Chaetacme aristata* in Natal. Anatomical studies showed the presence of specialized epidermal cells. Counts revealed 12 specialized cells/mm^2 on the lower epidermis and 34 specialized cells/mm^2 on the upper epidermis.

Based on the present state of knowledge it appears that most water absorbed by leaves enters directly through the cuticle. The rate of absorption is markedly influenced by the internal water balance of the leaf and the potential gradient across the cuticle. However, other factors associated with cuticle hydration also may be involved (Slatyer, 1960). The amount of moisture absorbed through leaves appears to be quite limited. Vaadia and Waisel (1963), for example, working with Aleppo pine (*Pinus halepensis*), concluded that cuticular transpiration exceeds cuticular water absorption at night because the potential gradient is greater during the day. However, assuming a foliar absorption rate similar to a cuticular transpiration rate of 10 mg/gm fresh weight, water absorption during a dewy night of 10 hr would be 100 mg H_2O/gm fresh weight. This amount would last for about 20 min. of moderate transpiration during the day. It was concluded that water absorbed may improve hydrature of leaves and other shoot tissues, but that it appeared doubtful that the small quantities of water absorbed would move into the stem, roots, and soil. Theoretically, such negative transpiration should be possible, but would require much longer periods than the relatively short time at night when absorption of water by leaves is physically possible (Slatyer, 1960; Vaadia and Waisel, 1963). Because of the small amounts of water absorbed and its slow transport, its value may be greatest for plant survival. However, restoration of turgidity of droughted leaves for a portion of the night and early morning may allow plants to continue growth for a longer time than would otherwise be possible. Fog appears to provide greater amounts of water than dew.

Exploitation of dew as a source of water has been proposed by Gindel (1966) who irrigated seedlings of *Pinus halepensis, Pinus brutia, Cupressus horizontalis, Eucalyptus gomphocephala,* and *Tamarix aphylla,* with polyethylene collecting sheds in the Judean Hills of Israel. It was considered possible to plant xerophytes any place in Israel where there is enough dew and mist to cause a flow of water from a collecting shed into the planting pit provided so that available moisture can be concentrated at the root system. Moisture condensed on a 5-m² area and concentrated on a 1-m² area was sufficient to maintain the tree seedlings investigated during the summer dry season and until the advent of winter rains.

3. Water Vapor

Direct uptake of atmospheric water vapor by a leaf or stem is theoretically possible when the free energy or water potential of the leaf is less than that of the air; this is the reverse of the situation during transpiration. Slatyer (1956) demonstrated under laboratory conditions the absorption of water vapor by needles of 1-yr-old shortleaf pine (*Pinus echinata*) seedlings in both saturated and unsaturated air (water potential − 15 bars; relative humidity 98%). However, more water was absorbed by the roots of these seedlings and transported to the needles under the same potential gradient. This suggested that the cutinized epidermis of needles was much less permeable to water than the suberized epidermis of roots in young *Pinus echinata* trees. Similar results were obtained by Stone *et al.* (1950), who found that Coulter pine (*Pinus coulteri*) seedlings grown in soil unwatered for 10 months could absorb vapor from air with a relative humidity below 90%, indicating that a water potential of approximately −230 bars existed at the leaf surface.

Obviously, the only plants likely to benefit from absorption of water vapor are those that can survive the extreme water stress necessary to produce the water potential gradient needed for absorption from air that is only slightly below full saturation. Several desert and Mediterranean-type woody plants develop such low internal water potential under drought conditions (Stone et al., 1950; Went, 1955). An adaptation for absorption through the cuticle also may be significant as indicated by Wood (1925), who found no absorption of atmospheric vapor by the cutinized leaves of sugar eucalyptus (Eucalyptus corynocalyx), kurrajong bottletree (Sterculia diversifolia), and Acacia decussata, whereas the leaves of saltbush (Atriplex vesicarium) absorbed considerable atmospheric moisture. Wood (1925) suggested that the rapid absorption of moisture by Atriplex vesicarium might be due to the high salt concentration of its tissues and the resulting osmotic concentration of the cell sap. A similar adaptation has been suggested for Tamarix aphylla, a salt excretor. Highly hygroscopic salt crystals cover the branches and absorb water vapor from the atmosphere even at 80% relative humidity. The moisture accumulated on leaf surfaces can be absorbed, and give T. aphylla an advantage over other tree species which do not excrete salt (Waisel, 1960).

B. SOIL WATER

1. Factors Influencing Soil Moisture Availability

a. Texture, Porosity, and Colloidal Content. The water-holding capacity of a soil is determined chiefly by the size distribution of mineral particles, the manner in which the mineral particles are aggregated, the kind of clay minerals present, and the amount of organic matter incorporated in the soil (Kramer, 1969; Pritchett, 1979). Mineral particles are the predominate components of most soils, except in an organic soil such as peat. The relative amounts of sand, silt, and clay determine soil texture. The simplest soil is a sand, which by definition contains less than 15% silt and clay. Sandy soils are generally loose and noncohesive. Clay soils represent the other extreme in complexity. They contain 40% clay particles and less than 45% sand or silt. These soils are typically aggregated into granules or other structures. Because of their small size (<0.002 mm diameter) and plate-like structure, clay particles have a much greater surface area than do the solid particles of silt or sand. The extensive surface alone enables clay soils to hold more water than sandy soils. For example, Day et al. (1967) state that the area of solid surface available to bind water ranges from less than 1000 cm²/gm for coarse sand to more than

1,000,000 cm²/gm for clay soils. The latter figure includes large amounts of interlayer spaces which accommodate only thin layers of water.

Pore space is that fraction of soil volume occupied by air and water. It usually amounts to about 50% of soil volume. Total pore space is less important than the proportion of capillary to noncapillary pore space (Bauer, 1956). The noncapillary pores drain freely after rain or irrigation and become filled with air, whereas the capillary pores contain water that remains after free drainage is completed. The large noncapillary pores typical of sandy soils result in better drainage and aeration, but they also result in a lower water-holding capacity than in soils of heavier texture, which have a large proportion of capillary pores. An ideal soil has pore space about equally divided between large and small pores so that adequate drainage and aeration are permitted while adequate water-holding capacity is available (Bauer, 1956). Differences in amounts of capillary pore space in an old field and in an adjacent mixed hardwood forest in the South Carolina Piedmont are shown in Fig. 8. The large percentage of noncapillary pore space increases the rate of infiltration and decreases surface runoff during heavy rains (Hoover, 1949). Pore volume of forested soils is normally greater than that of soil used for agricultural purposes, because continuous tillage results in reduction of organic matter and noncapillary pores.

Pore space properties of soils can be influenced markedly by organic matter content. Organic matter affects the aggregation of particles in clay soils and tends to create pore space properties characteristic of sandy soils. Conversely, organic matter in sandy soils tends to impart moisture and cation retention characteristics of a sandy soil. Sodium ions particularly influence pore size by causing dispersion of soil aggregates and reduction in pore space (Kramer, 1969).

Different texture, porosity, organic matter, and other soil properties at various depths result in distinguishable soil profiles. Consequently, the water holding capacity and availability of water to roots is not uniform throughout the volume of soil that tree roots penetrate (Will and Stone, 1967). The amount of water available at different depths of a deep pumice soil that supports Monterey pine (*Pinus radiata*) in New Zealand is shown in Table III.

i. Moisture constants. The force with which water is held in soils is expressed in thermodynamic terms as the water potential (ψ_{soil}). The principal component forces which contribute to the ψ_{soil} are matric potential (ψ_m), osmotic potential (ψ_s), and pressure potential (ψ_p). Matric potential is due to the texture, structure, and colloidal content that influence adsorption and capillarity. Osmotic potential is produced by the ions and other

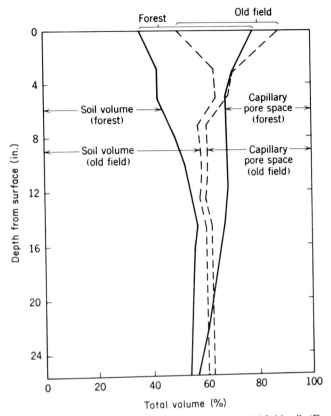

Fig. 8. Capillary and noncapillary pore space in forest and old field soils (From Hoover, 1949.)

TABLE III

ESTIMATED AVAILABLE MOISTURE STORAGE IN A KAINGAROA SILTY SAND PROFILE IN NEW ZEALAND[a]

Layer	Depth (cm)	Available moisture (% vol.)	Available moisture (cm)
I. Taupo top soil	0.0–30.5	30	9.2
II. Taupo ash	30.5–61.0	36	10.9
III. Taupo block member	61.0–91.5	44	13.5
IV. Taupo lapilli	91.5–152.5	29	17.8
V. Rotongaio sands	152.5–167.5	22	3.3
VI. Old surface soil I	167.5–198.0	33	10.2
VII. Ash	198.0–228.5	41	12.4
VIII. Old surface soil II	228.5–274.3	31	14.2

[a] Modified from Will and Stone (1967).

solutes in the soil solution while any pressure component contributes to the pressure potential. The ψ_{soil} is a summation of the component potentials and can be written $\psi_{soil} = \psi_m + \psi_s + \psi_p$. It is expressed in terms of Gibbs free energy and most often in units of bars. Pure free water has a potential of 0 bars. Since the free energy of water in soil is less than that of pure free water, soils typically have a negative potential (Rose, 1966; Slatyer, 1967). Water tends to move from a zone of less negative (high free energy) to one of more negative (low free energy) water potential and this is true for movement of water within the soil as well as movement from soil into roots. The amount of movement depends on the difference in ψ between the two zones.

Besides thermodynamic terms, several other terms are used to describe soil water characteristics of significance to trees. Field capacity is the soil water content after gravitational water has drained away and further downward movement is limited or has stopped. Water content at field capacity is influenced by soil texture and structure and the depth of the soil profile. Water potential normally varies between -0.1 to -0.3 bars. The upper limit of availability of soil moisture to roots is taken as field capacity (Pritchett, 1979).

The permanent wilting percentage is given as the lower limit of available soil water and is the soil moisture content at which plants remain permanently wilted unless water is added to the soil. Briggs and Shantz (1911, 1912) and Richards and Wadleigh (1952) measured the soil water content and soil water potential at permanent wilting for a large number of species. Similar values for these parameters at permanent wilting among a wide range of plants resulted in the wide acceptance that permanent wilting occurred at a soil water potential of -15 bars (Kramer, 1969; Pritchett, 1979).

Slatyer (1956) strongly criticized the concept of permanent wilting percentage. He emphasized that wilting occurred when there was a dynamic equilibrium between water potential of soil and water potential of the plant root. Consequently, soil water potential at wilting could be expected to vary as much as plant water potential, which can range from -5 to -200 bars. Nevertheless, the permanent wilting point of -15 bars is useful since the amount of water retained in soils at potentials less than -15 bars is extremely limited (Fig. 9). Water in soils with a ψ_{soil} of -50 bars or less is considered hygroscopic water and is tightly adsorbed on the soil particles, particularly the colloids, as a nonliquid and can move only in the vapor phase (Pritchett, 1979; Wilde, 1958).

ii. Hydraulic conductivity. The rate of water movement through soil and to roots, or its hydraulic conductivity, varies with soil texture, porosity, and moisture content. Movement is along water potential gradients. In

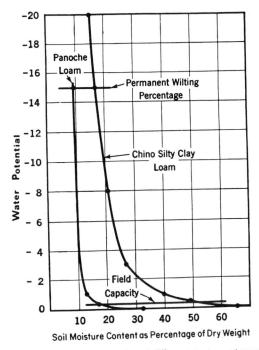

Fig. 9. Soil moisture content of two soils of different texture at permanent wilting and field capacity. (From Kramer, 1969.)

saturated soils gravity controls the ψ gradient, whereas in soils at field capacity or below, matric potential and root water potential are the principal influences on potential gradients. Under unsaturated conditions, noncapillary pores are primarily filled with air, and water movement is in thin films surrounding the soil particles and in the capillary pores. As the soil water content and ψ_{soil} decrease, the hydraulic conductivity decreases very rapidly. For example, hydraulic conductivity is approximately 10,000 times greater at field capacity than at the permanent wilting point. At a water potential of approximately -15 bars the continuity of liquid films is broken and water moves in the vapor phase. Vapor phase movement is very slow, approximately 10^{-7} cm/sec. However, diffusion of water vapor from one soil area to another of lower moisture content occurs under the driving force of a vapor pressure gradient. In very dry soils under conditions of nocturnal cooling the result may be condensation of moisture in the upper soil layers. Although considerable research has been done on hydraulic conductivity of agronomic soils, little information exists concerning hydraulic conductivity of forest soils (Hinckley et al., 1978; Kramer, 1969; Rose, 1966; Slatyer, 1967).

iii. Groundwater. Noncapillary water held in the soil profile by an impervious stratum constitutes the groundwater. The volume of groundwater is highly dynamic, flowing horizontally and fluctuating in depth under the influence of climate and biotic factors. Following heavy rain or snow melt the groundwater level may rise several feet. During drought, evaporation and absorption by roots may depress groundwater level several feet. A very important source of water for many trees is the moist zone above the groundwater called the capillary fringe. In clay or silt soils the capillary fringe may attain a height in excess of 1 m, whereas in sandy soils it seldom exceeds 0.3 m. The capillary fringe is sufficiently aerated to allow for active root growth. Survival of trees during drought often depends on water absorbed by roots in the capillary fringe (Pritchett, 1979).

Groundwater located at a suitable depth, such as 1 m, provides an increased supply of moisture and often is very beneficial to tree growth. This is especially true in sandy soils, which retain only small amounts of capillary water. A striking example of the value of groundwater was shown by Dosen *et al.* (1950). They found that red pine (*Pinus resinosa*) and white pine (*Pinus strobus*) trees on a sandy outwash with groundwater at a depth of 120 cm reached an average height of 10.6 m at the age of 22 years, whereas trees growing at higher elevations with groundwater 401 cm deep attained an average height of only 4.9 m at the same age (Fig. 10). Utilization of groundwater is much less important in fine-textured soils because of a large supply of capillary water (Pritchett, 1979; Wilde, 1958).

Many woody plants, called phreatophytes, utilize groundwater in such quantities that they are considered a problem in arid regions because

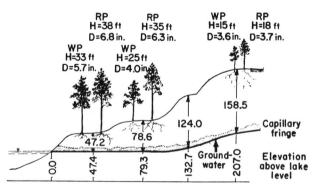

Fig. 10. Effect of groundwater on the growth of 22-yr-old plantation of red and white pine on a coarse sandy soil. WP = white pine; RP = red pine; H = average height; D = average diameter breast height. (After Dosen *et al.*, 1950.)

they deplete groundwater. Species of *Salix, Populus, Platanus, Nyssa, Picea, Larix,* and *Taxodium,* to list only a few, have the capacity to tolerate water saturated soils for prolonged periods. Roots of many of these species may penetrate to considerable depths to utilize groundwater (Bowie and Kam, 1968; Jaworski, 1968; Sebenik and Thames, 1967).

b. Characteristics of Woody Plants. i. Root absorption capacity. Although absorption by the above-ground parts of plants is an important source of water under some conditions, the root system constitutes the region through which the greatest amount of water enters woody plants. Spatial distribution of root systems, morphological characteristics, and rhizosphere associations show considerable diversity both within and between species. These factors vary in response to soil type, soil environment, competition from roots of nearby plants, and age of the root system. These characteristics and variations markedly influence the water absorbing capacity of roots (Hinckley *et al.,* 1978; Slatyer, 1967).

The roots of a woody plant constitute a dynamic and heterogeneous system that is continually growing into unprobed regions of soil and changing its water permeability through differentiation, suberization, and secondary growth while other parts of the system die and are shed. Seldom are two regions of the root system exposed to the same soil ψ. Movement of water from the soil through the root is predominately a passive process driven by a water potential gradient. This gradient usually originates in the shoots rather than in the roots as a result of transpirational water loss. Consequently, the ψ in roots is normally determined by the ψ in leaves that is transmitted through the cohesive water columns of the transpiration stream. For example, Kaufmann (1968) found that when the ψ of needles of loblolly pine (*Pinus taeda*) was −9 bars, root ψ was −5 bars, and soil ψ was −4 bars. When needle ψ was −18 bars, ψ in roots and the soil was −16 bars. Much higher potentials can exist in roots of xerophytic plants subjected to severe water deficits (Kramer, 1969). The rate of flow of water through a root system is a function of the resistances encountered in the root and the integration of all the water potential gradients that exist among the ramifications of the root system and the soil in which they are in contact (Hinckley *et al.,* 1978; Kramer, 1969; Slatyer, 1967).

The zone of most rapid water absorption lies behind the meristematic region of the root tip and ahead of the region where suberization occurs. Consequently, as a root grows, this zone moves through the soil. Root hairs commonly develop in the zone of most rapid water uptake, and although their water permeability appears to be no greater than that of other unsuberized epidermal cells, they may dramatically increase the absorbing surface of the root system (Rosene, 1943, 1956). They may also improve soil–root contact, particularly when a gap between the root and soil occurs. Gaps of 10 μm or more may result when roots grow down channels left by decay of other roots or burrowing of soil organisms. Contraction of drying soil or diurnal contraction of roots also may cause separation. Root hairs extending across the gap would aid water uptake (Newman, 1976b).

The rate of absorption is much less behind the region of root hair development where suberization and secondary growth occur. Nevertheless, considerable quantities of water are absorbed through suberized roots (Addoms, 1946; Chapman and Parker, 1942; Hayward *et al.*, 1942). When the proportion of the total root system of a tree is considered, it becomes evident that the actively growing unsuberized roots may constitute only a small portion of the total system. Consequently, the older, less permeable portions of a root system may absorb considerable amounts of water. In the winter most absorption must occur through such roots (Kramer, 1969).

The flow of water through a root is along the path of least resistance and, hence, is mostly apoplastic, involving cell walls and intercellular spaces in the epidermis and cortex. Movement across the endodermis, if still intact, is symplastic and represents the main source of resistance to water entry in young roots. Resistance to movement in the xylem is relatively low since these elements are especially adapted for water conduction. Vessel elements offer much less resistance to water transport than the tracheids, however (Baxter and West, 1977; Nobel, 1974; Zimmerman and Brown, 1971). Root resistance has been shown to increase with root age, length of the root, low soil oxygen or high carbon dioxide, and temperature fluctuations (Baxter and West, 1977; Busscher and Fritton, 1978; Newman, 1976a,b).

Water absorption by mycorrhizae has been studied very little, although it often is mentioned along with nutrient absorption as a function of the symbiotic relationship (Bowen, 1973). Hatch (1937) suggested that the survival benefit of mycorrhizal fungi to trees was due to increase in the absorbing surface of the root caused by increased diameter and branching of mycorrhizal rootlets, to growth of hyphae into soil, and to the greater longevity of mycorrhizae. Similar proposals have been presented by Cromer (1935), Goss (1960), Lobanow (1960), and Theodorou and Bowen (1970). Higher osmotic potentials of mycorrhizal than non-mycorrhizal roots of Norway spruce (*Picea abies*) have been found by Uhlig (1973), although the greater water absorbing capacity of mycorrhizal roots was attributed to their larger surface area. Safir *et al.* (1971) have shown approximately 50% less resistance to water movement through mycorrhizal than non-mycorrhizal soybeans. A similar effect on resistance to water transport in trees has not been demonstrated. Another possibility is that the mycelia of mycorrhizae bridge the gap that may occur between root and soil and aid in water absorption as has been proposed for root hairs.

ii. Root system morphology. Since water movement in soils toward roots is usually relatively slow, the water available to a tree is influenced by the volume of soil with which its roots are in contact. Genetic potential is an important factor controlling the spread and extent of root systems. Three morphological types tend to exist—taproot, heartroot, and plateroot (Fig. 11). A taproot system is common in *Gleditsia* and *Quercus* species. Its full development occurs in permeable, adequately moist, but well-aerated soils. Heartroots also develop in permeable, well-aerated soils but characterize species of *Acer, Fagus,* and *Abies.* The plateroot system commonly occurs in *Picea* and *Tsuga* (Wilde, 1958).

Inherent differences in patterns of root development are especially noticeable during early seedling growth and can significantly affect establishment and survival. Seedlings of baldcypress (*Taxodium distichum*) and yellow birch (*Betula alleghaniensis*) have a shallow plateroot system and can survive droughts only when rooted in moist soils. Species such as *Quercus* and *Carya* develop taproots that provide water even after the surface soil has become dry. Roots of bur oak (*Quercus macrocarpa*), which grow on dry sites, penetrated 1.7 m in the first season, whereas roots of basswood (*Tilia americana*) penetrated only 0.3 m into the same soil (Holch, 1931; Toumey, 1929). Ecotypic variation was found in seedlings of Douglas-fir (*Pseudotsuga menziesii*) with respect to drought tolerance. Survival was strongly correlated with early bud-burst and the associated ability of the root system to penetrate to a depth of 30 cm or more (Heiner and Lavender, 1972).

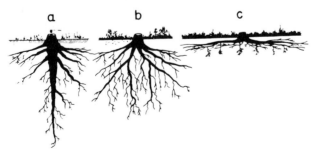

Fig. 11. Major types of tree root systems: (a) taproot; (b) heartroot; (c) plateroot. (From Wilde, 1958, with permission of Ronald Press.)

Albertson and Weaver (1945) concluded that survival of species of *Quercus, Gleditsia, Juglans,* and *Maclura* in the prairie region of the United States during the severe drought of the 1930's was related to their deeply penetrating tap root system. Drought resistance of Scotch pine (*Pinus sylvestris*) in the U.S.S.R. also has been attributed to deeply penetrating root systems (Abaturov *et al.,* 1976; Sudnitsyn *et al.,* 1976).

Many examples of woody plants that show a deep and ramifying root system exist. For example, Weaver and Kramer (1932) described a 65-yr-old bur oak (*Quercus macrocarpa*) with a taproot that penetrated to 4.3 m and had 30 or more large branch roots. The branch roots extended outward for 6.1 to 18.3 m. According to Herre (1954), the conifer-like *Welwitschia mirabilis* growing in the Namib desert may root to depths of 18 m. Roots of *Prosopis* were reported at an extraordinary 53 m below the soil surface in Arizona (Phillips, 1963). Several woody species including *Jacaranda decurrens, Attalea exigua,* and *Aidira* spp. have roots that penetrate the soil to 10 m or more in the Campos Cerrados savanna in Brazil (Rawitscher *et al.,* 1943). Many of the deep rooting species are phreatophytes, using groundwater as a source of moisture.

The size of root systems is often much reduced when grown in competition with that of other trees. The reduced root growth can be attributed to (1) competition for limited water and nutrients, (2) release of toxic substances from roots or other organs associated with plants, or (3) production of toxic substances during decomposition. The result is a reduction of the water absorbing surface (Kramer, 1969). In open-grown trees it is common to find lateral roots extending two to three times beyond the radius of the crown, although the majority of the absorbing roots may occur within the soil under the crown. Hough *et al.* (1965) investigated lateral root spread by placing ^{131}I in the soil and measuring radioactivity in the surrounding trees. It was absorbed as far away as 17 m in longleaf pine (*Pinus palustris*) and 16.5 m in turkey oak (*Quercus laevis*). The severe effect of competition on root development is emphasized by the observation that a 4-month-old loblolly pine (*Pinus taeda*) seedling growing in the absence of competition developed several times more roots than a 1-yr-old seedling growing in a forest (Kozlowski and Scholtes, 1948).

The inherent growth patterns of root systems often are modified by environmental and edaphic factors (Fig. 12). Root development is stimulated in the region of soil where texture, structure, and colloidal content provide adequate moisture. *Welwitschia mirabilis* may penetrate desert sands 18 m to groundwater (Herre, 1954), but may develop a root system no more than 1.5 m deep if the plant occurs in flat depressions where fog condenses (Oppenheimer, 1960). Litwak *et al.* (1963) established that the vertical taproot of the shrub spiny burnet (*Poterium spinosum*), which may

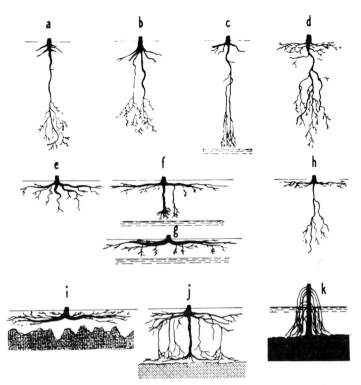

Fig. 12. Modification of root systems of trees by site. (a,b) Taproots and heartroots with reduced upper laterals; patterns encountered in coarse sandy soils underlain by fine-textured substrate; (c) taproot terminated by long tassels, a morphology induced by extended capillary fringe; (d) a collar of superficial laterals and a deep network of fibrous roots delineating the interlayer of porous material; (e) flattened heartroot formed in a lacustrine clay over a sand bed; (f) plateroot development in a soil with a reasonably deep groundwater table; (g) plateroot formed in organic soils with a shallow groundwater table; (h) bimorphic system of plateroot crown and a heartroot or taproot, a pattern characteristic of leached soils with a surface enriched in organic matter; (i) plateroot of Angiosperms in strongly leached soil with raw humus; (j) two parallel plateroots connected by vertical joiners in a hardpan podzol; (k) pneumatophores of mangrove trees on tidal lands. (From Wilde, 1958, with permission of Ronald Press).

grow down 40 cm in the first season in Israel, later branches out into the fissures of the underlying rock. There it uses the water stored in soil that fills the cracks. The shrub grows poorly, or is lacking, on flat soils covering unfissured rock. Some unfissured but porous limestone and sandstone rocks have been reported to store considerable quantities of water that is available to roots. Zohary and Orshansky (1951) found that dwarf shrubs of the Mediterranean rocky heaths could grow on moist, bare rocks.

Roots of *Quercus ithaburensis,* Palestine oak (*Quercus calliprinos*) and Palestine pistache (*Pistacia palaestina*) penetrated the solid rock of either porous or softened character and absorbed moisture. Penetration of the rocks was aided by root excretion of acid compounds (Oppenheimer, 1957, 1960).

The infrequent and limited precipitation of arid regions is usually retained in the uppermost soil layers. Hence, desert plants are often extremely shallow-rooted (Oppenheimer, 1960). Extremely long horizontal roots extending to as much as 27 m were developed by mesquite (*Prosopis spicigera*), and the root system of beancaper (*Zygophyllum dumosum*) reached a radius of 4 m (Evenari, 1938). When studying root penetration of common pistache (*Pistacia vera*) and almond trees in the dry mountains of the U.S.S.R., Zalensky (1940) found flat rooting that contrasted with that of pistache and almond trees grown with more rainfall.

An unusual stalagmiform root has been reported in Tuart eucalyptus (*Eucalyptus gomphocephala*). The root consists of a dense mount of fibrous roots that protrude as much as 12 cm from the floor of limestone caves. Localized abundance of water at sites beneath drip-points in the cave ceiling stimulates root development (Lamont and Lange, 1976).

iii. Crown morphology. Crown shape, density, foliage type, and persistence of foliage have a striking influence on the amount of annual precipitation that reaches the soil under a tree. Varying amounts of precipitation falling on crowns of different species are intercepted and evaporated back to the atmosphere whereas some of the precipitation may flow down the stem of the tree to the soil. Depending on the quantity of precipitation and the density of foliage, some precipitation may fall uninterrupted through the crown or drip from leaves. Because of the greater surface area and persistence of needles, species of gymnosperms generally intercept a greater proportion of annual precipitation than deciduous angiosperms (Table IV) (Büsgen and Münch, 1929; Penman, 1963; Zimmermann and Brown, 1971).

Büsgen and Münch (1929) discuss the importance of crown form to the distribution of stem flow and drip around individual trees. Young trees, which usually have acute branch angles, tend to channel stem flow toward the center of the crown and down the main bole. Older trees or those with more horizontal branches tend to favor movement of water toward the outside of the crown where it drips from leaves and twigs. These crown morphologies result in water being channeled to the soil zone that generally contains the greatest number of absorbing roots. Quantitative measurements have been made on mulga acacia (*Acacia aneura*), a tree of arid central Australia, that has branches which point diagonally upward and stiff leaf-like phyllodes oriented nearly vertically.

TABLE IV

PARTITIONING OF 403 MM OF RAIN FALLING ON 55-YR-OLD TREES[a]

| Species | Total rainfall on crown (liters) | Intercepted by crown | | |
		Stem flow (%)	Evaporated (%)	Through fall (%)
Beech	26,081	12.8	21.8	65.4
Oak	24,273	5.7	20.7	73.6
Maple	36,901	6.0	22.5	71.5
Spruce	12,044	1.4	58.8	39.8

[a] Modified from Büsgen and Münch (1929).

Of the rain falling on the tree crown, approximately 40% flowed down the trunk and penetrated the ground near its base (Slatyer, 1962).

c. Community Structure. Rhythmic arrangements of trees, shrubs, and grasses into dense bands (vegetation arcs) separated by nearly bare ground have been observed in several arid and semiarid parts of the world. The vegetation arcs occur on gentle slopes with a tendency for sheetflow of water and follow the contour of the land (Fig. 13). Hence,

Fig. 13. Aerial view of an *Acacia aneura* community in Australia showing marked grove–intergrove patterns. (From Slatyer, 1961.)

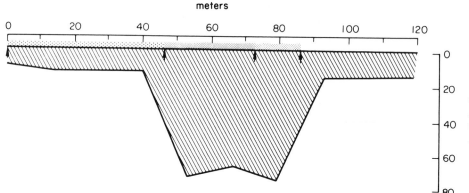

Fig. 14. Depth of water penetration into soil under a vegetation arc 3 days after rain in Somalia. The ground sloped downward to the left. Dots = vegetated areas; crosshatch = wet soil. (From Boaler and Hodge, 1964.)

they collect runoff from the intervening bare ground and more rainwater enters the soil near the base of plants than further away (Fig. 14).

The vegetation arcs were first appreciated by Gillett (1941) in Somalia and subsequently studied by Macfadyen (1950), Gilliland (1952), Greenwood (1957), and Boaler and Hodge (1964). Individual arcs were found to vary in width from 15–70 m and consisted of species of *Acacia* mixed with grasses. The intervening ground was almost completely bare and was three to four times the width of the vegetation band.

Similar distribution of trees also has been observed in Sudan (Worrall, 1960) and in central Australia (Litchfield and Mabbutt, 1962; Slatyer, 1961). In areas where mulga acacia (*Acacia aneura*) forms narrow grove patterns in Australia, the intergrove soil is a very shallow red earth (0–3.5 cm) but the grove soil is much deeper (0–24 cm) and occupies a narrow trench in the hardpan. Silt content of the intergrove soil is notably higher than that of the grove soil and causes it to seal to a greater extent. The trench thus acts as a trap for water draining from the intergroves upslope.

C. INTERNAL WATER RESERVOIRS

1. Tissue Capacitance

Another important source of water for trees is the water stored in plant tissues. More than three-fourths of green plant mass and a significant portion of the weight of wood are water. If the resistances between storage tissue and xylem are lower than those between xylem and soil, sufficient water may be withdrawn from tissues to meet transpirational requirements for considerable periods (Hinckley *et al.*, 1978).

Both diurnal and seasonal fluctuations in tissue water content have been measured and related to transpirational losses and available soil moisture (Chalk and Bigg, 1956; Chaney and Kozlowski, 1969d; Clark and Gibbs, 1957; Gibbs, 1939; Kramer and Kozlowski, 1979). A comprehensive coverage of seasonal changes in stem water content in various hardwood trees was given by Gibbs (1958). He reported a seasonal change of over 40% in water content (dry weight basis) of yellow birch (*Betula alleghaniensis*), for example, compared to a maximum diurnal change of 10%. Sapwood water content of eastern redcedar (*Juniperus virginiana*) was found to have fluctuated from a high of 130% (dry weight basis) in May to a low of 87% in September in Missouri (McGinnes and Dingeldein, 1969). Little variation occurred in the heartwood. A summertime low of 50% of saturated water content was found in the outer 2 cm of sapwood of 50 to 60 m tall Douglas-fir (*Pseudotsuga menziesii*) trees (Waring and Running, 1978). Wintertime highs approached a fully saturated condition. Rapid variations of as much as 30% per week were noted during the summer in response to rain storms that promoted partial recharge of the sapwood. Sapwood water content and its relationship to soil water were used to model variations in transpiration of Douglas-fir (*Pseudotsuga menziesii*) (Running *et al.*, 1975; Waring and Running, 1976).

Both diurnal and seasonal dimensional fluctuations in tree foliage, fruits, branches, and boles have been extensively documented and often used as quantitative measures of tissue water changes (Chaney and Kozlowski, 1969a,b,c; Dobbs and Scott, 1971; Fritts, 1958; Hinckley *et al.*, 1974; Hinckley and Bruckerhoff, 1975; Holmes and Shim, 1968; Kozlowski, 1972; Lassoie, 1973; Leikola, 1969; Pereira and Kozlowski, 1978; Roberts, 1976; Turner and Waggoner, 1968). A consequence of transpirational water loss is the development of a potential gradient not only in the conducting tissues from the leaves into the roots but also between conducting and nonconducting tissues. If the ψ gradient is adequate to overcome the resistance to flow, dehydration of parenchyma cells, cambial initials and undifferentiated derivatives, phloem and sapwood may occur. It is this loss of water to the conducting xylem that results in detectable contractions in tissues of trees. Under favorable conditions the water withdrawn from storage tissues during the day is replenished at night by water absorbed and transported from the soil. If soil water is limited, then tissue replenishment and expansion to the original size are incomplete. Since mature xylem elements are rigid cells that undergo negligible elastic deformations when subjected to hydrostatic stresses normal in trees, the fluctuations in dimensions are due almost entirely to hydration changes of

TABLE V

POTENTIAL HOURS OF TRANSPIRATION USING WATER STORED IN VARIOUS TISSUES OF
CONIFERS, HARDWOODS, AND HERBACEOUS PLANTS[a]

Storage tissue	Potential transpiration supply (hr)		
	Conifers	Hardwoods	Herbaceous
Roots	14.0	4.9	2.8
Stem—sapwood	50–180	12.2	1.5
Stem—extensible cells	1.0	6.7	1.3
Foliage	1.1	2.4	0.3

[a] From Hinckley et al. (1978).

the extensible living cells (Molz and Klepper, 1973; Waring and Running, 1976). Generally, more diurnal stem shrinkage occurs in or at the base of tree crowns than lower on the bole. In addition, shrinkage in the morning and recovery at night begin earlier at midcrown positions than at other stem heights. These patterns result from greater transpiration rates near midcrown and earlier development of steep ψ gradients (Dobbs and Scott, 1971; Hinckley and Bruckerhoff, 1975; Hellkvist et al., 1974; Kozlowski, 1972).

The potential contribution to transpiration of water from various internal reservoirs in conifers, hardwoods and herbaceous species is shown in Table V. The amount of water stored in conifer sapwood is impressive. Turner and Waggoner (1968) emphasized that the water reservoirs in the bole of red pine (Pinus resinosa) would be expected to allow transpiration to be maintained for longer periods of stress than in crops such as corn or tobacco which have little stored water. Both conifers and hardwoods utilize some water from xylem elements to help supply transpirational demands. However, a high proportion of sapwood to heartwood is a structural feature of many conifers and the wood normally contains more water than that of hardwoods. Stewart (1967) states that the average moisture content (dry weight basis) of sapwood of ring porous, diffuse porous, and nonporous species is <75%, approximately 100%, and >130%, respectively. Water content is generally higher in sapwood than in heartwood (Ovington, 1956).

A single 80-m tall Douglas-fir tree has been estimated to store more than 4000 liters of water, and a forest stand of Douglas-fir can have more than 290 tons/ha of water available in tissues (75% in sapwood) (Running et al., 1975; Waring and Running, 1978). This compares with about 40 tons/ha of water stored in tissues of a stand of oaks (Fig. 15). Roberts (1976) observed that sapwood capacitance accounted for about 5% of transpirational loss from 16-m tall Scotch pines (Pinus sylvestris) in the

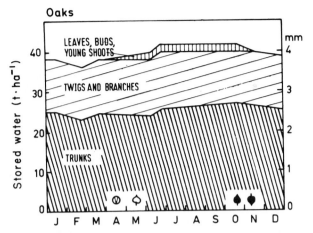

Fig. 15. Water storage in trunks, branches, twigs, young shoots, buds, and leaves of a stand of oaks. Phenological symbols from left to right: bud-swelling, leafing out, yellowing of leaves, leaf fall. (From Larcher, 1975, with permission of Springer–Verlag.)

East Anglia region of England during the autumn. The contribution of stored water was probably slightly larger since the study did not account for water stored in extensible stem, leaf, or root tissues. In addition, the study was conducted in the early autumn when available internal water is normally low compared to spring and early summer. Cermák *et al.* (1976) estimated that the storage reserves of freely accessible water above 1.3 m of a 6-m tall sweet cherry (*Prunus avium*) accounted for about 3.6% of daily transpiration.

Although the amount of water stored in foliage is small (Table V), it represents a reserve that may be critically important during dry periods. Water contents of leaves of various plants under conditions of saturation are shown in Fig. 16. The "available water" zone indicates the water that could be lost in cuticular transpiration or used in metabolic processes after stomatal closure without damage to cell protoplasm. The value of water in cell walls of leaves as a buffer against loss of water from protoplast during temporary adverse conditions also was emphasized by Gaff and Carr (1961).

The importance of sapwood as a water reservoir in areas where periodic drought is common was emphasized by Running *et al.* (1975). Large sapwood volume coupled with small leaf area might also be advantageous in the winter or early spring when air temperature and radiation are adequate for growth, but frozen soil or low temperatures limit water uptake by roots. Pressure bomb measurements on three populations of ribbon eucalyptus (*Eucalyptus viminalis*) showed the lowest values of wall water

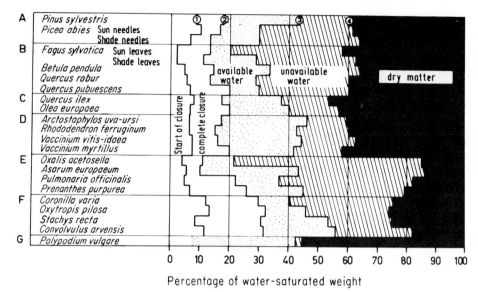

Fig. 16. Water content of leaves at various plant conditions and availability of the water. (1) Water deficit as percentage of saturated weight at onset of stomatal closure; (2) saturation deficit after complete stomatal closure; (3) saturation deficit at the appearance of the first signs of desiccation injury; (4) percentage of dry weight. The species are grouped as follows: A—conifers; B—deciduous trees; C—evergreen trees; D—dwarf shrubs; E—herbaceous sciophytes; F—herbaceous heliophytes; G—poikilohydric fern. (From Larcher, 1975, with permission of Springer–Verlag.)

content in the least drought-resistant population (Ladiges, 1975). Many trees and woody shrubs are adapted to endure drought conditions through phenological and anatomical mechanisms that reduce transpirational losses. The succulent condition that provides for a large volume to surface ratio and hence a large water storage capacitance is most common in species of cactus and *Euphorbia* (Bannister, 1976; Parker, 1968).

A specialized water storage organ occurs on the roots of the shrub *Ceiba parvifolia,* which grows in the thorn forest of Mexico. The roots produce large spherical swellings called "camotes" which may be 30 cm in diameter (Fig. 17). Water stored during the rainy season in the parenchymatous tissue is extracted for flowering and fruiting during the dry season (Moseley, 1956).

2. Internal Competition

Transpiration from leaves, stems, buds, flowers, and fruits occurs at varying rates because of anatomical and physiological differences in these tissues. Consequently, complex water potential gradients develop within the crown of trees that provide for redistribution of water stored in tissues. In addition, water absorbed by roots is not equally proportioned among all aerial organs. Tissues with low water potentials are the most competitive sinks for absorbed water. Lower shaded leaves often produce less carbohydrate than more exposed leaves and are thus less able to compete osmotically for water. The demand of developing peach fruits for carbohydrates and the resulting gradient for water movement to the fruit increased water stress and reduced growth in branches of fruiting peach trees (Chalmers and Wilson, 1978).

Redistribution of water from fruits to leaves during the day is common and has been shown to occur in a variety of fruits including lemons, oranges, walnuts, cherries, cotton bolls, avocados, pears, apples, and fruits and cones of forest trees (Kozlowski, 1972). Expansion and contraction of fruits and leaves are indicative of these hydration changes. Tukey (1962), for example, reported strong correlation between the amount of daily contraction of apple fruits and factors that influenced transpiration. When moisture absorbed by roots did not meet transpiration requirements and water deficits developed in trees, water was drawn from the fruits. When transpiration declined the fruits expanded. Similar results were shown by Chaney and Kozlowski (1971), who found diurnal fluctuations in fruit diameter and leaf thickness of Calamondin orange trees related to transpiration from leaves and internal redistribution of water from fruits to leaves. The percentage of moisture content of leaves on excised fruit-bearing branches was higher than it was in leaves on branches without

Fig. 17. A partially grown "camote" (about 10 cm in diameter) attached to its mother root. (Moseley, 1956.)

Fig. 18. Turgid leaves on Calamondin orange branch with attached fruit and wilted and curled leaves on branch without fruit. (From Chaney and Kozlowski, 1971.)

fruits. Five days after excision, leaves on branches without fruits were visibly wilted and curled, whereas those on fruit-bearing branches were turgid (Fig. 18). A similar internal redistribution of water has been demonstrated in lemon (Bartholomew, 1923). Furr and Taylor (1933) concluded that there was a ready cross-transfer of water in mature lemon trees. They showed that diurnal water deficits occurred regularly in trees even in moist soil and that leaf transpiration caused translocation of water from fruits.

It should not be assumed that water is always drawn from fruits. For example, water was translocated into rather than out of ripening fruits of black cherry (*Prunus serotina*) (Chaney and Kozlowski, 1969d). Rokach (1953) found that young oranges do not lose water during periods of stress. This observation agrees with that of Anderson and Kerr (1943), who found that leaves were unable to remove water from young cotton bolls, but older bolls showed daily midday shrinkage. Generally, younger plant parts obtain or retain water at the expense of older parts. However, the ultimate driving force for internal redistribution of water is a water potential gradient.

Water in tissues of trees or that absorbed by their roots may be translocated to adjacent trees through root grafts. Natural root grafting among trees of the same species is very common. Interspecific grafts occur but are much less common (Graham and Bormann, 1966; Kozlowski, 1971). The age of trees and their density markedly influence the degree of root grafting. Kuntz and Riker (1955) observed numerous grafted roots among 4-yr-old red pine (*Pinus resinosa*) seedlings in a nursery bed. *Pinus resinosa* plantations, however, at a 1.8 × 1.8 m spacing did not have grafted roots until 15 years old (Armson and Van Den Driessche, 1959). The proportion of root-grafted Norway spruce (*Picea abies*) trees in a Danish plantation increased from 3–5% at the time of the first thinning to at least 25–35% by the time the trees were 40–60 years old (Holmsgaard and Scharff, 1963).

Root grafts are a functional union between cambium, phloem, and xylem tissues (Fig. 19). Transfer of water as well as food, hormones,

Fig. 19. Root grafts in *Abies balsamea* trees growing approximately 1.5 feet apart. (From Kozlowski and Cooley, 1961.)

minerals, pesticides, and fungal spores has been found to occur through grafted roots (Eis, 1972; Kozlowski, 1971). Bormann (1966) suggested that transfer of materials translocated in the phloem was not important because xylem transport involves cross-grain movement. Water and minerals which move in the xylem tend to follow the grain of the wood to the crown, and hence root-grafted trees could not direct large amounts of water and minerals from adjacent individuals. Stone and Stone (1975), however, found measurable water flow capacity between grafted roots excised from *Pinus resinosa*. The rate of flow was a function of the tension applied. The resistance to cross-xylem transfer of water was discounted, and a mechanism for variable sharing among roots at the graft junction resulting from the anatomy and arrangement of tracheids in the fusion tissue was proposed. Such a mechanism could allow for transfer of available water among several interconnected trees.

As with tissue capacitance and with dew, fog and atmospheric vapor sources of water, the absolute amount of water available to a tree through root grafts may be small. However, at certain times this limited amount of water may be critical for growth or survival.

REFERENCES

Abaturov, Y. D., Bogatyrev, Y. G., and Orlov, A. Y. (1976). The role of deep roots in the life of young Scots pine stands on loose sandy soils. *Lesovedenie* **4**, 18–25.

Addoms, R. M. (1946). Entrance of water into suberized roots. *Plant Physiol.* **21**, 109–111.

Albertson, F. W., and Weaver, J. E. (1945). Injury and death or recovery of trees in prairie climate. *Ecol. Monogr.* **15**, 393–433.

Anderson, D. B., and Kerr, T. (1943). A note on the growth behavior of cotton bolls. *Plant Physiol.* **18**, 261–269.

Armson, K. A., and Van Den Driessche, R. (1959). Natural root grafts in red pine (*Pinus resinosa* Ait.). *For. Chron.* **35**, 232–241.

Arvidsson, I. (1958). Plants as dew collectors. *In* "International Union of Geodesy and Geophysics," Vol. 2, pp. 481–484. International Association of Scientific Hydrology, General Assembly of Toronto, Gentbrugge.

Bannister, P. (1976). "Introduction to Physiological Plant Ecology." Blackwell, Oxford.

Bartholomew, E. T. (1923). Internal decline of lemons. II. Growth rate, water content, and acidity of lemons at different stages of maturity. *Am. J. Bot.* **10**, 117–126.

Bauer, L. D. (1956). "Soil Physics," 3rd ed. Wiley, New York.

Baxter, P., and West, D. (1977). The flow of water into fruit trees. I. Resistance to water flow through roots and stems. *Ann. Appl. Biol.* **87**, 95–101.

Berry, F. A., Bollay, E., and Beers, N. R., eds. (1945). "Handbook of Meteorology." McGraw-Hill, New York.

Billings, W. D. (1970). "Plants, Man and the Ecosystem," 2nd ed. Wadsworth, Belmont, California.

Boaler, S. B., and Hodge, C. A. H. (1964). Observations on vegetation arcs in the northern region, Somali Republic. *J. Ecol.* **52**, 511–544.

Bormann, F. H. (1966). The structure, function, and ecological significance of root grafts in *Pinus strobus* L. *Ecol. Monogr.* **36**, 1–26.

Bowen, G. D. (1973). Mineral nutrition of ectomycorrhizae. *In* "Ectomycorrhizae: Ecology and Physiology" (G. C. Marks and T. T. Kozlowski, eds.), pp. 151–205. Academic Press, New York.

Bowie, J. E., and Kam, W. (1968). Use of water by riparian vegetation, Cottonwood Wash, Arizona. *U. S., Geol. Surv., Water Supply Pap.* **1858.**

Breazeale, E. L., and McGeorge, W. T. (1953). Exudation pressure in roots of tomato plants under humid conditions. *Soil Sci.* **75,** 293–398.

Breazeale, E. L., McGeorge, W. T., and Breazeale, W. F. (1950). Moisture absorption by plants from an atmosphere of high humidity. *Plant Physiol.* **25,** 413–419.

Brierley, W. G. (1934). Absorption of water by the foliage of some common fruit species. *Proc. Am. Soc. Hortic. Sci.* **32,** 277–283.

Briggs, L. J., and Shantz, H. L. (1911). A wax seal method for determining the lower limit of available soil moisture. *Bot. Gaz. (Chicago)* **51,** 210–219.

Briggs, L. J., and Shantz, H. L. (1912). The relative wilting coefficients for different plants. *Bot. Gaz. (Chicago)* **53,** 229–235.

Brooks, C. E. P. (1928). Influence of forests on rainfall. *Q. J. R. Meteorol. Soc.* **54,** 1–13.

Büsgen, M., and Münch, E. (1929). "The Structure and Life of Forest Trees" (Engl. transl. by T. Thomson). Wiley, New York.

Busscher, W. J., and Fritton, D. D. (1978). Simulated water flow through the root xylem. *Soil Sci.* **125,** 1–6.

Byers, H. R. (1953). Coast redwoods and fog drip. *Ecology* **34,** 192–193.

Cannon, W. A. (1901). Relation of redwoods and fog. *Torreya* **1,** 137–139.

Cermák, J., Palat, M., and Penka, M. (1976). Transpiration flow rate in a full-grown tree of *Prunus avium* L. estimated by the method of heat balance in connection with some meteorological factors. *Biol. Plant.* **18,** 111–118.

Chalk, L., and Bigg, J. M. (1956). The distribution of moisture in the living stem in Sitka spruce and Douglas-fir. *Forestry* **29,** 5–21.

Chalmers, D. J., and Wilson, I. B. (1978). Productivity of peach trees: Tree growth and water stress in relation to fruit growth and assimilation demand. *Ann. Bot. (London)* [N.S.] **42,** 285–294.

Chaney, W. R., and Kozlowski, T. T. (1969a). Seasonal and diurnal expansion and contraction of fruits of forest trees. *Can. J. Bot.* **47,** 1033–1038.

Chaney, W. R., and Kozlowski, T. T. (1969b). Seasonal and diurnal expansion and contraction of *Pinus banksiana* and *Picea glauca* cones. *New Phytol.* **68,** 873–882.

Chaney, W. R., and Kozlowski, T. T. (1969c). Diurnal expansion and contraction of leaves and fruits of English Morello Cherry. *Ann. Bot. (London)* [N.S.] **33,** 991–999.

Chaney, W. R., and Kozlowski, T. T. (1969d). Seasonal and diurnal changes in water balance of fruits, cones and leaves of forest trees. *Can. J. Bot.* **47,** 1407–1417.

Chaney, W. R., and Kozlowski, T. T. (1971). Water transport in relation to expansion and contraction of leaves and fruits of Calamondin orange. *J. Hortic. Sci.* **46,** 71–81.

Chapman, H. D., and Parker, E. R. (1942). Weekly absorption of nitrate by young bearing orange trees growing out of doors in solution cultures. *Plant Physiol.* **17,** 366–376.

Clark, J., and Gibbs, R. D. (1957). Studies in tree physiology. IV. Further investigations of seasonal changes in moisture contents of certain Canadian forest trees. *Can. J. Bot.* **35,** 219–253.

Cooper, W. C. (1917). Redwoods, rainfall and fog. *Plant World* **20,** 179–189.

Critchfield, H. J. (1966). "General Climatology," 2nd ed. Prentice-Hall, Englewood Cliffs, New Jersey.

Cromer, D. A. N. (1935). The significance of the mycorrhiza of *Pinus radiata. Bull. For. Bur. Aust.* **16,** 1–19.

Day, P. R., Bolt, G. H., and Anderson, D. M. (1967). Nature of soil water. *In* "Irrigation of

Agricultural Lands'' (R. M. Hagan, ed.), pp. 193–208. Am. Soc. Agron., Madison, Wisconsin.

Deacon, E. L., Priestley, C. H. B., and Swinbank, W. C. (1958). Evaporation and the water balance. *Arid Zone Res.* 10, 9–34.

DeForest, H. (1923). Rainfall interception by plants: An experimental note. *Ecology* 4, 417–419.

Dobbs, R. C., and Scott, D. R. M. (1971). Distribution of diurnal fluctuations in stem circumference of Douglas-fir. *Can. J. For. Res.* 1, 80–83.

Dosen, R. C., Peterson, S. F., and Pronin, D. T. (1950). Effect of ground water on the growth of red pine and white pine in central Wisconsin. *Trans. Wis. Acad. Sci., Arts Lett.* 40, 79–82.

Duvdevani, S. (1947). An optical method of dew estimation. *Q. J. R. Meteorol. Soc.* 73, 282–296.

Duvdevani, S. (1964). Dew in Israel and its effect on plants. *Soil. Sci.* 98, 14–21.

Eis, S. (1972). Root grafts and their silvicultural implications. *Can. J. For. Res.* 2, 111–120.

Ekern, P. C. (1964). Direct interception of cloud water on Lanaihale, Hawaii. *Soil Sci. Soc. Am., Proc.* 28, 419–421.

Evenari, M. (1938). Root conditions of certain plants in the wilderness of Judaea. *J. Linn. Soc. London, Bot.* 51, 383–388.

Fritschen, L. J., and Doraiswamy, P. (1973). Dew: An addition to the hydrologic balance of Douglas fir. *Water Resour. Res.* 9, 891–894.

Fritts, H. C. (1958). An analysis of radial growth of beech in a central Ohio forest during 1954–1955. *Ecology* 39, 705–720.

Furr, J. R., and Taylor, C. A. (1933). The cross transfer of water in mature lemon trees. *Proc. Am. Soc. Hortic. Sci.* 30, 45–51.

Gaff, D. F., and Carr, D. J. (1961). The quantity of water in the cell wall and its significance. *Aust. J. Biol. Sci.* 14, 299–311.

Gates, F. (1914). Winter as a factor in the xerophily of certain evergreen ericads. *Bot. Gaz. (Chicago)* 57, 445–489.

Gibbs, R. D. (1939). Studies in tree physiology. I. General introduction. Water contents of certain Canadian trees. *Can. J. Res.* 17, 460–482.

Gibbs, R. D. (1958). Patterns in the seasonal water content of trees. *In* ''The Physiology of Forest Trees'' (K. V. Thimann, ed.), pp. 43–69. Ronald Press, New York.

Gillett, J. B. (1941). The plant formations of western British Somaliland and the Harar Province of Abyssinia. *Kew Bull.* 2, 37–199.

Gilliland, H. B. (1952). The vegetation of eastern British Somaliland. *J. Ecol.* 40, 91–124.

Gindel, I. (1965). Irrigation of plants with atmospheric water within the desert. *Nature (London)* 207, 1173–1175.

Gindel, I. (1966). Attraction of atmospheric moisture by woody xerophytes in arid climates. *Commonw. For. Rev.* 45, 297–321.

Goss, R. W. (1960). Mycorrhizae of ponderosa pine in Nebraska grassland soils. *Nebr., Agric. Exp. Stn., Res. Bull.* 192, 1–47.

Graham, B. F., Jr., and Bormann, F. H. (1966). Natural root grafts. *Bot. Rev.* 32, 255–292.

Greenwood, J. E. G. W. (1957). The development of vegetation patterns in Somaliland Protectorate. *Georg. J.* 123, 465–473.

Haines, F. M. (1952). The absorption of water by leaves in an atmosphere of high humidity. *J. Exp. Bot.* 3, 95–98.

Haines, F. M. (1953). The absorption of water by leaves in fogged air. *J. Exp. Bot.* 4, 106–107.

Harlow, W. H., and Harrar, E. S. (1969). ''Textbook of Dendrology,'' 5th ed. McGraw-Hill, New York.

Harrold, L. L. and Dreibelbis, F. R. (1951). Agricultural hydrology as evaluated by monolith lysimeters. *U.S. Dept. Agric. Tech. Bull. 1050.*

Hatch, A. B. (1937). The physical basis of mycotrophy in the genus *Pinus*. *Black Rock For. Bull.* **6,** 1–168.

Haynes, R. J., and Goh, K. M. (1977). Review of physiological pathways of foliar absorption. *Sci. Hortic.* **7,** 291–302.

Hayward, H. E., Blair, W. M., and Skaling, P. E. (1942). Device for measuring entry of water into roots. *Bot. Gaz. (Chicago)* **104,** 152–160.

Heiner, T. D., and Lavender, D. P. (1972). "Early Growth and Drought Avoidance in Douglas-fir Seedlings," Res. Pap. No. 14. Forest Research Laboratory, Oregon State University, Corvallis.

Hellkvist, J., Richards, G. P., and Jarvis, P. G. (1974). Vertical gradients of water potential and tissue water relations in Sitka spruce trees measured with the pressure chamber. *J. Appl. Ecol.* **11,** 637–668.

Herre, M. (1954). *Welwitschia mirabilis* Hook. F. from seed to seed in the botanic garden of the University of Stellenbosch, C. P. *J. S. Afr. Bot.* **20,** 23–24.

Hinckley, T. M., and Bruckerhoff, D. N. (1975). The effects of drought on water relations and stem shrinkage of *Quercus alba. Can. J. Bot.* **53,** 62–72.

Hinckley, T. M., Chambers, J. L., Bruckerhoff, D. N., Roberts, J. E., and Turner, N. (1974). Effect of mid-day shading on net assimilation rate, leaf surface resistance, branch diameter, and xylem potential in a white oak seedling. *Can. J. For. Res.* **4,** 296–300.

Hinckley, T. M., Lassoie, J. P., and Running, S. W. (1978). Temporal and spatial variations in the water status of forest trees. *For. Sci. Monogr.* **20.**

Hofmann, G. (1958). Dew measurement by thermodynamical means. *In* "International Union of Geodesy and Geophysics," Vol. 2, pp. 443–445. International Association of Scientific Hydrology, General Assembly of Toronto, Gentbrugge.

Holch, A. E. (1931). Development of roots and shoots of certain deciduous tree seedlings in different forest sites. *Ecology* **12,** 259–298.

Holmes, J. W., and Shim, S. Y. (1968). Diurnal changes in stem diameter of Canary Island pine trees (*Pinus canariensis* C. Smith) caused by soil water stress and varying microclimate. *J. Exp. Bot.* **19,** 219–232.

Holmsgaard, E., and Scharff, O. (1963). Levende stodi rodgransevoksninger. *Forstl. Forsoegsvaes. Dan.* **28,** 99–150.

Hoover, M. D. (1949). Hydrologic characteristics of South Carolina Piedmont forest soils. *Soil Sci. Soc. Am., Proc.* **14,** 353–358.

Hough, W. A., Woods, F. W., and McCormack, M. L. (1965). Root extension of individual trees in surface soils of a natural longleaf pine-turkey oak stand. *For. Sci.* **11,** 223–242.

Jaworski, J. (1968). Evapotranspiration of plants and fluctuation of the groundwater table. *Int. Assoc. Sci. Hydrol. Publ.* **83,** 730–739.

Kaufmann, M. R. (1968). Water relations of pine seedlings in relation to root and shoot growth. *Plant Physiol.* **43,** 281–288.

Kerfoot, O. (1968). Mist precipitation on vegetation. *For. Abstr.* **29,** 8–20.

Kittredge, J. (1948). "Forest Influences." McGraw-Hill, New York.

Kozlowski, T. T. (1971). "Growth and Development of Trees," Vol. 2. Academic Press, New York.

Kozlowski, T. T. (1972). Shrinking and swelling of plant tissues. *In* "Water Deficits and Plant Growth" (T. T. Kozlowski, ed.), Vol. 3, pp. 1–64. Academic Press, New York.

Kozlowski, T. T., and Cooley, J. C. (1961). Root grafting in northern Wisconsin. *J. For.* **59,** 105–107.

Kozlowski, T. T., and Scholtes, W. H. (1948). Growth of roots and root hairs of pine and hardwood seedlings in the Piedmont. *J. For.* **46,** 750–754.

Kramer, P. J. (1969). "Plant and Soil Water Relationships: A Modern Synthesis." McGraw-Hill, New York.

Kramer, P. J., and Kozlowski, T. T. (1979). "Physiology of Woody Plants." Academic Press, New York.

Kuntz, J. E., and Riker, A. J. (1955). The use of radioactive isotopes to ascertain the role of root grafting in the translocation of water, nutrients and disease-inducing organisms. *Proc. Int. Conf. Peaceful Uses Atomic Energy, 1st 1955* Vol. 12, pp. 144–148.

Ladiges, P. Y. (1975). Some aspects of tissue water relations in three populations of *Eucalyptus viminalis* Labill. *New Phytol.* **75**, 53–62.

Lamont, B. B., and Lange, B. J. (1976). Stalagmiform roots in limestone caves. *New Phytol.* **76**, 353–360.

Larcher, W. (1975). "Physiological Plant Ecology" (translated by M. A. Biederman-Thorson). Springer-Verlag, Berlin and New York.

Lassoie, J. P. (1973). Diurnal dimensional fluctuations in a Douglas-fir stem in response to tree water status. *For. Sci.* **19**, 251–255.

Leikola, M. (1969). The effect of climatic factors on the daily hydrostatic variations in stem thickness of Scots pine, *Ann. For. Fenn.* **6**, 171–181.

Litchfield, W. H., and Mabbutt, J. A. (1962). Hardpan in soils os semi-arid Western Australia. *J. Soil Sci.* **13**, 148–159.

Litwak, M., Kupernik, G., and Orshan, G. (1963). The role of competition as a factor in determining the distribution of dwarf shrub communities in the Mediterranean territory in Israel. *J. Ecol.* **51**, 467–480.

Lloyd, F. E. (1905). The artificial induction of leaf formation in the ocotillo. *Torreya* **5**, 175–179.

Lobanow, N. W. (1960). "Mykotrophie der Holzpflanzen." VEB Dsch. Verlag Wiss., Berlin.

Macfadyen, W. A. (1950). Soil and vegetation in British Somaliland. *Nature (London)* **165**, 121.

McGinnes, E. A., Jr., and Dingeldein, T. W. (1969). Selected wood properties of eastern redcedar (*Juniperus virginiana* L.) grown in Missouri. *Mo. Agric. Exp. Stn., Res. Bull.* **960**, 2–19.

Marloth, R. (1903). Results of experiments on Table Mountain for ascertaining the amount of moisture deposited from the southeast clouds. *Trans. S. Afr. Philos. Soc.* **14**, 403–408.

Marloth, R. (1905). Results of further experiments on Table Mountain for ascertaining the amount of moisture deposited from the southeast clouds. *Trans. S. Afr. Philos. Soc.* **16**, 97–105.

Masson, H. (1952). "Dew and the Possibility of Using It." UNESCO, NS, Arid Zone Research, 100, Paris.

Means, T. H. (1927). Fog precipitation by trees. *J. For.* **25**, 1015–1016.

Meidner, H. (1954). Measurements of water intake from the atmosphere by leaves. *New Phytol.* **53**, 423–426.

Meigs, P. (1966). "Arid Zone Research. Geography of Coastal Deserts." UNESCO, Paris.

Merriam, R. A. (1973). Fog drip from artificial leaves in a fog wind tunnel. *Water Resour. Res.* **9**, 1591–1598.

Michaelis, G., and Michaelis, P. (1934). Ökologische Studien an der alpinen Baumgrenze. III. Über die winterlichen Temperaturer der Pflanzlichen Organe, insbesondere der Fichte. *Bot. Zentralbl.* **52B**, 333–377.

Milthorpe, F. L. (1960). The income and loss of water in arid and semiarid zones. *In* "Plant-Water Relationships in Arid and Semi-Arid Conditions," Reviews of Research, pp. 9–36. UNESCO, Paris.

Miranda, F., and Sharp, A. J. (1950). Characteristics of the vegetation in certain temperate regions of eastern Mexico. *Ecology* **31**, 313–333.

Molz, F. J., and Klepper, B. (1973). On the mechanism of water-stress-induced stem deformation. *Agron. J.* **65**, 304–306.

Monteith, J. L. (1957). Dew. *Q. J. R. Meteorol. Soc.* **83**, 322–341.

Monteith, J. L. (1963). Dew: Facts and fallacies. *In* "The Water Relations of Plants" (A. J. Rutter and F. H. Whitehead, eds.), pp. 37—56. Blackwell, Oxford.

Moseley, M. F., Jr. (1956). The anatomy of the water storage organ of *Ceiba parvifolia*. *Trop. Woods* **104**, 61–79.

Newman, E. I. (1976a). Water movement through root systems. *Philos. Trans. R. Soc. London* **273**, 463–478.

Newman, E. I. (1976b). Water relations. *In* "Plant Structure, Function and Adaptation" (M. A. Hall, ed.), pp. 157–196. Macmillan, New York.

Nobel, P. S. (1974). "Biophysical Plant Physiology." Freeman, San Francisco, California.

Oberlander, G. T. (1956). Summer fog precipitation in San Francisco peninsula. *Ecology* **37**, 851.

Oppenheimer, H. R. (1957). Further observations on roots penetrating into rocks and their structure. *Bull. Res. Counc. Isr., Sect. D* **6**, 18–31.

Oppenheimer, H. R. (1960). Adaptation to drought: Xerophytism. *In* "Plant-Water Relationships in Arid and Semi-Arid Conditions," Reviews of Research, pp. 105–138. UNESCO, Paris.

Ovington, J. D. (1956). The form, weights and productivity of tree species grown in close stands. *New Phytol.* **55**, 289–304.

Parker, J. (1968). Drought-resistance mechanisms. *In* "Water Deficits and Plant Growth" (T. T. Kozlowski, ed.), Vol. 1, pp. 195–234. Academic Press, New York.

Penman, H. L. (1963). "Vegetation and Hydrology." Commonw. Agric. Bur., Farnham Royal, England.

Pereira, J. S., and Kozlowski, T. T. (1978). Diurnal and seasonal changes in water balance of *Acer saccharum* and *Betula papyrifera*. *Physiol. Plant.* **43**, 19–30.

Phillips, W. S. (1963). Depth of roots in soil. *Ecology* **44**, 424.

Pritchett, W. L. (1979). "Properties and Management of Forest Soils." Wiley, New York.

Rawitscher, F., Ferri, M. G., and Rochid, M. (1943). Profundidade dos solos e vegetacao em campos cerrados do Brasil meridional. *An. Acad. Bras. Sci.* **15**, 267–294.

Richards, L. A., and Wadleigh, C. H. (1952). Soil water and plant growth. *In* "Soil Physical Conditions and Plant Growth" (B. T. Shaw, ed.), pp. 73–251. Academic Press, New York.

Roberts, J. (1976). An examination of the quantity of water stored in mature *Pinus sylvestris* L. trees. *J. Exp. Bot.* **27**, 473–479.

Rokach, A. (1953). Water transfer from fruits to leaves in the Shamouti orange tree and related topics. *Palest. J. Bot., Rehovot Ser.* **8**, 146–151.

Rose, C. W. (1966). "Agricultural Physics." Pergamon, Oxford.

Rosene, H. F. (1943). Quantitative measurement of the velocity of water absorption in individual root hairs by a microtechnique. *Plant Physiol.* **18**, 588–607.

Rosene, H. F. (1956). The water absorptive capacity of root hairs. *Congr. Int. Bot., Rapp. Commun., 8th, 1954* Vol. 11, pp. 217–218.

Running, S. W., Waring, R. H., and Rydell, R. A. (1975). Physiological control of water flux in conifers. *Oecologia* **18**, 1–16.

Russell, M. B., and Hurlbut, L. W. (1959). The agricultural water supply. *In* "Water and Its Relation to Soils and Crops" (M. B. Russell, ed.), pp. 6–19. Academic Press, New York.

Safir, G., Boyer, J. S., and Gerdemann, J. W. (1971). Mycorrhizal enhancement of water transport in soybean. *Science* **172**, 581–583.

Schönherr, J. (1976). Water permeability of isolated cuticular membranes: The effect of cuticular waxes on diffusion of water. *Planta* **131**, 159–164.

Sebenik, P. G., and Thames, J. L. (1967). Water consumption by phreatophytes. *Prog. Agric.* **14**, 10–11.

Slatyer, R. O. (1956). Absorption of water from atmospheres of different humidity and its transport through plants. *Aust. J. Biol. Sci.* **9**, 552–558.

Slatyer, R. O. (1960). Absorption of water by plants. *Bot. Rev.* **26**, 331–392.

Slatyer, R. O. (1962). Methodology of a water balance study conducted on a desert woodland (*Acacia aneura* F. Meull) community in Central Australia. *Arid Zone Res.* **16**, 15–26.

Slatyer, R. O. (1967). "Plant-Water Relationships." Academic Press, New York.

Spalding, V. M. (1906a). Absorption of atmospheric moisture by desert shrubs. *Bull. Torrey Bot. Club.* **33**, 367–375.

Spalding, V. M. (1906b). Biological relations of desert shrubs. II. Absorption of water by leaves. *Bot. Gaz. (Chicago)* **41**, 262–282.

Stewart, C. M. (1967). Moisture content of living trees. *Nature (London)* **214**, 138–140.

Stone, E. C. (1957). Dew as an ecological factor. I. A review of the literature. *Ecology* **38**, 407–413.

Stone, E. C., Went, F. W., and Young, C. L. (1950). Water absorption from the atmosphere by plants growing in dry soil. *Science* **111**, 546–548.

Stone, E. C., Shachori, A. Y., and Stanley, R. G. (1956). Water absorption by needles of ponderosa pine seedlings and its internal redistribution. *Plant Physiol.* **31**, 120–126.

Stone, J. E., and Stone, E. L. (1975). The communal root system of red pine: Water conduction through root grafts. *For. Sci.* **21**, 255–261.

Sudnitsyn, I. I., Shein, E. V., and Gael, A. G. (1976). Effect of the concentration of Scots pine roots on the absorption of soil moisture. *Lesovedenie* **4**, 9–17.

Theodorou, C., and Bowen, G. D. (1970). Mycorrhizal responses of radiata pine in experiments with different fungi. *Aust. For.* **34**, 183–191.

Toumey, J. W. (1929). Initial root habit in American trees and its bearing on regeneration. *Proc. Int. Bot. Congr., 4th, 1926* Vol. 1, pp. 713–728.

Tukey, L. D. (1962). Factors affecting rhythmic diurnal enlargement and contraction in fruits of the apple (*Malus domestica* Bork). *Proc. Int. Hortic. Congr., 16th, 1962* Vol. 3, pp. 328–336.

Turner, N. C., and Waggoner, P. E. (1968). Effects of changing stomatal width in a red pine forest on soil water content, leaf water potential, bole diameter and growth. *Plant Physiol.* **43**, 973–978.

Twomey, S. (1957). Precipitation by direct interception of cloud water. *Weather* **12**, 120–122.

Uhlig, S. K. (1973). The water exchange capacity of mycorrhizal and non-mycorrhizal roots of Norway spruce. *Wiss. Z. Tech. Univ., Dresden* **22**, 393–395.

Vaadia, Y., and Waisel, Y. (1963). Water absorption by the aerial organs of plants. *Physiol. Plant.* **16**, 44–51.

Vogelmann, H. W. (1973). Fog precipitation in the cloud forests of eastern Mexico. *BioScience* **23**, 96–100.

Vogelmann, H. W., Siccama, T., Leedy, D., and Ovitt, D. C. (1968). Precipitation from fog moisture in the Green Mountains of Vermont. *Ecology* **49**, 1205–1207.

Waggoner, P. E., Begg, J. E., and Turner, N. C. (1969). Evaporation of dew. *Meteorology* **6**, 227–230.

Waisel, Y. (1960). Ecological studies on *Tamarix aphylla* (L.) Karst. II. The water economy. *Phyton (Buenos Aires)* **15**, 19–28.

Walter, H. (1955). The water economy and hydrature of plants. *Annu. Rev. Plant Physiol.* **6,** 239–252.

Waring, R. H., and Running, S. W. (1976). Water uptake, storage and transpiration by conifers: A physiological model. *Ecol. Stud.* **19,** 189–202.

Waring, R. H., and Running, S. W. (1978). Sapwood water storage: Its contribution to transpiration and effect upon water conductance through the stems of old growth Douglas-fir. *Plant, Cell, & Environ.* **1,** 131–140.

Weaver, J. E., and Kramer, J. (1932). Root system of *Quercus macrocarpa* in relation to the invasion of prairie. *Bot. Gaz. (Chicago)* **94,** 51–85.

Went, F. W. (1955). Fog, mist, dew and other sources of water. *U.S., Dep. Agric., Yearb. Agric.* pp. 103–109.

White, G. (1788). "The Natural History of Selbourne." Aldine, Chicago, Illinois.

Wilde, S. A. (1958). "Forest Soils." Ronald Press, New York.

Will, G. M., and Stone, E. L. (1967). Pumice soils as a medium for tree growth. I. Moisture storage capacity. *N. Z. J. For.* **12,** 189–199.

Wood, J. G. (1925). The selective absorption of chlorine ions and the absorption of water by the leaves in the genus Atriplex. *Aust. J. Exp. Biol. Med. Sci.* **2,** 45–56.

Worrall, G. A. (1960). Tree patterns in the Sudan. *J. Soil Sci.* **11,** 63–67.

Zalensky, O. V. (1940). [Distribution and ecological peculiarities of pistachio (*Pistacia vera* L.) and almond (*Amygdalus communis* L.) in the western Kopet-Dagh. (Turkemenian S.S.R.).] *Bot. Zh.* **25,** 20–37.

Zimmerman, M. H., and Brown, C. L. (1971). "Trees-Structure and Function." Springer-Verlag, Berlin and New York.

Zohary, M., and Orshansky, G. (1951). Ecological studies on lithophytes. *Palest. J. Bot.* **5,** 119–128.

CHAPTER 2

CONIFEROUS FORESTS AND PLANTATIONS

D. Whitehead

FOREST RESEARCH INSTITUTE, NEW ZEALAND FOREST SERVICE, PRIVATE BAG,
ROTORUA, NEW ZEALAND

and

P. G. Jarvis

DEPARTMENT OF FORESTRY AND NATURAL RESOURCES,
UNIVERSITY OF EDINBURGH, EDINBURGH, SCOTLAND

Water Deficits and Plant Growth, Vol. VI
Copyright © 1981 by Academic Press, Inc.
All rights of reproduction in any form reserved.
ISBN 0-12-424156-5

I. INTRODUCTION

In transpiration water moves from the source of high potential in the soil to the sink of low potential in the atmosphere through the vegetation, along a flow pathway which has both flow resistance and capacitance. Forest is characterized by substantial spatial separation between source and sink, the transfer pathway being potentially of large resistance because of its length and large capacitance because of its bulk (Fig. 1). As a consequence of the frictional resistances to flow in the system, there are drops in potential along the pathway, the largest drops in potential occurring where the resistances are largest (Philip, 1957, 1966). As a result of capacitance in the system, the flow of water is rarely in a steady state and exhibits flow divergence from one level to another. The vapor flow from the canopy usually exceeds the uptake of water from the soil in daytime and at night the reverse occurs. The sheer height of the trees, too, causes a reduction in water potential in the foliage at the top and also leads to enhancement of evaporation from the foliage because of the considerable aerodynamic surface roughness which results (Jarvis and Stewart, 1979).

In the last few years there has been considerable interest in the concepts of flow resistance and capacitance in trees and they will be considered in detail in this chapter. Capacitance in particular will be discussed in some detail as the water storage capacity of trees is considerable (Jarvis, 1975), and development of the concepts and their consequences is an area of current interest.

In managed forest, particularly plantation forest, respacing is often carried out during canopy closure and a few years after closure commercial thinnings are removed. These operations bring about changes in the

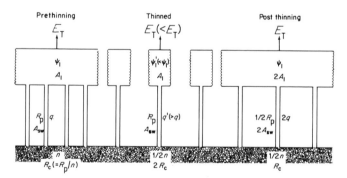

Fig. 1. A diagram to show the relationships between leaf ψ, sapwood cross-sectional area, and transpiration rate of a stand of trees before and after thinning. q is the flow of water through an individual tree with a cross-sectional area of sapwood, A_{sw}, a flow resistance, R_p, and a leaf area A_l, in a stand containing initially n trees ha^{-1}, with a liquid flow resistance from soil to canopy of R_c. (Redrawn from Jarvis, 1975.)

number and size of the stems connecting the canopy to the soil (Fig. 1) and hence cause rapid changes in the resistance and capacitance of this part of the pathway. In the following season this would be expected to result in somewhat lower leaf water potential (ψ_l) (i.e., post-thinning stress) because of the greater penetration of radiation into the crowns of the remaining trees. Later as the canopy fills in to the initial leaf area index (LAI), the volume flow of water per stem increases considerably and would be expected to lead to much lower values of ψ_l unless the resistance of each stem also falls in proportion (Jarvis, 1975). The main pathway for water transport in the stem is the earlywood of the sapwood. Thus a lower resistance implies a larger, or more effective cross-sectional area of functional sapwood. This chapter will therefore also discuss the interrelationships and homeostasis between the anatomy, morphology and amount of sapwood and leaf area. We aim to establish the properties of the soil–plant–atmosphere continuum which determine the rate and variation in rate of transpiration and the diurnal and seasonal changes in water content and ψ throughout the tree. Second, we wish to establish the basis for the developmental homeostasis and behavioral feedback that result in the containment of ψ_l within relatively narrow limits over a wide range of evaporative demand and transpiration rates.

II. TRANSPIRATION

A. GENERAL THEORY

The energy "available" for the exchanges of sensible and latent heat by a leaf or canopy is

$$A = Q_n - \Delta S = \lambda E_T + C, \tag{1}$$

where Q_n is the flux of net radiation and ΔS is the net change in energy stored in the leaf or vegetation and includes changes of sensible heat and metabolic energy in the biomass and soil and changes in the sensible and latent heat content of the air contained in the vegetation. λE_T is the flux of latent heat in transpiration and C the flux of sensible heat. The transpiration rate can be written as the diffusion equation

$$E_T = \frac{c_p \rho_a}{\lambda \gamma} [e_s(T_1) - e_a] g_w, \tag{2}$$

where c_p is the specific heat of dry air of density ρ_a, λ is the latent heat of vaporization of water, γ is the psychrometric constant, $e_s(T_1)$ is the saturated water vapor pressure of air at leaf temperature (T_1), e_a is the water vapor pressure of the ambient air, and g_W is the transfer conductance for water vapor between the leaves and bulk air.

Similarly the convective flux of sensible heat is

$$C = c_p \rho_a (T_L - T_a) g_H \tag{3}$$

where T_a is the air temperature and g_H is the transfer conductance for sensible heat between leaves and bulk air.

Assuming that

$$\frac{1}{g_W} = \frac{1}{g_c} + \frac{1}{g_a}, \tag{4}$$

where g_c is stomatal or canopy conductance and g_a is the boundary layer conductance for water vapor, and that

$$g_a \simeq g_H, \tag{5}$$

equations (1) to (4) can be combined to give an expression for transpiration rate that does not include leaf temperature (Monteith, 1965; Jarvis, 1981a):

$$E_T = \frac{sA + c_p \rho_a D g_a}{\lambda[s + \gamma(1 + g_a/g_c)]} \tag{6}$$

where D is the water vapor saturation deficit of the air (VPD) and s is the rate of change of saturation vapor pressure (e_s) with respect to air temperature, i.e., de_s/dT_a. The fluxes and conductances in these equations can be expressed on a unit leaf area basis or on a unit ground area basis depending on whether transpiration from leaves or canopies is being considered.

For aerodynamically rough canopies, such as coniferous plantations and forests, and for small leaves, such as coniferous needles, g_a is usually large in relation to g_c ($g_a/g_c \simeq 5$ to 20) and sufficiently large such that $c_p \rho_a D g_a$ is large in relation to sA (also up to 20 times) (Jarvis and Stewart, 1979; James and Jarvis, 1981).

Consequently, for conifers equation (6) can be simplified to the following (McNaughton and Black, 1973; Jarvis and Stewart, 1979; Jarvis, 1981a):

$$E_T = \frac{c_p \rho_a}{\gamma \lambda} D g_c. \tag{7}$$

This equation is similar to equation (2) for the condition of equal leaf and air temperatures. In conifers this is a good approximation in the forest because the small size of the leaves and large boundary layer conductance result in a close coupling between leaf and air temperatures so that leaf temperatures seldom differ from air temperatures by more than 0.5°C (Jarvis et al., 1976; Jarvis, 1981a; Grace, 1980).

Equation (7) shows that transpiration from a coniferous forest stand depends strongly on the VPD and the canopy conductance and that in the forest these are the variables of primary importance. That equation (7) is an adequate approximation in these circumstances has recently been elegantly demonstrated by Tan et al. (1978) (see Fig. 2). In growth rooms and assimilation chambers, on the other hand, the speed of air movement is generally much less (Van Bavel, 1973) so that g_a is much smaller than in the forest canopy and comparable in size or less than g_c. Consequently, in these situations as well as in still air conditions in the field, transpiration must be regarded as dependent on the additional variables contributing to equation (6), particularly the flux density of available energy and air temperature (Jarvis, 1981a). The flux density of available energy is largely the net radiation, since the storage terms can usually be neglected except near dawn and dusk (Jarvis et al., 1976). Apart from the effect of air temperature on VPD, temperature also has a strong influence on s, which has values of 0.45, 0.83, 1.45, and 2.44 mbar °C^{-1} at 0, 10, 20, and 30°C, respectively. The other physical coefficients, c_p, ρ_a, λ, and γ are only weakly dependent on temperature (Monteith, 1973).

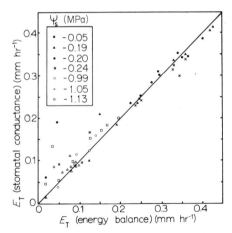

Fig. 2. The relation between transpiration rate calculated from stomatal conductance using equations (7) and (8) and transpiration rate determined from the energy balance or Bowen ratio method, for a stand of *Pseudotsuga menziesii* with an understory of *Gaultheria shallon*. The equation for the points is: E_T (stomatal conductance) = E_T (energy balance) 0.903 + 0.028, with $r^2 = 0.95$ for $n = 52$. ψ_s is soil water potential. (Redrawn from Tan et al., 1978; copyright 1978 by the Ecological Society of America.)

The rate of transpiration from leaves or canopy can readily be determined from equation (2) given leaf temperature, from equation (6), given the net radiation absorbed and boundary layer conductance, usually a function of windspeed (Jarvis et al., 1976; Jarvis and Stewart, 1979), or from equation (7). In all these cases the leaf or canopy conductance for water vapor transfer is required. Alternatively, in the absence of information on leaf or canopy conductance, the transpiration rate can be found by solving equation (3) for C and substituting into equation (1). This requires measurements of canopy–air temperature differences that are difficult to obtain with sufficient accuracy for conifers because they are usually very small. Since leaf temperatures are generally difficult to measure routinely in the field anyway, equations (6) and (7) provide the best estimates of transpiration, other than estimates based on micrometerological methods, such as Bowen ratio and eddy correlation methods (e.g., Tanner, 1968).

Determination of transpiration rate by solution of equations (2), (6), or (7) requires estimate or measurement of either stomatal or canopy conductance for water vapor transfer. Stomatal (or leaf) conductance can be measured in practice with diffusion porometers; several are now available for use on conifers. The conductance of all or part of the canopy is then derived as

$$g_c = \Sigma \, (\bar{g}_j L_j),\tag{8}$$

where $\bar{g}_j$ is the average leaf conductance of a cohort of the population of leaves of leaf area index L_j. Alternatively, canopy conductance can be calculated from the Bowen ratio, β, derived from micrometeorological measurements, as

$$g_c = \frac{g_a g_i}{g_i \left(\dfrac{s}{\gamma}\beta - 1\right) + g_a (\beta + 1)},\tag{9}$$

where g_i, the reciprocal of the "isothermal" or climatological resistance (Monteith, 1965; Stewart and Thom, 1973), is

$$g_i = \frac{A}{D} \cdot \frac{\gamma}{\rho_a c_p}.\tag{10}$$

Estimation of canopy conductance and transpiration rate using equation (8) requires measurement of the partial leaf area indices of different cohorts of foliage, as well as measurements of stomatal conductance. The accuracy of measurement of L_j is generally not high and is the largest source of error in determining transpiration in this way (Roberts *et al.*, 1980). Estimation of canopy conductance from the Bowen ratio using equation (9) is also subject to substantial error because the gradients of temperature and humidity over coniferous forest are very small and consequently difficult to measure with sufficient accuracy (Jarvis *et al.*, 1976).

B. Transpiration Rate

With one exception reported, mean hourly maximum transpiration rates from coniferous forest stands do not exceed 0.7 mm hr^{-1} (Table I). These data are for days of exceptionally good weather and in the absence of water stress. In general, maximum mean hourly rates are about 0.3 mm hr^{-1}. The figures in Table I include transpiration from ground vegetation and evaporation from the soil. In the case of the *Picea* stands, the contribution from these sources is negligible, but in the more open stands of *Pinus* and *Pseudotsuga menziesii* a substantial part of the total evaporative water loss is derived from the ground vegetation. Tan *et al.* (1978) have shown that as water stress increased the proportion of transpiration from

TABLE I

MAXIMUM MEAN HOURLY AND DAILY TRANSPIRATION RATES FOR
CONIFEROUS FOREST STANDS

Genus	Transpiration rate, E_T		Source
	mm hr^{-1}	mm day^{-1}	
Pinus sylvestris	0.33	3.0	Stewart and Thom (1973)
[+ *Pteridium aquilinum*]	0.48	2.7	McNeil and Shuttleworth (1975)
	0.57	3.8	Tajchman *et al.* (1979)
	0.38	2.3	Waring *et al.* (1979)
Pinus radiata	1.2	7.5	Denmead (1969)
	0.50	3.3	Moore (1976)
	0.22	1.2	D. Whitehead and R. K. Brownlie (unpublished data)
Pinus contorta	0.42	3.5	Gay (1971)
Picea abies	0.68	4.1	Tajchman (1972)
	0.50	2.8	Roberts (1978)
Picea sitchensis	0.52	3.0	Milne (1979)
	0.50	3.0	James and Jarvis (1981)
Pseudotsuga menziesii	0.52	4.8	McNaughton and Black (1973)
(+ *Gaultheria shallon*)	0.42	3.1	Tan *et al.* (1978)
	0.54	3.7	Spittlehouse and Black (1979)

the understory of *Gaultheria shallon* rose from 45 to 70%. Similarly, Roberts *et al.* (1980) have shown that the proportion of the transpiration from the *Pteridium aquilinum* understory in a stand of *Pinus sylvestris* rose from 15 to 60% in late summer.

The high rates of transpiration reported by Denmead (1969) for three days of measurements on *Pinus radiata,* near Canberra, Australia, are anomalous and need confirmation. Although *Pinus radiata* grows very fast, there is no evidence in the literature to suggest that canopy conductances are remarkably high in comparison with those of other species (see Section II,C) and VPD's did not exceed 14 mbar on the day with the highest transpiration rates.

Again, with the exception of Denmead's data for *Pinus radiata,* maximum reported daily rates of transpiration do not exceed 7 mm (Table I) and are generally about one-third of that amount.

Equation (6) suggests that a very wide range in rates of transpiration might be attainable in practice depending upon the particular combination of environmental variables, particularly available energy, VPD, temperature and windspeed, and canopy conductance. Similarly, equation (7) suggests that very high rates of transpiration might be attained at large VPD's. For example, a VPD of 25 mbar in combination with a canopy conductance of 0.02 m sec^{-1} might be expected to result in a transpiration rate from a stand of 2.6 mm hr^{-1}. However, such high transpiration rates

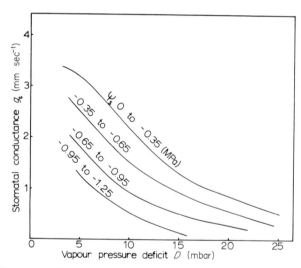

Fig. 3. The relationship between stomatal conductance, g_s, and vapor pressure deficit, D, for *Pseudotsuga menziesii* at four different ranges of soil water potential, ψ_s. The smooth curves have been redrawn from stomatal resistance data of Tan *et al.* (1977).

are never found. The rates of transpiration are constrained within the relatively narrow limits shown in Table I by closure of stomata associated with increases in VPD, and, to a lesser extent, as a result of water stress. Stomatal closure in response to increasing VPD of the air, now widely reported in many species, is also found in *Picea sitchensis* (Grace et al., 1975; Watts et al., 1976; Watts and Neilson, 1978), *Picea engelmannii* (Kaufmann, 1975), *Pinus sylvestris* (Ng, 1978; Jarvis, 1981b), *Pinus radiata* (Bennett and Rook, 1978; Benecke, 1980), and *Pseudotsuga menziesii* (Tan and Black, 1976; Tan et al., 1977) (Fig. 3). This effect causes the transpiration rate to reach a plateau at a particular VPD rather than continuing to increase with increasing VPD (Watts and Neilson, 1978; Landsberg and Butler, 1980). In addition, the analysis by Landsberg and Butler (1980) leads to the conclusion that the effect of humidity in decreasing stomatal conductance causes the proportion of available energy dissipated as latent heat to fall as the available energy increases, so that transpiration rates in high energy environments are likely to be similar to those in low energy environments.

C. TRANSFER CONDUCTANCES

1. Boundary Layer Conductance

The boundary layer conductance of coniferous shoots depends not only on windspeed but also on the mutual shelter that one needle gives another (Landsberg and Thom, 1971). Equations in the literature describing dependence on these variables were reviewed by Jarvis et al. (1976, Table VI). The shelter factor would be expected to vary with the size and arrangement of needles. The dense grouping of needles in some pines, for example, might be expected to lead to considerably more aerodynamic interference among needle boundary layers than occurs in the shoots of spruce with more regularly distributed needles. However, at the present time, detailed information on such interactions is lacking. The data summarized by Jarvis et al. (1976) indicate boundary layer conductances for water vapor (on a projected needle area basis) ranging from 0.06 to 2.0 m sec^{-1} at a windspeed of 1 m sec^{-1} depending on the species, needle density, and direction of air flow. For typical mid-canopy shoots of *Picea sitchensis* boundary layer conductances for water vapor are 0.07 and 0.1 m sec^{-1} at windspeeds of 0.4 and 1 m sec^{-1}, respectively.

Figures for the boundary layer conductance for canopies were also reviewed by Jarvis et al. (1976, Table V). A simple, practical generalization emerged that for coniferous forest $g_a \simeq 0.1\, u(h)$, where $u(h)$ is the half-hourly or hourly average windspeed at the top of the canopy in

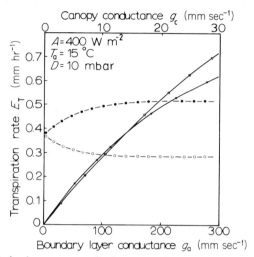

Fig. 4. The relation between transpiration rate and canopy g_c (solid lines) and boundary layer g_a conductance (dashed lines) for the conditions shown. Each pair of lines covers the range of likely values for the other conductance for coniferous forest: ∇, $g_a = 300$ mm sec^{-1}; $\blacktriangledown$, $g_a = 100$ mm sec^{-1}; $\bullet$, $g_c = 20$ mm sec^{-1}, $\bigcirc$, $g_c = 10$ mm sec^{-1}.

m sec^{-1}. This generalization is based on 15 studies on stands of 10 species by 13 sets of investigators in 6 countries. At the present time there is no systematic information on the effects of species, height, spacing, thinning regime or any other stand characteristic on boundary layer parameters. For the practical purpose of solving equation (6), g_a can be taken as 0.2 m sec^{-1}. However, as has been often pointed out, the rate of transpiration from coniferous forest and foliage is largely independent of the boundary layer conductance because it is an order of magnitude larger than the canopy or stomatal conductance (e.g., McNaughton and Black, 1973; Jarvis and Stewart, 1979; Jarvis, 1981a). This is illustrated in Fig. 4.

2. Stomatal and Canopy Conductance

Although a number of species of conifers seem to have rather low stomatal conductances of less than 5 mm sec^{-1} for water vapor on a projected leaf area basis (e.g., Pinus halepensis, Pinus ponderosa, Picea engelmannii, Tsuga heterophylla), values in other species can reach 10 mm sec^{-1} (e.g. Pinus radiata, Picea sitchensis, Pseudotsuga menziesii) (see Jarvis et al., 1976, Table XII; Jarvis, 1981b). Probably many of the low values reported are for plants regularly undergoing stress of some kind.

The responses of conifer stomata to environmental variables (light, temperature, CO_2, ψ_l, and VPD) have recently been reviewed by Jarvis (1980) and will not be discussed here. It is, however, pertinent to point out that in several recent studies of stomatal conductance in forest canopies, it has proved possible to account for approximately three-quarters of the short-term variations in stomatal conductance of populations of leaves of particular age and position by variation in the incident photon flux density and ambient VPD, with variations in ψ_l contributing significantly to the variation during stress periods (Jarvis, 1976; Watts *et al.*, 1976; Running, 1976; Tan and Black, 1976; Tan *et al.*, 1978; D. Whitehead, P. G. Jarvis, and R. H. Waring, unpublished data; Beadle *et al.*, 1982).

Maximum canopy conductance of coniferous forest canopies is about 0.02 m sec^{-1} for *Picea sitchensis, Picea abies,* and *Pseudotsuga menziesii,* and about 0.01 m sec^{-1} for *Pinus sylvestris* (Stewart and Thom, 1973; McNaughton and Black, 1973; Jarvis *et al.*, 1976; Tan and Black, 1976; Milne, 1979; James and Jarvis, 1981). Canopy conductance has been shown to decrease in response to increases in VPD and decreases in soil water potential ψ_s (Tan and Black, 1976; Jarvis, 1981a). Often, better correlations are obtained between the overall canopy conductance and particular environmental variables than with the stomatal conductance of populations of needles of certain age and position within the canopy.

Canopy conductance derived from measurements of stomatal conductance on leaf populations within the canopy, using equation (8), depends on the availability of satisfactory estimates of LAI of the different cohorts of leaves at different levels. The measurement of total and partial LAI's is both labor-intensive and prone to large errors. The largest single error in the calculation of transpiration rate from equations (6) or (7) lies in the accurate determination of LAI for the calculation of canopy conductance (Roberts *et al.*, 1980). The agreement between the two independent estimates of transpiration rate shown in Fig. 2 is, therefore, remarkable.

At the present time, the only sure way of determining the LAI of a stand is the harvesting of a number of whole trees consisting of a representative sample of the height or dbh (diameter at breast height = 1.3 m) size classes (e.g., Whitehead, 1978). However, considerable hope is attached to the regressions between leaf area and basal area of sapwood that recently have been appearing in the literature (e.g., Grier and Waring, 1974; Waring *et al.*, 1977; Whitehead, 1978; Rogers and Hinckley, 1979) and that appear to be reasonably species specific over a wide range of habitats (see Section III,B,4). If this specificity is confirmed, it will be possible to obtain estimates of LAI for a stand from measurements of sapwood basal area made with an increment borer. Such estimates of leaf area based on highly significant and specific regressions are likely to be at least as good as those obtained from a small (< 10) sample of harvested trees. Although substantial amounts of data on leaf weight have been gathered in the past, their use in deriving leaf areas is limited because the specific leaf area typically changes by a factor of two from top to bottom of a coniferous canopy.

Estimation of the partial LAI's of leaves of particular cohorts at particular nodes or stem heights is much more difficult on any kind of routine basis. Tan *et al.* (1978) were apparently successful in estimating the LAI of each of three layers of a canopy of *Pseudotsuga menziesii* by sampling an average branch at each whorl on four trees for leaf area. Study of the phenology and demography of needle populations and needle area has hardly begun, although demographic models of needle populations and area are required for simulations of canopy processes lasting more than a few days. At the moment, careful nondestructive studies of individual needles in the canopy, coupled with destructive sampling of branches of needles, are needed to form the basis of such models (Jarvis *et al.*, 1981).

III. WATER MOVEMENT IN THE TREE

A. ORIGIN OF WATER POTENTIAL

1. Simple Resistance Model

a. The Flow System. As water is lost by evaporation from leaves, ψ in the cell walls falls and, because of the hydraulic nature of the pathway for water movement through the plant, a wave of changing ψ moves down through the system. There is only a small drop in potential between average ψ of the leaf mesophyll and the xylem pressure potential, because the solute concentration of xylem water is low in the extremities of the plant and the flow resistance between mesophyll cells and xylem is small. In trees larger drops in potential occur in the xylem in the twigs and branches as well as in the trunk and roots (Jarvis, 1976). These differences in ψ drive the water flow against the frictional resistances in the pathway and the gravitational pull. Even when absorption by roots is equal to water losses by leaves, there will be a substantial drop in potential in the tree to overcome these resistances to flow (Jarvis, 1975).

The cohesion theory for water movement in plants (see Preston, 1961; Greenidge, 1957; Kozlowski, 1961; Zimmermann, 1971) has been used to explain the common occurrence of ψ's below -2.0 MPa in conifer foliage without fracture of the water columns in the xylem. The hydraulic nature of the water pathway in tall trees is demonstrated by changes in ψ_l closely following changes in stem water potential after the stem is severed from the root system under water. Roberts (1977) found that an increase of about 0.3 MPa in ψ_l occurred in trees of *Pinus sylvestris* within 30 min.

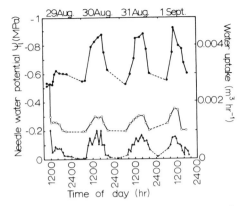

Fig. 5. The changes in needle water potential in cut and control trees and the rates of water uptake during four days. $\bigcirc$ = cut trees, $\bullet$ = control trees, $\blacktriangle$ = water uptake. (Redrawn from Roberts, 1977.)

of the tree being severed (Fig. 5) and that similar but smaller changes in ψ_l occurred in *Picea abies* within the same time after cutting (Roberts, 1978). The resistance to water movement through stems of conifers occurs in the tracheid lumens and the bordered pit structures (Section III,B,2).

A water deficit in nearby cells results from the fall in ψ in the xylem since water moves to maintain equilibrium between cell ψ and xylem ψ. The size of the water deficit therefore depends on the water content—ψ relationship for the tissue, the location of the tissue in relation to the sources of water, and the flow rate through the pathway (Jarvis, 1975).

b. Definition of Resistance. In a one-dimensional, rigid system where the difference in ψ between soil and leaves is linearly dependent on the volume flow rate of water, q, it is customary to apply an Ohm's law analogy (Huber, 1924; van den Honert, 1948; Richter, 1973) so that

$$\psi_l = \psi_s - \sum_{i=s}^{i=l} (q_i R_i) - h \rho_w g, \qquad (11)$$

where q_i and R_i are partial, series-linked flow rates and resistances in the pathway from soil to leaf; $h\rho_w g$ is the gravitational pull on the column of water (0.01 MPa over a vertical distance of 1 m). If the dimensions of q_i are $m^3 \ sec^{-1}$, the dimensions of the resistance R_i are $kg \ m^{-4} \ sec^{-1}$ (or Pa $sec \ m^{-3}$) (Table II). The product $(q_i \cdot R_i)$ has the dimensions of $kg \ m^{-1} \ sec^{-2}$, which are the dimensions of ψ when expressed as pressure (Pa or N m^{-2}).

TABLE II

DEFINITIONS OF FLOW RESISTANCE AND CONDUCTANCE[a]

Name	Symbol	Definition	Units	Name	Symbol	Units
Resistance	R	$\Delta\psi/q$	Pa sec m^{-3}	Conductance	K, G	m^3 Pa^{-1} sec^{-1}
Resistivity	ρ	$(\Delta\psi A_s)/(ql)$	Pa sec m^{-2}	Conductivity	σ	m^2 Pa^{-1} sec^{-1}
Relative resistance	R_r	$(\Delta\psi A_s)/(ql\eta)$	m^{-2}	Relative conductivity or permeability	k	m^2

[a]From Jarvis (1975).

In a multibranched plant system it is difficult to measure the partial flows and resistances in each individual part. Consequently, total evaporation from a plant or forest canopy, E_T (m^3 sec^{-1}), is often used in equation (11) in place of q_i, as follows:

$$\psi_l = \psi_s - E_T R_p - h\rho_w g \qquad (12)$$

and

$$R_p = (\psi_s - \psi_l - h\rho_w g)/E_T. \qquad (13)$$

R_p is then the *resistance* of the entire pathway of parallel and series linked conduits from soil to leaves, its value depending in part on the degree of branching in the system. The reciprocal of resistance is *conductance, K,* having the dimensions of m^3 Pa^{-1} sec^{-1} (Table II). If E_T is expressed per unit area (ground, leaf, or sapwood cross-section), the dimensions of the resistance and conductance are kg m^{-2} sec^{-1} (Pa sec m^{-1}) and m^2 sec kg^{-1} (m Pa^{-1} sec^{-1}), respectively. It is desirable to standardize these units (Jarvis, 1975).

c. Gradient of Water Potential. It follows from equation (12) that if no flow of water through the plant system is occurring, for example, during the night or when the foliage is wet, the gradient of ψ is equal only to the hydrostatic gradient so that the difference between ψ_s and ψ_l in a 10-m tall tree would be expected to be 0.1 MPa. The fact that values of ψ_l near zero are hardly ever recorded even before dawn or under conditions of no transpiration has been discussed by Ritchie and Hinckley (1975). Nonetheless the predawn ψ_l is useful in indicating (a) the value of ψ_s actually experienced by the plant, and (b) the value of ψ_l at which the tree begins each daylight period (Hinckley et al., 1978). Consequently predawn values of ψ_l have been used in models which describe the daily and seasonal fluctuations in stomatal conductance and ψ_l (e.g., Reed and Waring, 1974; Running et al., 1975).

By taking samples of foliage at different heights, gradients of ψ_l equal to the hydrostatic gradient have been measured in tall trees (Table III). Scholander et al. (1965) measured such gradients in an 82-m tall tree of Sequoia sempervirens, a 79-m tall tree of Pseudotsuga menziesii, and a 90-m tall tree of Sequoiadendron giganteum. Zimmermann (1971) and Bolton (1976) both considered that these examples provided evidence that the gradient of ψ_l in tall trees was equal to the hydrostatic gradient. Tobiessen et al. (1971) measured a gradient of less than the hydrostatic gradient in a 90-m tall Sequoiadendron giganteum tree and Hinckley and Ritchie (1970) reported inverted gradients in Abies amabilis. Hellkvist et al. (1974) measured gradients of ψ_l equal to the hydrostatic gradient in the early morning and also reported inversion of the gradient in the middle of the day at the top of trees of Picea sitchensis.

It was first pointed out by Richter (1973) that the measurements of ψ_l on transpiring leaves at the ends of branches could lead to apparent positive or negative gradients of varying magnitude depending upon the rate of transpiration in different parts of the crown. As shown in Section II, the rate of transpiration depends on the local absorption of radiation, the windspeed, and stomatal conductance in different parts of the crown. Thus, not too much significance should be given to apparently anomalous

TABLE III

GRADIENTS OF WATER POTENTIAL IN SEVERAL SPECIES[a,b]

Species	Gradient (MPa m^{-1})	Height of tree (m)	Source
Abies amabilis	0.02–0.15	8.0	Hinckley and Ritchie (1970)
A. amabilis	0.03–0.12	1.5	Kotar (1972)
Chamaecyparis obtusa	0.01–0.02	16.0	Morikawa (1974)
Juniperus communis	0.18–0.32	—	Wiebe et al. (1970)
Picea abies	0.02–0.04	20.0	Halbwachs (1970)
P. sitchensis	0.10–0.20	11.5	Hellkvist et al. (1974)
Pinus monticola	0.02–0.03	45.0	Cline and Campbell (1976)
P. resinosa	0.01–0.08	15.0	Waggoner and Turner (1971)
P. sylvestris	0.03	—	Halbwachs (1970)
Pseudotsuga menziesii	0.01–0.01	79.0	Scholander et al. (1965)
P. menziesii	0.04–0.12	14.0	Tan et al. (1978)
Sequoiadendron giganteum	0.008	90.0	Tobiessen et al. (1971)
S. giganteum	0.01	90.0	Rundel and Stecker (1977)
Sequoia sempervirens	0.01–0.01	82.0	Scholander et al. (1965)

[a] From Hinckley et al. (1978).

[b] The two numbers in the gradient column refer to measurements made under no flow and during flow.

gradients based on measurements of ψ_l: gradients of the appropriate size and direction are only to be expected along an unbranched flow continuum, e.g., in the stem (Jarvis, 1975). Stomatal closure in shoots in the upper part of the canopy leading to a reduced upward flow of water can easily result in apparently inverted gradients.

Gradients of ψ_l much larger than the hydrostatic gradient have been found under conditions of normal transpiration in a number of conifers (see Table III). The steepest gradients were reported by Wiebe et al. (1970) in *Juniperus communis* (0.1 to 0.32 MPa m^{-1}) and by Hellkvist et al. (1974) (0.1 to 0.2 MPa m^{-1}) in *Picea sitchensis*. However, as pointed out above, it is difficult to interpret such gradients when the points of measurement do not lie on the same flow pathway. To get measurements on the same flow pathway, samples of foliage can be enclosed in humid plastic bags (Hellkvist et al., 1974), or transducers of various kinds can be inserted into the xylem to measure the gradients, e.g., osmotic tensiometers (Richards, 1973), plaster-of-Paris resistance blocks (Spomer, 1968), and thermocouple psychrometers (Wiebe et al., 1970). Making holes in the tree may damage the conducting system as a result of air entry into the tracheids (Section III,B,1,c) and, after a short time, a layer of resin may be formed between wood and sensor in resinous species. Thus, care should be used in the interpretation of results from these methods.

Hellkvist et al. (1974) used the humid bag method to measure gradients of ψ in the stem and along branches of *Picea sitchensis*. They found gradients of 0.09 to 0.28 MPa m^{-1} in the branches from the lowest to the highest levels in the canopy: gradients in the secondary branches were twice as large as in the primary branches. Gradients in roots of 0.2 MPa m^{-1} were also measured, ψ rising with increasing distance from the trunk but in a nonlinear manner.

Although the pressure chamber has enabled rapid measurements of foliage ψ (Scholander et al., 1965; Waring and Cleary, 1967; Roberts and Fourt, 1977), because of the lack of techniques for directly measuring ψ in the flow pathways without damage to the system, it is difficult to make meaningful comparisons between the wide range of gradients reported by various investigators using different techniques on different species.

2. Relative Size of Resistances in Parts of the Tree

The overall resistances for trees of several species in forest stands are given in Table IV. It appears that the resistance of *Picea sitchensis* is appreciably larger than the resistance of the other species.

Much work on herbaceous plants suggests that the major part of the resistance to water movement occurs in the roots (see Jarvis, 1975). When transpiring *Pinus sylvestris* trees were cut off from their roots under water,

TABLE IV
Estimates of Flow Resistance for Different Species

Species	Stems ha⁻¹	LAI (projected)	$R_1{}^a$ (MPa sec m⁻³)	$R_2{}^b$ (MPa sec m⁻¹)	$R_3{}^c$ (MPa sec m⁻¹)	Source
Picea sitchensis	4320	10.2	1.0×10^7	21.0×10^7	2.3×10^7	Hellkvist et al., (1974)
Pinus resinosa	1880	$(5.5)^d$	0.2×10^7	6.0×10^7	1.1×10^7	Waggoner and Turner (1971)
Pinus sylvestris	$(740)^d$	$(2.6)^d$	0.6×10^7	22.3×10^7	8.5×10^7	Roberts (1977)
	3281	3.1	0.2×10^7	1.8×10^7	0.6×10^7	D. Whitehead, P. G. Jarvis and R. H. Waring (unpublished data)
	608	2.4	0.1×10^7	3.2×10^7	1.3×10^7	
Pseudotsuga menziesii	840	$(3.6)^d$	0.2×10^7	9.7×10^7	2.7×10^7	Nnyamah et al. (1978)
	e	e	—	—	1.1×10^7	Waring and Running (1978)

[a] $\Delta\psi_l/q$
[b] $\Delta\psi_l A_t/q$ — A_t is area of foliage per tree.
[c] $\Delta\psi_l A_g/q$ — A_g is area of ground per tree.
[d] Not given in the original source, but estimated or found elsewhere.
[e] Data not given in the source.

TABLE V

MEAN RESISTANCE OF PARTS OF THE PATHWAY IN *Pinus sylvestris*[a]

| Tree part | Resistance | |
	MPa sec m^{-1}	%
Root[b]	11.4×10^7	51
Stem[b]	8.4×10^7	38
Branch[b]	2.5×10^7	11
Total	22.3×10^7	
Stem + Branch[c]	5.7×10^7	

[a]From Roberts (1977).
[b]Resistance estimated from plant water potential and transpiration rates.
[c]Resistance estimated from a tree-cutting experiment.

Roberts (1977) found that ψ_1 rose about 0.6 MPa at midday, suggesting that a major part of the flow resistance was in the roots. In experiments with large trees cut off from their roots and standing in water Roberts (1977, 1978) measured water uptake and ψ_1 over several days. At midday there was a difference in ψ_1 between a control and a cut *Pinus sylvestris* tree of 0.6 MPa at an uptake rate of 0.001 m^3 hr^{-1} (Fig. 5). Roberts concluded that the root resistance accounted for 51% of the total tree resistance and that the resistance in the stem and branches was equal to the root resistance for whole trees (Table V). In *Picea abies* the difference between ψ_1 in control and cut trees was only 0.2 MPa at midday, suggesting that in this species a larger resistance occurred in the stem than in the roots (Roberts, 1978). This was in agreement with conclusions reached by Jarvis (1976) from a comparison between ψ_1 and transpiration rates for plantations of *Pinus sylvestris* with 800 stems ha^{-1} and *Picea sitchensis* with 4100 stems ha^{-1}. At equal transpiration rates per area of ground there was five times as large a volume flow through the pine trees as through the spruce trees for about the same drop in ψ (Fig. 6). Jarvis concluded that resistance to flow in the pine tree stems was about one-fifth of the resistance in individual spruce stems. Hellkvist *et al.* (1974) had previously concluded that the resistance in stems of *Picea sitchensis* was larger than in pine and two to three times the resistance in the roots. Nnyamah *et al.* (1978) also concluded that stem resistance was one to two times larger than root resistance in *Pseudotsuga menziesii*. Running (1979) calculated the root resistance in young *Pinus contorta* trees to be 67% of the total tree resistance at a root temperature of 7°C and the root resistance increased to 93% of the total at 0°C (see Section V,B).

Hellkvist *et al.* (1974) also inferred values of flow resistance in stems and branches of *Picea sitchensis* from measurements of ψ. The gradient of ψ in the primary branches was twice that in the stem and calculations suggested that the contributions of stem:primary:secondary branches to overall resistance were in the proportions 500:10:1. However, these values were calculated on the basis of assumed, rather than measured, volumes of water flowing through the different segments of the pathway.

Opinions differ as to the proportion of the resistance in the living cells of the leaf. Richter (1973) reported that the gradient of ψ in transpiring and non-transpiring twigs of *Taxus baccata* was practically the same, suggesting that there was negligible resistance to water flow in the leaf tissue. Tyree *et al.* (1975) carefully infiltrated the air spaces in leaves of *Tsuga canadensis* with water and then selectively removed some leaves. They concluded that most ($>\frac{2}{3}$) of the resistance to water flow occurred in the xylem up to the leaf tissue, rather than between the tracheids and the living cells in the leaves.

Although equation (11) predicts a linear dependence of flow rate on water potential with a constant plant resistance, in many herbaceous species the relationship between E_T and ψ_1 is not linear and implies a plant resistance that decreases with increasing flow rate (see reviews by Weatherley, 1970; Boyer, 1974; Jarvis, 1975). Such a flow-dependent resistance may therefore be a mechanism by which flow rate through the plant is maintained without decreasing ψ_1. A decreasing resistance with increasing flow rate has been reported for several conifers but in completely uncon-

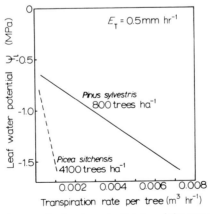

Fig. 6. The relation between leaf water potential ψ_1 and the rate of transpiration per tree for two stands of different species and density transpiring at the same rate of $E_T = 0.5$ mm hr^{-1} per unit ground area. (Redrawn from data given by Jarvis, 1976.)

trolled conditions in the field (Roberts, 1977; Nnyamah *et al.*, 1978; Waring and Running, 1978). A careful laboratory study of this phenomenon in conifers is needed to see if it is of significance.

In drying soil, a substantial proportion of the total resistance is found in the soil. Nnyamah *et al.* (1978) concluded that the bulk soil to root resistance in a *Pseudotsuga menziesii* plantation was constant and small in comparison with the root resistance, when the soil was wet. In dry soil a contact (or perirhizal) resistance between soil and root accounted for up to three-quarters of the bulk soil to xylem resistance. Sands and Theodorou (1978) found with potted seedlings of *Pinus radiata* that the soil resistance became the major part of the total soil to leaf resistance as the soil dried. In seedlings with mycorrhizae the ratio of soil to plant resistance was 0.8 at ψ_s of -0.05 MPa and the ratio increased to 7.4 when ψ_s reached -1.2 MPa, on the assumption that the plant resistance remained constant.

However, observations in the field by Sucoff (1972), Hinckley and Ritchie (1973), and Reed and Waring (1974), that up to 80% of the rooting zone soil water can be depleted before there is any effect on predawn ψ_l, suggest that the rate of water movement through the perirhizal zone usually does not limit water uptake by the roots. The perirhizal resistance, although a function of soil hydraulic conductivity and rooting density (Reicosky and Ritchie, 1976; Taylor and Klepper, 1978), is not usually a large resistance in comparison with the root resistance in conifers.

3. Conductivity and Relative Conductivity

a. Definition of Conductivity and Relative Conductivity. Two sets of observations lead to the desirability of isolating simple geometrical properties of the pathway, such as length and area, from the resistance or conductance. First, the water potential of the foliage in herbaceous and woody plants is generally maintained within narrow limits and rarely falls much below -3.0 MPa in mesophytes. Secondly the cross-sectional area of conducting tissue per unit of foliage varies among species and is, for example, ten times larger in conifers than in herbaceous plants. This ratio, "the Huber value," is larger in young stems near to the tops of trees than in older stems near the base (Zimmermann, 1971, quoting work of Huber, 1928, with *Abies concolor*). These observations suggest effective feedback between the volume of water that flows through the plant and the size and properties of the conducting pathway and implies homeostasis between leaf area and cross-sectional area of the xylem. This is recognized in the pipe model theories of Shinozaki *et al.* (1964a,b), the leaf area/sapwood area models of Waring (Running *et al.*, 1975; Waring *et al.*, 1977; Waring and Running, 1978), and the stand density/ψ hypothesis of Jarvis (1975) (see Section VI,A). All are based on the idea that the cross-sectional area of conducting tissue in tree stems is a variable which plays a part in the conservation of ψ within limits beneficial to growth.

Continuing the Ohm's law analogy, it is useful to define *conductivity* for the pathway as:

$$\sigma = \frac{q}{\Delta\psi} \frac{l}{A_s} = \frac{1}{R_p} \frac{l}{A_s}, \tag{14}$$

where $\Delta\psi$ is the drop in potential across a segment of pathway of conducting tissue, of length l, and cross-sectional area A_s. The *resistivity*, ρ, is the reciprocal of the conductivity. To compare species it is more useful to include cross-sectional area of sapwood (A_{sw}) than that of tracheid lumens in calculations of pathway properties and water flow since this is a useful and easily measurable variable.

Since viscosity varies with temperature, pressure, and solute concentration, it has also been found useful to extract it from the resistivity. Isolation of the viscosity η leads to the definition of *relative conductivity k*, following Heine (1971) as

$$k = \frac{q\eta}{\Delta\psi} \cdot \frac{l}{A_s}. \tag{15}$$

The dimensions and interrelationships of these properties are summarized in Table II.

Wood technologists have been interested in flow of liquids through wood for the purposes of infiltration with preservatives and an equation identical to (15) is used based on Darcy's law (Kelso *et al.*, 1963; Bailey and Preston, 1970; Siau, 1971). In such cases the area used in equation (15) is usually the total cross-sectional area and k is often referred to as *permeability*.

However, when defining a property of the conducting system itself, the conducting area should be used and this can be identified using radioactive tracers (Heine, 1970; Heine and Farr, 1973) or biological stains that do not diffuse throughout the wood, such as toluidene blue (Booker, 1977). In deciduous hardwoods, probably only the elements formed during the current or previous season conduct water (Greenidge, 1955a; Peel, 1965), but in conifers the earlywood in all the sapwood rings conducts (Swanson, 1975; Kozlowski *et al.*, 1966, 1967; Booker and Kininmonth 1978), whereas the heartwood, drywood, latewood and undifferentiated wood do not conduct (Plate I).

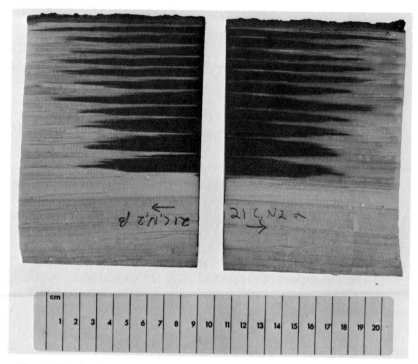

Plate I. The dye flow pattern through a specimen of stem of 28-year-old *Pinus radiata* from just below the crown. Only the sapwood earlywood conducts; the latewood and undifferentiated wood show virtually no conduction. Penetration time 55 sec with toluidine blue solution. [Photo from Booker and Kininmonth (1978) with permission.]

b. Measurement of k. Usually, measurements of relative conductivity are made in the laboratory on cores or segments from green stem. The apparatus is designed to apply a constant pressure difference of water across a specimen so that k is calculated under conditions of steady state flow (Huber and Merz, 1958a; Kelso *et al.*, 1963; Peel, 1965; Comstock, 1965; Banks, 1968; Petty and Preston, 1969; Siau, 1971; Puritch, 1971; Lin *et al.*, 1973; Gregory, 1977a).

A linear relationship between pressure difference across a wood specimen and the flow rate of water passing through it is taken as evidence that flow in the wood complies with the use of equation (15) (Petty and Preston, 1969; Puritch, 1971; Gregory, 1977a; Bolton and Petty, 1978b). In some cases departures from this relationship have been found and attributed to air blockage, the "Jamin" effect (Kelso *et al.*, 1963). To prevent this problem occurring the ends of the specimens should be carefully trimmed with a microtome or a chisel and the water used should be distilled, filtered through a 100-nm filter, and deaerated by ultrasonic cavitation under vacuum, and the specimens should be saturated before measurements are made (Booker, 1977). Only then is the flow constant with time at a constant pressure difference and the relationship between flow and pressure difference linear. Booker (1977) further suggests that many of the measurements made without deaerated water (e.g., Farmer, 1918; Huber and Merz, 1958a,b; Peel, 1965; Lin *et al.*, 1973) have produced values of k that are too low because of air blockage in the conducting elements. Table VI contains values of k for different species from studies in which precautions were taken to avoid air blockages. Values of k for conifer stem wood are about $\frac{1}{5}$ to $\frac{1}{3}$ of those for hardwoods (Cowan and Milthorpe, 1968, quoting Huber, 1956; Heine, 1971, quoting Farmer, 1918).

Bailey and Preston (1969) considered that radial flow of water through *Pseudotsuga menziesii* sapwood was negligible compared to axial flow (ratio of 1:5000) and that there was no conduction through ray parenchyma cells in a radial direction and very little through ray tracheids because the pits were often encrusted and the area of the longitudinal faces was small. Kininmonth (1970) found that the ratio of radial to axial k in the sapwood of *Pinus radiata* was 1:2000. This ratio is much smaller than in hardwoods (Kininmonth, 1971) but still very highly in favor of axial flow. Erickson and Balatinecz (1964) used a technique in which styrene was forced through wood of *Pseudotsuga menziesii* followed by a polymerizing liquid so that the styrene solidified and could be examined. The styrene moved from tracheids into ray tracheids and radially in ray

TABLE VI
MEAN AXIAL RELATIVE CONDUCTIVITIES FOR DIFFERENT SPECIES

Species	Relative conductivity k^a (m^2)	Source
Pseudotsuga menziesii	9.9×10^{-12}	Erickson and Crawford (1959)
Pinus radiata	6.8×10^{-12}	Kininmonth (1970)
	5.5×10^{-12}	Booker and Kininmonth (1978)
Pinus sylvestris	6.7×10^{-12}	Gregory (1977b)
Pinus contorta	3.4×10^{-12}	R. E. Booker (unpublished data)
	2.1×10^{-12}	W. R. N. Edwards (unpublished data)
Tsuga canadensis	5.4×10^{-12}	Comstock (1965)
Tsuga heterophylla	5.3×10^{-12}	Erickson and Crawford (1959)
Picea sitchensis	4.0×10^{-12}	W. R. N. Edwards (unpublished data)

[a] From measurements made on stem sections in the laboratory with precautions to prevent air blockage during measurements.

tracheids and, occasionally, from ray tracheids into longitudinal tracheids, but the ray parenchyma was almost impermeable. However, we also know that water and solutes do move between parenchyma ray cells (Wodzicki and Brown, 1970).

Relative conductivity increases toward the top of the tree (Comstock, 1965; Puritch, 1971; Booker and Kininmonth, 1978) (Table VII) and decreases across the sapwood toward the heartwood (Puritch, 1971; Markstrom and Hann, 1972). In many species k decreases linearly with increasing wood density (Comstock, 1965; Gregory, 1972; Booker and Kininmonth, 1978) and changes in k with position in the tree stem can be partly attributed to differences in wood density resulting from different growth rates. In *Pinus radiata,* Booker and Kininmonth (1978) found that $k = (20.475 - 0.0329 \, \rho_{ds})10^{-12}$ (m²), where ρ_{ds} is the dry density of sapwood (kg m⁻³). The value of k is reduced by a fall in sapwood water content (see Section IV,C) and is also reduced when the sapwood is infected by fungal attack (Puritch and Petty, 1971; Puritch, 1971; Coutts, 1977; Gregory, 1977b) and insect attack (Coutts, 1969).

Siau (1971) considered that k in pines was about ten times higher than in spruces and cedars. Table VI shows that the value of k for *Pseudotsuga menziesii* is higher than for pines and the values for hemlocks and spruce are only slightly lower than those for pines. In view of the arguments previously discussed in Sections III,A,1 and III,A,2, it is interesting to note that although the sapwood cross-sectional area is larger in pines than in firs, hemlocks, or spruces, the values of k in these species are not larger than in pines.

Zimmerman (1978) recently discussed the hydraulic architecture of several diffuse-porous trees. Following Huber (1928), Zimmermann expressed the conductivity of different parts of the stems of the trees on the

TABLE VII

AVERAGE VALUES OF PERMEABILITY AT DIFFERENT HEIGHTS IN TWO TREES OF *Pinus radiata*[a]

| Tree | Height (m) | Number of rings of | | | Average sapwood k (m²) |
		Sapwood	Heartwood	Total	
1	0.6	15	>7	>22	5.09×10^{-12}
	9.9	12	6	18	7.95×10^{-12}
	19.5	10	4	14	8.02×10^{-12}
2	0.6	17	>3	>20	2.54×10^{-12}
	5.9	14	7	21	4.89×10^{-12}
	17.4	12	2	14	5.74×10^{-12}

[a] Average heartwood permeability was zero. (From Booker and Kininmonth, 1978.)

basis of unit fresh weight of foliage, and called this "leaf-specific conductivity" or LSC. It is unfortunate that the choice of this parameter does not enable species to be compared readily when they have different specific leaf areas and cross-sectional areas of conducting sapwood. Nonetheless, the results are interesting and provocative. Because values of LSC of the stem were higher than values for the branches, Zimmermann deduced that vascular junctions were hydraulic constrictions. Values of LSC throughout the tree were related to vessel diameter so the values were lower in branches and lower parts of the stem than in the upper parts of the stem. Since the ratios of conducting area to leaf weight were similar, Zimmermann concluded that the pressure drop across the upper lengths of the stem was less than the drop across the lower lengths of the stem for the same flow rates per unit leaf weight. This suggests that the peripheral parts of the tree are more vulnerable to water stress than the trunk. He also made the curious observation that the flow rate of dilute potassium chloride through the vessels was higher than that for tap water, which was, in turn, higher than that for distilled water. There is no adequate explanation for this at present.

B. THE SAPWOOD PATHWAY

1. Properties of Sapwood

a. Structure and Tracheid Size. Tracheid cells occupy most of the volume of coniferous sapwood, with the remaining volume occupied chiefly by ray parenchyma and ray tracheids. The sapwood of *Pinus strobus* consists of approximately 93% tracheids, 6% ray cells, and 1% resin canals (Panshin *et al.*, 1964). Tracheids are vertically orientated, elongated dead cells, and the rays are mainly horizontally orientated, living parenchymatous cells; every tracheid is in contact with a ray cell (Carlquist, 1975). The tracheids give wood its mechanical strength and their lumens are the pathway for water movement.

On average, tracheids in conifers are about 3500 μm long and 3.3 μm wide, giving softwoods their characteristic properties; in contrast, in hardwoods the average fiber length is 1550 μm and vessel width 20–300 μm (Siau, 1971). In the early part of the growing season tracheids are wider and have thinner walls (earlywood) than those formed toward the end of the growing season (latewood), especially in temperate conifers. Tracheid dimensions differ widely amongst species (Carlquist, 1975) and with position in the tree (Bannan, 1965).

The development of tracheids can be split into three phases: cell production from the meristematic cambium; cell expansion; and cell-wall thickening. Skene (1969) showed that cell division took place over a period of 4 weeks in *Pinus radiata* and was followed by a period of 1.5 to 3 weeks during which the cells expanded radially. The deposition of secondary wall lasted 3 to 4 weeks but this occurred 8 to 10 weeks later in the season. There were only about 4 weeks during the winter when no cell division or radial growth took place. The duration of these processes is likely to be specific for species, latitude, and climate (Barnett, 1971) and position in the tree (Denne, 1979). Barnett (1971) compared the rates of cell production by the cambium in *Pinus radiata* in New Zealand and *Pinus strobus* in the United States, where winter temperatures were colder than in New Zealand. In *Pinus radiata* cells were produced from the cambium throughout the year, although only phloem cells were produced during mid-winter, whereas in *Pinus strobus* cell production ceased for 12 weeks during the winter. In Britain, cell production by the cambium in *Picea sitchensis* lasted for about 16 weeks (Denne, 1979). Thus, there are differences associated with climatic factors and partly with genetic variation.

There has been considerable work on the effects of environmental variables on tracheid growth. Periods of drought reduced tracheid cell production in *Sequoiadendron giganteum* (Zahner et al., 1964) and tracheid diameter in *Pinus resinosa* (Larson, 1963; Glerum, 1970), *Pinus radiata* (Shepherd, 1964; Harris et al., 1978), and *Picea glauca* (Glerum, 1970), but did not affect tracheid length in *Sequoiadendron giganteum* (Zahner et al., 1964). Decreased photoperiod produced short tracheids in *Pinus resinosa* (Larson, 1964), *Larix decidua* (Wodzicki, 1964), *Picea sitchensis,* and *Pinus sylvestris* (Denne and Smith, 1971). Ford et al. (1978) correlated cell production with daily input of solar radiation and Denne (1979) with light intensity in *Picea sitchensis.* In controlled environments, low light intensity reduced cell diameter and cell wall thickening in the same species (Denne, 1974). In *Pinus radiata* cell production and radial diameter were inversely related to temperature (Jenkins and Shepherd, 1972), and night temperature was more important than day temperature; tracheid length and diameter were proportional to night temperature and cell wall thickness was inversely proportional to night temperature. Richardson (1964) found the same effects with *Pseudotsuga menziesii* and *Sequoia sempervirens.* After water stress was relieved (Shepherd, 1964; Glerum, 1970; Harris et al., 1978) or short days extended (Larson, 1963), tracheid diameter increased, giving the appearance of a "false ring." The effect of several drying and rewetting cycles on growth of *Pinus radiata* was so dramatic that the large, thin-walled cells produced during the wet periods subsequently collapsed during drought periods (Barnett, 1976).

The amount of carbohydrate in the developing zone in the cambium is thought to determine the rate of tracheid growth and development

(Richardson, 1964). Nicholls and Wright (1976) related ring width in *Pinus radiata* to the number of potential growth days. Whitmore and Zahner (1967) showed that water stress caused a decrease in the amount of glucose incorporated into developing tracheid cells in *Pinus sylvestris*. Smith (1975) described substrate availability and Zahner (1968) competition between individual cells for substrate in the meristematic zone as important factors in the development process. Ford *et al.* (1978) postulated that an increased rate of production of xylem cells caused a larger queue of cells in the development stages, so that individual cells spent a longer time differentiating, with the result that they expanded to larger diameters with thicker walls. Denne (1976) suggested that the effects of night temperature and light intensity on substrate availability were not large enough to account for the observed seasonal changes in tracheid dimensions and concluded, in agreement with Larson (1963) and Brown (1970), that growth regulators were an essential part of the processes. A change in balance between the growth regulators that have been shown to affect the processes of xylem development (auxins, gibberellins, and cytokinins) (Denne and Wilson, 1977) is probably the trigger for the processes involving carbohydrate metabolism (Brown, 1970). Abscisic acid, known to be present in xylem sap (Davidson and Young, 1973) and to affect tracheid size in *Pinus radiata* (Jenkins and Shepherd, 1974) and *Abies balsamea* (Little, 1975), is also probably involved in these processes. Wodzicki (1964) has also postulated a stimulus in the form of an inhibitor induced in the foliage. Roberts (1969) reviewed xylem differentiation extensively in relation to hormones and should be referred to for further details. However, Barnett (1979) concluded that although it is certain that the stimuli affecting xylem cell growth and differentiation are induced in the stems the exact processes are still not understood. In general those environmental variables which cause bud break, rapid shoot growth, and leaf development give rise to increased levels of auxin and so lead to the production of larger diameter tracheid cells (Brown, 1971). At the end of the maturation phase the cytoplasmic material is removed, the cells die and their lumens become a functional part of the water conducting system. These processes have important positive roles in xylem formation (Barnett, 1979).

Carlquist (1975) argued that tracheid dimensions are a result of a compromise between mechanical strength and water-conducting efficiency. He pointed out that long tracheids are desirable for both high mechanical and conducting efficiency; in addition, thick-walled, narrow tracheids with few small-diameter pits provide maximum mechanical strength whereas thin-walled, wide tracheids with a large number of large diameter pits give maximum conducting efficiency. In support of this hypothesis, Carlquist pointed out that tracheid diameter changes during maturation but tracheid length does not. He cited Bannan (1965), who showed that tracheids in the roots were generally longer than in stems, thus enabling more efficient water conduction in the roots. In relating mechanical strength of tracheids to the capacity to withstand water tensions, Carlquist supposed that species such as *Agathis australis* and *Podocarpus minor,* which never experience severe ψ because they grow in climates where there is always an adequate water supply, develop tracheids which are thin-walled, long, and wide in the earlywood, the necessary mechanical strength being provided by the thick-walled, narrow tracheids in the latewood. In contrast short, thick-walled, narrow tracheids are found in species that regularly experience severe stress and where mechanical strength is important, such as the shrub *Microstrobus niphophilus,* and in the upper parts of tall trees where values of ψ are lowest. Rundel and Stecker (1977) confirmed that tracheid diameter in the crown of an 89-m tall *Sequoiadendron giganteum* tree was linearly related to ψ_l measured at dawn. There was, however, no correlation between ψ_l and tracheid length.

b. Bordered Pits. In conifer sapwood, water flows from one tracheid to the next through bordered pit pairs. The pits occur mainly on the long tapered radial walls of secondary tracheids. The pits in latewood are fewer in number and smaller than those in earlywood (Siau, 1971). Using a suspension of carbon particles in liquid, Bailey in 1913 (see Bailey, 1958) showed with the light microscope that these particles accumulated around the bordered pits. As a result, he deduced the pit structure, and subsequently the electron microscope has been used to elucidate its fine structure (Frey-Wyssling *et al.,* 1956; Liese, 1965; Tsoumis, 1965; Côté, 1958; Thomas and Nicholas, 1966; Petty, 1970). The pit chamber varies in diameter between species from about 6 to 30 μm (Siau, 1971). In the Pinaceae the center of the pit chamber is covered by a membrane or margo that supports the torus, which is thickened with primary cell wall material and usually considered to be impermeable to water. The membrane consists of strands or fibrils formed mainly of cellulose and hemicelluloses (Bauch *et al.,* 1968). In addition, the strands contain large amounts of pectin that can be degraded by pectinase enzymes (Tschernitz and Sachs, 1975) and some phenolic compounds in some species, possibly induced by reaction to bacteria in species prone to wetwood formation (Bauch and Berndt, 1973). Although Sachs (1963) showed that the margo contained lignin, Bamber (1961) had previously shown that lignin was absent. It is likely, however, that the strands become lignified during heartwood formation (Bauch *et al.,* 1968). The structure of the strands is still uncertain, but it is known that they are covered with extractive coatings, although these coatings provide no mechanical strength (Bolton, 1976).

The formation of pits occurs in the primary wall stage (Bauch *et al.*, 1968) probably as a result of redistribution of cell wall material caused by an unidentified agent that penetrates the radial wall of the tracheid during its enlargement. This gives the appearance of a thick circular rim (Barnett and Harris, 1975) upon which the margo fibrils and torus are deposited (Thomas, 1970).

c. Pit Aspiration. Bordered pits in secondary xylem are able to act as valves to prevent the spread of bubbles of air or water vapor in the conducting system when the water column is broken by injury or when bubble formation occurs as a result of cavitation or freezing in the water column (Section IV,C,3,b). When the torus moves across and seals the pit chamber, the pit is said to be aspirated and water flow ceases in the tracheid (Bailey and Preston, 1969). Pit aspiration has been of considerable interest to wood technologists concerned with the flow of preservatives through dried wood. Solvent-exchange drying with organic solvents can prevent pit aspiration (Thomas and Nicholas, 1966; Hart and Thomas, 1967; Liese and Bauch, 1967) and resaturation of dried wood with organic solvents has been shown to reverse aspiration (Thomas and Nicholas, 1966), suggesting that the surface tension of the liquid is important in preventing aspiration. Petty (1972) concluded that if the surface tension of the liquid always exceeds the mechanical resistance of the pit membrane to movement, aspiration occurs, and Bolton and Petty (1977a) showed that if the surface tension of the saturating liquid was progressively increased, then the degree of pit aspiration increased as well. Comstock and Côté (1968), Thomas and Kringstad (1971), and Petty (1972) recognized that a capacity of the liquid to promote adhesion of the pit to the border by hydrogen bonding was also essential for aspiration.

Bolton and Petty (1977b) found that after solvent-exchange drying of *Pinus sylvestris* wood with ethanol, there was permanent displacement of the membrane but only small changes in the size of the pit membrane pores. This could account for the observed differences in the dimensions of the pit structures between air and solvent-dried wood specimens (Petty and Puritch, 1970).

In the living tree only the earlywood conducts water. Phillips (1933) showed that in the heartwood of *Pinus sylvestris, Pseudotsuga menziesii* and *Pinus nigra* the earlywood pits were aspirated but the latewood ones were not. Harris (1954) concluded that conduction of water ceased in *Pinus radiata* sapwood when 50% of the pits were aspirated. The use of dyes with fresh sapwood also reveals that the latewood does not conduct. Harris (1961) showed this in *Pseudotsuga menziesii, Pinus nigra* var *maritima,* and *Pinus radiata* using safranin, and Booker and Kininmonth (1978) in *Pinus radiata* using toluidene blue. Harris (1961) associated nonconduction in the latewood with the low water content. Petty (1970) showed that embolism occurred only in the latewood but earlywood pits were aspirated following measurements of gas permeability in *Picea sitchensis*.

Gregory and Petty (1973) calculated that in earlywood of *Pinus sylvestris* the pressure difference needed to cause full displacement of the pit membrane and thus pit aspiration (about 0.033 MPa) was much less than the pressure required to move a meniscus of water through the largest margo pore (about 0.1 MPa). The consequence of this is that if a bubble of air or water vapor forms and expands in an earlywood pit, menisci form in the margo pores and remain, and the bubble is prevented from moving into the next tracheid. Consequently, although the pressure difference across the pit exerted by the transpiration stream is sufficient to cause aspiration, the bubble is prevented from moving into the next tracheid. Hart and Thomas (1967) had reached the same conclusion earlier. In latewood, however, because of the different dimensions of the pits and the larger diameter margo pores (Petty, 1972), the pressure difference needed to cause full displacement of the pit membrane was much larger (7.3 MPa) than the pressure required to move a meniscus through the margo pores (about the same as in earlywood). Therefore, in latewood the pits do not aspirate and the bubble of air or water vapor easily spreads throughout the whole conducting system. Using a more complex model, Bolton and Petty (1978a) reached the same conclusion but calculated that the pressure differences required to aspirate pits in earlywood and latewood of *Pinus sylvestris* were 0.56 and 68 MPa, respectively (Fig. 7). These values are an order of magnitude larger than those previously calculated by Gregory and Petty (1973) because Gregory and Petty used the total area of the membrane in their calculations, whereas Bolton and Petty used only the effective area of the membrane in calculating the pressure drop across it. If embolism occurs in earlywood through injury or freezing, the pits will aspirate, since sufficient pressure to move the membrane is present in the transpiration stream. However, in latewood the pressure required for aspiration is larger than that provided by the transpiration stream and the pits remain open, with the result that a bubble will expand throughout the conducting system. It would seem that this occurs in latewood each year since the latewood does not conduct water in living trees.

Without the introduction of a bubble into a tracheid, the likelihood of aspiration occurring is extremely small. Bolton (1976) discussed the maximum possible flow rates through an earlywood and a latewood tracheid required to create a pressure difference large enough to cause aspiration. Since the critical pressure differences across a tracheid of 0.079 and 22 MPa for earlywood and latewood, respectively, are unlikely to be reached

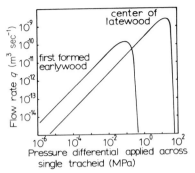

Fig. 7. The relation between the volume flow rate through the tracheid lumen/bordered pit system and the magnitude of the applied pressure difference in *Pinus sylvestris*. (Redrawn from Bolton and Petty, 1978a.)

in natural conditions, the chances of aspiration are remote. Bolton calculated that a gradient of ψ of 0.01 MPa m^{-1} is equivalent to pressure differences across earlywood and latewood tracheids of 0.0053 and 0.0032 MPa, respectively; even if the gradient was an order of magnitude larger (Section II,A,1,c), the critical values would still not be reached.

d. Heartwood Formation. As a tree grows the center part of the sapwood undergoes physiological and biochemical changes and becomes heartwood. The parenchyma cells die (Stewart, 1966) and all free water is lost so that heartwood has a low water content (Büsgen and Münch, 1929; Chalk and Bigg, 1956; Clark and Gibbs, 1957; Etheridge, 1958; Hillis *et al.*, 1962; Stewart, 1967). The tori in the bordered pits become lignified (Bauch *et al.*, 1968), the pit membranes encrusted (Krahmer and Côté, 1963), and the cells contain excretory products including polyphenolic materials (Chattaway, 1952; Bamber, 1961; Hillis *et al.*, 1962; Stewart, 1966) and pinosylvin, which is resistant to fungal infection (Jorgensen, 1961). Heartwood does not conduct water (Harris, 1961; Bailey and Preston, 1970; Puritch, 1971; Booker and Kininmonth, 1978), although Clark and Gibbs (1957) recorded small seasonal changes of about 10% in heartwood water content. Evidence from *Pinus radiata* suggests that the formation of heartwood occurs mainly during the dormant season (Harris, 1954; Shain and Mackay, 1973) and is not always confined to complete growth rings (Harris, 1954; Carrodus, 1972).

The physiological changes occurring in the ray cells during the transition from sapwood to heartwood include hydrolysis of starch (Frey-Wyssling and Bosshard, 1959; Rudman, 1966; Nečasaný, 1964; Hillis *et al.*, 1962; Shain and Mackay, 1973), decreased respiration rate (Nečasaný,

1964) and removal of minerals (Bamber, 1976). The pH of the cell sap also changes (Nečasaný, 1964). Stewart (1966) considered that the phenolic extractives were transported along the rays into the heartwood but there is evidence to suggest that the extractives are produced at the heartwood boundary (Hillis *et al.*, 1962; Dietrichs, 1964; Hasegawa and Shiroya, 1964).

A zone, variously called a dry zone (Büsgen and Münch, 1929), an intermediate zone (Chattaway, 1952), or a transition zone, (Harris, 1954; Stewart, 1966) has been identified surrounding the heartwood in several species. This zone is paler in color (Harris, 1954; Shain, 1979) than heartwood and includes living parenchyma cells that are metabolically active (Bamber, 1976; Coutts and Rishbeth, 1977) and contain highly active respiratory enzymes (Shain and Mackay, 1973). Stewart (1967) showed that the transition zone cells also had higher water contents than those in the sapwood and heartwood.

The formation of dry zones in the sapwood has also been shown following infection by fungi (Coutts, 1976; Gregory, 1977b; Shain, 1979), and reaction zones similar to heartwood have been found following mechanical damage (Bamber, 1976) and insect attack (Puritch and Johnson, 1971). The formation of reaction zones is thought to offer defense against subsequent fungal attack (Shain, 1979).

There are a number of hypotheses on the formation of heartwood, nearly all of them viewing the process as one of degradation. Chattaway (1952) proposed that polyphenols and other extractives were formed as a result of increased metabolic activity and starch breakdown at the heartwood border, whereas Stewart (1966) suggested that the extractives were transported towards the heartwood in the rays until they were prevented from further movement, built up to toxic levels, and killed the cells. On the other hand, Frey-Wyssling *et al.* (1956) saw the process of degeneration in the ray cells beginning at the cambium and the gradual death of the cells further toward the heartwood boundary with the production of polyphenols at the end of the process. Carrodus and Triffett (1975) suggested that polyphenols were produced from phosphoenol-pyruvic acid, which they showed occurred in *Acacia mearnsii* as a result of the high concentrations of carbon dioxide in the transition zone (Carrodus, 1971). Rudman (1966) considered that trees with high growth rates produced fewer "by-products" from metabolic processes and so did not convert as much carbohydrate into waste products and produced less heartwood than slowly growing trees. Contrary to the idea that heartwood is formed as a result of processes of degradation, Bamber (1976) argued that heartwood formation is an active growth process maintained to regulate the amount of sapwood and is initiated by a "heartwood initiating substance," possibly ethylene, which moves centripetally along the rays from the cambium until its progress is blocked.

Stewart (1966), Hillis *et al.* (1962), and Nečasaný (1964) argued that water was lost from the ray cells in the heartwood as a result of cell death and was not the cause of death. The process by which the water is removed has been a matter of debate. Harris (1953, 1954) argued that removal of water occurred under conditions of high ψ in the dying cells by metabolic processes, whereas Rudman (1966) and Stewart (1966) suggested that the water was removed by gradients of ψ between the dead cells and the transpiration stream. Coutts (1977) put forward the idea that the dead cells released substances into the adjacent tracheids, which caused embolism and resulted in the water moving from the dead cells into the tracheids and the transpiration stream. Shain (1979) suggested that the water from the dead parenchyma cells moves into the living cells of the transition zone under the influence of an osmotic gradient, as the osmotic potential in the dead cells increases following the conversion of soluble molecules into less soluble ones, and that the transfer of water is facilitated by increased permeability of the cell membranes caused by ethylene, which has been shown to be present in xylem sap (Davidson and Young, 1973) and to accumulate in the transition zone in *Pinus radiata* (Shain and Hillis, 1973).

2. Flow of Water in Tracheids

The flow of water through conifer sapwood is confined to the lumina of the tracheids so that the dimensions of the tracheids directly affect the flow rates of water in the tree. Petty and Preston (1969) and Petty (1970) identified the tracheid lumina and the pit margo pores as components of the pathway which offer resistance to water flow. Having shown that the flow of water through the tracheids and margo pores was laminar (Reynolds number <10), Petty (1970, 1974) described the resistance to water flow using a modified Hagen–Poiseuille equation. For a volume flow rate of water, q, moving in response to a pressure gradient, ΔP, along a length of tissue, l, through N tracheids of equivalent radius, r, with m pores in the margo of depth, t, and equivalent radius, a, the total resistance, R is:

$$R = \frac{\Delta P}{q} = \frac{8\eta l}{\pi N r^4} + \frac{8\eta(t + 1.15a)}{\delta \pi m a^4}. \tag{16}$$

The first term on the right defines the resistance to flow of the tracheid lumina, the second term that of the pores in the margo. The δ allows for overlap of the tracheids ($= 0.75 l_{tr}/l$ where l_{tr} is the mean tracheid length and 0.75 accounts for tracheid overlap) and the addition of $1.15a$ to the length t of the margo pores is known as the Couette correction, which becomes important when the radius of a capillary is of the same order as its length. η is the viscoscity of the sap, which Petty (1974) took to be equal to that for water. Because of the short length of the tracheids, kinetic energy losses may be important at low flow rates when the Reynolds number exceeds the length to width ratio of the tracheid $(0.1 < Re > 1.0)$ and a correction should be applied if the flow is not linearly proportional to the applied pressure (Siau and Petty, 1979). Petty and Puritch (1970) calculated for air-dried wood of *Abies grandis* that the resistance in the tracheid lumina accounted for 13% of the total, the pores in the margo being responsible for the other 87%. However, in solvent-dried wood the resistance in the lumina was larger, accounting for 39% of the total. Bolton and Petty (1975) identified the pit aperture as a third component in the resistance to water flow and attributed 3% of the resistance to this in air-dried wood of *Picea sitchensis*, the remaining resistance being attributable to pores in the margo (81%) and the tracheid lumina (16%). After submersion of the wood in water for several months at about 30°C to allow bacterial action to destroy the pit membranes, the average distribution of the resistances to liquid flow was pit apertures, 21%, and tracheid lumina, 79%. Johnson (1979) also showed that treating sapwood of *Pseudotsuga menziesii*, *Abies grandis*, and *Tsuga heterophylla* with bacteria, especially *Bacillus polymyxa*, increased axial conductivity as a result of degradation of the pit membranes. Bolton and Petty (1978a) extended their model to include changes in pit geometry as the pit membrane deflects under pressure, and also applied kinetic energy corrections in their calculations. Using data obtained from light and electron micrographs of *Pinus sylvestris* (Table VIII), they affirmed that the critical Reynolds number is never exceeded in earlywood so that flow is laminar. However, flow in latewood might be turbulent when there are large pressure differences across the tracheids, when kinetic energy corrections might also be important. In agreement with earlier findings, Bolton and Petty concluded that a significant part of the resistance to water flow at low pressure differences was in the tracheid lumina with a small but distinct contribution made by pit apertures in latewood.

As the tracheids develop, the resistance offered by the pores in the margo changes. During the first year of development, Mark and Crews (1973) showed that the margo became more perforate in *Picea engelmannii* and *Pinus contorta* and so the resistance to flow decreased. In latewood the margo is denser (Liese, 1965; Banks, 1968; Puritch and Johnson, 1971) and offers more resistance than in earlywood. The lysis of tori after infection of *Abies grandis* with the fungus *Fomes annosus* reduced resistance to water flow and subsequently caused the pits to become embolized

TABLE VIII

ASSUMED VALUES OF DIMENSIONAL AND PHYSICAL QUANTITIES OF TRACHEIDS IN *Pinus sylvestris*[a]

Quantity	First-formed earlywood	Center of latewood
Radius of margo strands (m)	1.50×10^{-8}	5.00×10^{-8}
Maximum membrane displacement (m)	2.00×10^{-6}	2.00×10^{-6}
Pit membrane diameter (m)	1.70×10^{-5}	8.00×10^{-6}
Fractional porosity of margo	2.00×10^{-1}	4.50×10^{-1}
Tracheid lumen diameter (m)	3.00×10^{-5}	8.00×10^{-6}
Tracheid lumen diameter (m)	3.50×10^{-5}	2.50×10^{-5}
Maximum diameter of pit (m)	7.00×10^{-6}	1.20×10^{-5}
Minimum pit aperture radius (m)	2.70×10^{-6}	1.50×10^{-6}
Constant for rectangular capillaries	1.78	2.15
Margo strand length (m)	4.25×10^{-6}	2.00×10^{-6}
Mean tracheid lenght (m)	4.00×10^{-3}	2.40×10^{-3}
No. of margo pores per pit	2.00×10^{2}	5.00×10^{1}
No. of margo strands per pit	1.00×10^{2}	5.00×10^{1}
Half no. of pits per tracheid	5.00×10^{1}	1.00×10^{1}
Margo pore radius (m)	1.50×10^{-7}	2.00×10^{-7}
Pit membrane thickness (m)	2.10×10^{-7}	2.80×10^{-7}

[a] From Bolton and Petty (1978).

(Coutts, 1976), in a similar reaction to that produced after ponding samples of *Picea sitchensis* (Bolton and Petty, 1975). The attack of the balsam woolly aphid, *Adelges piceae* on *Abies grandis* caused incrustation of the margo pores and increased their resistance to water flow (Puritch, 1971; Puritch and Johnson, 1971; Puritch and Petty, 1971); incrustation of the margo pores also occurs in heartwood (Krahmer and Côté, 1963; Lin *et al.*, 1973; Petty and Preston, 1969; Puritch, 1971).

3. Amount of Sapwood

Accurate determination of the cross-sectional areas of sapwood and heartwood in tree stems is necessary because the sapwood is involved in the flow of water, and the heartwood is of interest to wood technologists because it is often resistant to fungal attack.

Many organic and inorganic chemicals have been used to stain the cross-section of heartwood in tree discs, mostly as a result of reaction with the polyphenols (Kutscha and Sachs, 1962; Sandermann *et al.*, 1967; Beszterda and Raczowski, 1976). Koch and Krieg (1938) and Nicholls (1965) used diazotized benzidine, which turned the heartwood of several species dark red. The same effect was found using diazotized tolidine by

Shain (1967) with *Pinus taeda* and Whitehead (1978) with *Pinus sylvestris*. Wellwood (1955) successfully used perchloric acid to stain the heartwood of *Pseudotsuga menziesii*. Sandermann and Schmit (1965) used a concrete solution on discs of several woods and were able to differentiate heartwood from sapwood in the concrete relief picture since heartwood extractives prevented complete setting of the concrete. Biological stains have been introduced into tree stems and their distribution subsequently examined to identify the cross-sectional conducting area in the sapwood. Greenidge (1955b) and Kozlowski and Winget (1963) used acid fuchsin in several species; Talboys (1955) suggested that basic dyes were preferable since they diffused less from the vascular tissues into surrounding cells; Harris (1961) identified transport pathways in *Pseudotsuga menziesii, Pinus nigra,* and *Pinus radiata* using safranin and Booker and Kininmonth (1978) used toluidene blue in *Pinus radiata*.

The diameter of sapwood in tree stems has been shown to vary widely between species, growing conditions, age, and height within individual trees. Büsgen and Münch (1929) generalized that sapwood diameter is larger in dominant trees. Smith *et al.* (1966) accounted for the high variability in amount of sapwood in samples from 58 *Pseudotsuga menziesii* trees by correlation with diameter at breast height, crown width, radial growth rate during the previous 10 years, and tree age. They also showed that suppressed trees had less sapwood than dominant ones. The amount of sapwood was most closely related to tree vigor and growth rate in *Thuja plicata* (Wellwood and Jurazs, 1968). Wellwood (1955) also showed that the amount of sapwood in *Pseudotsuga menziesii* was more closely related to crown class in individual trees than to site index. There was more sapwood in coastal provenances of *Pseudotsuga menziesii* than in those from the interior, and the amount increased with tree diameter and decreased with elevation in the coastal provenances (Lassen and Okkonen, 1969). Panshin *et al.* (1964) showed that generally the width of the sapwood zone in *Pinus palustris* was narrower than that in *Pinus elliotti* and *Pinus taeda*.

A number of authors have expressed the amount of sapwood as the percentage of the total diameter occupied by the sapwood and referred to it as "sapwood diameter." We define "sapwood diameter" as

$$d_s = 100(l_t - l_h)/l_t , \qquad (17)$$

where l_t is the average diameter of the tree and l_h is the average diameter of the heartwood. Then $(l_t - l_h) = l_s$ is the average total width of sapwood across the average diameter. The cross-sectional area of the sapwood is given by

$$A_{sw} = \frac{\pi}{4}(l_t^2 - l_h^2).$$ (18)

In *Tsuga heterophylla* trees ranging in diameter at breast height from 38 to 51 cm, d_s was 40%, and in 61-cm diameter trees of *Abies amabilis* and *Abies grandis*, d_s was 40% (Eades, 1958). Wellwood and Jurazs (1968) measured d_s of only 7% in 103-yr-old, 60-cm diameter trees of *Thuja plicata*. Wellwood (1955) measured d_s of 21% in 59 to 78-yr-old trees of *Pseudotsuga menziesii* of average diameter 29 cm and Smith *et al.* (1966) found d_s of 19% in 85-yr-old, 40-cm diameter trees of the same species. Carrodus (1972) showed that d_s varied from 43 to 83% in 30-cm diameter, 30-yr-old trees of *Pinus radiata*, and Harris (1954) measured a range of d_s of 3 to 19% in the same species from trees harvested from different environments. Whitehead (1978) measured d_s of 69% in trees of *Pinus sylvestris* growing at 3281 stems per hectare and d_s of 79% in trees growing at 608 stems per hectare on the same site. Lassen and Okkonen (1969) ranked five species in order of amounts of sapwood, so that for 50-cm diameter trees the values of d_s were 10% in *Larix occidentalis* and in *Thuja plicata*, 20% in *Picea engelmannii*, 32% in *Pinus contorta*, and 52% in *Pinus ponderosa*. From these widely ranging values we may conclude that the amount of sapwood depends on the species, age, and growing conditions and is related to the volume of water flowing in the stems.

The amount and proportion of sapwood also varies within a tree but a rather unclear picture results from the different forms of expression used by different authors. Büsgen and Münch (1929) stated that the amount of sapwood falls off rapidly from the base to the lower part of the stem, less rapidly in the crown, and diminishes rapidly towards the top of the tree. Panshin *et al.* (1964) argued that the volume of sapwood (i.e., the cross-sectional area A_{sw}) remained constant with height, although the width (i.e., l_s) changed with position in the tree. Wellwood and Jurazs (1968) showed that above the first 7 m in tall *Thuja plicata* trees the width of the sapwood (i.e., l_s) did not vary much but, in contrast, Wellwood (1955) found d_s of 60% at the top, 44% in the middle, and 43% at the base in *Pseudotsuga menziesii*. Smith *et al.* (1966) confirmed that l_s decreased and d_s increased with height in the same species.

The number of growth rings which remain as sapwood also varies widely. Smith *et al.* (1966) showed that the maximum number of sapwood rings was 18 in 85-yr-old *Pseudotsuga menziesii* trees, whereas there were 150 to 200 in *Pinus ponderosa*.

The amount of heartwood appears to have a low heritability of 37% in *Pinus radiata* and little gain would be expected in attempting to select for low or high amounts in a breeding program (Nicholls, 1965).

4. Relation with Foliage Area

In predicting foliage biomass of forest stands, considerable emphasis has been placed on empirical relationships between the amount of foliage present on individual trees and the following properties of the stem: tree girth (Ovington and Madgwick, 1959); tree diameter (Hall, 1965; Stiell, 1966; Loomis *et al.*, 1966; Curtis and Reukema, 1970; Keays, 1971; Moir and Francis, 1972; Gary, 1976; Gholz *et al.*, 1976; Albrektsson, 1976; Hanley, 1976; Brown, 1978; Snell and Brown, 1978); diameter squared (Shinozaki *et al.*, 1964b; Tadaki, 1966); diameter transformed logarithmically (Kittredge, 1944; Rutter, 1955; Tadaki, 1963; Kinerson *et al.*, 1974; Baskerville, 1972; Fujimori *et al.*, 1976; Grier and Logan, 1977; Madgwick *et al.*, 1977; Alban *et al.*, 1978; Gholz *et al.*, 1979; Harrington, 1979); basal area (Tadaki, 1963; Shinozaki *et al.*, 1964a; Tadaki *et al.*, 1969; Newbould, 1967; Ovington *et al.*, 1967; Miller *et al.*, 1976); current annual increment (Zavitovski and Dawson, 1978); and cylindrical stem volume (Tadaki, 1963). However, as Madgwick (1970) and Satoo (1970) pointed out, the relationship changes with age (Tadaki, 1966) and season (Loomis *et al.*, 1966) and rarely holds for more than one stand of one species and so cannot be generalized.

Büsgen and Münch (1929) and Smith *et al.* (1966) noted that crown size was related to sapwood thickness within a species, Grier and Waring (1974) showed that the weight of foliage on *Pseudotsuga menziesii, Abies procera* and *Pinus ponderosa* trees was linearly related to the cross-sectional area of sapwood at 1.3 m above ground level, and Running (1979) also showed a linear relationship in *Pinus contorta*. However, the slopes of the lines for these species are different (Fig. 8).

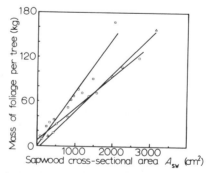

Fig. 8. The relation between mass of foliage per tree and the cross-sectional area of the sapwood A_{sw} at 1.3 m for *Pinus ponderosa*, $\bigcirc$; *Pseudotsuga menziesii*, $\square$; and *Abies procera*, $\triangle$. (Redrawn from Grier and Waring, 1974.)

The conversion of foliage weight to area is complicated by large changes in specific foliage area (ratio of area to weight) between species (Jarvis and Jarvis, 1964; Gholz *et al.*, 1976), with height in individual tree crowns (Tadaki, 1963, 1970; Tadaki *et al.*, 1969; Lewandowska and Jarvis, 1977), and growing treatment (Brix, 1967; Tucker and Emmingham, 1977; Del Rio and Berg, 1979). Mellor and Tregunna (1972) showed that specific foliage area was constant for a given species in the seedling stage. Beets (1977) expressed the relationships between surface area, A_1, fresh volume, V_1, and dry weight, W_{dl}, of a pine fascicle in the form

$$A_1 = k_f\sqrt{V_1 l_f} = k_f\sqrt{\rho_{dl} \cdot W_{dl} \cdot l_f} \tag{19}$$

where l_f is the length of the fascicle, ρ_{dl} ($= W_{dl}/V_1$) is the density and k_f is a constant that is determined by the geometrical shape of the fascicle. Since volume is reasonably simple to measure using liquid displacement techniques (Kozlowski and Schumacher, 1943; Ronco, 1969; Huxley, 1971; Beets, 1977; Burdett, 1979), surface area can be readily obtained using equation (19). Beets (1977) showed that k_f was constant for a large size range of fascicles on *Pinus radiata* and that ρ_{dl} was constant for all fascicles on one tree within the same age class. Since the geometric and arithmetic means for V_1 and W_{dl} are very close for a sample of fascicles (Beets, 1977), the total foliage area on a tree can be calculated from the total dry weight of foliage and measurements of ρ_{dl} and l_f on a sample from each needle age class.

Linear relationships between sapwood basal area and foliage area for a range of tree sizes have been found for a number of broad-leaved species (Dixon, 1971; Waring *et al.*, 1977; Rogers and Hinckley, 1979), and for *Pinus sylvestris* (Whitehead, 1978), (Fig. 9), *Pinus contorta* (D. Whitehead, J. W. Leverenz, and W. R. N. Edwards, unpublished data), *Picea sitchensis* (D. Whitehead, W. R. N. Edwards, and D. C. Malcolm, unpublished data), *Pinus radiata* (P. W. Beets, unpublished data), *Tsuga mertensiana* (R. H. Waring, unpublished data), and *Juniperus occidentalis* (Gholz, 1979). At the moment, however, there is insufficient evidence to determine how consistent the relationship is for a particular species on a range of sites.

Polster (1967) related transpiration rate from the canopy to the green foliage mass and Kline *et al.* (1976) showed that transpiration rates were linearly related to sapwood basal area in *Pseudotsuga menziesii*, and Jordan and Kline (1977) found the same in several species in an Amazonian tropical rain forest.

The Shinozaki *et al.* (1964a) model considered a tree stem to consist of an assemblage of unit pipe systems containing pipes of unit thickness

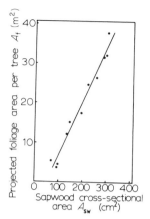

Fig. 9. The relation between the area (projected) of foliage per tree, A_f, and the cross-sectional area of the sapwood, A_{sw}, at 1.3 m for *Pinus sylvestris*. $A_f = 0.137 \times A_{sw} - 7.00$, $r^2 = 0.97$ (Redrawn from Whitehead, 1978.)

through which water flowed to supply a unit amount of leaves. However, as they later pointed out (Shinozaki *et al.*, 1964b), the proportionality between the total stem cross-sectional area and the amount of leaves above did not hold at positions near the base of the tree below the lowest living branches. Extending the analogy, they attributed this departure from proportionality in the lower parts of the tree to disused pipes that were originally connected to living branches but had become redundant after the lower branches died, presumably as a result of the formation of heartwood. Jarvis (1975) proposed that following thinning in a forest canopy the increase in foliage area would be proportional to the increase in sapwood basal area in the remaining trees. Both hypotheses suggest a homeostatic relationship between the cross-sectional area of sapwood, which is a characteristic of the water supply pathway to the foliage, and the foliage area from which water is transpired.

It is clear from earlier considerations of the main variables affecting transpiration from conifers (i.e., Section II) that the rate of transpiration from a stand of conifers is strongly dependent on the leaf area. This is because transpiration E_T from rough canopies is approximately proportional to the VPD, D, and the canopy conductance, g_c, as shown in equation (7). Since the canopy conductance is proportional to the LAI (equation 8), the transpiration rate is also proportional to LAI, except in still air. It is therefore reasonable to suppose a close developmental interrelationship between the extent of the foliage area, which is a characteristic of the loss system, and the size of the supply system. Nonetheless, it is not unexpected that relationships between sapwood basal area and foliage area will vary considerably with species, site factors, such as nutritional status and climate, and the degree of dominance of individual trees, all of which may influence transpiration rate by affecting properties of the liquid flow pathway in the plant and behavior of the stomata.

These concepts can be expressed quantitatively by replacing foliage area with sapwood cross-sectional (or basal) area in the Darcy equation (equation 15) and assuming that flow rate in tree stems is equal to the transpiration rate from the foliage. If a linear relationship exists between foliage area A_f and sapwood basal area A_{sw}, then if the relationship passes through zero, the slope of the relationship is

$$S = \frac{A_f}{A_{sw}}. \tag{20}$$

Substituting equation (20) into (15) with $q = \bar{E}_T/(\rho_w n)$, and rearranging for S gives

$$S = \frac{kA_f n \rho_w}{\eta \bar{E}_T} \frac{\Delta\bar{\psi}}{l}, \tag{21}$$

where the bar indicates an average value, n is the number of trees per unit area and ρ_w is the density of water. We would, therefore, expect S to be different for various species and sites depending on the conductivity of the sapwood, the transpiration rate, and the gradient of ψ with height in the stems. The higher the average transpiration rate at a site, the smaller the leaf area that would be supplied by unit sapwood basal area. Since the partitioning of energy by coniferous forest (i.e., β) varies strongly with the climatological conductance g_i (see Section II,A and Jarvis et al., 1976), lower values of the slope, S, would be expected at inland sites in continental climates than at coastal sites in oceanic climates.

From equations (6) and (9), the average transpiration rate at a site can be written as

$$\bar{E}_T = \frac{\bar{A}(\bar{s}/\gamma + \bar{g}_a/\bar{g}_i)}{\lambda(\bar{s}/\gamma + 1 + \bar{g}_a/\bar{g}_c)}. \tag{22}$$

Replacing $\bar{E}_T$ in equation (21) with the above, we see more clearly that S depends upon properties of the canopy as well as the climate:

$$S = \frac{kA_f n \lambda \rho_w(\bar{s}/\gamma + 1 + \bar{g}_a/\bar{g}_c)}{\eta \bar{A}(\bar{s}/\gamma + \bar{g}_a/\bar{g}_i)} \cdot \frac{\Delta\bar{\psi}}{l}. \tag{23}$$

Thus the slope S depends on a wood property, k, canopy properties, LAI ($= nA_f$), $\bar{g}_c$, and $\bar{g}_a$, and a property of the climate, $\bar{g}_i$ (see equation 10).

Sufficient information from a range of climates to test this hypothesis properly is not available. However, the variation in values of S found by Whitehead (1978) for several stands of *Pinus sylvestris* at different places does support this hypothesis. He found a smaller value for S for Thetford Forest in southeast England ($\bar{g}_i \simeq 25$ mm sec^{-1}) than for Roseisle Forest in northeast Scotland ($\bar{g}_i \simeq 34$ mm sec^{-1}) or Devilla Forest in central Scotland ($\bar{g} \simeq 50$ mm sec^{-1}).

C. Variations in Water Potential

1. Dependence on Evaporation

Typically, in conifers as in other species, ψ_l falls diurnally from a high value in the morning to a low value in the middle of the day and rises again in the afternoon (Fig. 10) (Waring and Cleary, 1967; Waggoner and Turner, 1971; Heth, 1974; Hellkvist *et al.*, 1974; Reed and Waring, 1974; Cline and Campbell, 1976; Rook *et al.*, 1977; Roberts, 1976a; Running, 1976; Sato and Morikawa, 1976; Watts, 1977; Aussenac and Granier, 1978; Tan *et al.*, 1978; Waring *et al.*, 1979) and similar changes occur in root water potential (ψ_{root}) (Hellkvist *et al.*, 1974; Nnyamah *et al.*, 1978).

With increasing transpiration rate E_T during the early part of the day, the difference ($\psi_s - \psi_l$) would be expected to increase in proportion to the

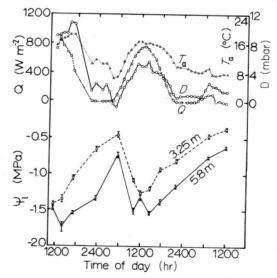

Fig. 10. The diurnal course of leaf water potential at two heights above the ground in a canopy of *Picea sitchensis*. Each point is the mean of five measurements and two standard errors are shown. Concurrent changes in solar irradiance, Q, air temperature, T_a, and vapor pressure deficit, D, are shown. (Redrawn from Hellkvist *et al.*, 1974, Fig. 5.)

flux of water through the stem (equation 11). Consequently, it would be expected that ψ_l would be more or less inversely proportional to E_T (equation 12). However, a large number of empirical correlations have been tried between ψ_l and a number of weather variables, including irradiance Q_s, net radiation Q_n, vapor pressure deficit, D, and air temperature, T_a, with varying degrees of success. For example, Barker (1973) used a multiple regression model and accounted for 86% of the variation in ψ_l in *Abies concolor* and 76% in *Pinus ponderosa* by Q_s, $D^{1/2}$, height of tree, and stomatal aperture. Pereira and Kozlowski (1976) found that in *Abies balsamea* 70% of the variation in ψ_l could be accounted for by the square root of the value of irradiance received 2 hr previously, $Q^{1/2}$ (−2 hr), since severe water deficit occurred on the sampling days. In *Pinus ponderosa* 73% of the variation in ψ_l was accounted for by a model containing log q, calculated from $D \cdot g_s$ and $Q_s^{1/2}$ (−2 hr) at low water stress, and 93% of the variation in water-stressed plants. Kaufmann (1975) related ψ_l to Q_n in *Picea engelmannii* at temperatures above 7°C. Hellkvist (1973) showed that ψ_l was linearly related to Q_s and T_a in *Pinus sylvestris* and Hellkvist and Parsby (1976) extended this work to show that ψ_l could be linearly related to potential transpiration rate E_o. However, the slopes of the relations between ψ_l and E_o were different for the two clones studied and changed with time of year.

Sucoff (1972) was perhaps the first to show that ψ_l could be linearly related to E_T, in *Pinus resinosa*. Kaufmann (1979) concluded that the relationship between ψ_l and E_T in *Picea engelmannii* was linear at low flux densities through the trees and, as soil water was depleted, the intercept and the slope of the relation between ψ_l and E_T changed as a result of stomatal closure. Kaufmann (1977b) also showed that the slope of the relation between ψ_l and E_T was influenced by root temperature in *Pinus radiata* (see Section V,B). Jarvis (1976) showed a linear relationship between ψ_l and E_T in *Pinus sylvestris* and *Picea sitchensis* but the slope of the relationship was much steeper for individual trees of *Pinus sylvestris* than for *Picea sitchensis* (see Fig. 6 in Section III,A,2). Waring and Running (1978) found linear relationships between ψ_l and E_T in *Pseudotsuga menziesii* trees and the relationships became steeper as the season progressed (Fig. 11). D. Whitehead, P. G. Jarvis, and R. H. Waring (unpublished data) showed that the slope of the linear relationship between ψ_l and E_T for *Pinus sylvestris* varied with tree spacing. On a plot with 608 stems ha⁻¹, the slope of the relationship was 2.2 times steeper than on a plot with 3281 stems ha⁻¹. Nnyamah *et al.* (1978) also found a linear relationship between ($\psi_S - \psi_{root}$) and E_T in trees of *Pseudotsuga menziesii* and the slope of the relationship was steeper on a plot with 840 stems ha⁻¹ than on a plot with 1840 stems ha⁻¹. Thus, there is considerable evidence for a linear relationship between ψ_l (and ψ_{root}) and E_T in conifers, in contrast to the wide range

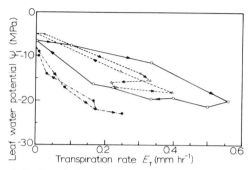

Fig. 11. Hysteresis in the relation between leaf water potential, ψ_l, and transpiration rate, E_T, of *Pseudotsuga menziesii* on days with different sapwood relative water contents: $\theta_s = 0.90$, $\bigcirc$; $\theta_s = 0.78$, $\triangle$; $\theta_s = 0.56$, $\bullet$. Each point is the average from three trees. (Redrawn from Waring and Running, 1978.)

of nonlinear relationships found in herbaceous species (see Jarvis, 1975). However, some nonlinear relationships have also been found. Kaufmann (1975) found a nonlinear relation between ψ_l and E_T in *Picea engelmannii;* Morikawa and Sato (1976) also found a nonlinear relationship between ψ_l and q, measured using the heat-pulse velocity method in the upper part of the crown of *Chamaecyparis obtusa;* and Nnyamah et al. (1978) obtained a curved relation between $(\psi_s - \psi_l)$ and E_T. Some of these observations of nonlinearity may be related to the hysteresis discussed in the following section.

2. Hysteresis in ψ_1/E_T Relationships

Hysteresis is not infrequently found in plots of ψ_l in relation to transpiration rate because ψ_l reaches a minimum later in the day than does maximum transpiration rate (Fig. 11). Kaufmann (1977a) found hysteresis in plots of ψ_l against E_T with 2-yr-old seedlings of *Pinus radiata* during a 9-day drying cycle. However, the hysteresis was minimal after rewatering the stressed plants and in the unstressed controls. Jarvis (1976) showed that marked hysteresis between ψ_l and E_T could occur for large *Pinus sylvestris* trees and commented that the degree of hysteresis changed with season and type of weather. Waring and Running (1978) found that the amount of hysteresis between ψ_l and E_T from large *Pseudotsuga menziesii* trees became less as the summer progressed. In the late summer when predawn values of ψ_l had reached -0.61 MPa no hysteresis was evident (Fig. 11). Running (1979) also showed hysteresis between ψ_l and E_T in trees of *Pinus contorta*.

Jarvis (1976) considered the possibility that this hysteresis might be the result of errors in estimation of E_T but thought it more likely that hysteresis was the result of changes in resistance of the liquid flow pathway during the day, coupled with changes in the amount of water stored within the tree. In support of this suggestion, Waring and Running (1978) showed that the degree of hysteresis became less as the relative water content of the sapwood fell from 0.9 to 0.56 (Fig. 11).

Hysteresis can also be found if ψ_l is plotted against Q_s, Q_n, or D. For example, Hellkvist and Parsby (1976) found hysteresis in plots of ψ_l against D or Q_s in *Pinus sylvestris* and the hysteresis was less marked in cloudy conditions. In such cases only a part of the driving force for transpiration is being taken as the independent variable (see Section II) and consequently some hysteresis will result if radiation and VPD reach their maxima at different times of the day. For example, Landsberg *et al.* (1975) showed that the very marked hysteresis in plots of ψ_l in apple trees against Q_n or D disappeared when ψ_l was plotted against E_T. Hellkvist and Parsby (1976) accounted for some of the hysteresis in ψ_l that they observed in *Pinus sylvestris* by the lag of the peak in D behind the peak in Q_s.

3. Lags in ψ and Shrinkage

Lags between changes in ψ in the upper parts of tree stems or foliage and the lower parts of the stem or roots have also been recorded. Richards (1973) showed a phase lag of 2.8 hr between changes in ψ_{stem} at 5.7 m and 0.2 m in the trunk of *Picea sitchensis* and Morikawa (1974) measured similar lags between ψ_l and ψ of the stem phloem tissue of *Chamaecyparis obtusa*. Nnyamah *et al.* (1978) showed a 3 to 4-hr lag between the start of recovery of ψ in twigs and roots on some occasions during the summer in 7 to 9 m tall *Pseudotsuga menziesii* trees. These lags, or phase shifts, in ψ at the base of the stem, compared with the mid-canopy level, are also accompanied by smaller changes in ψ (Hellkvist *et al.*, 1974). Both attenuation and phase shift in the development of water potentials and water deficits can be explained as the result of significant water storage capacitance in the wood and tissues of the tree (Powell and Thorpe, 1977; Jarvis, 1975; Landsberg *et al.*, 1976).

Swelling and shrinkage of the cambium and phloem of the stem follow changes in ψ in the adjacent xylem and hence also exhibit phase shifts and attenuation at the base of the stem as compared with in the crown. Consequently, greater shrinkage of these tissues tends to occur in the mid-crown rather than lower down, and shrinkage begins several hours earlier at the top than at the bottom of the tree in the morning and expansion at the top begins earlier than at the bottom during the late afternoon (Dobbs and Scott, 1971; Kozlowski, 1972, 1979; Richards, 1973; van Laar, 1967; Hinckley and Bruckerhoff, 1975; Lassoie, 1979). For these reasons Pereira and Kozlowski (1976) recorded a lag of 3–4 hr between diameter at the base and ψ_l in the crown of *Pinus resinosa* and similar observations were made by Zaerr (1971) on *Pseudotsuga menziesii*. Similarly, Waggoner and Turner (1971) found that shrinkage of cambium and phloem at 1.4 m above ground level lagged 3 hr behind the radiation received by the mid-crown in *Pinus resinosa*, and Holmes and Shim (1968) showed a lag of 1 hr between sunrise and the start of shrinkage in the stem of *Pinus canariensis* in the summer and ¼ hr in mid-winter.

4. Transpiration Lags

Initially, when transpiration begins, water moves into the leaves from the nearest available sources. In the nearby xylem, ψ falls and water moves from the surrounding cells into the xylem as ψ tends to equilibrate throughout the tissue (Jarvis, 1975). The result is a wave of decreasing ψ moving down the plant (Jarvis, 1976), with significant flow beginning in the top parts of the tree first (Huber, 1952). Several studies comparing the flow of water in tree stems near the base, using the heat-pulse method, with transpiration rate from the foliage, have shown apparent lags in the flow rate at the base of the stem. For example, Morikawa (1974) found that the diurnal rise and fall in sap flux lagged several hours behind transpiration rate from a young *Chamaecyparis obtusa* tree. Swanson (1972) found much the same thing in *Pinus halepensis:* sap flux lagged behind transpiration rate by 1–2 hr in the morning and continued longer when transpiration had fallen in the evening. Following a thunderstorm that caused transpiration to fall to zero, Swanson (1967) observed that the sap flux declined more slowly in pine and spruce trees. Reynolds (1966) also found variations in the relationship between sap flux and transpiration rate in 20-yr-old spruce trees and Waggoner and Turner (1971) showed that peak sap flux lagged behind the peak value of net radiation by 2–3 hr in *Pinus resinosa*. Waring and Roberts (1979) compared the rate of water uptake of a large cut off *Pinus sylvestris* tree standing in water with the velocity of water movement higher up the stem, measured using a radio-active isotope, and found lags between the two methods at the start and end of the day. There was also a delay of 2 to 3 hr after the onset of rain before and rate of water uptake fell to zero. The movement of the isotope also lagged behind transpiration measured using equation (6) (Waring, *et al.*, 1980).

Lags between sap flux at the base of the stem and transpiration rate from the crown can also be interpreted in terms of changes in the amount of water stored in the tissues in the stem. Since liquid flow rate through the tree depends on differences in ψ, the progressive removal of water from storage down the stem results in the appearance of lags between sap flux and transpiration rates (Weatherley, 1970; Jarvis, 1975).

In conclusion, therefore, the use of equation (12) is not valid in plants in which there is significant storage of water, and simple resistance models of flow of water through trees can be misleading if absorption and transpiration of water only are considered without regard to changes in stored water. In order to construct more adequate models of water flow that will take into account movement of water in and out of storage, it is necessary to know the volumes of stored water, the capacitances of the tissues, and the resistances to water movement from store to xylem. In the following section, therefore, we consider the diurnal and seasonal variation in water content of different parts of the tree.

IV. CHANGES IN WATER STORAGE

A. Tissue Capacitance

A hydraulic liquid flow system with a capacity to store and exchange water associated with the flow pathways must necessarily lead to the appearance of lags when the rate of flow in and out of the system is changed (Jarvis, 1975). The solid line in Fig. 12 shows the changes in liquid flow rate which would result from changes in transpiration rate if the tree were completely inelastic and without storage capacity. The dashed lines show the flow of water at various points in a tree with storage capacity. Uptake from the soil begins at the same time as transpiration in the morning but much more slowly because of the extraction of water from stores within the plant. Water is withdrawn from tissues during the first part of the day when transpiration begins, and the tissues are later recharged by uptake from the soil continuing after transpiration has

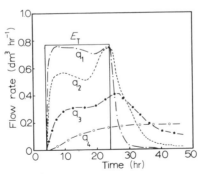

Fig. 12. Simulation of the flow of water in a tree where transpiration rate is changed abruptly. E_T is the loss of water from the leaves in transpiration; q_1 is the flow from twigs to leaves; q_2 is the flow from the top of the stem to the twigs; q_3 is the flow half way up the stem at 4 m; and q_4 is the flow in the stem at ground level. (From W. R. N. Edwards, H. Talbot, J. J. Landsberg, and P. G. Jarvis, unpublished.)

ceased at the end of the day. Hence transpiration rate, E_T, is made up of a number of partial flows from stores within the plant and the soil.

We can describe the movement of water out of storage in the tree in the following way (Landsberg et al., 1976). If ψ_t is the total potential of water in a tissue and ψ_x the pressure potential of water in the xylem, then the volume flow of water from the tissue into the xylem is given by

$$q_{tx} = (\psi_t - \psi_x)/R_{tx} , \tag{24}$$

where R_{tx} is the lateral resistance to water movement between the tissue and the xylem (Fig. 13). This flow of water from tissue to xylem is also equal to the rate of change of water content of the tissue,

$$q_{tx} = \frac{dV_w}{dt} = V_{wt} \frac{d\theta_t}{dt} , \tag{25}$$

where V_w is the volume of water in the tissue, V_{wt} the volume of water in the turgid tissue, and θ_t the relative water content of the tissue. The volume of water that is exchanged between tissue and xylem between time t_1 and t_2 is therefore

$$\Delta V_w = V_{wt} \Delta\theta_t = \int_{t_1}^{t_2} q_{tx} \cdot dt. \tag{26}$$

We may define the *capacitance* of the tissue as the change in its water content per unit change in water potential as follows

$$C_t = \frac{dV_w}{d\psi_t} = V_{wt} \frac{d\theta_t}{d\psi} \tag{27}$$

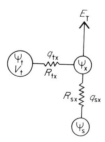

Fig. 13. A diagram showing one node of a model of flow, resistance, and water storage in a tree. E_T is transpiration rate; ψ_x, ψ_s, and ψ_t are water potentials in the xylem, a source, and a tissue, respectively, and q and R are the volume flows and resistances between the locations. V_t is the volume of water in the tissue.

The relationship between θ_t and ψ_t is given by the moisture characteristic or pressure–volume curve for the tissue.

Then from equations (26) and (27) the change in volume of water in the tissue is

$$\Delta V_w = C_t \, \Delta \psi_t . \tag{28}$$

From equations (25) and (27), the volume flow out of the tissue is

$$q_{tx} = \frac{dV_w}{dt} = C_t \, \frac{d\psi_t}{dt} . \tag{29}$$

From equations (24) and (29), the change in water potential of the tissue is

$$\frac{d\psi_t}{dt} = \frac{\psi_t - \psi_x}{R_{tx} \cdot C_t} . \tag{30}$$

The product $(R_{tx} \cdot C_t)$ is the time constant for exchange between tissue and xylem (Landsberg *et al.*, 1976). By analogy with electrical theory, equation (30) leads to

$$\frac{d\psi_t}{d\psi_x} = 1 - e^{-t/(R_{tx}C_t)} . \tag{31}$$

These equations demonstrate explicitly that the changes in potential and water content of a tissue that exchanges water with the xylem will exhibit phase lags and reductions in amplitude relative to the xylem, dependent upon the flow resistance between tissue and xylem and on the capacitance of the tissue. Thus, tissue capacitance needs to be included in models of water flow through the tree. This conclusion was arrived at intuitively by Weatherley in 1970.

If an axial resistance is now considered in conjunction with the lateral resistance (Fig. 13), then

$$E_T = q_{sx} + q_{tx} \tag{32}$$

$$= \frac{\psi_s - \psi_x}{R_{sx}} + C_t \, \frac{d\psi_t}{dt} . \tag{33}$$

Nodal modules of this kind can be added together to give a model of water flow through the tree which extends beyond the rigid pipe concepts of Gradmann and van den Honert, to include the water capacitance of the associated tissues. At the present time further developments in this area must, however, await information on tissue capacitances, lateral (or radial) flow resistances between tissues, and the xylem and axial resistances of segments of the pathway.

The capacitance of needles is quite well known (e.g., Jarvis and Jarvis, 1963a,b; Roberts and Knoerr, 1977; Waring and Running, 1978), but knowledge of the capacitance of other tissue is very sketchy. Capacitance of stem sapwood has been determined for *Pseudotsuga menziesii* (Waring and Running, 1978) and *Pinus sylvestris* (Waring *et al.*, 1979), but not, as far as we know, for any tissues of roots or branches.

B. WATER IN TISSUES

1. Living Cells in Foliage, Stem Cambium, Phloem, and Roots

Because living cells and tissues change considerably in volume as their water content changes, it is not appropriate to define the water contents in terms of volume fractions. Instead the water content of living tissues is usually expressed in terms of the dimensionless relative water content (Weatherley, 1950, 1965; Barrs, 1968; Čatský, 1974) as

$$\theta_t = \frac{V_w}{V_{wt}} = \frac{W_{ft} - W_{dt}}{W_{tt} - W_{dt}} \tag{34}$$

where V_w is the volume of water in a tissue that has the capacity to hold the volume V_{wt} when fully turgid and W_{tt}, W_{ft}, and W_{dt} are the turgid, fresh, and dry weights of the tissue, respectively. The water deficit is then $(1 - \theta_t)$. The volume of water in the tissue at any time is given by

$$V_w = V_{wt}\theta_t = \frac{W_{ft} - W_{dt}}{\rho_w} \tag{35}$$

where ρ_w is the density of water.

However, not all of the water present in a tissue is freely available for osmotic exchange within the cells or for loss from the tissue. A part of the water is held in small capillaries or bound to the cellulose matrix in the cell walls. Pressures required to remove this water exceed -30 MPa and are considerably larger than is likely to be encountered under normal physiological conditions (Slatyer, 1966; Jarvis and Slatyer, 1970). Thus, it is useful to distinguish between that fraction of water present that can be regarded as unavailable or bound, B_t, and the fraction which is available for exchange, $(\theta_t - B_t)$.

The volume of freely exchangeable water present in a tissue at any time is therefore

$$V_{wa} = V_{wt}(\theta_t - B_t) = \frac{W_{tt} - W_{dt}}{\rho_w} (\theta_t - B_t). \tag{36}$$

The change in volume of available water in the tissue over a period of time is

$$\Delta V_{wa} = V_{wt}\Delta(\theta_t - B_t). \tag{37}$$

Estimates of B_1 have been made for leaves of various species using a variety of techniques but values are largely unknown for other tissues. Slavík (1963) calculated B_1 for leaves of *Nicotiana* sp. to be 0.2 to $0.4\theta_1$ using a cryoscopic method, and Warren Wilson (1967) used a gravimetric vapor pressure method to calculate B_1 as $0.2\theta_1$ for *Zea mays* $0.3\theta_1$ for *Helianthus annuus* and $0.4\theta_1$ for *Brassica napus* leaves. Hellkvist *et al.* (1974) derived B_1 from pressure–volume curves in *Picea sitchensis* and showed that B_1 changed throughout the year. In autumn B_1 ($= 0.5\theta_1$) was higher than it was in the summer ($= 0.2\theta_1$) and seasonal differences have also been reported for several conifers by Kramer (1955). In interpreting pressure–volume curves of leaves, it has also proved useful to express the content of available water on a dimensionless scale of 0–1 analogously to relative water content. The available (or free) water content F_1 is then related to θ_1 by (Hellkvist *et al.*, 1974):

$$F_1 = (\theta_1 - B_1)/(1 - B_1). \tag{38}$$

2. Branch, Stem, and Root Sapwood

In the dead cells of the sapwood, changes in water content are associated with changes in the fraction of the volume occupied by water and by air or water vapor: changes in the volume of the tissue are negligible.

It is convenient to illustrate the interrelationships between the three phases of solid material, liquid water, and gas in wood in a phase diagram (Fig. 14). The solids that form the cell wall consist of cellulose, lignin, hemicelluloses, and extractives [45–50%, 25–30%, 20–25%, and 0–10% by weight, respectively (Dinwoodie, 1975)]. Some of the water in the sapwood is bound on to the cellulose walls and the remainder can be regarded as freely available. The gas phase consists largely of air saturated with water vapor.

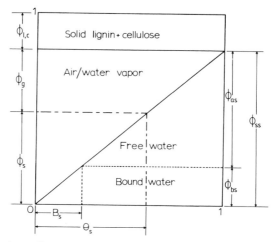

Fig. 14. A phase diagram to demonstrate the interrelations between solid material, bound water, available water, and air in sapwood. ϕ is volume fraction and θ relative water content. The subscripts are: l,c, lignin and cellulose; g, air and water vapor; s, water; as, available water; bs, bound water; and ss, water in water-saturated wood. (Redrawn from Waring and Roberts, 1979.)

In a sample of sapwood of volume V_{fs}, fresh weight W_{fs}, and dry weight W_{ds}, the volume of freely available water, V_{as}, is given by

$$V_{as} = V_{fs}(\phi_s - \phi_{bs}) , \qquad (39)$$

where ϕ_s is the volume fraction of water and ϕ_{bs} is the volume fraction of bound water. The volume fraction of water can be conveniently expressed as:

$$\phi_s = \frac{V_w}{V_{fs}} = \frac{W_{fs} - W_{ds}}{V_{fs}\rho_w} = \frac{\rho_{fs} - \rho_{ds}}{\rho_w} \qquad (40)$$

where V_w is the total volume of water in the sample and $\rho_{fs}(= W_{fs}/V_{fs})$, $\rho_{ds}(= W_{ds}/V_{fs})$, and ρ_w are the densities of wet wood, oven-dry wood, and water, respectively.

If the sample of wood has relative water content θ_s, part of which is the bound water content, B_s, then from the phase diagram the volume fraction of bound water is:

$$\phi_{bs} = B_s \cdot \phi_{ss} \qquad (41a)$$

or

$$\phi_{bs} = \frac{B_s}{\theta_s} \cdot \phi_s , \qquad (41b)$$

where ϕ_{ss} is the volume fraction of water in saturated wood and

$$\theta_s = \phi_s / \phi_{ss} .$$

Substituting equations (40) and (41) into equation (39) gives

$$V_{as} = V_{fs} \left[\frac{\rho_{fs} - \rho_{ds}}{\rho_w} - B_s \cdot \phi_{ss} \right] \tag{42a}$$

or

$$V_{as} = V_{fs} \left[\frac{\rho_{fs} - \rho_{ds}}{\rho_w} \right] \left[1 - \frac{B_s}{\theta_s} \right] \tag{42b}$$

In saturated wood the solid and liquid phases only are present and the volume fraction of water, ϕ_{ss}, is given by:

$$\phi_{ss} = \frac{V_{fs} - V_{l,c}}{V_{fs}} , \tag{43}$$

where $V_{l,c}$ is the volume of the solids, largely lignin and cellulose, and is given by:

$$V_{l,c} = \frac{W_{ds}}{\rho_{l,c}} = \frac{V_{fs} \cdot \rho_{ds}}{\rho_{l,c}} , \tag{44}$$

where $\rho_{l,c}$ is the density of the solid materials.
Substituting for $V_{l,c}$ from equation (44) into equation (43) gives

$$\phi_{ss} = 1 - \frac{\rho_{ds}}{\rho_{l,c}} \tag{45}$$

and substituting for ϕ_{ss} from equation (45) in equation (42a) or for θ_s in equation (42b) gives

$$V_{as} = V_{fs} \left[\frac{\rho_{fs} - \rho_{ds}}{\rho_w} - B_s \left(1 - \frac{\rho_{ds}}{\rho_{l,c}} \right) \right] \tag{46}$$

From equation (44) the volume fraction of solids is

$$\phi_{l,c} = \frac{\rho_{ds}}{\rho_{l,c}} \tag{47}$$

and thus the volume fraction of the gas phase,

$$\phi_g = 1 - \phi_s - \phi_{l,c} \tag{48}$$

can be obtained from equations (40) and (47).

θ_s ($= \phi_s/\phi_{ss}$) can be conveniently calculated from equations (40), (44) and (45) as either:

$$\theta_s = \frac{W_{fs} - W_{ds}}{\rho_w(V_{fs} - W_{ds}/\rho_{l,c})} \tag{49a}$$

or

$$\theta_s = \frac{\rho_{l,c}(\rho_{fs} - \rho_{ds})}{\rho_w(\rho_{l,c} - \rho_{ds})} \tag{49b}$$

The bound water content of sapwood is expected to be a fixed amount and Siau (1971) found it to be 30% of the oven-dry weight of the wood. Using a vapor pressure equilibration method, Waring and Running (1978) and Waring et al. (1979) found B_s to be about $0.2\theta_s$ in both *Pseudotsuga menziesii* and *Pinus sylvestris* (Fig. 15).

The density of the solids, predominantly lignin and cellulose, $\rho_{l,c}$, can be regarded as constant at 1530 kg m^{-3} (Siau, 1971; Skaar, 1972). The density of dry sapwood varies between 350 and 500 kg m^{-3} depending on species and position in the tree, and the density of water saturated sapwood is in the range 870 to 1020 kg m^{-3} (Hamilton, 1975).

This set of equations, especially equations (40), (48), and (49) provide a convenient means of analyzing and expressing changes in sapwood water relations.

Many studies with a range of conifers have shown that the water content expressed as a fraction of the dry weight of sapwood is greater at the top of a tree than lower down the stem (Luxford, 1930; Fielding, 1952; Ovington, 1956; Johnston, 1970) and generally decreases toward the center of the tree (Swanson, 1967; Markstrom and Hann, 1972, but cf. McGinnes and Dingeldein, 1969; Stewart, 1967). However, this expression of water content does not take into account variation in the dry density of wood in different parts of the tree and may be misleading (Chalk and Bigg, 1956). θ_s increases with height in trees by about 10 to 15% (Waring and Running, 1978; Waring and Roberts, 1979; Waring et al., 1979).

The convenient method of sampling cores to show seasonal changes in θ_s has been criticized because some water may be squeezed out of the sample during the boring procedure or pulled out of adjacent tracheids as the tension in the sapwood is released when the hole is made. McDermott

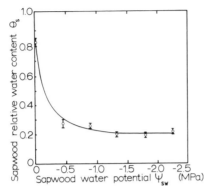

Fig. 15. The relation between sapwood relative water content, θ_s, and sapwood water potential, ψ_{sw}, for *Pinus sylvestris,* determined by vapor equilibration over salt solutions at constant temperature. The equation for the curve is $\theta_{sw} = 0.2072 + \exp[4.679\ (\psi_{sw} - 0.111)]$, $r^2 = 0.99$. (Redrawn from Waring *et al.,* 1979.)

(1941) showed that the measured water content of twigs of *Pinus taeda* and *Pinus echinata* was 6 to 15% lower in sections cut in two places with an interval of 30 sec between cuts than when the cuts were made simultaneously. Although Yerkes (1967) showed no differences in water content between cores extracted from *Pinus ponderosa* sapwood using borers with diameters of 0.5 and 1.3 cm, Swanson (1970) observed 40 to 50% higher water contents in samples from *Pinus contorta* sapwood when a corer was hammered very quickly into the wood compared with the conventional twist boring method. To minimize this problem, Waring and Running (1978) and Waring *et al.* (1979) took their samples in the early part of the day when tensions in the sapwood tracheids were low. In both cases, values of θ_s close to 1.0 were reported in the winter suggesting that water was not squeezed from the cores during extraction.

Nondestructive methods using the attentuation of x-rays or gamma radiation have been used to measure the total density of tree stems; this includes the dry density of the wood and the water content. Kühn and Handl (1973) used x-rays to determine whether 250-yr-old lime trees were rotten in the center of the bole and later used americium 241 as a source of gamma radiation (Handl *et al.,* 1974) to measure variations in dry density in spruce wood but did not measure fresh specimens. Woods *et al.* (1965) used iodine 131 as a source of gamma radiation to measure the water contents of blocks of wood of *Pinus taeda* between saturation and ovendryness. Olszyczka (1979) used americium 241 to measure seasonal changes in the water content of stems of *Pinus contorta* and showed that θ_s in a tree severed from its roots fell by 25% in 3 weeks. From measurements on wooden blocks in the laboratory, careful determination of the length of the pathway through which the radiation beam passed, and measurement of dry wood density, Olszyczka (1979) concluded that the technique should be able to detect changes in θ_s of 2%. These methods need more exploration to determine their usefulness in measuring diurnal and seasonal changes in θ_s in forest trees. By scanning across the stem at regular intervals, this method provides a means of determining daily changes in water content at different depths (W. R. N. Edwards, personal communication).

C. CHANGES IN STORAGE

1. Amounts

a. Foliage. Reynolds (1966) considered the amount of water stored in foliage to be trivial compared with the amount evaporated from the soil, but other work has shown that appreciable amounts of water are stored in the foliage in coniferous canopies. Having measured the water contents of foliage for several different species in southeastern England, Ovington (1956) showed that the maximum amount stored in a plantation of 47-yr-old *Picea abies* was equivalent to 8.1 mm.

Equation (37) shows that the water available in foliage for transpiration depends on the total volume of water storage, the changes in relative water content, and the bound water fraction. The minimum values of θ_1 found under normal physiological conditions in experiments where trees have been allowed to become severely water stressed are usually between 0.7 and 0.84. The minimum value of θ_1 in droughted 3-yr-old seedlings of *Pinus sylvestris* was 0.83 (Rutter and Sands, 1958), and the minimum in forest trees of the same species on plots where water had been withheld was 0.88 (Rutter, 1967). Hellkvist (1973) measured a minimum θ_1 in two clones of *Pinus sylvestris* of 0.75, and Waring *et al.* (1979) reported a minimum θ_1 of 0.8 in detached branches of the same species. Running (1979) measured a minimum θ_1 of 0.6 in detached branches of *Pinus contorta,* and Ashby (1966) a minimum of 0.75 in suppressed trees of *Pinus echinata* following a 48-hour drought period. Hellkvist *et al.* (1974) found a minimum of 0.87 in the winter in *Picea sitchensis* and Johnston (1964) and Oppenheimer (1968) measured θ_1 values of between 0.78 and 0.75 in the late summer in droughted *Pinus radiata* trees. Using a gravimetric vapor pressure equilibration method Jarvis and Jarvis (1963a,b) showed that θ_1 fell to between 0.7 and 0.8 in *Pinus sylvestris* and *Pinus abies* over the physiological range of water potential from 0 to -2.0 MPa.

The total amount of water stored in the foliage and available for transpiration depends on the volume of foliage present and hence on the LAI, as well as on changes in θ_1. A canopy with large LAI may be expected to store a larger amount of water that is available for transpiration than a canopy with small LAI but will also tend to have higher transpiration rates (Section II,A). As the age of the trees increases, the ratio of the amount of water in the foliage to the total amount in the whole tree decreases in accordance with the decreasing proportion of leaf biomass in the total (Ovington, 1956). Running (1979) showed that 33% of the total water was present in the foliage of 2-yr-old seedlings of *Pinus contorta* compared to only 4% in forest trees ranging from 10 to 60 years old.

Using maximum changes in θ_1 and the amounts of foliage on forest plantations, estimates of the total water storage capacity of forest canopies have been made. Jarvis (1975) calculated for a young, dense forest with 4000 stems ha^{-1} and a projected LAI of 10 that the amount of water stored in the foliage of 0.4 mm was sufficient to supply transpiration demand for 1.1 hr, and Waring and Running (1978) calculated for a 50- to 60-m tall plantation of old growth *Pseudotsuga menziesii*, with a projected LAI of 8.4, that the water stored in the foliage was equivalent only to 0.03 mm, less than 0.1% of the total amount stored in the whole trees and sufficient to supply transpiration for only a few minutes. This estimate agrees with calculations by Roberts (1976a), which showed that less than 0.1 mm of water were supplied from the foliage of 43-yr-old, 16-m tall trees of *Pinus sylvestris* during a tree cutting experiment in which individual trees were allowed to dry out during a summer period. However, Waring *et al.* (1979) calculated that larger quantities of water were stored in *Pinus sylvestris* foliage on four plots in a spacing experiment. The values ranged from 0.4 to 0.5 mm (Table IX) and were sufficient to supply transpiration for several hours. Furthermore, Reynolds (1966) calculated from

TABLE IX

THE WATER STORAGE CAPACITIES OF DIFFERENT TISSUES IN FOUR STANDS OF *Pinus sylvestris*[a]

Tissue	Plot 1		Plot 2		Plot 3		Plot 4	
	mm	%	mm	%	mm	%	mm	%
Foliage	0.4	4.0	0.5	2.0	0.5	3.0	0.5	3.0
Cambium and phloem	0.1	1.0	<0.1	0.2	<0.1	0.3	0.1	0.4
Stem sapwood	7.2	53.0	15.6	74.0	9.8	67.0	9.6	63.0
Branch sapwood	3.4	27.0	3.1	14.0	2.8	19.0	3.4	22.0
Root sapwood	1.3	11.0	2.0	9.0	1.6	11.0	1.7	11.0
Total	12.4		21.2		14.7		15.3	

[a] Water storage capacity is the volume of water in the fully turgid tissue × the possible change in relative water content and is equal to the maximum amount of "available" water. There were 608, 3281, 1725, and 1178 trees ha^{-1} on Plots 1, 2, 3, and 4, respectively. From Waring *et al.* (1979).

Johnston's (1964) data for *Pinus radiata* that possibly 0.4 mm of water were stored in the foliage of 9-yr-old, 10-m tall trees. Running (1979) measured the water content in the foliage of 16-m tall *Pinus contorta* trees varying in age from 10 to 60 yr in a stand with 2742 trees ha^{-1} and a projected LAI of approximately 3.0. He calculated that approximately 0.15 mm were stored in the foliage, forming 4% of the total available water in the trees.

b. Cambium and Phloem. More than 90% of the changes in diameter of tree stems are accounted for by changes in the volume of the cambium and phloem cells surrounding the sapwood (Richards, 1973; Lassoie, 1979). Lorio and Hodges (1968) measured the relative water contents of this tissue and showed that in droughted trees θ_t fell to 0.6–0.7 whereas in control trees the value remained above 0.8, suggesting that changes in θ_t are close to those observed for foliage. Stewart *et al.* (1973) showed that the developing phloem tissue lost water at a slower rate than developing xylem tissue in stems of *Eucalyptus regnans,* but the water content of the cells decreased to half the turgid value between dawn and midday.

The amount of water available from storage in the cambium and phloem can be calculated from such changes in θ_t and the volume of the tissue using equation (36), or it can be calculated from the changes in stem diameter. Jarvis (1975) calculated the change in the volume of water in the cambium and phloem, $\Delta V_{c,p}$, in a tree stem from equation (50), where A_s is the cross-sectional area of the stem, $\bar{d}$ is the average diameter of the stem, Δd is the maximum change in the diameter observed, l is the length of the stem, and α is the fraction of available water:

$$\Delta V_{c,p} = \Delta A_s l \alpha \approx \pi \bar{d} \frac{\Delta d}{2} l \alpha. \qquad (50)$$

The thickness of cambium and phloem in trees is usually about 3 to 4 mm but increases with height in the stem (Lassoie, 1979), and the total volume of the tissue is larger in large trees than in small ones (Waring *et al.,* 1979). Taking an average change in cambium and phloem thickness of 300 μm (Lassoie, 1979), Jarvis (1975) calculated that in a 20-yr-old conifer stand with 4000 trees ha^{-1}, 0.4 mm of water were stored in these tissues, sufficient to supply transpiration demand for 1 hr. A figure of 0.1 mm was calculated for total storage in cambium and phloem in *Pinus sylvestris* (Table IX) by Waring *et al.* (1979). Compared with sapwood, the total amount of water available for transpiration from cambium and phloem cells is small and comparable with that in the foliage.

c. Sapwood. The sapwood in conifer stems, branches and roots contains by far the largest amount of water storage in trees. Stewart (1967) considered it to be a supplementary reservoir of water available, especially at times of peak transpiration rates, and referred to it as the dominant, major reservoir. In 1930 Luxford measured the water content of an "ordinary" sized redwood tree that was 60-cm in diameter and 61-m high and concluded that the stem contained 17 m³ of water. Fielding (1952) calculated that 1.6 m³ of water were stored in 30-yr-old trees of *Pinus contorta*, each of total volume 2.3 m³. Ovington (1956) measured the water content of a number of coniferous species in Britain and showed that up to 15 mm of water were stored in the stem, comprising more than 70% of the total water stored in the aerial parts of the trees.

The amount of water available in sapwood of *Pseudotsuga menziesii*, calculated from equation (46) by Waring and Running (1978), was 0.71 m³ per m³ of sapwood when the sapwood was fully saturated ($\theta_s = 1.0$, $\rho_{ds} = 450$ kg m⁻³). If $B_s = 0.2$ then 0.57 m³ of water were available per m³ of sapwood. In a 50- to 60-m tall, 450-yr-old stand, this was equivalent to 20 mm stored in the stem sapwood and 1.4 mm in the branch sapwood, sufficient to supply transpiration demands for several days. Similar quantities were calculated by Jarvis (1975) for a young, dense stand and an older, thinned stand with 4000 and 1000 trees ha⁻¹, respectively. Waring *et al.* (1979) calculated that 7.2 and 15.6 mm were stored in the stem sapwood of trees of *Pinus sylvestris* in two plots with spacings of 608 and 3281 stems ha⁻¹, respectively (Table IX). An additional 3.3 and 3.0 mm were stored in branch sapwood and 1.3 and 2.0 mm in root sapwood on the two plots, respectively, amounting to 12 and 21 mm in total, sufficient to supply transpiration needs for several days. These quantities were more than 95% of the total water storage in the trees. Nnyamah and Black (1977) concluded that in a 20-yr-old, 7- to 9-m tall stand of *Pseudotsuga menziesii* with 840 stems ha⁻¹, 2% of the total water transpired came from stem sapwood during a four week drying summer period. This was equivalent to providing the transpiration demand for 12 hr at a rate of about 2.3 mm day⁻¹. In contrast, Roberts (1976a) examined the possibility that the discrepancy over a 19-day period between two independent measurements of transpiration could be accounted for by internal water reservoirs in a 43-yr-old, 16-m tall stand of *Pinus sylvestris* with 740 stems ha⁻¹. He concluded from measurements of water content changes in cut trees at the end of the summer that only 1.6 mm of water were available from the sapwood. However, Waring *et al.* (1979) calculated that the values of θ_s during Roberts' experiments were already low and changed by only 1 to 10%, suggesting that the sapwood was already substantially depleted before the experiments were begun.

d. Roots. Ovington (1956) did not include the root systems in his estimates of the dry weights and water contents of various coniferous forest species in Britain. If his dry weights are increased by about 20% to account for the roots (Rutter, 1975; Albrektsson, 1978), the water stored in the root systems is of the order to 6 to 7 mm. Waring and Running (1978) calculated that 5.3 mm of water were stored in root systems in a stand of 450-year-old *Pseudotsuga menziesii*, and Waring *et al.* (1979) indirectly estimated the size of the root systems in four stands of *Pinus sylvestris* and calculated that between 1.3 and 2.0 mm of water were stored there (Table IX). Calculations from Running's (1979) data suggest that the roots in a *Pinus contorta* stand stored 0.34 mm of water.

These estimates are based on the storage capacity of the sapwood alone in the root systems. In addition, it is likely that the many small, fine roots are able to supply significant amounts of water. The average root diameter in *Pseudotsuga menziesii* was only 0.8 mm (Nnyamah *et al.*, 1978); more than 90% of the total root length was less than 0.5 mm in diameter in *Pinus sylvestris* (Roberts, 1976b); and the lengths, weights, and diameters of the fine roots changed seasonally in an 11-yr-old plantation of *Picea sitchensis* (Ford and Deans, 1977). Taking the fine root system into consideration, Jarvis (1975) calculated that in a dense conifer stand with 4000 stems ha^{-1} up to 5 mm of water might be stored in the whole root system, sufficient to supply transpiration demand for up to 14 hr, and Roberts (1976a) estimated that root systems might supply up to 0.46 mm day^{-1} toward transpiration in a *Pinus sylvestris* stand.

2. Diurnal and Seasonal Changes in Storage of Water

a. Foliage. Typically, diurnal changes in θ_l of conifers are about 10%, with the maximum occurring during the night and the minimum around midday (Rutter and Sands, 1958; Harms and McGregor, 1962; Jameson, 1966; Rutter, 1967, 1975; Waring *et al.*, 1979), but the fluctuations are much less when the trees are droughted and maximum θ_l falls to about 0.8 (Johnston, 1964).

These diurnal changes in θ_l represent changes in storage of water of about 1 mm (Rutter, 1975). Reynolds (1966) calculated from Johnston's (1964) data for *Pinus radiata* that the diurnal change in water stored in the foliage was equivalent to 0.15 mm day^{-1}, and Waring *et al.* (1979) calculated that the water stored in the foliage of *Pinus sylvestris* changed by an amount equivalent to 0.2 mm day^{-1}.

Seasonally, θ_l is generally higher in winter and early spring than it is in late spring and summer (Rutter, 1967; Kramer and Kozlowski, 1979), although seasonal changes are often smaller than diurnal ones (Clausen and Kozlowski, 1965; Doley, 1967) and are usually less than 5% over the growing season (Jameson, 1966; Pharis, 1967; Gary, 1971; Hellkvist, 1973).

Older needles have lower water contents because of the increased thickness of cell walls (Kozlowski and Clausen, 1965; Clausen and Kozlowski, 1965; Ashby, 1966; Pharis, 1967; Gary, 1971; Hellkvist, 1973) with 1 to 10% lower values of θ_1.

Seasonal changes in θ_1 lead to estimates of changes in stored water similar to the estimates given in Section IV,C,1,a.

b. Cambium and Phloem. Diurnal changes in cambium and phloem thickness range from 0 to 1600 μm, varying with species, size of tree, and changes in the weather (Kozlowski and Winget, 1964; van Laar, 1967; Kozlowski, 1972; Richards, 1973; Hinckley *et al.*, 1978; Lassoie, 1979). Changes are usually larger in the midcrown region than in other parts of the tree stem (Roberts, 1976a; Lassoie, 1979) (Fig. 16), and the changes in this region often occur up to 2 hr before equivalent changes at the base of the tree (Dobbs and Scott, 1971; Richards, 1973), because water is first withdrawn in the midcrown region and later from the lower stem tissue.

Seasonal changes in cambium and phloem thickness reflect the rate of tree growth (Kozlowski, 1972) and fluctuations are often smaller diurnally because of summer drought (Lassoie, 1979).

Lassoie (1979) calculated that the cambium and phloem tissues of *Pseudotsuga menziesii* trees were capable of providing daily 0.04 m^3 of water per tree, which was up to 5% of the total water transpired daily. From the maximum diurnal change in stem thickness recorded in *Pinus sylvestris* trees, Waring *et al.* (1979) estimated that these tissues provided 0.07 mm per day. This was only ca. 10% of the amount available daily from the sapwood.

c. Sapwood. Diurnal changes in θ_s are poorly known because of difficulties of measurement. They are probably less than 10% (Kramer and Kozlowski, 1979). Isolated measurements of larger diurnal changes have been made but require confirmation. Swanson (1970) recorded decreases in

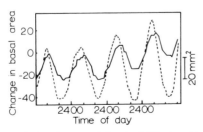

Fig. 16. Diurnal change in basal area at 1.3 m (solid line) and in midcrown (dashed line) of a dominant tree of *Pseudotsuga menziesii* at the end of the rapid growth period and before the onset of summer drought. (Redrawn from Lassoie, 1979.)

water content in *Pinus contorta* between early morning and sunset of about 20%. Waring and Running (1978) measured a change in θ_s from 0.81 to 0.61 in a 3- to 5-m tall *Pinus sylvestris* tree during the first clear day following winter snow.

Seasonal changes in sapwood water content have been well documented from measurements of fresh and dry weights of discs cut from tree stems or from cores removed from sapwood using "Pressler" increment borers. The most comprehensive studies spanning 20 yr by Gibbs (1935, 1958), and Clark and Gibbs (1957) show that for a very wide range of coniferous species, sapwood water contents are usually between 15 to 50% lower in the late summer than they are in the winter. Seasonal changes of the same magnitude for other species have also been reported by Büsgen and Münch (1929), Fielding (1952), Parker (1954), Peck (1959), Yerkes (1967), McGinnes and Dingeldein (1969), and Markstrom and Hann (1972).

Chalk and Bigg (1956) showed that in the outer sapwood of *Picea sitchensis,* θ_s decreased by up to 0.4 between early summer and autumn, reaching a value as low as 0.31 in one tree, and even lower values were found in the inner sapwood in the autumn. In *Pseudotsuga menziesii* in Oregon, United States, θ_s in the outer sapwood was almost completely saturated in winter and, as rates of transpiration increased throughout the summer, fell to 0.5 by autumn, before gradually increasing again (Fig. 17). Similar patterns of seasonal change in θ_s have been found in *Pinus sylvestris* in eastern Scotland (Waring *et al.,* 1979) and in *Pinus contorta* in

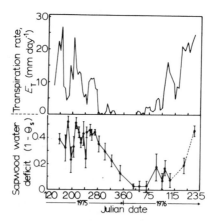

Fig. 17. The relation between transpiration rate, E_T, and sapwood water deficit, $(1 - \theta_s)$, at 1.3 m in *Pseudotsuga menziesii* through a cycle of depletion and recharge of sapwood water. Values of $(1 - \theta_s)$ are means from 5 to 10 trees and two standard errors are shown. (Redrawn from Waring and Running, 1978.)

Colorado, United States (Running, 1979), although the changes were not as large. The minimum θ_s observed by Waring *et al.* (1979) was 0.7 and by Running (1979) was 0.6 after a long summer drought.

These observed changes in θ_s represent large diurnal and seasonal changes in the amount of water stored in the sapwood. Reynolds (1966) calculated that a change in water content of 20% applied to the biomass data from Ovington (1956) would represent a change of 0.6 mm of water in a closed stand. Waring and Running (1978) concluded that a maximum of 1.7 mm day^{-1} could be provided from the sapwood to contribute to transpiration demand and Waring *et al.* (1979) estimated that between 1 and 1.5 mm of water could be available each day for transpiration, i.e., 30 to 50% of the total water transpired, throughout the summer from the sapwood. Seasonally, much larger quantities of water are available and estimates of these amounts have been given in Section IV,C,1,c.

d. Roots. Diurnal shrinkage and swelling of fine roots has been studied in herbaceous plants but few data are available for conifers. In herbaceous plants roots may shrink in diameter diurnally by up to 40% (Huck *et al.*, 1970; Faiz, 1973; Cole and Alston, 1974), leading to changes in relative water content of more than 40% (Jarvis, 1975). Roberts (1976a) proposed that the maximum daily change in relative water content of the fine roots of *Pinus sylvestris* was 20%. Hellkvist *et al.* (1974) recorded diurnal changes in ψ_{root} of about 0.2 MPa in *Picea sitchensis*, although the changes were not related to time of day or weather. If these changes in ψ represent the same changes in water content as in foliage, the changes in relative water content would be about 5%. This is less than in foliage, but as the quantity of water stored in roots is probably larger than in foliage, diurnal changes in the amount of water in the roots may be larger than in foliage.

Little is known about seasonal changes in root water content. Kozlowski (1972) cited observations of D. T. MacDougal on the diurnal changes in root diameter of *Pinus radiata* throughout a season; as the growing season progressed, root shrinkage began earlier in the day, coinciding with the beginning of stem shrinkage. However, we know of no quantitative information on which changes in the amount of water in tree root systems could be based.

3. Mechanisms

a. Shrinkage. The living cells in the leaves, roots, cambium, and phloem are thin-walled so that changes in their water content and ψ are accompanied by deformation in cell size when transpiration from the leaves begins. Water is lost by evaporation from the terminal mesophyll cells and consequently is withdrawn from the adjacent cells because they are at a higher potential. There are three parallel pathways of water movement from cell to cell, viz., through the cell walls (apoplastic), through the symplasm, or from cell vacuole to cell vacuole (Weatherley, 1970). It is most probable that water moves through all these pathways, most of it following the pathways of least resistance primarily through the cortical and stelar cell walls, with passage into the symplasm and through the vacuoles largely restricted to the endodermis in roots and needles.

A wave of falling ψ passes across the mesophyll tissue to the xylem and subsequently through the entire plant to the roots. Because of the water storage capacity of the tissues along the pathway, there is a delay, or time lag, between the start of transpiration from the leaves and the first detectable sign of water absorption by the roots. This has commonly been observed as a time lag in the contraction of the cambium and phloem of the stem, which begins earlier in the upper stem than in the lower part of the bole (Dobbs and Scott, 1971; Hinckley and Bruckerhoff, 1975; Roberts, 1976a; Lassoie, 1979). Changes in ψ at the base of the stem and in the roots are smaller in amplitude and occur after changes in ψ_l (Hellkvist et al., 1974).

The elasticity of the cell walls determines the change in ψ consequent on a change in water content. The turgor pressure and water potential of a cell with more rigid walls will fall further in response to a change in water content than the turgor pressure and water potential in a cell with less rigid walls (Cheung et al., 1975). Thus, the volume of water removed from a tissue, and hence the degree of deformation of the tissue, in response to a change in ψ, depends upon the bulk elastic modulus of the tissue.

For shoots and leaves, the bulk volumetric elastic modulus can be determined from pressure–volume curves obtained with the pressure chamber:

$$\epsilon = \frac{dP}{dV/V} \tag{51}$$

where dP and dV are the changes in turgor potential and volume in cells of bulk-averaged turgor potential, P, and volume V. Hellkvist et al. (1974) found that ϵ increased linearly with turgor potential in Picea sitchensis whereas Cheung et al. (1976) found asymptotic relationships in several species (Fig. 18). ϵ can vary from very low values of ca. zero, when the turgor potential is zero, up to several hundred MPa at full turgor depending on the species, type of cell, or tissue (Dainty, 1976; Tyree, 1976). The leaf tissue of conifers is more rigid than in hardwoods with values of ϵ up to twice the value for hardwoods (Table X). Hence, a substantial reduction in ψ is associated with only small losses of water. For example, a

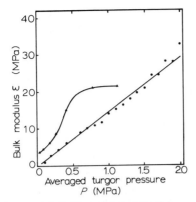

Fig. 18. The relation between the bulk modulus of elasticity, ϵ, and the volume averaged turgor potential, P, in *Abies concolor*, ▲, and *Picea sitchensis*, ●. (Redrawn from Cheung *et al.*, 1976, and Hellkvist *et al.*, 1974, respectively.)

water potential of -1.0 MPa is associated with relative water contents of 0.93, 0.94, 0.89, and 0.95 in shoots of *Picea abies, Picea sitchensis, Pinus sylvestris,* and *Ginkgo bioloba*, respectively, whereas much lower relative water contents are obtained in less rigid leaves (Jarvis and Jarvis, 1963b; Hellkvist *et al.*, 1974; Cheung *et al.*, 1975). Values of ϵ for other tissues in the tree are not known. In living tissues ϵ has an important influence on processes such as extension growth and turgor control, as well as on the movement of water. Rigid cell walls and tissues other than in xylem or fibers might be disadvantageous in growing tissues where low plasticity and elasticity of cell walls might retard growth (Tyree, 1976).

TABLE X

THE MAXIMUM BULK ELASTIC MODULUS ϵ_{max} MEASURED BY THE PRESSURE CHAMBER TECHNIQUE OF SOME GYMNOSPERMS AND SOME ANGIOSPERMS[a]

Species	ϵ_{max} (MPa)
Gymnosperms	
Abies concolor	20
Picea sitchensis	30
Ginkgo biloba	26
Angiosperms	
Acer saccharum	15
Fraxinus pennsylvanica	17
Populus balsamifera	14

[a]From Cheung *et al.* (1976) and Hellkvist *et al.* (1974).

b. Cavitation in Sapwood Tracheids. Whereas living cells expand and contract with changes in water content and ψ, changes in water content of the more rigid, dead xylem tracheids are associated with changes in the volume fractions of water and air or vapor. Once formed in a tracheid, an air bubble will rapidly expand, releasing the tension inside the lumen and causing a pressure difference across the pit membrane, with the result that the torus moves across the pit aperture and conduction ceases in the tracheid (see Section III,B,1,c). The pit is said to be aspirated and the tracheid embolized or cavitated. The cohesion theory of water movement up tree stems postulates that tensions exist in the water columns in the tracheid system and that once the tensions are broken, collapse of the columns will result (see reviews by Greenidge, 1957; Zimmermann, 1971). Although pure water is able to withstand tensions up to 10 MPa (Scholander *et al.,* 1964), quantities of dissolved gases in xylem sap [e.g., higher concentrations of carbon dioxide and lower concentrations of oxygen than in air (Carrodus and Triffett, 1975; Hook and Scholtens, 1978)] reduce the tension that can be withstood. The mechanism of aspiration in earlywood tracheids is seen as an adaptive function to prevent the spread of air bubbles through the entire system once a bubble has been formed (Carlquist, 1975). The unresolved questions still remaining are:

1.   Is cavitation a normal seasonal occurrence in sapwood associated with the observed decreases in sapwood water content during the summer months (Section IV,C,2,c)?

2.   Is cavitation in conifer sapwood reversible, thereby accounting for the increase in sapwood water content during the winter months? In deciduous hardwood trees, root pressure could effect "refilling" of the conducting system during winter but in conifers this is less likely.

Bolton (1976) came to the conclusion that under normal physiological flow rates through individual tracheids, the pressure drop across the tracheid lumen-bordered pit system would be insufficient to cause pit aspiration (see Fig. 7). However, he assumed a maximum gradient of ψ within the stem of only 0.01 MPa m^{-1}. If he is correct in his conclusion, cavitation would only be likely to occur when a bubble of air was introduced into a tracheid by a separate process. From where or how this bubble would originate is not clear. Oertli (1971) argued that the presence of dissolved gases in xylem water is irrelevant since, first, nuclei for spontaneous bubble formation are absent in the columns and, second, the free energy required to form a bubble of critical size for it to grow is very high and not likely to be provided by chemical processes ordinarily occurring in the xylem. Preston (1961) postulated that air could be drawn into water-filled tracheids from adjacent air-filled cells, but this is unlikely since the pressure difference required to force a bubble through the pores in the pit membrane of about 1.4 μm in radius is approximately 0.1 MPa (Gregory and Petty, 1973), and a pressure difference of this size is unlikely to occur. However, cavitation does occur in the latewood each year and spreads throughout because the pits do not aspirate (Section III,B,1,c) with the result that latewood does not conduct water.

When freezing takes place in conifer stems, all of the free water, but not the bound water, can be converted to ice (Lybeck, 1959). Sucoff (1969) demonstrated that when sap froze in 10-cm long segments of 3-yr-old seedlings of *Thuja occidentalis*, *Picea glauca*, and *Pinus sylvestris*, undissolved air formed bubbles and some water migrated to nonfrozen parts of the stem. The result was that the frozen tracheids cavitated and remained so until thawing took place, when the air bubbles redissolved before the pits deaspirated and conduction recommenced. Hammel (1967), in contradiction, suggested that cavitation did not occur during the freezing process. He measured the resistance to water flow in *Tsuga canadensis* twigs which had been completely frozen and found no increase in resistance after freezing and rethawing the twigs. He postulated that the bordered pits closed and isolated the frozen tracheids as a result of positive pressure in the tracheid, consequent to expansion and freezing of the sap. Dainty (1976) pointed out that a high value of the bulk elastic modulus ϵ for xylem, such as 230 MPa (Hammel, 1967), has the consequence that the conversion of only 3% of water in the tracheids to ice would change the pressure in the tracheid from negative to positive. The high values for ϵ in conifer tissues (Table X) are, therefore, an important factor in preventing embolism during freezing.

Indirect evidence from work with herbaceous species has led to the inference that cavitation is a normal, regular event occurring daily. Milburn and Johnson (1966) constructed apparatus that amplified the sound in a xylem vessel when they supposed it cavitated. The result was that a "click" could be heard when the event took place. The number of clicks occurring increased as leaves, or vascular bundles, of several plants were allowed to dry out, and since the fundamental frequencies of the clicks for many species were the same, it was concluded that the same process was taking place (Milburn, 1973; Milburn and McLauchlin, 1974). The number of clicks is related to ψ_l and the process is reversible even when leaves are cooled to 1°C (Milburn, 1966), suggesting that the process is entirely physical. This phenomenon has also been found in branches of *Picea sitchensis* (J. A. Milburn, personal communication) and more recently in 2-yr-old trees of *Pinus radiata* (D. W. Sheriff, personal communication), the numbers of clicks increasing as the trees were allowed to dry out. It seems likely therefore that the same process of cavitation and its reversal also occurs in the tracheids of conifers during the diurnal and seasonal changes in ψ in the summer months.

Less information is available on the reversibility of cavitation that apparently occurs during the winter months (Section IV,C,2,c). Thomas and Kringstad (1971) showed that total deaspiration of all the pits in *Pinus taeda* wood occurred when specimens were soaked in water for 1 week after being dried to 5% water content during a 3-month period. However, after being dried for 7 months only 40% of the pits deaspirated in the specimens after soaking in water for 1 week. Thomas and Nicholas (1966) concluded that deaspiration of pits in *Pinus taeda* wood by solvent drying had no effect on the structure of the pit margo fibrils, but Petty (1972) found that some permanent creep occurred in the margo strands in *Pinus sylvestris* after deaspiration had taken place. The degree of creep was proportional to the time during which the specimens were kept in a dry condition.

Uptake of water by the foliage from rain or dew might account for some increase in water content of sapwood during the winter months but it is unlikely to be a fully adequate explanation because of the quantities of water involved. Stone (1958) showed that the uptake of atmospheric moisture by droughted seedlings of *Pinus jeffreyi, Pinus ponderosa,* and *Abies concolor* increased survival times by up to 4 weeks, and Johnston (1964) found that θ_1 in *Pinus radiata* needles increased by 6% following overnight rain after a period of drought. Leyton and Juniper (1963) showed that uptake of surface water by *Pinus sylvestris* needles was mainly through the adaxial needle surfaces normally enclosed in the sheath. Leyton and Armitage (1968) showed that water could be taken up over the whole surface of needles of *Pinus radiata*. The area normally in the sheath absorbed water to give an increase in fresh weight of 8.5%, absorption of water by the middle of the needles accounted for an increase of 11.2% of the fresh weight, and the tips absorbed the most water, giving an increase in fresh weight of 15.6% during 24 hr of submersion in water. The different uptake rates from the different parts of the needles were accounted for by differences in "wettability" of the areas. Cremer and Svensson (1979) found that it took 22 hr of misting to replace the water lost from storage in cut shoots of *Pinus radiata* during 2 hr of rapid transpiration. Although such a rate of uptake is probably too slow to support growth, except perhaps in areas of frequent fog, it may be sufficient to lead to normal refilling of cavitated tracheids. Vaadia and Waisel (1963) showed that the rates of uptake of water by *Pinus halepensis* needles decreased over a 3-hr-period and, using tritiated water, they demonstrated that the transport of the absorbed water out of the needle was very slow. Rutter (1975) concluded that the amounts of water taken up by canopy foliage were small, even during the night when the soil was dry and the maximum difference in vapor pressure between the air and foliage existed. Taking a canopy with a LAI of 5 and a fresh weight–to–leaf area ratio of 0.02 gm cm^{-2}, Rutter (1975) proposed that the maximum increase in fresh weight would be 10% as a result of absorption of water. This was equivalent to only 0.1 mm taken up by the foliage and is insignificant in comparison with transpiration rates of over 3 mm day^{-1} and diurnal changes in sapwood storage of up to 1.5 mm day^{-1}.

Nearly every tracheid is in contact with a living parenchyma cell, which compose up to 40% of the total sapwood cells. The ray cells could, therefore, be responsible for maintenance of the conducting system. Wodzicki and Brown (1970) measured the rates of transport of acid fuchsin dye in living and dead 2-yr-old stems of *Platanus occidentalis* and *Pinus densiflora* and showed that the rates were always higher in the living stems. They concluded that the ray cells have the capacity to assist the transport process, possibly acting as osmotic pumps, and this may be why the hydraulic conductivity of sapwood is higher with water containing potassium ions than with pure water (Zimmermann, 1978). The ray cells are known to store fats, nitrogenous compounds, and starch (Ziegler, 1964) and are involved in a secretory process which transports inorganic ions from symplasm to xylem vessels (Läuchli, 1972). There is, however, no direct evidence at present that ray cells play a part in the refilling of cavitated tracheids in conifer sapwood.

The occurrence of cavitation, its reversibility, and the process of refilling the tracheids with water pose problems to be answered by future research.

4. Effect of Changes in Water Content on Conductivity

It follows from the definition of relative conductivity (equation 15) that a reduction in the cross-sectional area available for water transport in a tissue will reduce its relative conductivity. It is likely in primary, extensible tissues that reductions of cross-sectional area for water conduction occur when the tissue shrinks. The diurnal shrinkage in roots (Section, IV,C,2,d) almost certainly leads to changes in the diameter of the conducting vessels which would decrease root conductivity during the middle of the day.

It is more likely, however, that diurnal and seasonal changes in conductivity result from cavitation of some of the conducting elements, which reduces the cross-sectional area available for water movement. Milburn and McLaughlin (1974) concluded that cavitation was an everyday occurrence in stems of herbaceous plants, and Byrne *et al.* (1977) showed that a high flow resistance in cotton roots was directly attributable to cavitation of some of the vessels. At values of ψ_l below -2 MPa the maximum potential drop required to maintain flow across a 1-cm segment of lateral root increased by nearly 200%.

In sapwood of conifers blockage of tracheids after cavitation also reduces k. Puritch (1971) examined the effects of the attack of the balsam woolly aphid *Adelges piceae* on water flow in sapwood of *Abies grandis*. In healthy trees the inner sapwood had a lower k than the outer sapwood and the difference was correlated with the volume fraction of the wood that was air-filled (refer to Fig. 14). In the infected trees conductivity was only 5% of that in the healthy trees, and even after allowance for an increase in the air-filled fraction, there was still a threefold decrease in conductivity. This was attributed to incrustation of the pit membrane pores (Puritch and Johnson, 1971), similar to the effects of blockage in heartwood tracheids (Côté, 1958; Krahmer and Côté, 1963; Bauch *et al.*, 1968). Mitchell (1967) showed that the conduction of acid fuchsin dye through sapwood in *Abies grandis* and *Abies lasiocarpa* infested with the balsam woolly aphid was prevented or restricted to a narrow zone in the earlywood.

Gregory (1977b) examined conductivity in *Pinus sylvestris* sapwood that had been infested with *Peridermium pini* and showed that the fraction of air-filled tracheids in the infected wood was ten times higher than in the healthy sapwood. In agreement with Puritch (1971), Gregory concluded that part of the reduction in k resulted from embolism in some of the tracheids and part from effects on pit membrane structure.

Based on Puritch's (1971) data, relative conductivity, normalized to 100% at saturation, is shown in relation to sapwood relative water content in Fig. 19. As θ_s decreases from 1.0 to 0.9, k falls to 58%; at $\theta_s = 0.2$ (bound water only) k is zero. Waring and Running (1978) found good agreement between values of k estimated *in situ* in large *Pseudotsuga menziesii* trees from measurements of transpiration rate and ψ, and the laboratory-determined relationship shown in Fig. 19. The values of k predicted from Fig. 19 at a particular θ_s agreed closely with the values estimated in the forest at the same θ_s. Their data strongly suggest that diurnal and seasonal changes in cavitation, as the result of changes in water content, are the cause of diurnal and seasonal changes in k. Clearly large, regular changes in k are to be expected if θ_s changes frequently to the extent indicated in Section IV,C,2,c. This is the only demonstration of a relationship between k and θ_s and it is not yet possible to consider differences which might exist between species.

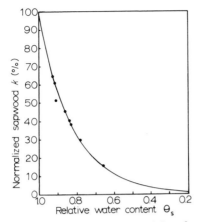

Fig. 19. The relationship between relative conductivity of sapwood normalized to the value at saturation k and sapwood relative water content, θ_s, for *Abies grandis*. The equation for the curve is $k(\%) = 0.4092 \exp(5.5\theta_s), r^2 = 0.97$. (Redrawn from data of Puritch, 1971, following Waring and Running, 1978.)

V. SOIL WATER

A. WATER AVAILABILITY

1. Soil Properties

The total potential of soil water is lowered because of strong attractive forces between the soil particles and water molecules (the matric potential), the presence of dissolved salts (the solute potential), and hydrostatic pressure resulting from the gravitational force of the soil above it (the pressure potential) (Gardner, 1968). The relationship between flow and potential is analogous to that in wood and has traditionally also been described by Darcy's law as follows:

$$q = -K_s \, d\psi_s / dz \qquad (52)$$

where q is the flow in the z direction in steady-state, isothermal conditions and K_s is the hydraulic conductivity of the soil. In saturated soil K_s is a constant as long as the structure is stable (Nye and Tinker, 1977). However, in an unsaturated soil K_s decreases by several orders of magnitude as the soil water content, θ'_s, decreases (Fig. 20).

It is also convenient to define the diffusivity of water in the soil, D_w, by

$$q = D_w \, d\theta'_s \rightarrow dz, \qquad (53)$$

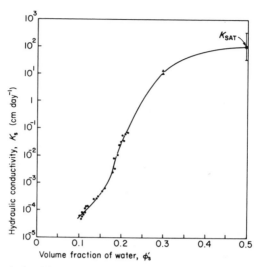

Fig. 20. The relationship between hydraulic conductivity of soil, K_s, and the volume fraction of water, θ'_s, after correction for the stone content in Dashwood gravelly sandy loam. (Redrawn from Spittlehouse, 1980.)

where $d\theta'_s/dz$ is the gradient of water content in the z direction (Slatyer, 1967). It follows from equations (52) and (53) that

$$D_w = -K_s \, d\psi_s/d\theta'_s, \qquad (54)$$

where the soil capacitance, $d\theta'_s/d\psi_s$ is found from the slope of the moisture characteristic curve. Since the energy required to remove water from the soil matrix is inversely related to pore size between the soil particles, and because the pores are irregularly shaped, pronounced hysteresis occurs in the relationship between matric potential and water content of the soil, i.e., the capacitance varies not only with water content but with previous wetting and drying history. Since the capacitance depends on a number of variables, theoretical treatments of the unsteady flow of water in soil are complex (Philip, 1974) and, as in wood, it is necessary to use empirical relationships between K_s, q, ψ_s, and θ'_s (Gardner, 1968). For example, semilogarithmic plots of K_s against θ'_s can be linear, e.g., for an alluvial sandy loam soil (Hillel et al., 1972), or nonlinear, e.g., for a gravelly loam, even after corrections have been made for the stone content of the soil (Fig. 20).

2. Root Distribution

The availability of water is related to K_s, and hence its relationship with θ'_s, but also depends strongly on the density and depth of the rooting system of the plants. The total quantity of water available increases with both θ'_s and with the volume of soil occupied by roots (Gardner, 1960; Cowan, 1965; Newman, 1969a,b). In a saturated soil, water is removed first from where it is most readily available, which is usually in the top layers of the profile because of the greater concentration of roots there (Woods and O'Neal, 1965). From an examination of the water content profile beneath a stand of *Pinus resinosa*, Waggoner and Turner (1971) showed that the upper layers of the profile lost more water initially when the profile was wet and only when it dried out did the lower levels lose water. Nnyamah and Black (1977) found almost exactly the same thing beneath a stand of *Pseudotsuga menziesii*. The zone of maximum water uptake correlated well initially with the distribution of root biomass but subsequently the amount of water extracted from a layer depended both on rooting density and on soil water content.

Depth of rooting is both genetically controlled and environmentally modified (Taylor and Klepper, 1978) and similar-sized trees may extend their root systems to different depths at the same site (Rutter, 1968). Waring and Cleary (1967) emphasized the importance of rooting by demonstrating that in a 25-m tall *Pseudotsuga menziesii* tree ψ_l fell to -2.0 MPa at midday, whereas in a 1-m tall tree at the same site ψ_l fell to nearly -4.0 MPa because the smaller tree evidently had a less extensive root system. The decline in rooting density with depth depends very much on the soil type. Roberts (1976b) showed that in an acid sandy soil overlying chalky drift at Thetford in eastern Britain, roots of 40-yr-old *Pinus sylvestris* trees were found in all layers down to a depth of 2 m but the quantity of roots decreased dramatically below 1 m (Table XI). In contrast, in a 36-yr-old stand of *Pseudotsuga menziesii* in western Washington few roots existed below a depth of 30 cm in a coarse, well-drained, sandy loam soil (Cole *et al.*, 1971). In a plantation of *Picea sitchensis,* growing at spacings of 1.5 m between both rows and trees on peat in southern Scotland, the largest concentration of fine roots was adjacent to the trees and the smallest concentration, 33% lower, was between the trees: 70% of the total length of roots occurred in the peat layer, the maximum depth of which was 0.2 m (Ford and Deans, 1977).

Water in the soil profile completely below the rooting depths cannot be regarded as readily available to the trees: only small quantities of up to a maximum of 0.2 to 0.3 mm day^{-1} (Nnyamah and Black, 1977) or 20–50 mm over the year (Rutter and Fourt, 1965; Rutter, 1967) may be available as the result of capillary rise into the rooting zone.

However, just a few roots penetrating to considerable depths may be able to maintain sufficient uptake for the avoidance of stress. Rutter (1967) failed to measure significant reductions in transpiration from a plot from

TABLE XI

The Quantity, Diameter, and Distribution of Tree Roots of *Pinus sylvestris* in Successive Horizons in a 43-Yr-Old Stand[a]

Depth (cm)	L_a (cm cm^{-2})	L_v (cm cm^{-3})	Root diameter (mm)
0–15	80.17 ± 6.08	5.26	0.278 ± 0.04
15–30	19.74 ± 1.38	1.25	0.325 ± 0.04
30–45	9.30 ± 0.79	0.61	0.475 ± 0.07
45–61	5.17 ± 0.79	0.34	0.508 ± 0.08
61–76	3.00 ± 0.40	0.19	0.553 ± 0.08
76–91	2.85 ± 0.48	0.18	0.531 ± 0.10
91–106	1.29 ± 0.35 (2.51)[b]	0.08 (0.16)	0.618 ± 0.21
106–122	1.38 ± 0.40 (3.11)	0.09 (0.20)	0.640 ± 0.18
122–137	0.94 ± 0.11 (2.05)	0.06 (0.13)	0.752 ± 0.24
138–153	1.06 ± 0.13 (2.75)	0.06 (0.18)	0.988 ± 0.21
153–168	0.56 ± 0.20 (1.52)	0.03 (0.09)	0.775 ± 0.32
168–183	0.69 ± 0.23 (2.84)	0.04 (0.18)	0.775 ± 0.32
Core sum	126.15		

[a] Average height 15 m and 740 stems ha^{-1}. L_a and L_v are total root lengths per unit area and volume of soil, respectively. From Roberts (1976b).

[b] Figures in parentheses represent the means of only those cores containing roots.

which water was withheld throughout the year in a plantation of *Pinus sylvestris,* even when the accumulated soil water deficit in the rooting zone (0–180 cm) reached 250 mm out of a possible 370 mm in a moderately dry year. The probable reason for this was the penetration of a few roots below the 180-cm depth since only about 30% of the extractable water remained in the root zone. In contrast, on soils with a maximum depth limited by sandstone to 70 to 85 cm, Black (1979) found that transpiration of *Pseudotsuga menziesii* began to fall at soil water deficits of about 50 to 60 mm when there was about 40% of the extractable water left in the root zone. The matric potentials at the bottom of the profile were then ca. −0.1 MPa. Rutter did find a fall in transpiration rate in an exceptionally dry year when the accumulated deficit reached 320 mm (out of 370 mm), and this also coincided with matric potentials at the bottom of the rooting zone of ca. −0.1 MPa.

If the soil is waterlogged at depth, the growth of roots penetrating to the water table is greatly reduced and the roots are frequently killed as the water table rises and falls. Coutts and Philipson (1978a) showed that roots of *Pinus contorta* were more tolerant of waterlogging than the roots of *Picea sitchensis,* and they subsequently showed that this was because of more effective transport of oxygen to the root tips (Coutts and Philipson, 1978b; Philipson and Coutts, 1978). Shallow, fluctuating water tables result in shallow-rooted unstable trees and reduce the volume of soil available for exploitation during dry periods.

B. SOIL TEMPERATURE

The uptake of water by plants is reduced at low soil temperature. Kramer (1942) showed that rates of water uptake at soil temperatures of 10°C were only 40% of the rates at 25°C in *Pinus taeda* and *Pinus caribaea* (natives of southeastern United States) and 60% in *Pinus resinosa* and *Pinus strobus* (natives of western United States). He attributed the differences amongst the species to adaptation to their native climates and the low rates of uptake at the low soil temperature to the increased viscosity of water and changes in the conductivity of the roots. Kuiper (1964) recalculated Kramer's (1942) data to correct for changes in viscosity at the lower temperatures and showed that there was a critical temperature, below which root conductivity fell substantially and that the critical temperature was much lower in *Pinus strobus* at 5°C than in many other species.

Lyons (1973) reviewed the effects of temperature on membrane structure and showed that in temperature-sensitive plants the cell membranes undergo a physical phase change from the normal, flexible, liquid-crystalline structure to a solid gel at a critical temperature. This leads to increased conductivity of the membrane, leakage of ions from the tissues, enzyme imbalance, and production of toxic metabolites which decrease energy production from mitochondrial respiration. Below the critical temperature, changes in membrane structure cause a change in the activation energy required to move water molecules across the structure, and this is reflected in a marked increase in the Q_{10} of the process (Slatyer, 1967). These changes are reversible if the root system is not exposed to temperatures below the critical temperature for long periods but after a certain time irreversible damage results.

Babalola *et al.* (1968) attributed the reduction in rates of transpiration and photosynthesis in *Pinus radiata* below 15°C directly to increase in viscosity of water. At a root temperature of 10°C the transpiration rates of 6-month-old seedlings in water culture were half the rates at 15°C. Rook and Hobbs (1976) likewise showed that rates of transpiration and photosynthesis in rooted cuttings of *Pinus radiata* at soil temperatures of 15°C were 300 and 125% higher than at 3°C. Kaufmann (1975) measured transpiration and ψ_l in *Picea engelmannii* growing in its native subalpine environment in Colorado, and found that at temperatures below 7.5°C the resistance to water flow in the roots increased. In contrast, Turner and Jarvis (1975) found that rates of transpiration and photosynthesis in seed-

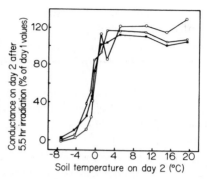

Fig. 21. The effect of soil temperature on rate of net photosynthesis (■), stomatal conductance (○), and mesophyll conductance (□) after 5.5 hr of irradiation on day 2 in an unhardened plant of *Picea sitchensis*. Values are expressed as percentages of the equilibrium values on day 1 at soil temperatures of 20°C. (Redrawn from Turner and Jarvis, 1975.)

lings of the more northerly *Picea sitchensis* were not reduced until soil temperatures fell below 1°C. This threshold is lower than the critical temperatures reported by Kramer (1942), Kuiper (1964), Babalola *et al.* (1968), and Rook and Hobbs (1976) but is in close agreement with the critical temperature of 3°C reported for *Pinus sylvestris,* another more northerly species, by Linder (1973).

Low soil temperature leads to low root conductivity and hence to low ψ_l and turgor potential (e.g., Anderson and McNaughton, 1973). Changes in root conductivity can only affect transpiration rates by bringing about changes in stomatal conductance (Slatyer, 1967). Once the critical temperature is reached, stomatal conductance falls sharply because of reduction in leaf turgor (Fig. 21) (Linder, 1973; Turner and Jarvis, 1975).

Tranquillini (1976) examined the changes in leaf water content with soil temperature in coniferous species growing above the "Krummholz" limit at 2230 m in the Austrian mountains. Leaf water content fell from 140 to 80% during the winter in *Pinus cembra* and from 140 to less than 50% in *Picea abies* in a severe winter. This low water content was less than the limit of 62% water content, below which needles were completely desiccated and fell off. The more drought resistant species, e.g., *Pinus cembra* and *Pinus mugo,* were better able to survive at higher altitudes than the more drought-susceptible *Picea abies*. Predawn ψ_l of *Juniperus virginiana* fell from −0.4 to −1.4 MPa when summer soil temperature fell below 16°C and winter soil temperature below 10°C (Lassoie and Dougherty, 1976). D. Whitehead, P. G. Jarvis, and R. H. Waring (unpublished data) also found that predawn values of ψ_l fell from −0.5 to −0.8 MPa in *Pinus sylvestris* trees during the winter when the stems were close to freezing temperatures. Zimmermann (1964) concluded that at stem temperatures of -1 to $-2°C$, the ascent of sap in conifers was reduced and ice formation took place in the stems. However, some species adapted to growing in very harsh environments are apparently able to withstand very low soil temperatures without any effects on their water relations. For example, Wolff *et al.* (1977) showed that in *Picea mariana* growing in conditions of "permafrost" and long photoperiods in Alaska, ψ_l did not fall below -1.8 MPa and the trees were apparently unaffected by the extremely low winter temperatures. With seedlings of *Pinus radiata* growing in pots in the greenhouse stomatal conductance was not affected but minimum diurnal ψ_l fell from -0.8 to -1.2 MPa as soil temperature was reduced from 20 to 6°C (Kaufmann, 1977a), and the slope of the relationship between E_T and ψ_l (resistance) increased as soil temperature was decreased from 19–25 to -1 to $+1°C$ in 2-yr-old seedlings of *Pinus radiata* (Fig. 22) (Kaufmann, 1977b). Running (1979) found the same effect when soil temperature was decreased from 22 to 0°C in seedlings of *Pinus contorta*.

The effects of low soil temperatures are reversible and plants can be acclimatized by preconditioning or hardening by exposure to low temperatures. Kuiper (1972) argued that the critical temperature, below which the potential energy barrier for water transport across cell membranes is much higher, strongly depends on the environmental conditions during growth and shifts to a lower value if roots are grown at lower temperatures. Turner and Jarvis (1975) showed differences in the response to low soil temperatures of seedlings of *Picea sitchensis* that had been hardened in

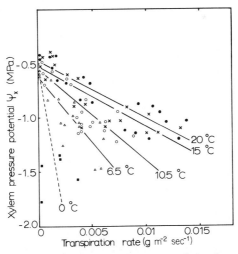

Fig. 22. The influence of root temperature on the relation between xylem pressure potential, ψ_x, and transpiration rate in seedlings of *Pinus radiata*. (Redrawn from Kaufmann, 1977b.)

comparison with seedlings that had not previously been exposed to low temperatures. The roots of the hardened plants were not frost-damaged by exposure to soil temperatures of $-7°C$ and the rate of photosynthesis and stomatal conductance of the hardened plants were not reduced until root temperatures were reduced below $-1°C$, as compared with $1°C$ in the unhardened plants. In the hardened seedlings, the osmotic potential was lower in the foliage and in the roots by 0.6 and 0.4 MPa, respectively, than in the unhardened plants. As a result, the critical value of ψ_l for stomatal closure was also 0.6 MPa lower in the hardened plants than in the unhardened plants, and consequently the hardened plants were able to withstand lower soil temperatures and root conductances before stomatal closure occurred. Thus, acclimation resulting in a decrease in osmotic potential in the foliage conferred extended tolerance to low soil temperatures.

The effects of low soil temperatures on survival and growth rates of coniferous species differ according to the species and the environment. The differences reported here between species in critical temperatures and the effects of hardening can presumably be explained by variations in the structure of cell membranes and their susceptibility to temperature changes. There is some evidence that abscisic acid is involved in these processes (Markhart *et al.*, 1979). Tranquillini (1976) concluded that considerable damage in species above the "Krummholz" zone in Austria occurred through the combined effects of frost and drought resulting from frozen soils. However, Turner and Jarvis (1975) concluded that low air temperatures and low irradiance were primarily responsible for the winter depression in growth of *Picea sitchensis* in eastern Scotland. Rook and Whitehead (1979) also concluded that low soil temperatures were probably not responsible for low rates of photosynthesis and growth in *Pinus radiata* in winter in New Zealand: when *Pinus radiata* seedlings were transferred from day/night air temperatures of $15°/10°C$ to $33°/28°C$ the rate of photosynthesis increased by 35% and the rate of respiration was halved within two days (Rook, 1969). Changes in rates of photosynthesis, transpiration, and respiration within several days of transfer from outdoor winter conditions to the laboratory at 25%C were also observed by Želawski and Kucharska (1967) in *Pinus sylvestris* seedlings.

C. INFLUENCE ON WATER POTENTIAL AND MOVEMENT

Water potential in the leaf depends on the soil water potential in a manner which can be approximated by equation (12). As ψ_s falls, ψ_l will also decrease, although the decrease will be modified to some extent by the capacitance of tree and soil, not included in equation (12). Thus, as discussed earlier (Section III,C,1), ψ_l will fluctuate diurnally following the diurnal changes in E_T. Since E_T is proportional to (Dg_c) (equation 7), and since g_c is a function of ψ_l (Beadle et al., 1979), ψ_s will eventually fall to a level which causes ψ_l to fall below the critical value for stomatal closure earlier and earlier in the day as the soil is slowly dried out (Gardner and Nieman, 1964).

The relationship between ψ_l and ψ_s as the soil dries will depend on the values of the resistances to water movement in the pathway from soil to leaf and the changes in these resistances. As Rutter (1975) pointed out, unless the soil component, R_{soil}, is small in comparison with R_p the reduction in conductivity in the soil which contributes substantially to R_{soil} will enhance the reduction in ψ_l. Jarvis (1975) and Nye and Tinker (1977) present concise reviews of the location of the resistance to water movement between soil and root. The application of models by Gardner (1960), Philip (1957), and Cowan (1965) to the flow of water in unsaturated soils suggested that the perirhizal resistance was large in comparison with the root resistance (Weatherley, 1970) but the inclusion of more realistic values for root lengths in the models led to the contrary suggestion that the perirhizal resistance was small and unlikely to be important in impeding plant water uptake (Newman, 1969a,b; Hansen, 1974; Reicosky and Ritchie, 1976; Taylor and Klepper, 1978), although it could become important when flow rates are high and rooting density is low (Williams, 1974; Nye and Tinker, 1977). It is more likely that the major resistance to water movement from soil to root is located at the interface between the soil and root where a vapor gap occurs when the roots shrink (Faiz, 1973; Faiz and Weatherley, 1977).

Various studies with conifers have established relationships between ψ_l and ψ_s (e.g., Wiebe et al., 1970) but the relationships are usually not linear. For example, exponential decreases in ψ_l as ψ_s decreases have been found (Sucoff, 1972; Sands and Theodorou, 1978) and transpiration rate has also been shown to decrease exponentially as ψ_s falls (e.g., Babalola et al., 1968; Jackson et al., 1973; Rook et al., 1977; Cremer, 1972). Nnyamah et al. (1978) found good linear relationships between rate of uptake of water by Pseudotsuga menziesii trees and the difference between root and soil ψ but the relationship was not so clear if soil to twig ψ difference was used.

A clear correlation between ψ_l and ψ_s would be expected in steady conditions of no transpiration with E_T zero in equation (12). This condition may obtain at dawn (Slatyer, 1967). Predawn ψ_l (see Section III,A,1,c) has been used as an indicator of soil water availability in coniferous forest, for example, at different forest sites, at different times throughout the year in Oregon (Waring and Cleary, 1967). Waring et al. (1978) were able to correlate the maximum LAI with predawn ψ_l, and a "temperature growth index" at sites in the Cascade and Siskiyou mountains in Oregon, and Barker (1973) calculated regressions of predawn water potential against environmental variables and stomatal aperture in *Abies concolor* and *Pinus ponderosa* to obtain estimates of soil water availability.

However, there are difficulties in relating ψ_l to ψ_s if transpiration is not actually zero and ψ_s is not uniform throughout the soil profile (Hinckley *et al.*, 1978). Furthermore, because of the shape of the $d\psi_s/d\theta'_s$ relationship, the availability of water does not change linearly as ψ_s (or ψ_l) decreases. Sucoff (1972) and Hinckley and Ritchie (1973) considered that predawn ψ_l did not decrease until 80% of the available water in the soil was removed, and Rutter (1967) concluded that 60% of the available water could be removed before transpiration was restricted in *Pinus sylvestris* trees.

VI. THE SOIL–PLANT–ATMOSPHERE CONTINUUM

A. THE HYPOTHESIS

In 1975, Jarvis put forward a hypothesis relating transpiration rate, liquid flow resistance, leaf area, sapwood basal area, and leaf water potential. This hypothesis is essentially described by Fig. 1 and has the following main features.

1. In the forest, water is transferred from the soil to the atmosphere through the stems of the trees which can be regarded as a number of parallel conduits linking the source and the sink.

2. The rate of transpiration from the canopy is proportional to the LAI, stomatal conductance, and VPD (equation 7).

3. The proportion of the total flow from soil to atmosphere to be found in individual stems depends on the number of stems per unit area; when there are many stems present, the flow through any one of them is smaller than when few are present, as seen in Fig. 6, for example.

4. Each stem has a flow resistance, R_i, and the flow resistance of the stand is, therefore

$$R_c = R_i/n \qquad (52)$$

where n is the number of stems per unit area.

5. If the canopy is complete, i.e., at ceiling LAI, the leaf area carried by a stem is inversely proportional to the number of stems per unit area.

6. The fall in ψ between soil and foliage depends on the flow rate

and flow resistance within the individual trees. (In its original form no account is taken of capacitance and storage capacity.)

7. The flow resistance depends to a large extent on the basal area A_{sw}, and relative conductivity, k, of the sapwood (equation 15).

8. The hypothesis is completed by the linear correlation found between sapwood basal area and foliage area on a tree, e.g., Fig. 9.

The relationships between leaf water potential and transpiration rate per tree replotted in Fig. 6 can be used to illustrate some of the implications of the above propositions. In both canopies ψ_l fell over the same range of values during the course of a number of days of different weather. In both cases minimum values of ψ_l of -1.6 MPa were reached at transpiration rates from the canopy of ca. 0.55 mm hr^{-1}, irrespective of stand density. Thus the flow resistances for the two stands were identical. However, because of the large differences in number of trees per unit area, the flow rates in the individual stems of the two stands were widely different, as shown in Fig. 6. Consequently, the flow resistance in the trees was much larger in the high-density stand than in the low-density stand. Since the values of ψ_l were identical at the same stand transpiration rates, the flow resistance of the trees was proportional to the number of trees present. Thus, halving the number of stems present would be expected to result in halving the stem resistance.

We would have expected this kind of situation for trees of the same species growing at different densities in the same locality but it is somewhat surprising for two such dissimilar species as *Picea sitchensis* and *Pinus sylvestris* growing at two very different sites: the two stands differ substantially in LAI and the sites differ in g_i. Substantial differences in relative conductivity between the species are also to be expected with the result that the relationship between leaf area and sapwood basal area, S, would be quite different for the two species (equation 23). In the event, therefore, these differences must compensate one another to lead to the same values of ψ_l at the same stand transpiration rates.

On the other hand, stands of the same species growing at different densities in the same locality would have broadly similar LAI's, values of k, and relationships between leaf area and sapwood basal area. The primary compensation expected to occur, therefore, is the proportional decrease in stem flow resistance, associated with a lower stand density and higher leaf area per tree, and resulting from an increase in sapwood basal area and possibly also k.

This hypothesis ignores a number of obviously important factors considered earlier in this chapter, such as the dependence of k on sapwood water content, and does not consider several possible interdependencies amongst the variables, such as correlation between relative conductivity and sapwood cross-sectional area. It is an outline of ideas which can be expanded to explore how changes in k, and other tree properties, particularly those associated with storage capacity and capacitance, may influence the movement and activity of water in the soil–plant–atmosphere continuum.

In this review we have tried to provide the background evidence for this hypothesis together with the ideas and equations to enable it to be expanded into a more realistic computer simulation of water flow in the soil–plant–atmosphere continuum.

B. Simulation

There have been a number of attempts to model water flow through trees and to predict leaf ψ_l using simple relationships like equation (12) which do not consider the water storage capacity and capacitance of the tissues. Some of these models have been very successful in predicting ψ_l in trees over short periods (e.g., Thompson and Hinckley, 1977), but in other cases it has been necessary to invoke diurnal and seasonal changes in resistance or conductivity to explain changes in water flow and potentials (Hellkvist and Parsby, 1976; Roberts, 1977; Kaufmann, 1977b; Waring and Running, 1978). Models that include storage capacity and capacitance have been developed for apple trees (Powell and Thorpe, 1977; Landsberg *et al.*, 1976), but the only published model we know of for conifers that includes changes in tissue water storage is that by Running *et al.* (1975) and Waring and Running (1976). This model includes changes in water storage in the sapwood and in the extensible tissues of the cambium and phloem and predicts transpiration rate and changes in the water content of the storage tissues.

In this chapter we have presented the concepts and equations we think are required for a water flow and potential model that includes water storage capacity and capacitance in the plant. The model we are currently using makes use of many of the equations and functions previously given, the most important being listed in Table XII.

More details are given by Jarvis *et al.* (1981). At the present time, the main usefulness of the model lies in providing a conceptual framework for our ideas and in describing the behavior of variables such as water potential, water content, and flow rates at different points in the tree where they are difficult to measure in practice. Fig. 23 shows the diurnal changes in these variables at the base of the stem and in the foliage. Clearly, as a result of resistance and capacitance there are substantial phase shifts and attenuation of these variables at the base of the stem as compared with in

TABLE XII

EQUATIONS AND FUNCTIONS GIVEN PREVIOUSLY WHICH ARE USED IN OUR MODEL OF
WATER FLOW IN THE TREE

Transpiration	Equations 6, 8
Water flow in the tree	Equations 12, 15, 20, 33; Figs. 9, 19
Tissue capacitance	Equations 25, 27, 29; Fig. 15
Tissue water content	Equations 35, 36, 37, 40, 48, 49

the foliage, in agreement with the few data available (e.g., Hellkvist *et al.,* 1974; Jarvis, 1975, 1976; Lassoie, 1979).

C. CONCLUSIONS

Our ideas about the water relations of trees have progressed substantially over the last 10 years as a result of the integration of a number of separate concepts into our thinking about the functioning of the complete soil–plant–atmosphere system. The concept of the soil–plant–atmosphere

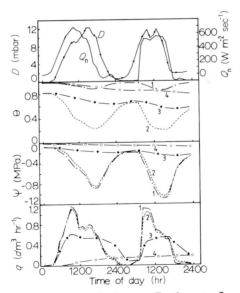

Fig. 23. The simulation of the transpiration rate, E_T, the water flow, q, water content, θ, and water potential, ψ, in the leaves, branches, top, and base of the stem (1, 2, 3, 4, respectively) of a tree of *Pinus contorta* over two sunny days with net radiation, Q_n, and vapor pressure deficit, D. (From W. R. N. Edwards, H. Talbot, J. J. Landsberg, and P. G. Jarvis, unpublished.)

system has been a motive force since 1957 (Philip, 1957). Similarly, the phenomena of changes in water storage have been familiar for many years (see Kozlowski, 1972), as have relationships between leaf mass and stem size (see Ovington, 1956), and simple models of flow resistance and potential also have a long history (see Huber, 1952). We are at the present time on the threshold of a major step forward which involves the full integration of these ideas into a far more comprehensive appreciation of the water relationships throughout the system. For this computer simulation is essential and the last 5 years have seen the first appearance of simulations of conifer water relations that attempt to embrace these concepts.

On the practical side things are less bright. There has been little advance in measurement and experimentation in tree–water relationships over the last 10 years because of technical inability to make the necessary measurements. To establish experimentally the relationships we can now predict from our grasp of the functioning of the system, we need to be able to measure water flow, potential and content at any point in the system. We are far from being able to do this, our ability to measure these variables being restricted to certain particular locations, e.g., water potential in the foliage by pressure chamber, flow at the base of the stem by heat-flux.

Consequently, we foresee difficulties in the next 10 years in the verification of the models which are being developed to embrace our ideas. Fig. 23 shows graphically what we expect to find but we do not have the techniques available to test fully the predictions made there, and if we expand our predictions to the twigs or roots, for example, the problem becomes more severe. Our final conclusion is, therefore, that there must be substantial advances in measurement techniques over the next 10 years if we are to be able to verify the predictions that arise from our developing ideas about the functioning of the system.

REFERENCES

Alban, D. H., Perala, D. A., and Schlaegel, B. E. (1978). Biomass and nutrient distribution in aspen, pine and spruce stands on the same soil type in Minnesota. *Can. J. For. Res.* **8**, 290–299.

Albrektsson, A. (1976). "The Amount and Distribution of Tree Biomass in Some Pine Stands (*Pinus sylvestris*) in Northern Gästrikland," Swed. Coniferous For. Proj., Uppsala, Intern. Rep. 38 (in Swedish).

Albrektsson, A. (1978). "Beskrivning av trädens biomassa i några tallbestand (*Pinus sylvestris*). "Swed. Coniferous For. Proj., Uppsala, Intern. Rep. 73.

Anderson, J. E., and McNaughton, S. J. (1973). Effects of low soil temperature on transpiration, photosynthesis, leaf relative water content, and growth among elevationally diverse plant populations. *Ecology* **54**, 1220–1233.

Ashby, W. C. (1966). Relative turgidity determinations for native and plantation shortleaf pine in southern Illinois. *Bot. Gaz. (Chicago)* **127**, 225–227.

Aussenac, G., and Granier, A. (1978). Quelques résultats de cinétique journalière du potential de sève chez les arbres forestiers. *Ann. Sci. For.* **35**, 19–32.

Babalola, O., Boersma, L., and Youngberg, C. T. (1968). Photosynthesis and transpiration of Monterey pine seedlings as a function of soil water suction and soil temperature. *Plant Physiol.* **43**, 515–521.

Bailey, I. W. (1958). The structure of tracheids in relation to the movement of liquids, suspensions, and undissolved gases. *In* "Physiology of Forest Trees" (K. V. Thimann, ed.), pp. 71–82. Ronald Press, New York.

Bailey, P. J., and Preston, R. D. (1969). Some aspects of softwood permeability. I. Structural studies with Douglas fir sapwood and heartwood. *Holzforschung* **23**, 113–120.

Bailey, P. J., and Preston, R. D. (1970). Some aspects of softwood permeability. II. Flow of polar and non-polar liquids through sapwood and heartwood of Douglas-fir. *Holzforschung* **24**, 37–45.

Bamber, R. K. (1961). Staining reaction of the pit membrane of wood cells. *Nature (London)* **191**, 409–410.

Bamber, R. K. (1976). Heartwood, its function and formation. *Wood Sci. Technol.* **10**, 1–8.

Banks, W. B. (1968). A technique for measuring the lateral permeability of wood. *J. Inst. Wood Sci.* **4**, 35–41.

Bannan, M. W. (1965). The length, tangential diameter, and length/width ratio of conifer tracheids. *Can. J. Bot.* **43**, 967–984.

Barker, J. E. (1973). Diurnal patterns of water potential in *Abies concolor* and *Pinus ponderosa*. *Can. J. For. Res.* **3**, 556–564.

Barnett, J. R. (1971). Winter activity in the cambium of *Pinus radiata*. *N. Z. J. For. Sci.* **1**, 208–222.

Barnett, J. R. (1976). Rings of collapsed cells in *Pinus radiata* stemwood from lysimeter-grown trees subjected to drought. *N. Z. J. For. Sci.* **6**, 461–465.

Barnett, J. R. (1979). Current research into tracheary element formation. *Curr. Adv. Plant Sci.* **11**, 33.1–33.13; commentaries in *Plant Sci.* No. 33.

Barnett, J. R., and Harris, J. M. (1975). Early stages of bordered pit formation in radiata pine. *Wood Sci. Technol.* **9**, 233–241.

Barrs, H. D. (1968). Determination of water deficits in plant tissue. *In* "Water Deficits and Plant Growth" (T. T. Kozlowski, ed.), Vol. 1, pp. 235–368. Academic Press, New York.

Baskerville, G. L. (1972). Use of logarithmic regression in the estimation of plant biomass. *Can. J. For. Res.* **2**, 49–54.

Bauch, J., and Berndt, H. (1973). Variability of the chemical composition of pit membranes in bordered pits of gymnosperms. *Wood Sci. Technol.* **7**, 6–19.

Bauch, J., Liese, W., and Scholz, F. (1968). Über die Entwickslung und stoffliche Zusammensetzung der Hoftupfelmembranen von Längstracheiden in Coniferen. *Holzforschung* **22**, 144–153.

Beadle, C. L., Jarvis, P. G., and Neilson, R. E. (1979). Leaf conductance as related to xylem water potential and carbon dioxide concentration in Sitka spruce. *Physiol. Plant.* **45**, 158–166.

Beadle, C. L., Neilson, R. E., Talbot, H., and Jarvis, P. G. (1981). Modelling the response of canopy conductance of *Pinus sylvestris* L. to the weather. *J. Appl. Ecol.* (in press).

Beets, P. (1977). Determination of the fascicle surface area for *Pinus radiata*. *N. Z. J. For. Sci.* **7**, 397–407.

Benecke, U. (1980). Photosynthesis and transpiration of *Pinus radiata* D. Don under natural conditions in a forest stand. *Oecologia* **44**, 192–198.

Bennett, K. J., and Rook, D. A. (1978). Stomatal and mesophyll resistances in two clones of *Pinus radiata* D. Don known to differ in transpiration and survival rate. *Aust. J. Plant Physiol.* **5**, 231–238.

Beszterda, A., and Raczowski, J. (1976). An attempt at identification of heartwood of Scots Pine (*Pinus sylvestris* L.) by means of a colour reaction with p-amino-N, N-dimethylaniline. *Zesz. Probl. Postepow Nauk Roln.* **185**, 167–172.

Black, T. A. (1979). Evapotranspiration from Douglas fir stands exposed to soil water deficits. *Water Resour. Res.* **15**, 164–170.

Bolton, A. J. (1976). Biological implications of a model describing liquid flow through conifer wood. *In* "Wood Structure in Relation to Biological and Technological Research" (P. Baas, A. J. Bolton, and D. M. Catlin, eds.), Leiden Bot. Ser. No. 3, pp. 222–237. Leiden Univ. Press, Leiden, The Netherlands.

Bolton, A. J., and Petty, J. A. (1975). Structural components influencing the permeability of ponded and unponded Sitka spruce. *J. Microsc. (Oxford)* **104**, 33–46.

Bolton, A. J., and Petty, J. A. (1977a). Variation of susceptibility to aspiration of bordered pits in conifer wood. *J. Exp. Bot.* **28**, 935–941.

Bolton, A. J., and Petty, J. A. (1977b). Influence of critical point and solvent exchange drying on the gas permeability of conifer sapwood. *Wood Sci.* **9**, 187–193.

Bolton, A. J., and Petty, J. A. (1978a). A model describing axial flow of liquids through conifer wood. *Wood Sci. Technol.* **12**, 37–48.

Bolton, A. J., and Petty, J. A. (1978b). The relationship between axial permeability of wood to dry air and to a non-polar solvent. *Wood Sci. Technol.* **12**, 111–126.

Booker, R. E. (1977). Problems in the measurement of longitudinal sapwood permeability and hydraulic conductivity. *N. Z. J. For. Sci.* **7**, 297–306.

Booker, R. E., and Kininmonth, J. A. (1978). Variation in longitudinal permeability of green radiata pine wood. *N. Z. J. For. Sci.* **8**, 295–308.

Boyer, J. S. (1974). Water transport in plants: Mechanism of apparent changes in resistance during absorption. *Planta* **117**, 187–207.

Brix, H. (1967). An analysis of dry matter production of Douglas-fir seedlings in relation to temperature and light intensity. *Can. J. Bot.* **45**, 2063–2072.

Brown, C. L. (1970). Physiology of wood formation in conifers. *Wood Sci.* **3**, 8–22.

Brown, C. L. (1971). Secondary growth. *In* "Trees Structure and Function" (M. H. Zimmermann and C. L. Brown, eds.), pp. 67–123. Springer-Verlag, Berlin and New York.

Brown, J. K. (1978). Weight and density of crowns of Rocky mountain conifers. *U. S., For. Serv., Res. Pap. INT* **197**, 1–56.

Burdett, A. N. (1979). A nondestructive method for measuring the volume of intact plant parts. *Can. J. For. Res.* **9**, 120–122.

Büsgen, M., and Münch, E. (1929). "The Structure and Life of Forest Trees" (transl. by T. Thompson), 3rd ed. Chapman & Hall, London.

Byrne, G. F., Begg, J. E., and Hansen, G. K. (1977). Cavitation and resistance to water flow in plant roots. *Agric. Meteorol.* **18**, 21–25.

Carlquist, S. (1975). "Ecological Strategies of Xylem Evolution." Univ. of California Press, Berkeley.

Carrodus, B. B. (1971). Carbon dioxide and the formation of heartwood. *New Phytol.* **70**, 939–943.

Carrodus, B. B. (1972). Variability in the proportion of heartwood formed in woody stems. *New Phytol.* **71**, 713–718.

Carrodus, B. B., and Triffett, A. C. K. (1975). Analysis of composition of respiratory gases in woody stems by mass spectrometry. *New Phytol.* **74**, 243–246.

Čatský, J. (1974). Water saturation deficit (relative water content). *In* "Methods of Studying Plant Water Relations" (B. Slavík, ed.), pp. 136–151. Springer-Verlag, Berlin and New York.

Chalk, L., and Bigg, J. M. (1956). The distribution of moisture in the living stem in Sitka spruce and Douglas fir. *Forestry* **29**, 5–21.

Chattaway, M. M. (1952). The sapwood-heartwood transition. *Aust. For.* **16**, 25–34.

Cheung, Y. N. S., Tyree, M. T., and Dainty, J. (1975). Water relations parameters on single leaves obtained in a pressure bomb and some ecological interpretations. *Can. J. Bot.* **53**, 1342–1346.

Cheung, Y. N. A., Tyree, M. T., and Dainty, J. (1976). Some possible sources of error in determining bulk elastic moduli and other parameters from pressure-volume curves of shoots and leaves. *Can. J. Bot.* **54**, 758–765.

Clark, J., and Gibbs, R. D. (1957). Studies in tree physiology. IV. Further investigations of seasonal changes in moisture content of certain Canadian forest trees. *Can. J. Bot.* **35**, 219–253.

Clausen, J. J., and Kozlowski, T. T. (1965). Use of the relative turgidity technique for measurement of water stress in gymnosperm leaves. *Can. J. Bot.* **43**, 305–316.

Cline, R. G., and Campbell, G. S. (1976). Seasonal and diurnal water relations of selected forest species. *Ecology* **57**, 367–373.

Cole, D. W., Gessel, S. P., and Dice, S. F. (1971). Distribution and cycling of nitrogen, phosphorus, potassium, and calcium in a second-growth Douglas-fir ecosystem. *In* "Symposium on Primary Productivity and Mineral Cycling in Natural Ecosystems" (H. E. Young, chairman), 2nd printing, pp. 197–232. Univ. of Maine Press, Orono.

Cole, P. J., and Alston, A. M. (1974). Effects of transient dehydration on absorption of chloride by wheat roots. *Plant Soil* **40**, 243–247.

Comstock, G. L. (1965). Longitudinal permeability of green eastern hemlock. *For. Prod. J.* **15** (10), 441–449.

Comstock, G. L., and Côté, W. A., Jr. (1968). Factors affecting permeability and pit aspiration in coniferous sapwood. *Wood Sci. Technol.* **2**, 279–291.

Côté, W. A., Jr. (1958). Electron microscope studies of pit membrane structure. Implication in seasoning and preservation of wood. *For. Prod. J.* **8** (10), 296–301.

Coutts, M. P. (1969). The mechanism of pathogenicity of *Sirex noctilo* on *Pinus radiata*. I. Effects of the symbiotic fungus *Amylosterum sp.* (Thelophoraceae). *Aust. J. Biol. Sci.* **22**, 915–924.

Coutts, M. P. (1976). The formation of dry zones in the sapwood of conifers. I. Induction of drying in standing trees and logs by *Fomes annosus* and extracts of infected wood. *Eur. J. For. Pathol.* **6**, 372–381.

Coutts, M. P. (1977). The formation of dry zones in the sapwood of conifers. II. The role of living cells in the release of water. *Eur. J. For. Pathol.* **7**, 6–12.

Coutts, M. P., and Philipson, J. J. (1978a). Tolerance of tree roots to waterlogging. I. Survival of Sitka spruce and Lodgepole pine. *New Phytol.* **80**, 63–69.

Coutts, M. P. and Philipson, J. J. (1978b). Tolerance of tree roots to waterlogging. II. Adaptation of Sitka spruce and Lodgepole pine to waterlogged soil. *New Phytol.* **80**, 71–77.

Coutts, M. P., and Rishbeth, J. (1977). The formation of wetwood in Grand fir. *Eur. J. For. Pathol.* **7**, 13–22.

Cowan, I. R. (1965). Transport of water in the soil-plant-atmosphere system. *J. Appl. Ecol.* **2**, 221–239.

Cowan, I. R., and Milthorpe, F. L. (1968). Plant factors influencing the water status of plant tissues. *In* "Water Deficits and Plant Growth" (T. T. Kozlowski, ed.), Vol. 1, pp. 137–193. Academic Press, New York.

Cremer, K. W. (1972). Immediate resumption of growth by radiata pine after five months of minimal transpiration during drought. *Aust. For. Res.* **6**(1), 11–16.

Cremer, K. W., and Svensson, J. G. P. (1979). Changes in length of *Pinus radiata* shoots reflecting loss and uptake of water through foliage and bark surfaces. *Aust. For. Res.* 9(3), 163–172.

Curtis, R. O., and Reukema, D. L. (1970). Crown development and site estimates in a Douglas-fir plantation spacing test. *For. Sci.* 16, 287–301.

Dainty, J. (1976). Transport of water across cell membranes. *In* "Handbuch der Pflanzen physiologie" (U. Lüttge and M. G. Pitman, eds.), Vol 2, Part A, p. 12–35. Springer-Verlag, Berlin and New York.

Davidson, R. M., and Young, H. (1973). Abscisic acid content of xylem sap. *Planta* 109, 95–98.

Del Rio, E., and Berg, A. (1979). Specific leaf area of Douglas-fir reproduction as affected by light and needle age. *For. Sci.* 25, 183–184.

Denmead, O. T. (1969). Comparative micrometeorology of a wheat field and a forest of *Pinus radiata*. *Agric. Meteorol.* 6, 357–372.

Denne, M. P. (1974). Effects of light intensity on tracheid dimensions in *Picea sitchensis*. *Ann. Bot. (London)* [N.S.] 38, 337–345.

Denne, M. P. (1976). Effects of environmental change on wood production and wood structure in *Picea sitchensis* seedlings. *Ann. Bot. (London)* [N.S.] 40, 1017–1028.

Denne, M. P. (1979). Wood structure and production within the trunk and branches of *Picea sitchensis* in relation to canopy formation. *Can. J. For. Res.* 9, 406–427.

Denne, M. P., and Smith, C. J. (1971). Daylength effects on growth, tracheid development, and photosynthesis in seedlings of *Picea sitchensis* and *Pinus sylvestris*. *J. Exp. Bot.* 22, 347–361.

Denne, M. P., and Wilson, J. E. (1977). Some quantitative effects of indoleacetic acid on the wood production and tracheid dimensions of *Picea*. *Planta* 134, 223–228.

Dietrichs, H. H. (1964). Das Verhalten von Kohlenhydraten bei der Holzverkernung. *Holzforschung* 18, 14–24.

Dinwoodie, J. M. (1975). Timber—a review of the structure—mechanical property relationship. *J. Microsc. (Oxford)* 104, 3–32.

Dixon, A. F. G. (1971). The role of aphids in wood formation. I. The effect of the sycamore aphid *(Drepansiphum platonoides* (Schr.) (Aphidae), on the growth of sycamore, *Acer pseudoplatanus* (L.) *J. Appl. Ecol.* 8, 165–179.

Dobbs, R. C., and Scott, D. R. M. (1971). Distribution of diurnal fluctuations in stem circumference of Douglas-fir. *Can. J. For. Res.* 1, 80–83.

Doley, D. (1967). Water relations of *Eucalyptus marginata* Sm. under natural conditions. *J. Ecol.* 55, 597–614.

Eades, H. W. (1958). Differentiation of sapwood and heartwood in western hemlock by color tests. *For. Prod. J.* 8(3), 104–106.

Erickson, H. D., and Balatinecz, J. J. (1964). Liquid flow paths into wood using polymerization techniques - Douglas-fir and styrene. *For. Prod. J.* 14(7), 293–299.

Erickson, H. D., and Crawford, R. J. (1959). The effect of several seasoning methods upon the permeability of wood to liquids. *Proc. Am. Wood-Preserv. Assoc.* 55, 210–220.

Etheridge, D. E. (1958). The effects on variations in decay of moisture content and rate of growth in subalpine larch. *Can. J. Bot.* 36, 187–206.

Faiz, S. M. A. (1973). Soil-root water relations. Ph.D. Thesis, University of Aberdeen, U. K.

Faiz, S. M. A., and Weatherley, P. E. (1977). The location of the resistance to water movement in the soil supplying the roots of transpiring plants. *New Phytol.* 78, 337–347.

Farmer, J. B. (1918). On the quantitative differences in the water conductivity of the wood in trees and shrubs. *Proc. R. Soc. London, Ser. B* 90, 218–250.

Fielding, J. M. (1952). The moisture content of the branches of Monterey pine trees. *Aust. For.* **16**, 3–21.

Ford, E. D., and Deans, J. D. (1977). Growth of a sitka spruce plantation: Spatial distribution and seasonal fluctuations of lengths, weights and carbohydrate concentrations of fine roots. *Plant Soil* **47**, 462–485.

Ford, E. D., Robards, A. W., and Piney, M. D. (1978). Influence of environmental factors on cell production and differentiation in the early wood cells of *Picea sitchensis*. *Ann. Bot. (London)* [N. S.] **42**, 683–692.

Frey-Wyssling, A., and Bosshard, H. H. (1959). Cytology of the ray cells in sapwood and heartwood. *Holzforschung* **13**, 129–137.

Frey-Wyssling, A., Bosshard, H. H., and Mühlethaler, K. (1956). Die submikroskopische Entwicklung der Hoftüpfel. *Planta* **47**, 115–126.

Fujimori, T., Kawanabe, S., Saito, H., Grier, C. C., and Shidei, T. (1976). Biomass and primary production in forests of the major vegetation zones of the northwestern United States. *J. Jpn. For. Soc.* **58**, 360–373.

Gardner, W. R. (1960). Dynamic aspects of water availability to plants. *Soil Sci.* **89**, 63–73.

Gardner, W. R. (1968). Availability and measurement of soil water. *In* "Water Deficits and Plant Growth" (T. T. Kozlowski, ed.), Vol. 1, Chapter 5, pp. 107–135. Academic Press, New York.

Gardner, W. R., and Nieman, R. H. (1964). Lower limit of water availability to plants. *Science* **143**, 1460–1462.

Gary, H. L. (1971). Seasonal and diurnal changes in moisture contents and water deficits of Engelmann spruce needles. *Bot. Gaz. (Chicago)* **132**, 327–332.

Gary, H. L. (1976). Crown structure and distribution of mass in a lodgepole pine stand. *U. S., For. Ser., Res. Pap. RM* **165**, 1–20.

Gay, L. W. (1971). Forest climatology studies at Oregon State University. *Proc. Oreg. Acad. Sci.* **7**, 11–23.

Gholz, H. L. (1979). Limits on aboveground net primary production, leaf area, and biomass in vegetational zones of the Pacific northwest. Ph.D. Thesis, Oregon State University, Corvallis.

Gholz, H. L., Fitz, F. K., and Waring, R. H. (1976). Leaf area difference associated with old-growth forest communities in the western Oregon Cascades. *Can. J. For. Res.* **6**, 49–57.

Gholz, H. L., Grier, C. C., Campbell, A. G., and Brown, A. T. (1979). Equations for estimating biomass and leaf area of plants in the Pacific northwest. *Oreg. State Univ. For. Res. Lab, Res. Pap.* **41**, 1–39.

Gibbs, R. D. (1935). Studies of wood. II. On the water content of certain Canadian trees and on changes in water-gas system during seasoning and flotation. *Can. J. Res.* **12**, 727–760.

Gibbs, R. D. (1958). Patterns in the seasonal water content of trees. *In* "Physiology of Forest Trees" (K. V. Thimann, ed.), pp. 43–69. Ronald Press, New York.

Glerum, C. (1970). Drought ring formation in conifers. *For. Sci.* **16**, 246–248.

Grace, J. (1980). Some effects of wind on plants. *In* "Plants and Their Atmospheric Environment" (J. Grace, E. D. Ford, and P. G. Jarvis, eds.), pp. 31–56. Blackwell, Oxford (in press).

Grace, J., Malcolm, D. C., and Bradbury, I. K. (1975). The effect of wind and humidity on leaf: Diffusive resistance in Sitka spruce seedlings. *J. Appl. Ecol.* **12**, 931–940.

Greenidge, K. N. H. (1955a). Observations on the movement of moisture in large woody stems. *Can. J. Bot.* **33**, 202–221.

Greenidge, K. N. H. (1955b). Studies on the physiology of forest trees. III. The effects of dramatic interruption of conducting tissues on moisture movement. *Am. J. Bot.* **42,** 582–587.

Greenidge, K. N. H. (1957). Ascent of sap. *Annu. Rev. Plant Physiol.* **8,** 237–256.

Gregory, S. C. (1972). Some effect of the pathogenic fungus *Peridermium pini* (Pers.) Lev. on the water physiology of *Pinus sylvestris* Linn. Ph. D. Thesis, University of Aberdeen, U. K.

Gregory, S. C. (1977a). A simple technique for measuring the permeability of coniferous wood and its application to the study of water conduction in living trees. *Eur. J. For. Pathol.* **7,** 321–328.

Gregory, S. C. (1977b). The effect of *Peridermium pini* (Pers.) Lev. on water conduction in *Pinus sylvestris* L. *Eur. J. For. Pathol.* **7,** 328–338.

Gregory, S. C., and Petty, J. A. (1973). Valve action of bordered pits in conifers. *J. Exp. Bot.* **24,** 763–767.

Grier, C. C., and Logan, R. S. (1977). Old growth *Pseudotsuga menziesii* communities of a western Oregon watershed: Biomass distribution and production budgets. *Ecol. Monogr.* **47,** 373–400.

Grier, C. C., and Waring, R. H. (1974). Coniferous foliage mass related to sapwood area. *For. Sci.* **20,** 205–206.

Halbwachs, G. (1970). Vergleichende Unterschungen über die Wasserbewegungen in gesunden und fluorgeschädigten Holzwächsen. *Centralbl. Gesamte Forstwes.* **87,** 1–20.

Hall, G. S. (1965). Wood increment and crown distribution relationships in red pine. *For. Sci.* **11,** 438–448.

Hamilton, G. J. (1975). Forest mensuration handbook. *G. B., For. Comm., Bookl.* No. 39, p. 274.

Hammel, H. T. (1967). Freezing of xylem sap without cavitation. *Plant Physiol.* **42,** 55–66.

Handl, J., Kühn, W., and Schätzler, H. P. (1974). Study of the diagnosis and treatment of rot in living trees. *Annu. Rep. Programme Biol. Health Prot. Comm. Eur. Commun.* EUR **5332,** 715–716.

Hanley, D. P. (1976). Tree biomass and productivity estimated for three habitat types of northern Idaho. *Univ. Idaho, Coll. For. Wildl. Range Exp. Stn., Bull.* **14,** 1–15.

Hansen, G. K. (1974). Resistance to water transport in soil and young wheat plants. *Acta Agric. Scand.* **24,** 37–48.

Harms, W. R., and McGregor, W. H. D. (1962). A method for measuring the water balance of pine needles. *Ecology* **43,** 531–532.

Harrington, G. (1979). Estimation of above ground biomass of trees and shrubs in a *Eucalyptus populnea* F. Muell. woodland by regression of mass on trunk diameter and plant height. *Aust. J. Bot.* **27,** 135–143.

Harris, J. M. (1953). Heartwood formation in *Pinus radiata* (D. Don.). *Nature (London)* **172,** 552.

Harris, J. M. (1954). Heartwood formation in *Pinus radiata* (D. Don.). *New Phytol.* **53,** 517–524.

Harris, J. M. (1961). Water conduction in the stems of certain conifers. *Nature (London)* **189,** 678–679.

Harris, J. M., McConchie, D. L., and Povey, W. A. (1978). Wood properties of clonal radiata pine grown in soil with different levels of available nitrogen, phosphorus and water. *N. Z. J. For Sci.* **8,** 417–430.

Hart, C. A., and Thomas, R. J. (1967). Mechanism of bordered pit aspiration as caused by capillarity. *For. Prod. J.* **17** (11), 61–68.

Hasegawa, M., and Shiroya, M. (1964). The formation of phenolic compounds at the sapwood-heartwood boundary. *Proc. Meet. Sect. 41, For. Prod., Int. Union For. Res. Organ., 1963* Melbourne. Vol. 1, pp. 1–10.

Heine, R. W. (1970). Estimation of conductivity and conducting area of poplar stems using a radioactive tracer. *Ann. Bot. (London)* [N. S.] **34**, 1019–1024.

Heine, R. W. (1971). Hydraulic conductivity in trees. *J. Exp. Bot.* **22**, 503–511.

Heine, R. W., and Farr, D. J. (1973). Comparison of heat-pulse and radioisotope tracer methods for determining sap-flow velocity in stem segments of poplar. *J. Exp. Bot.* **24**, 649–654.

Hellkvist, J. (1973). The water relations of *Pinus sylvestris*. II. Comparative field studies of water potential and relative water content. *Physiol. Plant.* **29**, 371–379.

Hellkvist, J., and Parsby, J. (1976). The water relations of *Pinus sylvestris*. III. Diurnal and seasonal patterns of water potential. *Physiol. Plant.* **38**, 61–68.

Hellkvist, J., Richards, G. P., and Jarvis, P. G. (1974). Vertical gradients of water potential and tissue water relations in Sitka spruce trees measured with the pressure chamber. *J. Appl. Ecol.* **11**, 637–667.

Heth, D. (1974). Water potentials and water saturation deficits in *Cupressus sempervirens* L., *Cupressus atalantica* Gaussen and *Pinus halepensis* Mill. *La-Yaaran* **24** (3–4), 38–43.

Hillel, D., Krentos, V. D., and Stylianou, Y. (1972). Procedure and test of an internal drainage method for measuring soil hydraulic characteristics *in situ*. *Soil Sci.* **114**, 395–400.

Hillis, W. E., Humphreys, F. R., Bamber, R. K., and Carle, A. (1962). Factors influencing the formation of phloem and heartwood polyphenols. II. The availability of stored and translocated carbohydrate. *Holzforschung* **16**, 114–121.

Hinckley, T. M., and Bruckerhoff, D. N. (1975). The effects of drought on water relations and stem shrinkage in *Quercus alba*. *Can. J. Bot.* **53**, 62–72.

Hinckley, T. M., and Ritchie, G. A. (1970). Within-crown patterns of transpiration, water stress, and stomatal activity in *Abies amabilis*. *For. Sci.* **16**, 490–492.

Hinckley, T. M., and Ritchie, G. A. (1973). A theoretical model for calculation of xylem sap pressure from climatological data. *Am. Midl. Nat.* **90**, 56–69.

Hinckley, T. M., Lassoie, J. P., and Running, S. W. (1978). Temporal and spatial variations in the water status of forest trees. *For. Sci. Monogr.* **20**, 1–72.

Holmes, J. W., and Shim, S. Y. (1968). Diurnal changes in stem diameter of Canary Island pine trees (*Pinus canariensis* C. Smith) caused by soil water stress and varying microclimate. *J. Exp. Bot.* **19**, 219–232.

Hook, D. D., and Scholtens, J. R. (1978). Adaptation and flood tolerance of tree species. *In* "Plant Life and Anaerobic Environments" (D. D. Hook and R. M. M. Crawford, eds.), pp. 299–331. Ann Arbor Sci. Publ., Ann Arbor, Michigan.

Huber, B. (1924). Die Beurteilung des Wasserhaushaltes der Pflanze: ein Beitrag zur vergleichenden. *Jahrb. Wiss. Bot.* **64**, 1–120.

Huber, B. (1928). Weitere quantitative Untersuchungen über das Wasserleitungssystem der Pflanzen. *Jahrb. Wiss. Bot.* **67**, 877–959.

Huber, B. (1952). Tree physiology. *Annu. Rev. Plant Physiol.* **3**, 333–346.

Huber, B. (1956). Die Gefässleitung. *In* "Handbuch der Pflanzenphysiologie" (W. Ruhland, ed.), Vol. 3, pp. 541–582. Springer-Verlag, Berlin and New York.

Huber, B., and Merz, W. (1958a). Ueber die Bedeutung der Hoftüpfelverschlussen für die axiale Wasserleitfähigkeit von Nadelhölzen. I. Grunderscheinungen. *Planta* **51**, 645–659.

Huber, B., and Merz, W. (1958b). Ueber die Bedeutung der Hoftüpfelverschlussen für die axiale Wasserleitfähigkeit von Nadechölzen. II. Vergleichende Untersuchungen und pracktische Anwendungen. *Planta* **51**, 660–672.

Huck, M. G., Klepper, B., and Taylor, H. M. (1970). Diurnal variation in root diameter. *Plant Physiol.* **45**, 529–530.

Huxley, P. A. (1971). Leaf volume: A simple method for measurement and some notes on its use in studies of leaf growth. *J. Appl. Ecol.* **8**, 147–153.

Jackson, D. S., Gifford, H. H., and Hobbs, I. W. (1973). Daily transpiration rates of radiata pine. *N. Z. J. For. Sci.* **3**, 70–81.

James, G. B., and Jarvis, P. G. (1981). Carbon dioxide and water vapour exchanges by a Sitka spruce forest canopy. II. Evaporation and surface resistance. *Plant, Cell, & Environ.* (in press).

Jameson, D. A. (1966). Diurnal and seasonal fluctuations in moisture content of Pinyon and Juniper. *U. S., For. Serv., Res. Note RM* **67**, 1–7.

Jarvis, P. G. (1975). Water transfer in plants. *In* "Heat and Mass Transfer in the Plant Environment" (D. A. deVries and N. G. Afgan, eds.), Part 1, pp. 369–394. Scripta Book Co., Washington, D. C.

Jarvis, P. G. (1976). The interpretation of the variations in leaf water potential and stomatal conductance found in canopies in the field. *Philos. Trans. R. Soc. London, Ser. B* **273**, 593–610.

Jarvis, P. G. (1980). Stomatal response to water stress in conifers. *In* "Adaptation of Plants to Water and High Temperature Stress" (N. C. Turner and P. J. Kramer, eds.), pp. 105–122. Wiley (Interscience), New York.

Jarvis, P. G. (1981a). Stomatal conductance, gaseous exchange and transpiration. *In* "Plants and their Atmospheric Environment" (J. Grace, E. D. Ford, and P. G. Jarvis, eds.), pp. 175–204. Blackwell, Oxford.

Jarvis, P. G. (1981b). Stomatal control of transpiration and photosynthesis. *In* "Stomatal Physiology" (P. G. Jarvis and T. A. Mansfield, eds.), pp. 247–279, SEB Sem. Ser. Cambridge Univ. Press, London and New York.

Jarvis, P. G., and Jarvis, M. S. (1963a). Effects of several osmotic substances on the growth of *Lupinus albus* seedlings. *Physiol. Plant.* **16**, 485–500.

Jarvis, P. G., and Jarvis, M. S. (1963b). The water relations of tree seedlings. IV. Some aspects of the tissue water relations and drought resistance. *Physiol. Plant.* **16**, 501–516.

Jarvis, P. G., and Jarvis, M. S. (1964). Growth rates of woody plants. *Physiol. Plant.* **17**, 654–666.

Jarvis, P. G., and Slatyer, R. O. (1970). The role of the mesophyll cell wall in leaf transpiration. *Planta* **90**, 303–322.

Jarvis, P. G., and Stewart, J. B. (1979). Evaporation of water from plantation forest. *In* "The Ecology of Even-Aged Forest Plantations" (E. D. Ford, D. C. Malcolm, and J. Atterson, eds.), pp. 327–350. Inst. Terrestrial Ecology, NERC, Cambridge.

Jarvis, P. G., James, G. B., and Landsberg, J. J. (1976). Coniferous forest. *In* "Vegetation and the Atmosphere" (J. L. Monteith, ed.), Vol. 2, Case Studies, pp. 171–240. Academic Press, New York.

Jarvis, P. G., Edwards, W. R. N., and Talbot, H. (1981). Models of plant and crop water use. *In* "Mathematics and Plant Physiology" (D. A. Rose and D. H. Charles-Edwards, eds). Academic Press, New York.

Jenkins, P. A., and Shepherd, K. R. (1972). Influence of temperature on cambial activity and cell diameter in *Pinus radiata* D. Don. *J. Inst. Wood Sci.* **6**, 36–39.

Jenkins, P. A., and Shepherd, K. R. (1974). Seasonal change in levels of indoleacetic acid and abscisic acid in stem tissue of *Pinus radiata*. *N. Z. J. For. Sci.* **4**, 511–519.

Johnson, B. R. (1979). Permeability changes induced in three western conifers by selective bacterial inoculation. *Wood Fiber* **11**, 10–21.

Johnston, R. D. (1964). Water relations of *Pinus radiata* under plantation conditions. *Aust. J. Bot.* **12**, 111–124.

Johnston, W. D. (1970). Some variations in specific gravity and moisture content of 100 year old lodgepole pine trees. *Can., Dep. Fish. For., For. Serv., For. Res. Lab., Inf. Rep.* **A-X-29**, 1–19.

Jordan, C. F., and Kline, J. R. (1977). Transpiration of trees in a tropical rain forest. *J. Appl. Ecol.* **14**, 853–860.

Jorgensen, E. (1961). The formation of pinosylvin and its monomethyl ether in the sapwood of *Pinus resinosa* Ait. *Can. J. Bot.* **39**, 1765–1772.

Kaufmann, M. R. (1975). Leaf water stress in Engelmann spruce. Influence of root and shoot environments. *Plant Physiol.* **56**, 841–844.

Kaufmann, M. R. (1977a). Soil temperature and drought effects on growth of Monterey pine. *For. Sci.* **23**, 317–325.

Kaufmann, M. R. (1977b). Soil temperature and drying cycle effects on water relations of *Pinus radiata*. *Can. J. Bot.* **55**, 2413–2418.

Kaufmann, M. R. (1979). Stomatal control and the development of water deficit in Engelmann spruce seedlings during drought. *Can. J. For. Res.* **9**, 297–304.

Keays, J. L. (1971). Complete tree utilization, an analysis of the literature. Part II. Foliage. *Can., Dep. Fish. For., For. Serv., Inf. Rep.* **VP-X-70**, 1–94.

Kelso, W. C., Gertjejansen, C. O., and Hoosfeld, R. L. (1963). Effect of air blockage upon the permeability of woods to liquids. *Univ. Minn., Agric. Res. Stn., Tech. Bull.* **242**, 1–40.

Kinerson, R. S., Higginbotham, K. O., and Chapman, R. C. (1974). The dynamics of foliage distribution within a forest canopy. *J. Appl. Ecol.* **11**, 347–353.

Kininmonth, J. A. (1970). An evaluation of timber drying problems in terms of permeability and fine structure. Ph.D. Thesis, University of Melbourne, Melbourne, Australia.

Kininmonth, J. A. (1971). Permeability and fine structure of certain hardwoods and effects of drying. I. Transverse permeability of wood to microfiltered water. *Holzforschung* **25**, 127–133.

Kittredge, J. (1944). Estimation of the amount of foliage of trees and stands. *J. For.* **42**, 905–912.

Kline, J. R., Reed, K. L., Waring, R. H., and Stewart, M. L. (1976). Field measurement of transpiration in Douglas-fir. *J. Appl. Ecol.* **13**, 273–283.

Koch, J. E., and Krieg, W. (1938). Neues Verfahren zur Unterscheidung von Kern und Splint im Kiefernholz. *Chem. Ztg.* **62**, 140–141.

Kotar, J. (1972). Ecology of *Abies amabilis* in relation to its altitudinal distribution and in contrast to its common associate *Tsuga heterophylla*. Ph. D. Thesis, University of Washington, Seattle.

Kozlowski, T. T. (1961). The movement of water in trees. *For. Sci.* **7**, 177–192.

Kozlowski, T. T. (1972). Shrinkage and swelling of plant tissues. *In* "Water Deficits and Plant Growth" (T. T. Kozlowski, ed.), Vol. 3, pp. 1–64. Academic Press, New York.

Kozlowski, T. T. (1979). "Tree Growth and Environmental Stresses." Univ. of Washington Press, Seattle.

Kozlowski, T. T., and Clausen, J. J. (1965). Changes in moisture contents and dry weights of buds and leaves of forest trees. *Bot. Gaz. (Chicago)* **126**, 20–26.

Kozlowski, T. T., and Schumacher, F. X. (1943). Estimation of stomated foliar surface of pines. *Plant Physiol.* **18**, 122–127.

Kozlowski, T. T., and Winget, C. H. (1963). Patterns of water movement in forest trees. *Bot. Gaz. (Chicago)* **124**, 301–311.

Kozlowski, T. T., and Winget, C. H. (1964). Diurnal and seasonal variation in radii of tree stems. *Ecology* **45**, 149–155.

Kozlowski, T. T., Hughes, J. F., and Leyton, L. (1966). Patterns of water movement in dormant gymnosperm seedlings. *Biorheology* **3**, 77–85.

Kozlowski, T. T., Hughes, J. F., and Leyton, L. (1967). Dye movement in gymnosperms in relation to tracheid alignment. *Forestry* **40**, 209–227.

Krahmer, R. L., and Côté, W. A., Jr. (1963). Changes in coniferous wood cells associated with heartwood formation. *Tappi* **46**, 42–49.

Kramer, P. J. (1942). Species differences with respect to water absorption at low soil temperatures. *Am. J. Bot.* **29**, 828–872.

Kramer, P. J. (1955). Bound water. In "Handbuch der Pflanzenphysiologie" (W. Ruhland, ed.), Vol. 1, pp. 232–242. Springer-Verlag, Berlin and New York.

Kramer, P. J., and Kozlowski, T. T. (1979). "Physiology of Woody Plants." Academic Press, New York.

Kühn, W., and Handl, J. (1973). Zur Erkennbarkeit von Fäulen in stehenden Baumen durch Röntgendiagose. *Holzforschung* **27**, 173–178.

Kuiper, P. J. C. (1964). Water uptake of higher plants as affected by root temperature. *Meded. Landbouw hogesch. Wageningen* **64**, 4, 1–11.

Kuiper, P. J. C. (1972). Water transport across membranes. *Annu. Rev. Plant Physiol.* **23**, 157–172.

Kutscha, N. P., and Sachs, I. B. (1962). Color tests for differentiating heartwood and sapwood in certain softwood tree species. *U.S., For. Serv., For. Res. Lab., Rep.* **2246**, 1–13.

Landsberg, J. J., and Butler, D. R. (1980). Stomatal response to humidity: Implications for transpiration. *Plant, Cell, & Environ.* **3**, 29–33.

Landsberg, J. J., and Thom, A. S. (1971). Aerodynamic properties of a plant of complex structure. *Q. J. R. Meteorol. Soc.* **97**, 565–570.

Landsberg, J. J., Beadle, C. L., Biscoe, P. V., Butler, D. R., Davidson, B., Incoll, L. D., James, G. B., Jarvis, P. G., Martin, P. J., Neilson, R. E., Powell, D. B. B., Slack, E. M., Thorpe, M. R., Turner, N. C., Warrit, B., and Watts, W. R. (1975). Diurnal energy, water and CO_2 exchanges in an apple *(Malus pumila)* orchard. *J. Appl. Ecol.* **12**, 659–684.

Landsberg, J. J., Blanchard, T. W., and Warrit, B. (1976). Studies on the movement of water through apple trees. *J. Exp. Bot.* **27**, 579–596.

Larson, P. R. (1963). The indirect effect of drought on tracheid diameter in red pine. *For. Sci.* **9**, 52–62.

Larson, P. R. (1964). Some indirect effects of environment on wood formation. In "The Formation of Wood in Forest Trees" (M. H. Zimmermann, ed.), pp. 345–365. Academic Press, New York.

Lassen, L. E., and Okkonen, E. A. (1969). Sapwood thickness of Douglas-fir and five other western softwoods. *U. S. Dep. Agr., Forest Ser., Forest Res. Lab., Res. Pap., FPL 124*, 1–16.

Lassoie, J. P. (1979). Stem dimensional fluctuations in Douglas-fir of different crown classes. *For. Sci.* **25**, 132–144.

Lassoie, J. P., and Dougherty, P. M. (1976). Fall and winter gas exchange rates in eastern red cedar. *Plant Physiol. Supplement* **57**, 45.

Läuchli, A. (1972). Translocation of inorganic solutes. *Annu. Rev. Plant Physiol.* **23**, 197–218.

Lewandowska, M., and Jarvis, P. G. (1977). Changes in chlorophyll and carotenoid content, specific leaf area and dry weight fractions in Sitka spruce, in response to shading and season. *New Phytol.* **79**, 247–256.

Leyton, L., and Armitage, I. P. (1968). Cuticle structure and water relations of the needles of *Pinus radiata* (D. Don). *New Phytol.* **67**, 31–38.

Leyton, L., and Juniper, B. E. (1963). Cuticle structure and water relations of pine needles. *Nature (London)* **198**, 770–771.

Liese, W. (1965). The fine structure of bordered pits in softwoods. In "Cellular Ultrastructure of Woody Plants'' (W. A. Côté, Jr., ed.), pp. 271–290. Syracuse Univ. Press, Syracuse, New York.

Liese, W., and Bauch, J. (1967). On the closure of bordered pits in conifers. *Wood Sci. Technol.* **1**, 1–13.

Lin, R. T., Lancaster, E. P., and Krahmer, R. L. (1973). Longitudinal water permeability of western hemlock. I. Steady state permeability. *Wood Fiber* **4**, 278–289.

Linder, S. (1973). The influence of soil temperature upon net photosynthesis and transpiration in seedlings of Scots pine and Norway spruce. Some aspects of pigmentation, photosynthesis and transpiration in nursery-grown seedlings of Scots pine and Norway spruce. D. Phil. Thesis, University of Umeå, Sweden.

Little, C. H. A. (1975). Inhibition of cambial activity in *Abies balsamea* by internal water stress: Role of abscisic acid. *Can. J. Bot.* **53**, 3041–3050.

Loomis, R. M., Phares, R. E., and Crosby, J. S. (1966). Estimating foliage and branchwood quantities in shortleaf pine. *For. Sci.* **12**, 30–39.

Lorio, P. L., Jr., and Hodges, J. D. (1968). Oleoresin exudation pressure and relative water content of inner bark as indicators of moisture stress in loblolly pines. *For. Sci.* **14**, 392–398.

Luxford, R. F. (1930). Distribution and amount of moisture in virgin redwood trees. *J. For.* **28**, 770–772.

Lybeck, B. R. (1959). Winter freezing in relation to the rise of sap in tall trees. *Plant Physiol.* **34**, 482–486.

Lyons, J. M. (1973). Chilling injury in plants. *Annu. Rev. Plant Physiol.* **24**, 445–466.

McDermott, J. J. (1941). The effect of the method of cutting on the moisture content of samples from tree branches. *Am. J. Bot.* **28**, 506–508.

McGinnes, E. A., Jr., and Dingeldein, T. W. (1969). Selected wood properties of eastern redcedar (*Juniperus virginiana* L.) grown in Missouri. *Mo., Agric. Exp. Stn., Res. Bull.* **960**, 1–19.

McNaughton, K. G., and Black, T. A. (1973). Evapotranspiration from a forest: A micrometeorological study. *Water Resour. Res.* **9**, 1579–1590.

McNeil, D. D., and Shuttleworth, W. J. (1975). Comparative measurements of the energy fluxes over a pine forest. *Boundary-Layer Meteorol.* **9**, 297–313.

Madgwick, H. A. I. (1970). Biomass and productivity models of forest canopies. *Ecol. Stud.* **1**, 47–54.

Madgwick, H. A. I., Jackson, D. S., and Knight, P. J. (1977). Aboveground dry matter energy and nutrient contents of trees in an age series of *Pinus radiata* plantations. *N. Z. J. For. Sci.* **7**, 445–468.

Mark, W. R., and Crews, D. L. (1973). Heat-pulse velocity and bordered pit condition in living Englmann spruce and lodgepole pine trees. *For. Sci.* **19**, 291—296.

Markhart, A. H., III, Fiscus, E. L., Naylor, A. W., and Kramer, P. J. (1979). Effects of abscisic acid on root hydraulic conductivity. *Plant Physiol.* **64**, 611–614.

Markstrom, D. C., and Hann, R. A. (1972). Seasonal variation in wood permeability and moisture content of three Rocky Mountain softwoods. *U.S., For. Serv., Res. Note RM* **212**, 1–7.

Mellor, G. E., and Tregunna, E. G. (1972). The relationship between leaf area and dry weight of three conifer species grown on three sources of nitrogen. *Can. J. For. Res.* **2**, 377–379.

Milburn, J. A. (1966). The conduction of sap. I. Water conduction and cavitation in stressed leaves. *Planta* **69**, 34–42.

Milburn, J. A. (1973). Cavitation studies on whole *Ricinus* plants by acoustic detection. *Planta* **112**, 333–342.

Milburn, J. A., and Johnson, R. P. C. (1966). The conduction of sap. II. Detection of vibrations produced by sap cavitation in *Ricinus* xylem. *Planta* **69**, 43–52.

Milburn, J. A., and McLaughlin, M. E. (1974). Studies of cavitation in isolated vascular bundles and whole leaves of *Plantago major* L. *New Phytol.* **73**, 861–871.

Miller, H. G., Miller, J. D., and Pauline, O. J. L. (1976). Effect of nitrogen supply on nutrient uptake in Corsican pine. *J. Appl. Ecol.* **13**, 955–966.

Milne, R. (1979). Water loss and canopy resistance of a young Sitka spruce plantation. *Boundary-Layer Meteorol.* **16**, 67–81.

Mitchell, R. G. (1967). Translocation of dye in grand and subalpine firs infested by the balsam woolly aphid. *U. S., For. Serv., Res. Note PNW* **46**, 1–17.

Moir, W. H., and Francis, R. (1972). Foliage biomass and surface area in three *Pinus contorta* plots in Colorado. *For. Sci.* **18**, 41–45.

Monteith, J. L. (1965). Evaporation and environment. *Symp. Soc. Exp. Biol.* **19**, 205–234.

Monteith, J. L. (1973). "Principles of Environmental Physics." Arnold, London.

Moore, C. J. (1976). Eddy flux measurements above a pine forest: Notes and correspondence. *Q. J. Roy. Meteorol. Soc.* **102**, 913–918.

Morikawa, Y. (1974). Sap flow in *Chamaecyparis obtusa* in relation to water economy of woody plants. *Bull. Tokyo Univ. For.* **66**, 251–297.

Morikawa, Y., and Sato, A. (1976). Diurnal trends of xylem sap pressure in the crown and sap speed through the trunk in an isolated large *Chamaecyparis obtusa* tree. *J. Jpn. For. Soc.* **58**, 11–14 (in Japanese).

Nečasaný, V. (1964). Heartwood formation as a physiological ageing process. *Proc. Meet. Sec. 41, For. Prod., Int. Union For. Res. Organ., 1963* Vol. 1, pp. 1–9.

Newbould, P. J. (1967). Methods of estimating primary production of forests. *IBP Handb.* **2**, 1–62.

Newman, E. I. (1969a). Resistance to water flow in soil and plant. I. Soil resistance in relation to amount of root: Theoretical considerations. *J. Appl. Ecol.* **6**, 1–12.

Newman, E. I. (1969b). Resistance to water flow in soil and plant. II. A review of experimental evidence on the rhizosphere resistance. *J. Appl. Ecol.* **6**, 261–272.

Ng, P. A. P. (1978). Response of stomata to environmental variables in *Pinus sylvestris* L. Ph.D. Thesis, University of Edinburgh, United Kingdom.

Nicholls, J. W. P. (1965). Heritability of heartwood formation in *Pinus radiata* D. Don. *Nature (London)* **207**, 320.

Nicholls, J. W. P., and Wright, J. P. (1976). The effect of environmental factors on wood characteristics. III. The influence of climate and site on young *Pinus radiata* material. *Can. J. For. Res.* **6**, 113–121.

Nnyamah, J. U., and Black, T. A. (1977). Rates and patterns of water uptake in a Douglas-fir forest. *Soil Sci. Soc. Am., Proc.* **41**, 972–979.

Nnyamah, J. U., Black, T. A., and Tan, C. S. (1978). Resistance to water uptake in a Douglas-fir forest. *Soil Sci.* **126**, 63–76.

Nye, P. H., and Tinker, P. B. (1977). "Solute Movement in the Soil-Root System." Blackwell, Oxford.

Oertli, J. J. (1971). The stability of water under tension in the xylem. *Z. Pflanzenphysiol.* **65**, 195–209.

Olszyczka, B. (1979). Gamma ray determination of surface water storage and stem water content for coniferous forests. Ph.D. Thesis, University of Strathclyde, U. K.

Oppenheimer, H. R. (1968). Drought resistance of Monterey pine needles. *Isr. J. Bot.* **17**, 163–168.

Ovington, J. D. (1956). The form, weights and productivity of tree species grown in close stands. *New Phytol.* **55**, 289–304.

Ovington, J. D., and Madgwick, H. A. I. (1959). Distribution of organic matter and plant nutrients in a plantation of Scots pine. *For. Sci.* **5,** 344–355.

Ovington, J. D., Forrest, W. G., and Armstrong, J. S. (1967). Tree biomass estimation. *In* "Primary Productivity and Mineral Cycling in Natural Ecosystems" (H. E. Young, ed.), pp. 4–32. Univ. of Maine Press, Orono.

Panshin, A. J., de Zeeuw, C., and Brown, H. P. (1964). "Textbook of Wood Technology," 2nd ed., Vol. 1. McGraw-Hill, New York.

Parker, J. (1954). Available water in stems of some Rocky Mountain conifers. *Bot. Gaz. (Chicago)* **115,** 380–385.

Peck, E. C. (1959). The sap or moisture in wood. *U.S., For. Serv., For. Prod. Lab., Rep.* **768,** 1–13.

Peel, A. J. (1965). On the conductivity of the xylem in trees. *Ann. Bot. (London)* [N. S.] **29,** 119–130.

Pereira, J. S., and Kozlowski, T. T. (1976). Diurnal and seasonal changes in water balance of *Abies balsamea* and *Pinus resinosa. Oecol. Plant.* **11,** 397–412.

Petty, J. A. (1970). Permeability and structure of the wood of Sitka spruce. *Proc. R. Soc. London, Ser. B* **175,** 149–166.

Petty, J. A. (1972). The aspiration of bordered pits in conifer wood. *Proc. R. Soc. London, Ser. B* **181,** 395–406.

Petty, J. A. (1974). Laminar flow of fluids through short capillaries in conifer wood. *Wood Sci. Technol.* **8,** 275–282.

Petty, J. A., and Preston, R. D. (1969). The dimensions and number of pit membrane pores in conifer wood. *Proc. R. Soc. London, Ser. B* **172,** 137–151.

Petty, J. A., and Puritch, G. S. (1970). The effects of drying on the structure and permeability of the wood of *Abies grandis. Wood Sci. Technol.* **4,** 140–154.

Pharis, R. P. (1967). Seasonal fluctuations in the foliage-moisture content of well-watered conifers. *Bot. Gaz. (Chicago)* **128,** 179–185.

Philip, J. R. (1957). The physical principles of water movement during the irrigation cycle. *Proc. 3rd Int. Congr. Irrig. Drainage,* pp. 8.125–8.154.

Philip, J. R. (1966). Plant water relations: Some physiological aspects. *Annu. Rev. Plant Physiol.* **17,** 245–268.

Philip, J. R. (1974). Fifty years progress in soil physics. *Geoderma* **12,** 265–280.

Philipson, J. J., and Coutts, M. P. (1978). The tolerance of tree roots to waterlogging. III. Oxygen transport in lodgepole pine and Sitka spruce roots of primary structure. *New Phytol.* **80,** 341–349.

Phillips, E. W. J. (1933). Movement of the pit membrane in coniferous woods, with special reference to preservative treatment. *Forestry* **7,** 109–120.

Polster, H. (1967). Wasserhaushalt. *In* "Gehölzphysiologie" (H. Lyr, H. Polster, and H. J. Fiedler, eds.), pp. 163–196. Fischer, Jena.

Powell, D. B. B., and Thorpe, M. R. (1977). Dynamic aspects of plant-water relations. *In* "Environmental Effects on Crop Physiology" (J. J. Landsberg and C. V. Cutting, eds.), pp. 259–279. Academic Press, New York.

Preston, R. D. (1961). Theoretical and practical implications of the stresses in the water-conducting system. *Proc. Int. Bot. Congr., 9th, 1959* Vol. 2, pp. 1144–1149.

Puritch, G. S. (1971). Water permeability of the wood of Grand fir (*Abies grandis* (Dougl.) Lindl.) in relation to infestation by the balsam woolly aphid, *Adelges piceae* (Ratz.). *J. Exp. Bot.* **22,** 936–945.

Puritch, G. S., and Johnson, R. P. C. (1971). Effects of infestation by the balsam woolly aphid, *Adelges piceae* (Ratz.) on the ultrastructure of bordered-pit membranes of Grand fir, *Abies grandis* (Dougl.) Lindl. *J. Exp. Bot.* **22,** 953–958.

Puritch, G. S., and Petty, J. A. (1971). Effects of balsam woolly aphid, *Adelges piceae* (Ratz.), infestation on the xylem of *Abies grandis* (Dougl.) Lindl. *J. Exp. Bot.* **22**, 946–952.

Reed, K. L., and Waring, R. H. (1974). Coupling of environment to plant response: A simulation model of transpiration. *Ecology* **55**, 62–72.

Reicosky, D. C., and Ritchie, J. T. (1976). Relative importance of soil resistance and plant resistance in root water absorption. *Soil Sci. Soc. Am. J.* **40**, 293–297.

Reynolds, E. R. C. (1966). The internal water balance of trees. *In* "Physiology in Forestry," Suppl. to *Forestry*, pp. 32–44. Oxford Univ. Press, London and New York.

Richards, G. P. (1973). Some aspects of the water relations of Sitka spruce. Ph.D. Thesis, University of Aberdeen, U. K.

Richardson, S. D. (1964). The external environment and tracheid size in conifers. *In* "Formation of Wood in Forest Trees" (M. H. Zimmermann, ed.), pp. 367–388. Academic Press, New York.

Richter, H. (1973). Frictional potential losses and total water potential in plants: A reevaluation. *J. Exp. Bot.* **24**, 983–994.

Ritchie, G. A., and Hinckley, T. A. (1975). The pressure chamber as an instrument for ecological research. *Adv. Ecol. Res.* **9**, 165–254.

Roberts, J. (1976a). An examination of the quantity of water stored in mature *Pinus sylvestris* L. trees. *J. Exp. Bot.* **27**, 473–479.

Roberts, J. (1976b). A study of root distribution and growth in a *Pinus sylvestris* L. (Scots pine) plantation in East Anglia. *Plant Soil* **44**, 607–621.

Roberts, J. (1977). The use of tree cutting in the study of the water relations of mature *Pinus sylvestris* L. I. The technique and survey of the results. *J. Exp. Bot.* **28**, 751–767.

Roberts, J. (1978). The use of the 'tree cutting' technique in the study of the water relations of Norway spruce, *Picea abies* (L.) Karst. *J. Exp. Bot.* **29**, 465–471.

Roberts, J., and Fourt, D. F. (1977). A small pressure chamber for use with plant leaves of small size. *Plant Soil* **48**, 545–546.

Roberts, J., Pymar, C. F., Wallace, J. S., and Pitman, R. M. (1980). Seasonal changes in leaf area, stomatal and canopy conductances and transpiration from bracken (*Pteridium aquilinum* (L.) Kuhn.) below a forest canopy. *J. Appl. Ecol.* **17**, 409–422.

Roberts, L. W. (1969). The initiation of xylem differentiation. *Bot. Rev.* **35**, 201–250.

Roberts, S. W., and Knoerr, K. R. (1977). Components of water potential estimated from xylem pressure measurements in five species. *Oecologia* **28**, 191–202.

Rogers, R., and Hinckley, T. M. (1979). Foliage weight and area related to current sapwood area in oak. *For. Sci.* **25**, 298–303.

Ronco, F. (1969). Volumeter for estimating quantity of conifer foliage. *U. S., For. Serv., Res. Note RM* **133**, 1–2.

Rook, D. A. (1969). The influence of growing temperatures on photosynthesis and respiration of *Pinus radiata* seedlings. *N. Z. J. Bot.* **7**, 43–55.

Rook, D. A., and Hobbs, J. F. F. (1976). Soil temperatures and growth of rooted cuttings of radiata pine. *N. Z. J. For. Sci.* **5**, 296–305.

Rook, D. A., and Whitehead, D. (1979). Temperature and growth of *Pinus radiata. Proc. Agron. Soc. N. Z.* **9**, 109–112.

Rook, D. A., Swanson, R. H., and Cranswick, A. M. (1977). Reaction of radiata pine to drought. *In N. Z. Dep. Sci. Ind. Res., Inf. Ser.* **126**, 55–68.

Rudman, P. (1966). Heartwood formation in trees. *Nature (London)* **210**, 608–610.

Rundel, P. W., and Stecker, R. E. (1977). Morphological adaptation of tracheid structure to water stress gradients in the crown of *Sequoiadendron giganteum. Oecologia* **27**, 135–139.

Running, S. W. (1976). Environmental control of leaf water conductance in conifers. *Can. J. For. Res.* **6**, 104–112.

Running, S. W. (1979). Environmental and physiological control of water flux through *Pinus contorta*. Ph.D. Thesis, Colorado State University, Fort Collins.

Running, S. W., Waring, R. H., and Rydell, R. A. (1975). Physiological control of water flux in conifers. *Oecologia* **18**, 1–16.

Rutter, A. J. (1955). The relation between dry weight increase and linear measures of growth in young conifers. *Forestry* **28**, 125–135.

Rutter, A. J. (1967). Studies on the water relation of *Pinus sylvestris* in plantation conditions. V. Responses to variation in soil water conditions. *J. Appl. Ecol.* **4**, 73–81.

Rutter, A. J. (1968). Water consumption by forests. *In* "Water Deficits and Plant Growth" (T. T. Kozlowski, ed.), Vol. 2, pp. 23–84. Academic Press, New York.

Rutter, A. J. (1975). The hydrological cycle in vegetation. *In* "Vegetation and the Atmosphere" (J. L. Monteith, ed.), Vol. 1, pp. 111–154. Academic Press, New York.

Rutter, A. J., and Fourt, D. F. (1965). Studies in the water relations of *Pinus sylvestris* in plantation conditions. III. A comparison of soil water changes and estimates of total evaporation on four afforested sites and one grass-covered site. *J. Appl. Ecol.* **2**, 197–209.

Rutter, A. J., and Sands, K. (1958). The relation of leaf water deficit to soil moisture tension in *Pinus sylvestris* L. I. The effect of soil moisture on diurnal changes in water balance. *New Phytol.* **57**, 50–65.

Sachs, I. B. (1963). Torus of the bordered-pit membrane in conifers. *Nature (London)* **198**, 906–907.

Sandermann, W., and Schmit, G. (1965). New methods of the differentiation of sapwood, heartwood and growth rings. I. The concrete-relief analysis. *Holz Roh- Werkst.* **23**, 221–227. Translated from German into English by L. Plaszner, University of British Columbia Faculty of Forestry, Translation No. 43.

Sandermann, W., Hausen, B., and Simatupang, M. (1967). Initial experiments to differentiate sapwood and heartwood as well as the transition zone of spruce and other coniferous woods. *(Darmstadt) Papier* **21**(7), 349–354. Translated from German into English by Canadian Dep. Forestry and Rural Development, Translation No. 267.

Sands, R., and Theodorou, C. (1978). Water uptake by mycorrhizal roots of radiata pine seedlings. *Aust. J. Plant Physiol.* **5**, 301–309.

Sato, A., and Morikawa, Y. (1976). Diurnal trends in xylem sap pressure of *Cryptomeria japonica* trees growing on two slopes. *J. Jpn. For. Soc.* **58**, 321–327 (in Japanese).

Satoo, T. (1970). A synthesis of studies by the harvest method: Primary production relations in the temperate deciduous forests of Japan. *Ecol. Stud.* **1**, 55–72.

Scholander, P. F., Hammel, H. T., Hemmingsen, E. A., and Bradstreet, D. (1964). Hydrostatic pressure and osmotic potential in leaves of mangroves and some other plants. *Proc. Nat. Acad. Sci. U.S.A.* **52**, 119–125.

Scholander, P. F., Hammel, H. T., Bradstreet, E. D., and Hemmingsen, E. A. (1965). Sap pressure in vascular plants. *Science* **148**, 339–346.

Shain, L. (1967). Resistance of sapwood in stems of loblolly pine to infestation by *Fomes annosus*. *Phytopathology* **57**, 1034–1045.

Shain, L. (1979). Dynamic responses of differential sapwood to injury and infection. *Phytopathology* **69**, 1143–1147.

Shain, L., and Hillis, W. E. (1973). Ethylene production in xylem of *Pinus radiata* in relation to heartwood formation. *Can. J. Bot.* **51**, 1331–1335.

Shain, L., and Mackay, J. F. G. (1973). Seasonal fluctuation in respiration of ageing xylem in relation to heartwood formation in *Pinus radiata*. *Can. J. Bot.* **51**, 737–741.

Shepherd, K. R. (1964). Some observations on the effect of drought on the growth of *Pinus radiata* D. Don. *Aust. For.* **28,** 7–22.

Shinozaki, K., Yoda, K., Hozumi, K., and Kira, T. (1964a). A quantitative analysis of plant form—the pipe model theory. I. Basic analyses. *Jpn. J. Ecol.* **14,** 97–105.

Shinozaki, K., Yoda, K., Hozumi, K., and Kira, T. (1964b). A quantitative analysis of plant form—the pipe model theory. II. Further evidence of the theory and its application in forest ecology. *Jpn. J. Ecol.* **14,** 133–139.

Siau, J. F. (1971). "Flow in Wood," Syracuse Wood Sci. Ser. 1. Syracuse Univ. Press, Syracuse, New York.

Siau, J. F., and Petty, J. A. (1979). Corrections for capillaries used in permeability measurements of wood. *Wood Sci. Technol.* **13,** 179–185.

Skaar, C. (1972). "Water in Wood," Syracuse Wood Sci. Ser. 4. Syracuse Univ. Press, Syracuse, New York.

Skene, D. S. (1969). The period of time taken by cambial derivatives to grow and differentiate into tracheids in *Pinus radiata* D. Don. *Ann. Bot. (London)* [N. S.] **33,** 253–262.

Slatyer, R. O. (1966). Some physical aspects of internal control of leaf transpiration. *Agric. Meteorol.* **3,** 281–292.

Slatyer, R. O. (1967). "Plant-Water Relationships." Academic Press, New York. Experimental Botany Series of Monographs number 2.

Slavík, B. (1963). Relationship between osmotic potential of cell sap and the water saturation deficit during the wilting of leaf tissue. *Biol. Plant.* **5,** 258–264.

Smith, C. J. (1975). Substrate availability in relation to daylength effects on tracheid development in *Picea sitchensis* (Bong.) Carr. *Ann. Bot. (London)* [N. S.] **39,** 101–111.

Smith, J. H. G., Walters, J., and Wellwood, R. W. (1966). Variation in sapwood thickness of Douglas-fir in relation to tree and section characteristics. *For. Sci.* **12,** 97–103.

Snell, J. A. K., and Brown, J. K. (1978). Comparison of tree biomass estimators—DBH and sapwood area. *For. Sci.* **24,** 455–457.

Spittlehouse, D. L. (1980). Measurement and modelling of forest evapotranspiration. Ph.D. thesis, Univ. of British Columbia, Vancouver.

Spittlehouse, D. L., and Black, T. A. (1979). Determination of forest evapotranspiration using Bowen ratio and eddy correlation measurements. *J. Appl. Meteorol.* **18,** 647–653.

Spomer, G. G. (1968). Sensors monitor tensions in transpiration strength of trees. *Science* **161,** 484–485.

Stewart, C. M. (1966). Excretion and heartwood formation in living trees. *Science* **153,** 1068–1074.

Stewart, C. M. (1967). Moisture content of living trees. *Nature (London)* **214,** 138–140.

Stewart, C. M., Tham, S. H., and Rolfe, D. L. (1973). Diurnal variations in water in developing secondary stem tissues of Eucalypt trees. *Nature (London)* **242,** 479–480.

Stewart, J. B., and Thom, A. S. (1973). Energy budgets in pine forest. *Q. J. R. Meteorol. Soc.* **99,** 154–170.

Stiell, W. M. (1966). Red pine crown development in relation to spacing. *Can., Dep. Fish. For., For., Serv., Publ.* **1145,** 1–44.

Stone, E. C. (1958). Dew absorption by conifers. *In* "Physiology of Forest Trees" (K. V. Thimann, ed.), pp. 125–153. Ronald Press, New York.

Sucoff, E. I. (1969). Freezing of conifer xylem and the cohesion-tension theory. *Physiol. Plant.* **22,** 424–431.

Sucoff, E. (1972). Water potential in red pine: Soil moisture, evapotranspiration, crown position. *Ecology* **53,** 681–686.

Swanson, R. H. (1967). Seasonal course of transpiration of lodgepole pine and Engelmann spruce. *In* "Forest Hydrology" (W. E. Sopper and H. W. Lull, eds.), pp. 419–434. Pergamon, Oxford.

Swanson, R. H. (1970). The tree as a dynamic system in forest-water resource research. *In* "Proceedings of the Third Forest Microclimate Symposium" (J. M. Powell and C. F. Nolasco, eds.), pp. 34–39. Can. Dep. Fish For., For. Serv., Calgary.

Swanson, R. H. (1972). Water transpired by trees is indicated by the heat pulse velocity. *Agric. Meteorol.* **10**, 277–281.

Swanson, R. H. (1975). Velocity distribution patterns in ascending xylem sap during transpiration. *In* "Flow—Its Measurement and Control in Science and Industry" (R. B. Dowdell, ed.), Vol. 1, pp. 1425–1430. Instru. Soc. Am.

Swanson, R. H., and Lee, R. (1966). Measurement of water movement from and through shrubs and trees. *J. For.* **64**, 187–190.

Tadaki, Y. (1963). Studies on production structure of forest. IV. Some studies on leaf amount of stands and individual trees. *J. Jpn. For. Soc.* **45**, 249–256. (in Japanese).

Tadaki, Y. (1966). Some discussions on the leaf biomass of forest stands and trees. *Bull. Gov. For. Exp. Stn. (Jpn.)* **184**, 135–161.

Tadaki, Y. (1970). Studies on the production structures of forest. XVII. Vertical change of specific leaf area in forest canopy. *J. Jpn. For. Soc.* **52**, 263–268.

Tadaki, Y., Hatiya, K., and Tochiaki, K. (1969). Studies of the production structure of forest. XV. Primary productivity of *Fagus crenata* in plantation. *J. Jpn. For. Soc.* **51**, 331–339 (in Japanese).

Tajchman, S. J. (1972). The radiation and energy balances of coniferous and deciduous forests. *J. Appl. Ecol.* **9**, 359–375.

Tajchman, S. J., Hadrich, F., and Lee, R. (1979). Energy budget evaluation of the transpiration-pF relationship in a young pine forest. *Water Resour. Res.* **15**(1), 159–163.

Talboys, P. W. (1955). Detection of vascular tissues available for water transport in the hop by colourless derivatives of basic dyes. *Nature (London)* **175**, 510.

Tan, C. S., and Black, T. A. (1976). Factors affecting the canopy resistance of a Douglas-fir forest. *Boundary-Layer Meteorol.* **10**, 475–488.

Tan, C. S., Black, T. A., and Nnyamah, J. U. (1977). Characteristics of stomatal diffusion resistance in a Douglas-fir forest exposed to soil water deficits. *Can. J. For. Res.* **7**, 595–604.

Tan, C. S., Black, T. A., and Nnyamah, J. U. (1978). A simple diffusion model of transpiration applied to a thinned Douglas-fir stand. *Ecology* **59**, 1221–1229.

Tanner, C. B. (1968). Evaporation of water from plants and soil. *In* "Water Deficits and Plant Growth" (T. T. Kozlowski, ed.), Vol. 1, pp. 74–106. Academic Press, New York.

Taylor, H. M., and Klepper, B. (1978). The role of rooting characteristics in the supply of water to plants. *Adv. Agron.* **30**, 99–128.

Thomas, R. J. (1970). Origin of bordered pit margo microfibrils. *Wood Fiber* **2**, 285–288.

Thomas, R. J., and Kringstad, K. P. (1971). The role of hydrogen bonding in pit aspiration. *Holzforschung* **25**, 143–149.

Thomas, R. J., and Nicholas, D. D. (1966). Pit membrane structure in loblolly pine as influenced by solvent exchange drying. *For. Prod. J.* **16**(3), 53–56.

Thompson, D. R., and Hinckley, T. M. (1977). A simulation of water relations of white oak based on soil moisture and atmospheric evaporative demand. *Can J. For. Res.* **7**, 400–409.

Tobiessen, P., Rundel, P. W., and Stecker, R. E. (1971). Water potential gradient in a tall *Sequoiadendron. Plant Physiol.* **48**, 303–304.

Tranquillini, W. (1976). Water relations and alpine timberline. *Ecol. Stud.* **19**, 473–491.

Tschernitz, J. L., and Sachs, I. B. (1975). Observations on microfibril organization of Douglas-fir bordered pit-pair membranes by scanning electron microscopy. *Wood Fiber* **6**, 332–340.

Tsoumis, G. (1965). Light and electron microscopic evidence on the structure of the membrane of bordered pits in tracheids of conifers. *In* "Cellular Ultrastructure of Woody Plants" (W. A. Côté, Jr., ed.), pp. 305–317. Syracuse Univ. Press, Syracuse, New York.

Tucker, G. F., and Emmingham, W. H. (1977). Morphological changes in leaves of residual western hemlock after clear and shelterwood cutting. *For. Sci.* **23**, 195–203.

Turner, N. C., and Jarvis, P. G. (1975). Photosynthesis in Sitka spruce (*Picea sitchensis* (Bong.) Carr.). IV. Response to soil temperature. *J. Appl. Ecol.* **12**, 561–576.

Tyree, M. T. (1976). Physical parameters of the soil-plant-atmosphere system: Breeding for drought resistance characteristics that might improve wood yield. *In* "Tree Physiology and Yield Improvement" (M. G. R. Cannell and F. T. Last, eds.), pp. 329–348. Academic Press, New York.

Tyree, M. T., Caldwell, C., and Dainty, J. (1975). The water relations of hemlock (*Tsuga canadensis*). V. The localization of resistances to bulk water flow. *Can. J. Bot.* **53**, 1078–1084.

Vaadia, Y., and Waisel, Y. (1963). Water absorption by the aerial organs of plants. *Physiol. Plant.* **16**, 44–51.

Van Bavel, C. H. M. (1973). Towards realistic simulation of the natural plant climate. *In* "Plant Response to Climatic Factors," Proc. Uppsala Symp., 1970, Ecology and Conservation Vol. 5, pp. 441–446. Unesco, Paris.

van den Honert, T. H. (1948). Water transport in plants as a catenary process. *Discuss. Faraday Soc.* **3**, 146–153.

van Laar, A. (1967). The influence of environmental factors on the radial growth of *Pinus radiata*. *S. Afr. For. J.* **61**, 24–39.

Waggoner, P. E., and Turner, N. C. (1971). Transpiration and its control by stomata in a pine forest. *Conn., Agric. Exp. Stn., New Haven, Bull.* **726**, 1–87.

Waring, R. H., and Cleary, B. D. (1967). Plant moisture stress: Evaluation by pressure bomb. *Science* **155**, 1248–1254.

Waring, R. H., and Roberts, J. M. (1979). Estimating water flux through stems of Scots pine with tritiated water and phosphorous-32. *J. Exp. Bot.* **30**, 459–471.

Waring, R. H., and Running, S. W. (1976). Water uptake, storage and transpiration by conifers: A physiological model. *In* "Water and Plant Life" (O. L. Lange, L. Kappen, and E.-D. Schulze, eds.), pp. 190–202. Springer-Verlag, Berlin and New York.

Waring, R. H., and Running, S. W. (1978). Sapwood water storage: Its contribution to transpiration and effect upon water conductance through the stems of old-growth Douglas-fir. *Plant, Cell, & Environ.* **1**, 131–140.

Waring, R. H., Gholz, H. L., Grier, C. C., and Plummer, M. L. (1977). Evaluating stem conducting tissue as an estimator of leaf area in four woody angiosperms. *Can. J. Bot.* **55**, 1474–1477.

Waring, R. H., Emmingham, W. H., Gholz, H. L., and Grier, C. C. (1978). Variation in maximum leaf area of coniferous forests in Oregon and its ecological significance. *For. Sci.* **24**, 131–140.

Waring, R. H., Whitehead, D., and Jarvis, P. G. (1979). The contribution of stored water to transpiration in Scots pine. *Plant, Cell, & Environ.* **2**, 309–317.

Waring, R. H., Whitehead, D., and Jarvis, P. G. (1980). Comparison of an isotopic method and the Penman-Monteith equation for estimating transpiration from Scots pine. *Can. J. For. Res.* **10**, 555–558.

Warren Wilson, J. (1967). The components of leaf water potential. I. Osmotic and matric potentials. *Aust. J. Biol. Sci.* **20**, 329–347.

Watts, W. R. (1977). Field studies of stomatal conductance. In "Environmental Effects on Crop Physiology" (J. J. Landsberg and C. V. Cutting, eds.), pp. 173–189. Academic Press, New York.

Watts, W. R., and Neilson, R. E. (1978). Photosynthesis in Sitka spruce (Picea sitchensis (Bong.) Carr). IX. Measurements of stomatal conductance and $^{14}CO_2$ uptake in controlled environments. J. Appl. Ecol. 15, 245–255.

Watts, W. R., Neilson, R. E., and Jarvis, P. G. (1976). Photosynthesis in Sitka spruce (Picea sitchensis (Bong.) Carr.) VIII. Measurement of stomatal conductance and $^{14}CO_2$ uptake in a forest canopy. J. Appl. Ecol. 13, 623–638.

Weatherley, P. E. (1950). Studies in the water relations of the cotton plant. I. The field measurement of water deficits in leaves. New Phytol. 49, 81–97.

Weatherley, P. E. (1965). The state and movement of water in the leaf. Symp. Soc. Exp. Biol. 19, 157–184.

Weatherley, P. E. (1970). Some aspects of water relations. Adv. Bot. Res. 3, 171–206.

Wellwood, R. W. (1955). Sapwood-heartwood relationships: Second-growth Douglas fir. For. Prod. J 5(2), 108–111.

Wellwood, R. W., and Jurazs, P. E. (1968). Variations in sapwood thickness, specific gravity, and tracheid length in western red cedar. For. Prod. J. 18(12), 37–46.

Whitehead, D. (1978). The estimation of foliage area from sapwood basal area in Scots pine. Forestry 51, 35–47.

Whitmore, F. W., and Zahner, R. (1967). Evidence for a direct effect of water stress on tracheid cell wall metabolism in pine. For. Sci. 13, 397–400.

Wiebe, H. H., Brown, R. W., Daniel, T. W., and Campbell, E. (1970). Water potential measurements in trees. BioScience 20, 225–226.

Williams, J. (1974). Root density and water potential gradients near the plant root. J. Exp. Bot. 25, 669–674.

Wodzicki, T. (1964). Photoperiodism control of natural growth substances and wood formation in larch (Larix decidua D. C.). J. Exp. Bot. 15, 584–599.

Wodzicki, T. J., and Brown, C. L. (1970). Role of xylem parenchyma in maintaining the water balance of trees. Acta. Soc. Bot. Pol. 39, 617–621.

Wolff, J. O., West, S. D., and Viereck, L. A. (1977). Xylem pressure potential in black spruce in interior Alaska. Can. J. For. Res. 7, 422–428.

Woods, F. W., and O'Neal, D. (1965). Tritiated water as a tool for ecological field studies. Science 147, 148–149.

Woods, F. W., Hough, W. A., O'Neal, D., and Barnett, J. (1965). Gamma ray attenuation by loblolly pine wood: An investigation of integral counting. For. Sci. 11, 341–435.

Yerkes, V. P. (1967). Effect of seasonal moisture variation and log storage on the weight of Black Hills ponderosa pine. U.S., For. Serv., Res. Note RM 96, 1–8.

Zaerr, J. B. (1971). Moisture stress and stem diameter in young Douglas-fir. For. Sci. 17, 466–469.

Zahner, R. (1968). Water deficits and growth of trees. In "Water Deficits and Plant Growth" (T. T. Kozlowski, ed.), Vol. 2, pp. 191–254. Academic Press, New York.

Zahner, R., Lotan, J. E., and Baughman, W. D. (1964). Earlywood-latewood features of red pine grown under simulated drought and irrigation. For. Sci. 10, 361–370.

Zavitkovski, J., and Dawson, D. H. (1978). Structure and biomass production of 1- to 7-year old intensively cultured jack pine plantations in Wisconsin. U.S., For. Serv., Res. Pap. NC 157, 1–15.

Želawski, W., and Kucharska, J. (1967). Winter depression of photosynthetic activity in seedlings of Scots pine (Pinus sylvestris L.). Photosynthetica 1, 207–213.

Ziegler, H. (1964). Storage, mobilization and distribution of reserve material in trees. *In* "Formation of Wood in Forest Trees" (M. H. Zimmermann, ed.), pp. 303–320. Academic Press, New York.

Zimmermann, M. H. (1964). Effect of low temperature on ascent of sap in trees. *Plant Physiol.* **39**, 568–572.

Zimmermann, M. H. (1971). Transport in the xylem. *In* "Trees: Structure and Function" (M. H. Zimmermann and C. L. Brown, eds.), pp. 169–220. Springer-Verlag, Berlin and New York.

Zimmermann, M. H. (1978). Hydraulic architecture of some diffuse-porous trees. *Can. J. Bot.* **56**, 2286–2295.

CHAPTER 3

TEMPERATE HARDWOOD FORESTS

T. M. Hinckley* and R. O. Teskey*

SCHOOL OF FORESTRY, FISHERIES AND WILDLIFE, UNIVERSITY OF MISSOURI,
COLUMBIA, MISSOURI

F. Duhme

LEHRSTUHL FÜR LANDSCHAFTSÖKOLOGIE, TECHNISCHE UNIVERSITÄT MÜNCHEN,
FREISING-WEIHENSTEPHAN, FEDERAL REPUBLIC OF GERMANY

H. Richter

BOTANISCHES INSTITUT, UNIVERSITÄT FÜR BODENKULTER, VIENNA, AUSTRIA

*Present address: College of Forest Resources, University of Washington, Seattle, Washington 98195.

Water Deficits and Plant Growth, Vol. VI
Copyright © 1981 by Academic Press, Inc.
All rights of reproduction in any form reserved.
ISBN 0-12-424156-5

I. INTRODUCTION

Hardwood forests are widely distributed throughout the world, with their greatest dominance in three climatic regions: (1) the tropics; (2) the Mediterranean; and (3) the belts between the continuous coniferous regions of the upper latitudes, and between the mixed conifer–hardwood forest of wet areas and dry grasslands (Fig. 1). Hardwood species form a major component of these three regions and are also an important part of desert and arctic and alpine ecosystems. Their wide distribution subjects hardwood species to a wide range of environmental stresses.

Hardwood species are capable of growing in areas which receive less than 50 mm of rainfall per year (e.g., *Artemisia herba alba* in the Negev desert of Israel) and in areas that are flooded for most of the year (e.g., *Nyssa aquatica* in south central United States). Certain hardwood species can grow in soils with as little as 5% extractable bases (e.g., *Quercus marilandica*) and in soils with a pH as low as 3.5 (e.g., *Betula pubescens*) or as high as 8.5 (e.g., *Sorbus aria*). Temperatures between -2 and $-25°C$ limit the distribution of evergreen hardwoods (e.g., *Arbutus menziesii*), while temperatures between -41 and $-47°C$ limit deciduous hardwoods (e.g., *Quercus rubra*) which possess a supercooling mechanism for living cells (Weiser, 1970; Burke *et al.*, 1976). Other deciduous hardwoods (e.g., *Betula papyrifera*) are not limited by low temperatures.

Any analysis of the distribution or productivity of hardwood species must at one point or another deal with water deficits in trees. Water deficits either directly or indirectly influence many important physiological processes (Hsiao, 1973; Kozlowski, 1968, 1972, 1976, 1978). Periods of water deficits in temperate hardwoods are most common during the growing season when evaporative demand is high or soil moisture supplies are low. However, water deficits may be a major problem in developing tissue that is not drought-resistant. In addition, appreciable water deficits may

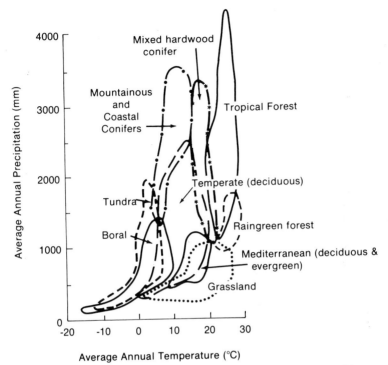

Fig. 1. Relationship between plant or community type and average annual temperature and precipitation. (Redrawn from Lieth, 1975.)

develop in deciduous species during the leafless period (Marchand and Chabot, 1978). Therefore, an understanding of how water deficits develop, their impact on physiological processes, and mechanisms by which trees resist the development of water deficits seems important and timely. This chapter will examine water relations of hardwood forest trees at several levels of biological organization and describe factors that make this forest type similar to or different from other forest types. The emphasis will be on whole-tree water relations. Most examples will be for deciduous hardwood trees, although examples of evergreen hardwoods and conifers will be used for comparison.

II. TEMPERATE HARDWOOD FORESTS

A. DISTRIBUTION

What determines the distribution of hardwoods as compared to conifers (Fig. 1)? For example, how does one explain the nearly continuous distribution of *Quercus-Carya* forests in central and eastern North America as well as the *Fagus-Quercus* forests of central, eastern, and southeastern Europe? Waring and Franklin's (1979) attempt to interpret the coniferous forest belt of the Pacific Northwest of the United States brought out the two major reasons for the distribution of the principal biomes of the world: vegetational history and environmental stresses. In central Europe, with its impoverished forest flora, the historical background is most important although it is frequently neglected (Frenzel, 1968).

In general, temperate hardwood forests are bound to climates with an almost even distribution of precipitation throughout the year (Walter, 1968). This precipitation can range between 50 and 200 cm per year. Precipitation in the growing season must normally exceed 20 to 30% of the total (Förster, 1979). Rainy winters with a prolonged summer drought would favor either conifers (Waring and Franklin, 1979) or sclerophyllous evergreen hardwoods. Occasional frost periods with temperatures well below $-15°C$ would exclude the latter. The classic latitudinal patterns of biome distribution from evergreen hardwoods to deciduous hardwoods to conifers are found only where summer droughts are moderate or infrequent. For example, in California, Oregon, and the eastern Mediterranean where summer droughts are frequent, the deciduous forest is absent and the evergreen hardwood forest is in direct contact with the conifer forest (Barbour and Major, 1977; Horvat *et al.*, 1974). However, before the success (or lack thereof) of deciduous hardwoods is ascribed to any one factor (i.e., water status), other factors such as nutrients, pathogens, insects, temperature, mechanical stresses, and the presence or absence of genetic diversity must be considered (Schulze *et al.*, 1977). Moreover, it can be assumed that water deficits not only limit productivity, but also influence both biome (Fig. 1) and species distribution (Fig. 2).

B. Hydrological Characteristics of the Temperate Hardwood Forest Ecosystem

The deciduous nature of most hardwoods of the temperate region determines their hydrological properties more than any other factor. Research in New Hampshire (Likens *et al.*, 1978), North Carolina (Swank and Douglass, 1974), and Tennessee (Henderson *et al.*, 1978; Luxmoore *et al.*, 1978) in the United States, in Belgium (Duvigneaud and Kestemont, 1977), and in West Germany (Benecke and Mayer, 1971; Benecke, 1976) provides much information on hydrological characteristics of hardwood forests as well as the differences between hardwood and coniferous forests.

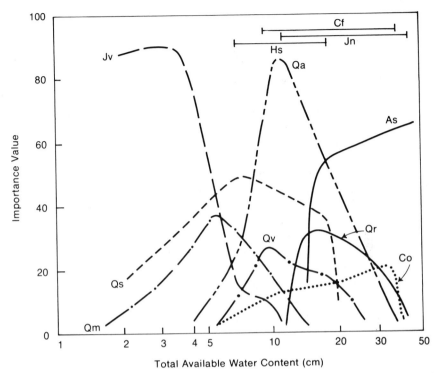

Fig. 2. Relationship between importance value [= (relative density + relative basal area + relative frequency)/3] and available water content for eight tree species from southern Illinois. Jv = *Juniperus virginiana*, Qa = *Quercus alba*, As = *Acer saccharum*, Qs = *Q. stellata*, Qm = *Q. marilandica*, Qv = *Q. velutina*, Qr = *Q. rubra*, Co = *Carya ovata*. Bars represent average water content (AWC) ranges that are occupied by other species such as *Cornus florida* (Cf), *Juglans nigra* (Jn), *Helianthus strumosus* (Hs). (After Fralish *et al.*, 1978). AWC (cm/100) soil horizon = AWC (%) × bulk density × depth of horizon (cm/100) where AWC (%) = 6.5 + 0.26 (% silt).

Both leaf area index and seasonal duration of leaf retention in a deciduous forest are only about half of those in a coniferous forest on comparable sites (Grier and Running, 1977; Gholz *et al.*, 1976). As a result, interception losses are less from a hardwood than from a coniferous forest (Rutter, 1975; Henderson *et al.*, 1977). Canopy storage capacities range between 1.0 and 2.1 mm of water in conifers and are approximately 1.0 mm in the summer and 0.5 mm in the winter for a deciduous hardwood forest (Rutter, 1975). For example, in the Appalachian mountains of North Carolina (210 cm rainfall per year), stream flow in a *Quercus-Carya* forest amounted to 1.48 × 10⁶ liter more per hectare per year than in a *Pinus strobus* forest (Swank and Douglass, 1974). Differences in interception appeared at least as important as differences in evapotranspiration. Interception losses can range between 20 and 85% of the total precipitation for conifers and between 10 and 50% for foliated hardwoods, depending on duration and intensity of individual rainfall events. Interception averaged 10 to 20% for a deciduous hardwood forest and 30% for a coniferous forest (Holch, 1931; Helvey and Patric, 1965; Benecke, 1976; Henderson et al., 1977). During the winter, interception for a deciduous hardwood forest averaged less than 10% (Rutter, 1975; Benecke, 1976). Stem flow is generally greater in hardwood trees than in conifers (Benecke, 1976).

Soil moisture depletion beneath fully stocked Quercus garryana stands occurred faster and to a greater depth than in a Pseudotsuga menziesii forest (Fig. 3). Leaf area index and seasonal duration of leaf retention are obviously important in determining transpirational losses. However, higher rates of transpiration (see Section III,A) and differences in root distribution and stomatal control (see Section IV) in Quercus garryana as compared to Pseudotsuga menziesii may play major roles in stand water balance. Individual hardwood trees (e.g., Quercus petraea) lost much more water than coniferous trees (e.g., Picea abies; Ladefoged, 1963). However, when water losses of a hardwood forest and a coniferous forest were expressed on a unit land area basis, there were only minor differences between these two types. This contrasts with results presented in Fig. 3 and may reflect differences in summer rainfall or potential evaporation between southern Sweden (wet) and Oregon (dry).

Differences in leaf area index and seasonal duration of leaf retention greatly influence the amount of rainfall intercepted, the timing of stand transpiration, and the microclimate of a forest. Wind and radiation levels, especially during the leafless period, are much greater within a hardwood forest than a coniferous forest (Monteith, 1975; Thompson and Hinckley, 1977a). In addition stem flow in the former is greater during the leafless period (Helvey and Patric, 1965). Higher temperatures at the forest floor in early spring mean higher rates of (1) growth of roots (Hinckley et al., 1979a); (2) respiration of microorganisms and roots (Garrett and Cox, 1973; Garrett et al., 1978); (3) decomposition of litter (Garrett and Cox, 1973); and (4) evaporation from the soil (Luxmoore et al., 1978). Additionally, there usually is a more abundant shrub and herb layer in a deciduous hardwood forest. Bud opening and leaf expansion occur earlier in the shrub and herb layer than in the overstory species (Hinckley et al., 1976). Therefore, this lower layer places another demand on soil water, especially that in the upper 30 cm of forest soil (Thompson and Hinckley, 1977a; Eliáš, 1978).

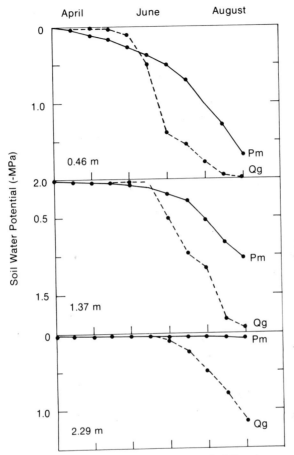

Fig. 3. Soil water potential at three depths in fully stocked *Pseudotsuga menziessi* (Pm) and *Quercus garryana* (Qg) stands near Corvallis, Oregon. (After Krygier, 1971.)

The reduced litter content of a hardwood forest generally results in a lower nutrient accumulation in the O soil horizon compared to that beneath a coniferous forest (Benecke and Mayer, 1971; Henderson *et al.*, 1978). The differences in litter content, soil evaporation rates, and soil temperatures may affect root distribution and growth and, hence, water uptake from the soil (see Section III,B). In summary, a hardwood forest intercepts less water and may deplete soil water faster and to a greater depth than a coniferous forest.

In addition to the above differences between a hardwood and a coniferous forest, in both forests, each stratum is exposed to different microclimates and soil environments. Water deficits may be greater in seedlings and herbs than in overstory species because the ameliorating influence of an overstory is cancelled by the relatively small volume of soil occupied (Thompson and Hinckley, 1977a; Eliáš, 1978). Therefore, a description of community or watershed water relations must integrate over time, species, and position in the stand what is known about water relations of single trees.

III. GENERAL WATER RELATIONS OF TREES

Water is moved through the soil–plant–atmosphere continuum because of differences in water potential between the soil and atmosphere. The rate at which water flows depends largely on the transpiring surface area, on stomatal aperture, and the gradient in water vapor pressure between the leaf and air. Stomatal aperture is influenced by a number of factors, including leaf water potential, which in turn depends on the potential of the soil water and the potential drop during water flow in the plant body. The soil is the ultimate source for almost all transpired water, but water stored in trees also is an important intermediate buffer and is utilized during periods of high transpiration (see Chapter 1 by Chaney in this volume; Powell and Thorpe, 1977; Federer, 1979). In addition, leaves may absorb some water from the atmosphere during dew formation or rainfall episodes (Rutter, 1975; Went and Babu, 1978).

The important components of the water transport system include properties of the root, xylem, and the leaf–air interface. These will be discussed separately, but it should be remembered that a balance between these components must exist (Jarvis, 1975; Grier and Running, 1977). For example, a tree that can transpire a large volume of water per unit foliage area per unit time must either have a conduction system capable of extremely high velocities of water flow, as in ring porous species, or, if it has a conducting system capable of only moderate velocities of water flow, it must use a larger cross-sectional area (i.e., diffuse porous and coniferous species). Coniferous species have a relatively slow flow rate but they use a large cross-sectional area for flow and have large amounts of stored water (Hinckley et al., 1978b; Waring and Franklin, 1979).

As will be illustrated in Section VII, the apparent balance between various components of the water transport system is affected by differences in drought avoidance and desiccation tolerance. For example, a tree with a small amount of stored water may close its stomata at a lower leaf water potential for stomatal closure than one with a high amount of stored water.

A. PATH OF WATER TRANSPORT

The general path of water transport through trees has been recognized since the time of Hales (1727) and has been described by several investigators (Kozlowski, 1961; Kramer, 1969; Jarvis, 1975; Kramer and Kozlowski, 1979). Differences in xylem and leaf morphology, root distribution, water storage capacity, leaf area index, and duration of leaf retention are the obvious factors that influence the specific patterns of water transport in trees.

Water is absorbed through the roots, crosses a membrane barrier at the endodermis, moves into the xylem of the root, and then moves upward through the xylem of the main stem, branches, and petioles, and finally through the vascular elements of each leaf. Water then moves along a cell wall pathway from the vascular bundle through the bundle sheath cells to the epidermis (Edwards and Meidner, 1978; Sheriff, 1979). Liquid water may then flow through the epidermis and cuticle or along the underside of the epidermis and a guard cell to a stomatal cavity where it evaporates into the atmosphere.

Upward water movement tends to be restricted to the outermost one or two annual rings in ring porous trees (e.g., *Quercus, Fraxinus, Carya*), to the last several annual rings in diffuse porous hardwoods (e.g., *Acer, Betula, Populus*) and to most of the sapwood in conifers (Kozlowski and Winget, 1964; Swanson, 1967; Chaney and Kozlowski, 1977; Lassoie *et al.*, 1977). Only small gradients of water potential are necessary to move water through large diameter vessels (Fig. 4). However, as vessels become air-filled, much larger gradients are required to maintain the same rate of sap flow through the remaining small vessels. Maximum sap velocities range from 6000 cm hr^{-1} in *Quercus macrocarpa* (ring porous) (Kuntz and Riker, 1955), to 450 cm hr^{-1} in *Acer saccharum* (diffuse porous) (Greenidge, 1958), to 80 cm hr^{-1} in *Pinus contorta* (Swanson, 1967). Although the rates of sap movement and specific conductivities are higher in ring porous than in diffuse porous or in coniferous species (Hinckley *et al.*, 1978b), the total amount of water lost in transpiration may not differ greatly because water moves through fewer elements in ring porous species than in diffuse porous species or conifers (Ladefoged, 1963; Kramer and Kozlowski, 1979).

During periods of long or severe water deficits when large diameter vessels may be cavitated, total water transport in conifers may exceed that in ring porous or diffuse porous trees. Many of the large diameter vessels that cavitate during the summer probably are not refilled following autumn rains. Furthermore, during a winter freeze most if not all of the large diameter vessels become cavitated (Hammel, 1967). The net effect is

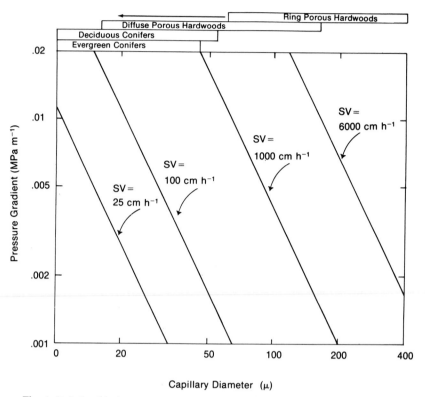

Fig. 4. Relationship between pressure gradient, capillary diameter, and peak sap velocities (SV) based on the Hagen–Poiseuille equation (Zimmermann and Brown, 1971, p. 190). Ranges of earlywood tracheid and vessel diameters are shown above the graph. (From Grosser, 1977.) The line with arrow indicates that diameters of latewood vessels can be small in ring porous hardwoods.

that some xylem increment is often necessary before bud opening and leaf expansion in ring porous trees. This newly formed xylem provides a low resistance pathway for water transport to the expanding leaves. For example, as much as 40% of the annual stem increment takes place before buds open in *Quercus alba* (Boggess, 1956; Dougherty *et al.*, 1979; Hinckley *et al.*, 1979) whereas bud opening and annual initiation of cambial growth usually coincide in many conifers (Lassoie, 1979). Also, injury to the outer annual rings by fire or insects will have a greater impact on hardwood trees than on conifer trees (assuming enough phloem tissue remains intact).

Carlquist (1975) and Zimmermann (1978b) reviewed the relationship between the structure of the vascular system and the ecological position occupied by a species. It seems that colonization of periodically dry sites by broad-leaved trees was possible only after the evolution of simple perforations in vessels. The transition from scalariform to simple perforations occurred independently in several systematic groups. Further expansion of a species into dry areas led to a reduction in vessel diameters and an increase in wall thickness. For every large vessel, a tree needs many small ones to move comparable amounts of water (Fig. 4). Air embolism in a large diameter vessel shuts off a far larger fraction of the total water flowing in the stem than embolism in one of numerous vessels of small diameters. Trees living in dry areas, such as evergreen oaks, generally have small diameter vessels. In contrast, oaks of temperate regions have vessels with large diameters.

As illustrated in Fig. 5, "leaf-specific conductivity" (conductivity of the xylem supplying water to a unit mass of leaves) varied greatly within *Acer saccharum* trees mostly, due to variations in vessel diameters (Zimmermann, 1978a). Highest conductivities were generally found in the stem and leaf, while hydraulic constrictions were observed at vascular junctions (e.g., branch-stem, petiole-branch) in *Acer pennsylvanicum, A. rubrum, A. saccharum, Betula papyrifera, Populus grandidentata* (Zimmer-

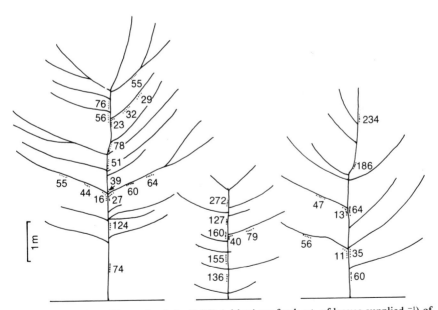

Fig. 5. Leaf-specific conductivity (LSC) (μl hr^{-1} gm fresh wt. of leaves supplied $^{-1}$) of three sugar maple (*Acer saccharum*) trees grown in open shade, under conditions of gravity flow. Conductivities are usually higher in stems than in branches and often increase toward the top of the tree. Hydraulic constrictions characterize the xylem pathway in its transition from stem to branch. Petiole junctions have a LSC between 1–3 and petioles between 5–10. (Redrawn from Zimmermann, 1978a.)

mann, 1978a) and *Populus deltoides* (Larson and Isebrands, 1978). Baxter and West (1977) noted that conductivity in the xylem of tree roots was about 5 times that in large branches and 10 times that in twigs of *Malus* spp. and *Prunus* spp. Zimmermann (1978a) concluded that water flow and gradients of water potential throughout a tree crown can be interpreted from these results (Fig. 5). These patterns of flow and gradients favor the main stem in situations of severe drought (i.e., leaf, twig, and then branch abscission; see Section VII) and provide a system whereby leaves at the top of the tree can compete with other leaves for water. Similar results have been noted in conifers (Hellkvist *et al.*, 1974; Zimmermann, 1978a), though leaf-specific conductivities were much lower and varied between 4.5 and 27.0 in *Tsuga canadensis* (Zimmermann, 1978a).

Important morphological features of the typical hardwood leaf include the following: (1) the blade or lamina is usually broad and flat; (2) a petiole supports the leaf; (3) the mesophyll is differentiated into distinct palisade and spongy parenchyma tissue; (4) the conducting tissue forms a very delicate web that provides a ready supply of water and inorganic nutrients and removes surplus assimilates; and (5) the stomata appear rather uniformly scattered across the surface of the epidermis with no particular orientation or alignment. The greater frequency of stomata, the greater pore diameters of stomata, the surface exposure of the stomata, and the more thorough vascular supply in hardwood leaves in comparison with conifer foliage account for hardwood leaves having greater potential for carrying on gas exchange (i.e., higher leaf conductances) and, therefore higher rates of photosynthesis and transpiration (Körner *et al.*, 1979). For example, maximum reported rates of transpiration in ring porous hardwoods (800 to 3500 mg water gm dry wt.$^{-1}$ hr^{-1}), diffuse porous hardwoods (300 to 500), and conifers (180 to 430) form a range very similar to the one shown for sap velocities in Fig. 5 (Lopushinsky, 1975; Lassoie and Chambers, 1976; Kramer and Kozlowski, 1979). Similar differences in maximum stomatal opening (Duhme, 1974; Körner *et al.*, 1979) have been reported for these broad species groups. Such data may suggest that hardwood forests transpire more water per unit foliage area than coniferous forests. However, when transpiration rates per unit land area are compared, they are rather similar for hardwood and coniferous forests (Ladefoged, 1963).

B. Water Absorption

1. Form and Growth of Roots

The form of a tree's root system is determined by the species and by the soil environment (Toumey, 1929; Weaver and Kramer, 1932; Biswell,

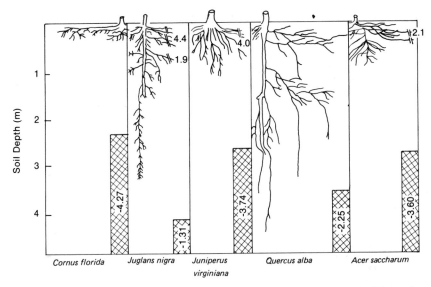

Fig. 6. Distribution of roots *Cornus florida, Juglans nigra, Juniperus virginiana, Quercus alba, Acer saccharum,* and minimum leaf water potential (cross-hatched areas) at predawn during a severe drought. (Redrawn from Biswell, 1935; Hinckley *et al.,* 1979; Sprackling and Read, 1979.) The numbers next to the root system (e.g., 4.4) correspond to the horizontal extension of the root, in meters.

1935; Yen, 1973; Yen *et al.,* 1978) (Fig. 6). The structure of large roots of hardwoods of Taiwan and the United States were classified into seven categories based on the relative distribution of roots 0.6 cm and greater in diameter (Yen, 1973; Yen *et al.,* 1978). These investigators concluded that although site can modify root distribution, root structure remains relatively fixed unless distinct physical barriers to growth are present (e.g., soil horizons with high bulk densities, high rock content, or saturated soil). This view contrasts with earlier work on roots of seedlings and saplings (Weaver and Kramer, 1932; Biswell, 1935). For example, the structure of *Acer negundo* and *Gleditschia triacanthos* roots varied widely on different sites (Biswell, 1935). By comparison, root structure of *Acer saccharum, Aesculus glabra, Carya ovata, Juglans nigra, Quercus rubra,* and *Q. macrocarpa* was relatively fixed. It appears that root structure varies among species, amount of overhead shade, and tree age (Holch, 1931; Biswell, 1935).

Most studies show that more than half of the tree roots (both fine roots, less than 2 mm in diameter, and large roots) occur in the upper 30 cm of soil (Braekke and Kozlowski, 1977; Hermann, 1977; Yen *et al.,* 1978). The better soil aeration and nutrient and water supply in upper soil layers account for the high concentration of tree roots close to the surface.

Deeper rooting patterns are advantageous under situations of recurring droughts (Scrivner *et al.*, 1973). For example, predawn water potentials (ψ^{leaf}) were 0.46 megapascals (MPa) more negative in *Acer saccharum* than in *Juglans nigra* during a minor drought in Missouri (Lucier *et al.*, 1978). During a year of extreme drought, the difference in predawn ψ^{leaf} in these species increased to 1.32 MPa (-2.80 and -4.12 MPa, respectively) (Hinckley *et al.*, 1979). These differences can be related in part to the relatively shallow rooting of *Acer saccharum* (Biswell, 1935) (Fig. 6).

Although site conditions do not always affect root form, they influence both the rate and periodicity of root growth. Except for a short period of rapid growth of flowers, leaves, and branches, root growth of a *Quercus alba* tree was regulated by soil temperature and soil water potential (Fig. 7) (Dougherty *et al.*, 1979; Hinckley *et al.*, 1979). Soil depth influenced root growth as it affected soil temperature (T^{soil}) and soil water potential (ψ^{soil}).

When soil temperatures are not limiting, root growth often is highly correlated with the amount of available soil moisture (Morrow, 1950; Stone, 1967; Larson and Whitmore, 1970; Stone and Jenkinson, 1970; Larson and Palashev, 1973). As soil water is depleted, both the rate at which root growth decreases and the minimum ψ^{soil} at which root growth stops vary with species. For example, the rate of root growth of *Quercus alba* seedlings was much greater than that of *Q. rubra* at -0.8 MPa ψ^{soil} (Larson, 1974).

2. Factors Affecting Water Absorption

Although suberized and unsuberized roots absorb water, their efficiency in water uptake varies greatly (Kramer and Bullock, 1966; Baxter and West, 1977). Even with small ψ^{soil} to ψ^{root} gradients, unsuberized roots often have higher rates of water absorption than suberized roots per unit root surface area (Chung and Kramer, 1975). As the potential gradient increases, differences in absorption between suberized and unsuberized roots are accentuated. Variation in rates of water absorption through suberized roots has been reported in these studies. This variation appears to be related to uneven distribution of water entry points in the suberized surface (e.g., cracks, checks) (Kramer and Bullock, 1966). Although data are limited, it appears that rates of water absorption are higher for suberized hardwood roots than for suberized conifer roots. For example, *Liriodendron tulipifera* and *Cornus florida* roots absorbed water faster than *Pinus echinata* roots (Kramer, 1946).

Suberization of roots would appear to limit water absorption, but it may be important to the process as ψ^{soil} decreases. Under drying soil conditions, suberization inhibits water transfer from the root to the soil,

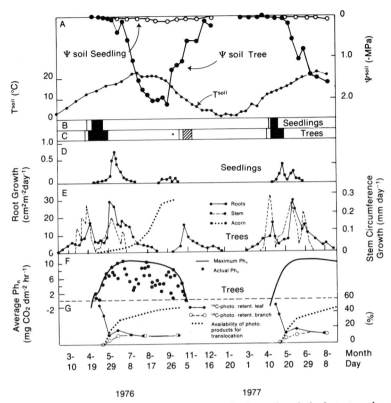

Fig. 7. Seasonal patterns of growth, photosynthesis, and carbohydrate translocation over a 2-yr period in *Quercus alba*. (From Dougherty, 1977; Hinckley *et al.*, 1979; Dougherty *et al.*, 1979; McLaughlin *et al.* 1979.) (A) Soil temperature (T^{soil}) and soil water potential (ψ^{soil}) averaged for a 1.2-m profile. (B) Time of bud opening of seedlings (vertical line) and period of maximum leaf growth (solid bar). (C) Same as B except for adult trees. *—initiation of leaf color; vertical line—initiation of leaf fall; cross-hatched area—period of maximum leaf fall. (D) Root growth (= average rate of root elongation × number of growing roots) of seedlings for entire profile. (E) Stem cambial and root (calculated as in D) growth for trees. Acorn growth on a cumulative basis is also shown. (F) Average (0500–1800 readings) rates of net photosynthesis (Ph_n) for foliage of the upper crown. The solid line is the boundary line (assumed to represent maximum average Ph_n), whereas the points are actual averages. (G) Patterns of translocation of photosynthate in foliage and branches for the entire crown of a tree. ^{14}C-photosynthate retention in the leaf and the branch are given.

thereby increasing the effectiveness of the root in conducting water. This may be especially important just behind the root tip, since high rates of water absorption would leave a zone of relatively dry soil behind the unsuberized growing root tip as it continues to penetrate wetter soil zones (Landsberg and Fowles, 1978).

Since water transport through growing roots involves crossing of a membrane barrier at the endodermis, factors that influence this barrier have a dramatic impact on water absorption. Important factors that influence absorption of water by roots include (1) soil moisture; (2) soil aeration; (3) soil temperature; (4) soil nutrient status; (5) the presence or absence of mycorrhizae; and (6) the ψ gradient within the tree (Kramer, 1969; Newman, 1976).

Soil moisture deficits probably have a more dramatic impact on soil resistance (reductions in hydraulic conductivity) and the soil-to-root resistance (increasing gaps between the soil and root due to root shrinkage) than on permeability of root membranes. However, since water deficits affect metabolic processes and severe water deficits can cause death of roots, soil water deficits ultimately affect root absorption. Rhizosphere resistance appears to be the major factor limiting water uptake from the soil by the plant (Tinker, 1976; Landsberg and Fowles, 1978). Root elongation may lead to higher rates of water uptake by lowering localized resistances in dry soil (Caldwell, 1976). Regular daily cycles of root resistance have been correlated with daily cycles of root growth (Bunce, 1978). Faster root growth during the night than during the day, as reported for *Quercus robur, Prunus avium,* and *Pyrus malus* (Head, 1965; Kazaryan and Khurshudyan, 1966; Lyr *et al.,* 1968), is closely related to the internal water status of a tree since higher ψ^{root} and, therefore, more optimal turgor and growth conditions occur at night.

A primary effect of flooding is decreased soil aeration. During flooding, the oxygen in water is rapidly depleted through respiration of roots and microorganisms (Ponnamperuma, 1972; Nuritdinov and Vartapetyan, 1976). The resulting anaerobic conditions adversely affect root growth and metabolism (Jackson and Campbell, 1975; Hook and Scholtens, 1978). In addition, decreased transpiration is often observed after flooding. Phung and Knipling (1976) found a 90 to 100% reduction in transpiration of *Citrus limon, Citrus reticulata,* and *Poncirus trifoliata* after 10 days of flooding. Reduced transpiration in other flooded hardwoods (e.g., *Eucalyptus camaldulensis, E. globulus, Populus deltoides, Salix nigra,* and *Ulmus americana*) does not seem to be the result of decreased water absorption or increased root resistance, but rather stomatal closure (Pereira and Kozlowski, 1977; Kozlowski and Pallardy, 1979). Stomatal closure of flooded trees does not appear to be a response to general leaf water deficits (Regehr *et al.,* 1975; Pereira and Kozlowski, 1977). It has been suggested that stomatal closure of flooded plants is a hormonally mediated response due to decreased hormone synthesis in the roots and interrupted transport between the root and shoot. Species that are highly resistant to flooding (*Fraxinus pennsylvanica, Nyssa aquatica, Taxodium distichum*) have roots that continue to grow when flooded (Hook *et al.,* 1970; Clemens *et al.,*

1978). These features reduce the effect of root anaerobiosis through physical and metabolic changes and, hence, they increase the supply of oxygen to the roots and prevent toxicity from the end products of anaerobic respiration (Hook and Scholtens, 1978). It is interesting to note that riparian and wetland communities are dominated by deciduous hardwoods, reflecting their higher flooding tolerance over conifers (Teskey and Hinckley, 1977). Exceptions include *Taxodium distichum* and bog species such as *Larix laricina, Thuja occidentalis, Picea mariana,* and *Pinus contorta.*

   Soil temperature affects the rate of water absorption through modification of the rate of root growth (Wilcox and Ganmore-Newman, 1975), membrane permeability, and water viscosity (Kuiper and Kuiper, 1978), but it has a minimal impact on nutrient uptake (Hermann, 1977). More research has been done on root permeability of conifers and agricultural crop species than evergreen or deciduous hardwoods, probably because of the relatively low frequency of low soil temperatures of most hardwood sites during the growing season (Teskey *et al.,* 1978). Obvious exceptions include high and middle elevation species (e.g., *Alnus sinuata* and *Fagus sylvatica*) that frequently begin to expand their shoots while there is still a snow pack, higher latitude species (e.g., *Populus nigra*), and evergreen hardwoods that grow in higher latitudes (e.g., *Arbutus menziesii*). Transpiration and net photosynthesis were reduced by only 4 and 1%, respectively, in *Salix* spp. from an elevation of 1830 m in Montana when soil temperature was lowered from 20 to 3°C (Anderson and McNaughton, 1973).

   Generalizations about causal relationships between availability of specific nutrient elements and root growth or permeability appear difficult because of the contradictory nature of the available data (Hermann, 1977). Nutrient disorders (i.e., imbalances, deficiences, and toxicities) can greatly affect root growth and membrane permeability (Proctor and Woodell, 1971). Under soil water deficits, nutrient deficiencies increase as nutrient transport decreases (Shaner and Boyer, 1976). High concentrations of aluminum and low concentrations of calcium and magnesium appeared to eliminate the mycorrhizal associations of *Quercus marilandica* roots (Reich and Hinckley, 1980). Mycorrhizae not only improve nutrient uptake (Hermann, 1977), but also increase absorption of water by increasing the area of contact between the root system and the soil (Kramer and Kozlowski, 1979).

C. Tissue Capacitance

   Most discussions of water transport through the soil–plant–air continuum have considered the soil as the only source of water even for meeting short-term or peak demands. Recently, the importance of plant tissues as an important source of water in some plants has been emphasized (Kozlowski, 1972, 1979; Waring and Running, 1978; Lassoie, 1981). Though plant tissues were recognized as sources of water as early as 1729 by Langley (cf., Bartholomew, 1926), the extent of available water in various tissues has not been adequately studied, especially in forest trees (Jarvis, 1975; Hinckley et al., 1978b). Tissue capacitance is much greater in conifers than in hardwoods (Table I).

In this chapter, stem shrinkage and expansion will be mentioned as essential aspects of diurnal and seasonal changes of water status of trees. However, stem shrinkage, or changes in the water content of the extensible tissue (i.e., phloem, cambial, and newly derived xylem cells) of the stem, represents only a part of the total capacitance system of a tree (Table I). Capacitance was discussed by Chaney in Chapter 1 of this volume and will not be covered here in detail.

D. Transpiration

As discussed in Section II,B, the typical deciduous hardwood forest has a lower leaf area index and a shorter period of leaf retention than a coniferous forest. However, transpiration rates of hardwoods per unit of foliage or per tree (Ladefoged, 1963; Hinckley et al., 1978b) and soil moisture depletion in hardwood stands (see Fig. 3) may be as great or

TABLE I

POTENTIAL HOURS OF WATER USED IN TRANSPIRATION THAT COULD COME FROM VARIOUS PLANT TISSUES IN CONIFERS AND HERBACEOUS PLANTS[a,b]

| Storage zone | Hours of potential transpiration supply | | | Hardwood capacitance $(m^3 MPa^{-1} \times 10^{-5})$ |
	Conifers	Hardwoods	Herbaceous plants	
Roots	14.0	4.9	2.8	—
Stem-sapwood	50–180	12.2	1.5	0.2
Stem-extensible	1.0	6.7	1.3	19.9
Foliage	1.1	2.4	0.3	1.9

[a] After Jarvis, 1975.

[b] Values for hardwoods come from Federer (1979) and from Powell and Thorpe (1977). Federer assumed that average daily transpiration would equal 3.40 mm. Powell and Thorpe (1977) determined foliage capacitance from pressure volume curves (see Fig. 10) and capacitance of extensible tissues of the stem from changes in stem diameter. They checked these values by cutting a tree down and measuring the change in weight as ψ^{leaf} dropped from -0.3 to -3.0 MPa. Extensible tissues of the stem include the phloem, cambium, and newly derived xylem cells.

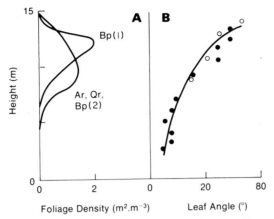

Fig. 8. Distribution (A) and orientation of foliage (B) in two adjacent stands: (1) pure *Betula papyrifera* (Bp) and (2) mixed *B. papyrifera, Acer rubrum* (Ar) and *Quercus rubra* (Qr). Open circles, Bp stand; solid circles, Ar, Bp, Qr stand. (Redrawn from Kinerson, 1979.)

greater than in conifers. Stomatal frequency, leaf morphology, leaf angle, stomatal action, and prevailing micrometeorological factors influence the rate of water loss from the leaf.

In a deciduous hardwood forest, most leaves are located near the point of canopy closure (Fig. 8A) (Thompson and Hinckley, 1977b; Kinerson, 1979). Variations in transpiration and hence in ψ^{leaf} throughout the tree crown probably reflect variations in leaf temperature (i.e., the vapor pressure difference between leaf and air) more than any other biotic or abiotic factor (Aubuchon *et al.*, 1978). In response to the greater radiation loads and higher potential foliage temperatures in the upper crown, sun leaves tend to be more dissected, have more stomata per unit leaf area, a greater mass-to-area ratio than foliage in the lower crown (shade leaves) as well as other xeromorphic characteristics (Huzulák, 1976; Aubuchon *et al.*, 1978; Kramer and Kozlowski, 1979). In addition, leaf angle decreases from the top to the bottom of a deciduous hardwood canopy (Kinerson, 1979; Fig. 8B). Similar but less dramatic changes in foliar morphology and needle angle are found in conifers (Hinckley *et al.*, 1978b). Other aspects of transpiration are discussed in Sections IV and V.

IV. STOMATAL APERTURE

The abiotic and biotic factors which regulate the opening and closing of stomata form a major focus in the study of plant and tree water relations. Only at the level of the guard cell can stomatal function be clearly defined; however, to study stomatal functioning at this level is difficult (Sheriff, 1979). Therefore, our discussion will be limited to indirect observation on stomatal function [i.e., leaf conductance (g^{leaf}), stomatal infiltration pressure].

The factors that have a major influence on g^{leaf} include (1) radiation, (2) humidity or vapor pressure gradient between the leaf and air, (3) leaf temperature, (4) external and internal CO_2 concentration, (5) leaf water balance, and (6) abscisic acid (Jarvis, 1976; Hinckley et al., 1978b). Except for alterations in the response pattern due to foliage age, stomata from all types of foliage respond very similarly to various factors. However, there appear to be differences in both the maximum and minimum g^{leaf} between broadly defined foliage types (Hinckley et al., 1978b; Körner et al., 1979).

A. RADIATION

Stomata in hardwood and conifer species open fully when exposed to 5 to 20% of full sunlight (100 to 400 μmoles m^{-2} sec^{-1}) (Fig. 9A) (Davies et al., 1974; Federer and Gee, 1976; Hinckley et al., 1978a,b; Pereira and Kozlowski, 1978; Kinerson, 1979). At light levels below saturation, Woods and Turner (1971) observed stomata in shade-tolerant species to open faster (e.g., 3 min. in Fagus grandifolia) and at lower light levels than in shade-intolerant species (e.g., 20 min. in Lirodendron tulipifera). However, Pereira and Kozlowski (1977) did not find consistent relationships between the speed of stomatal opening or closing and shade tolerance of forest trees.

Current theory suggests that two light systems are responsible for stomatal opening: (1) a blue light system associated with the tonoplast membrane and important in both membrane permeability and proton efflux from the guard cell; and (2) system reacting either to blue or red light which affects internal CO_2 concentration through photosynthesis (Sheriff, 1979). The net effect of both systems is to cause a change in the gradient in osmotic potential between the guard cell and surrounding subsidiary cells, which results in stomatal opening or closing. The change in osmotic potential is due to a concentration change in an organic acid or Cl$^-$ as the anion and K$^+$ as the balancing cation.

B. VAPOR PRESSURE GRADIENTS

As vapor pressure gradients increase, stomata close (Fig. 9B), but closure is sometimes in response to changes in bulk ψ^{leaf} (Davies and Kozlowski, 1974) and sometimes not (Lange et al., 1971). Until the work

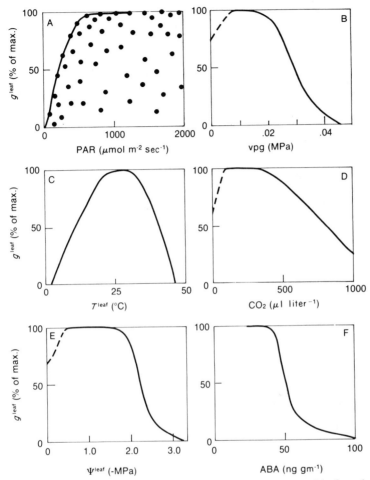

Fig. 9. Relationship between various abiotic and biotic factors and leaf conductance (g^{leaf}). (A) PAR—photosynthetically active radiation, (B) vpg—vapor pressure gradient; (C) T^{leaf}—leaf temperature; (D) CO_2—carbon dioxide concentration; (E) ψ^{leaf}–leaf water potential; (F)ABA—abscisic acid. Lines are drawn to indicate that a boundary line analysis was used (see Jarvis, 1976).

of Edwards and Meidner (1978) and Lösch and Schenk (1978), this relationship between humidity and stomatal action was difficult to ascertain. According to Edwards and Meidner (1978), stomata of plants such as *Prunus, Citrus,* and *Picea,* which show pronounced responses to changing humidity, have two distinctive characteristics: (1) the lack of a cuticle in the lower part of the guard cell; and (2) the water supply to the epidermis is relatively slow. Because of these two factors, the water relations of the stomatal apparatus are relatively isolated from that of the bulk leaf. Humidity (or bulk ψ^{leaf} and bulk leaf turgor pressure) influences stomatal activity by affecting the relationship between the pressure potential (turgor pressure) of the guard cell and the surrounding subsidiary cells. Any other relationship between stomatal action and water status is merely correlative.

C. Leaf Temperature

The response of g^{leaf} to leaf temperature shows that optimum values range from 22 to 34°C for deciduous hardwoods (Fig. 9C) (Phelps et al., 1976; Aubuchon et al., 1978; Kinerson, 1979; Reich and Hinckley, 1980). These optimum temperatures are higher than those cited for conifers (Hinckley et al., 1978b). Low or high temperatures may result in a delayed response of stomata to other environmental factors and directly inhibit the photosynthetic process (Larcher, 1975; Bauer, 1979). Separating direct and indirect effects of temperature on the stomatal mechanism has proved difficult (Wuenscher and Kozlowski, 1971).

D. CO_2 Concentration

Many investigators indicate that ambient or external CO_2 effects on stomatal action in a field situation are overridden by other factors or are not significant because ambient CO_2 concentration varies by 50 μl liter^{-1} around a mean of 330 μl liter^{-1} (Hinckley et al., 1978b; Sheriff, 1979) (Fig. 9D). In contrast, changes in internal CO_2 concentrations, as influenced by ambient concentrations as well as by relative rates of photosynthesis and respiration, affect stomatal activity (Raschke, 1975). High internal CO_2 concentrations induce stomatal closure and low concentrations cause opening.

E. Leaf Water Balance

As ψ^{leaf} decreases, stomata close due to water loss from guard cells and a decrease in guard cell potassium ion (K^+) concentration, which probably is associated with changes in abscisic acid (Ehret and Boyer, 1979). Three responses of stomata to water deficits are frequently cited: (1) abrupt closure (Fig. 9E); (2) gradual closure at a critical ψ^{leaf}; and (3) closure associated with predawn ψ^{leaf} (Dougherty et al., 1976; Federer and Gee, 1976; Phelps et al., 1976; Federer, 1977; Pereira and Kozlowski, 1977, 1978; Aubuchon et al., 1978; Eliáš, 1979a). The first response occurs when a critical turgor potential is reached. Since most investigators measure only ψ^{leaf}, a critical ψ^{leaf} instead of turgor pressure is most commonly recorded (Fig. 9E). In field studies, several critical ψ^{leaf} values for stomatal closure have been reported (Aubuchon et al., 1978), but usually only one turgor pressure value is reported. The second response, which commonly occurs under constant humidity conditions, represents a slight modification of the first response. The third response, stomatal control through predawn ψ^{leaf}, may result more from ψ^{leaf} affecting abscisic acid (ABA) levels, which in turn influence stomatal activity, than from ψ^{leaf} directly affecting stomata. This will be discussed in Section V,F.

Since turgor pressure provides a more direct index of stomatal activity than ψ^{leaf}, a pressure–volume curve (Fig. 10) is useful for a discussion of the relationship between leaf water status and stomatal action. A pressure–volume curve allows for (1) determination of osmotic potential and pressure potential at any relative water content or ψ^{leaf}; and (2) calculation of the elasticity of the cell wall and solute concentration changes (Tyree et al., 1978; Richter et al., 1981). These various parameters of a pressure–volume curve vary by species, drought preconditioning, and stage of leaf development. For example, Tyree et al. (1978) found that the turgor loss point decreased from −1.0 MPa to −2.3 MPa as *Acer saccharum* leaves expanded. When leaves were fully expanded, the turgor loss point could change by as much as 0.45 MPa over a 2-week period. The

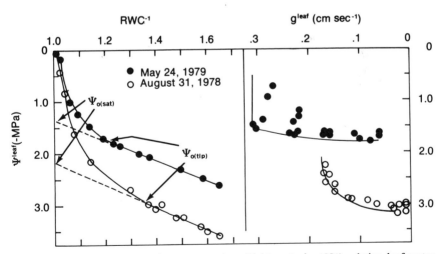

Fig. 10. Two pressure–volume curves (see Richter et al., 1981) relating leaf water potential (ψ^{leaf}) and the reciprocal of relative leaf water content (RWC^{-1}) for *Cornus sanguinea* leaves. $\psi_{o(sat)}$—saturated osmotic potential; $\psi_{o(tlp)}$—turgor loss point; ψ_o—osmotic potential. Leaf conductance (g^{leaf}) collected under field conditions is also shown and is enclosed by a boundary line.

turgor loss point is, at least for many species, very close to the critical turgor pressure for stomatal closure (Richter *et al.*, 1981), hence the critical ψ^{leaf} for stomatal closure in *Acer saccharum* might vary as much as 0.45 MPa over short periods. Aubuchon *et al.* (1978) observed a shift of -0.72 MPa in the critical ψ^{leaf} for stomatal closure in *Quercus alba*, whereas Hinckley *et al.* (1980) did not find a shift in *Crataegus monogyna*. These shifts may be due to active osmotic adjustment or changes in elasticity of the cell wall. The species as well as the stage of leaf development are the most important factors causing changes in the pressure–volume curve.

F. Abscisic Acid (ABA)

Davies and Lakso (1978) showed that as ABA levels in the foliage increased above a threshold, stomata of *Malus domestica* closed (Fig. 9F). Similar data were obtained for *Pseudotsuga menziesii* (Blake and Ferrell, 1977). Davies and Lakso (1978) noted that the increase in ABA was related linearly to changes in turgor pressure. Whether the increase in ABA causes stomatal closure directly or merely reinforces the effect of decreasing turgor pressures, the "first line" of defense (Mansfield *et al.*, 1978), is not clear. However, ABA released during the period of drought stress remains active in the guard cells for several days; therefore, full stomatal opening is delayed. Mansfield *et al.* (1978) term this response a "second line" of defense because a ceiling is imposed on the extent to which stomata can open.

Abscisic acid appears to be present in an inactive form in the chloroplasts or mesophyll cells of well-watered plants (Mansfield *et al.*, 1978). When ψ in the photosynthetic tissue decreases, ABA is released, moves to the guard cells, and induces a closing reaction by interfering with transport of potassium and hydrogen ions. The loss of stored ABA from the chloroplast induces synthesis of more ABA which stops only when ψ recovers. The rate of decrease in ABA varies with species and the extent to which the plant is stressed (Giles *et al.*, 1976). Stomatal closure that is correlated with predawn ψ^{leaf} (response 3) may indeed be an ABA-mediated response (e.g., *Juglans nigra*, Dougherty *et al.*, 1976; *Pseudotsuga menziesii*, Running, 1976).

Abscisic acid has also been used as an effective antitranspirant on deciduous hardwoods (Davies and Kozlowski, 1975; Goode *et al.*, 1978). The effectiveness of ABA depends on (1) how completely it penetrates the leaf; (2) how rapidly the amount applied to a leaf decreases below that needed for stomatal closure; and (3) the water use efficiency (ratio of total water vapor conductance to total carbon dioxide conductance) of the species involved.

G. OTHER FACTORS

At low wind velocities, boundary layer conductance limits diffusion of water vapor and CO_2. High wind velocities may cause increased evaporative losses and stomatal closure (Davies *et al.*, 1974) or stomatal dysfunction because of rapid and perhaps violent movement of the leaf (Pallardy and Kozlowski, 1979). Wind also causes particles and other leaves to abrade the surface of the leaf, resulting in tearing of the cuticular layer and erosion of leaf waxes (Grace, 1976). When wind speeds are low to moderate, leaf movement in *Populus* spp. reduces the boundary layer, promoting gas exchange by bulk air flow in response to pressure differences across the leaf and reducing leaf temperature (Shive and Brown, 1978; Day and Parkinson, 1979).

Mineral deficiencies, particularly of potassium and calcium, cause either complete or partial stomatal closure in response to environmental stresses. Responses to air pollutants are not uniform. Low dosages of ozone and sulfur dioxide, when absorbed by a leaf, appear to induce further stomatal opening, hence further foliar damage (Mudd and Kozlowski, 1975). Higher SO_2 dosages may induce stomatal closure (Noland and Kozlowski, 1979). Hydrogen fluoride induces closure (Halbwachs, 1970).

V. DIURNAL AND SEASONAL PATTERNS OF WATER STATUS

When soil moisture supplies are high, minimum ψ^{leaf} rarely is less than -2.5 ± 0.3 MPa in trees of temperate regions (Jarvis, 1975; Richter, 1976). The tree maintains a ψ^{leaf} higher than this value through either rapid water absorption, release of tissue water, or stomatal closure. This minimum ψ^{leaf} appears to be associated with the turgor loss point (Fig. 10). Turgor loss points change during leaf development from about -0.7 to about -2.5 MPa and minimum ψ^{leaf} reflects this range. Therefore, under conditions of adequate soil moisture, diurnal patterns of ψ^{leaf} and g^{leaf} are largely governed by the turgor loss point as influenced by leaf age and species. Trees from more xeric environments have lower turgor loss points; hence lower minimum ψ^{leaf} (Hinckley *et al.*, 1980). As soil moisture or tissue water capacitance decreases, or as the resistance to water flow increases (i.e., through cavitation of water conducting cells, root suberization, depletion of large-diameter soil capillaries), minimum ψ^{leaf} is achieved earlier during the day and tends to reach a plateau (Fig. 11B,C) near the turgor loss point. In trees of temperate regions, minimum ψ^{leaf} rarely decreases below the turgor loss point except when predawn ψ^{leaf} (i.e., effective soil in the root zone) is below this point. In trees from

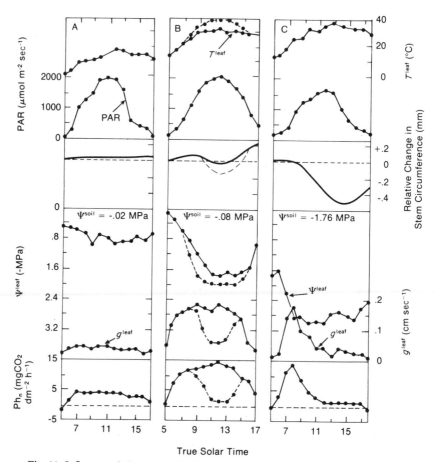

Fig. 11. Influence of photosynthetically active radiation (PAR), leaf temperature (T^{leaf}), soil water potential (ψ^{soil}), and day of the year (A: April 26; B: June 2; C: September 7) on relative change in stem circumference, leaf water potential (ψ^{leaf}), leaf conductance (g^{leaf}), and net photosynthesis (Ph_n) in *Quercus alba*. The dashed line (B) illustrates the impact of increasing T^{leaf}. (Redrawn from Hinckley *et al.*, 1979.)

warmer and drier environments (e.g., mediterranean regions), minimum ψ^{leaf} can decrease below the turgor loss point, especially when critical leaf temperatures are approached. Species that show such a response must have a relatively greater tolerance to low ψ^{leaf}.

When ψ^{leaf} decreases below the turgor loss point, small changes in ψ^{leaf} result in relatively large changes in relative leaf water content (RWC) (Fig. 10). Although ψ^{leaf} may not decrease much, RWC may decrease considerably and both may approach lethal levels.

Fig. 11B and C illustrate diurnal patterns of water status and other physiological processes in mature leaves of *Quercus alba*. Similar diurnal patterns have been observed in *Quercus pubescens* (Aussenac *et al.*, 1978), *Acer campestre, Carpinus betulus, Cornus mas, Q. cerris, Q. petraea* (Eliáš, 1979a,b,c), *Acer saccharum, Betula papyrifera* (Pereira and Kozlowski, 1978), and *Q. rubra* (Lassoie and Chambers, 1976). From these patterns, it appears that stomatal activity and leaf water status in mature leaves are influenced chiefly by the microclimate around the leaf (i.e., radiation level and atmospheric evaporative demand) and availability of soil moisture.

Maximum g^{leaf} and net photosynthesis probably change very little during the period between complete leaf maturation and initial leaf senescence in deciduous hardwoods (Figs. 7, 11) (Gee and Federer, 1972; Turner and Heichel, 1977; Hinckley *et al.*, 1978b, 1979; Roberts *et al.*, 1979). However, maximum g^{leaf}, minimum g^{leaf} (cuticular conductance), maximum net photosynthesis, and critical levels of leaf water status (e.g., $\psi^{leaf}_{0(sat)}$ and $\psi^{leaf}_{0(tlp)}$) do vary during the period of leaf development and senescence (Turner and Heichel, 1977; Tyree *et al.*, 1978; Dougherty *et al.*, 1979; Hinckley *et al.*, 1980). For example, stomata in immature leaves of *Quercus alba* did not respond, or responded only slightly, to increased light intensity and g^{leaf} remained at a rather constant value for the entire day (Fig. 11A). Predawn ψ^{leaf} was relatively negative (-0.6MPa) at this stage of development in spite of adequate soil moisture. This value was a result of higher cuticular conductances, incomplete development of the stomatal mechanism, and only partial development of vascular elements in ring porous species (Dougherty *et al.*, 1979). Similar changes in response of stomatal activity as needles developed were reported for *Pseudotsuga menziesii* (Lassoie, 1979).

Changes in stem circumference on a diurnal or seasonal basis provide an estimate of the use of tissue water reserves and the rate of cambial growth (Hinckley *et al.*, 1976, 1978b; Waring and Running, 1978; Lassoie, 1979, 1981). When fully expanded leaves were present and soil moisture reserves were high, diurnal changes in stem dimensions were small to moderate. These small changes reflect a relatively low resistance transport pathway between the soil and the leaf (Fig. 11B). As the soil dried, the resistances between the soil and leaf increased, and an increased volume of water moved radially from reservoirs such as the phloem to the vascular conducts. Therefore, the amount of stem shrinkage increased (Fig. 11C). Stem shrinkage decreased as the supply of tissue water dropped (Hinckley and Bruckerhoff, 1975).

Diurnal and seasonal changes in transpiration can be most easily described by the diurnal and seasonal patterns of g^{leaf} and vapor density gradients (Hinckley and Bruckerhoff, 1975; Landsberg *et al.*, 1975). Diurnal and seasonal patterns of net photosynthesis reflect changes in microclimate and g^{leaf} and are discussed in Section VI,B.

VI. WATER DEFICITS AND PHYSIOLOGICAL PROCESSES

The physiological response of trees to water deficits vary from very sensitive (cell elongation) to moderately sensitive (photosynthesis and respiration) to relatively insensitive (irreversible membrane damage) (Hsiao, 1973). These responses represent not only the short-term interactions between the environment and the tree's organs, but also the long-term interactions between the organs which result in the allocation of water, nutrients, and carbon. Such factors as shoot-to-root ratios, the complexity of the tree's transport system, and tree's size influence the response of a particular organ to an environmental factor. Many of the interactions are governed or triggered by specific hormones and the balances among hormones. Unfortunately, most of the experimental evidence on hormonal control of physiological processes in entire plants, particularly trees, is so fragmentary and indirect that extreme caution should be used in extrapolating results in an effort to interpret these interactions (King, 1976).

A. GROWTH

Growth of tissues involves division, enlargement, and differentiation of cells. Cell enlargement, a turgor-related process, is particularly sensitive to plant water deficits (Hsiao, 1973). Other aspects of the overall growth process are directly affected by water deficits and also indirectly affected through hormone balances and availability of photosynthetic products.

Large trees that do not lose leaves during periods of mild or moderate water deficits [e.g., species that exhibit "fixed growth" of shoots (Kramer and Kozlowski, 1979) such as *Acer saccharum* and species of *Carya, Fagus, Quercus*] produce their total complement of leaves in one short flush. This large mass of foliage does not change significantly from the time of full leaf expansion to leaf senescence (Dougherty, 1977). Under special circumstances, when appreciable defoliation occurs as a result of late frost after buds open or insects attack, all trees can produce another crop of leaves from opening of previously dormant buds (Borchert, 1978). Seedlings, saplings, and stump sprouts of species exhibiting fixed growth of shoots can flush several times in a year as long as light intensity and soil moisture supplies are high (Hoffman, 1972; Borchert, 1975).

Root growth of hardwoods and conifers of the temperate zone occurs from early spring to late fall (Lyr et al., 1968; Stone and Jenkinson, 1970; Riedacker, 1976; Hinckley et al., 1979; Fig. 7E, 12). Environmental factors prescribe limits for the amount of root growth but their effects are mediated through endogenous factors (Fig. 7D,E). The rate of root elongation is regulated by soil temperature and ψ^{soil}, whereas root initiation also depends strongly on apically produced hormones (Richardson, 1953; Webb, 1976). Root and shoot growth are interrelated since stress, which limits shoot activity, can induce a decrease or cessation of root growth (Hodgkinson and Becking, 1977; Eliasson, 1978). Similarily, hormone production and water and nutrient uptake by the root system influence physiological processes in the shoot. Periods of root growth may occur throughout the year in adult trees and seedlings, depending upon environmental conditions and available food reserves (Ovington and Murray, 1968; Cohen et al., 1978; Wargo, 1979). The amounts of stored food, inferred from root growth were small in Populus tremula and Acer saccharum seedlings. The reserves were depleted after 3 days when photosynthesis was reduced by low light levels or by drought (Eliasson, 1968; Parker, 1970; Webb, 1976). However, in stem-girdled canopy trees of Liriodendron tulipifera, root growth was reduced but continued for two years (Edwards and Ross-Todd, 1979).

   Growth processes in a mature deciduous hardwood species (Quercus alba) are illustrated in Fig. 7. Root growth was initiated in late February or early March as soil temperature increased above 3.5°C. At that time, starch reserves in ray parenchyma cells of roots began to decrease (Wargo, 1979). Cambial growth of the stem began some 35 days later, and, after an additional 14 to 20 days, buds opened and a major period of growth was initiated. During this relatively short but intense period of growth of leaves, branches, and flowers, the growth of the root and stem was inhibited (Figs. 7E and 11A). When leaves had attained approximately 90% of their maximum size and had begun to export carbohydrates, growth of the root and stem increased again. Two to four weeks later, storage of starch resumed in ray parenchyma cells of roots (Wargo, 1979). When predawn ψ^{leaf} decreased below approximately -0.25 MPa, cambial growth in the branches, at midcrown, and at the base of the tree was no longer detectable. Following cessation of cambial growth, progressive stem shrinkage was observed. Similar results were reported for Quercus alba by Fritts (1958), for Populus tremuloides and Quercus ellipsoidalis by Kozlowski and Winget (1964), and for Acer saccharum and Betula papyrifera by Pereira and Kozlowski (1978). However, cell division continued in Quercus alba until predawn ψ^{leaf} decreased below -0.6 MPa since stem circumference was greater immediately following soil rehydration than before (Hinckley and Bruckerhoff, 1975).

Even when predawn ψ^{leaf} dropped to -2.2 MPa, growth of acorns continued, indicating their high sink strength (McLaughlin et al., 1979) (Fig. 7E). In contrast to 1976 (a dry year), growth of acorns during years of high rainfall was rapid, few acorns abscised prematurely and their final size was large (Hinckley et al., 1976).

During mid-July, root cambial growth was initiated in Acer saccharum (Wargo, 1979). Cambial growth of the root occurs when starch reserves were completely refilled and soil moisture was adequate. Since the entire sapwood of the root is used to transport water and nutrients and only the outer one or two xylem rings in ring porous and several annual xylem rings in diffuse porous species are functional in the stem, cambial growth of roots is of low priority and is greatly reduced during years of drought or defoliation (Wargo, 1979). Under most conditions elongation of roots, important in the absorption of water, is reduced less than cambial growth of roots (Wargo, 1979). In Quercus alba (Fig. 7E), root growth continued until ψ^{soil} reached -1.3 MPa or predawn ψ^{leaf} was between -1.5 and -2.0 MPa (Teskey et al., 1978). Some 10 days after autumn rain (October, 1976), root growth resumed and continued until soil temperatures decreased to 2.5°C.

Patterns of growth and phenology similar to those for Quercus alba have been reported for other deciduous hardwood trees (Lyr and Hoffmann, 1967; Longman and Coutts, 1974). However, patterns for seedlings (Fig. 7B,D) and saplings (Fig. 12) (Hoffmann, 1972; Riedacker, 1976) differ from those for large trees. Growth patterns of Quercus alba seedlings (Fig. 7B,D) were similar to growth patterns of Quercus robur saplings (Fig. 12), but both differed from the growth pattern of dominant Quercus alba and Q. robur trees (Fig. 7C,E). In addition, root growth of trees exhibiting fixed growth of shoots was different from root growth of trees with free growth (Fig. 12).

B. PHOTOSYNTHESIS AND RESPIRATION

As previously discussed (see Figs. 7 and 11), water deficits can decrease photosynthesis either (1) indirectly by altering total conductance to CO_2 ($g_{CO_2}^{total} = g_{CO_2}^{leaf} + g_{CO_2}^{residual}$); or (2) indirectly by affecting photosynthetic and respiratory mechanisms. Water deficits cause a decrease in photosynthesis through stomatal closure ($g_{CO_2}^{stomata}$), reduction in cellular water content ($g_{CO_2}^{residual}$), increase solute content ($g_{CO_2}^{residual}$), change in membrane structure and permeability (both $g_{CO_2}^{residual}$ and the photosynthetic mechanism), reduction in enzyme levels through a reduction in protein synthesis, increase in degradative enzymes, and reduction in chlorophyll content (Hsiao, 1973; Boyer, 1976). Maximum $g_{CO_2}^{leaf}$ does not

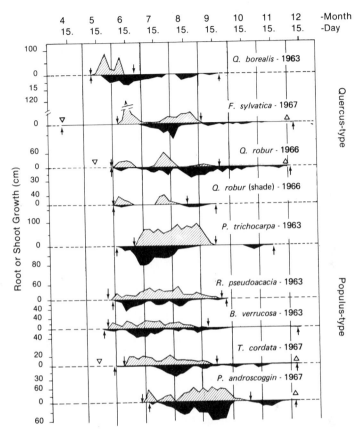

Fig. 12. Seasonal patterns of shoot growth (above the line) and root growth (below the line) for hardwood saplings (6 to 8 years old). The distance between arrows represents complete period of shoot or root growth. Observations were made over a 4-year period, during which periodic droughts occurred. (Redrawn from Hoffmann, 1972.)

vary greatly among deciduous hardwoods (Wuenscher and Kozlowski, 1970; Phelps *et al.*, 1976; Körner *et al.*, 1979), but maximum $g_{CO_2}^{residual}$ does. Though changes in $g_{CO_2}^{leaf}$ are regarded as having a dominant impact on photosynthesis, changes in $g_{CO_2}^{residual}$ or in the photosynthetic mechanism may be as important, particularly during recovery following drought (Boyer, 1976; Bunce, 1977).

As ψ^{leaf} reaches a critical value, photosynthesis decreases (Fig. 13). The value at which this decrease occurs is highly species dependent and usually reflects the typical ψ^{leaf} values found in that environment (Bunce *et al.*, 1979; Chabot and Bunce, 1979). Figure 7F illustrates that potential daily average net photosynthesis did not decrease appreciably in mature

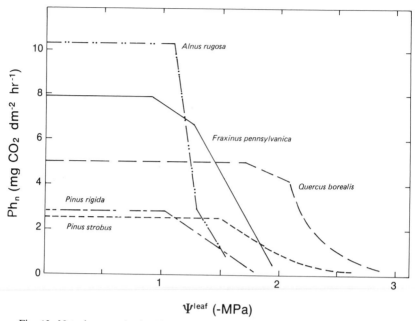

Ψ^{leaf} (-MPa)

Fig. 13. Net photosynthesis (Ph_n) as a function of leaf water potential (ψ^{leaf}) (redrawn from Chabot and Bunce, 1979).

Quercus alba leaves in spite of a prolonged and intense drought. In addition, even when ψ^{soil} decreased to less than -2.0 MPa, average daily net photosynthesis remained quite close to potential net photosynthesis. Obviously the capacity of stomata to open fully even when ψ^{leaf} approached -2.6 MPa was critical in this tree (Fig. 11B,C).

Dark respiration of leaves and root respiration decreased in several deciduous hardwood species with decreasing ψ^{leaf} (Jarvis and Jarvis, 1965; Bunce and Miller, 1976; Brandle *et al.*, 1977; Dougherty, 1977). In addition, photorespiration often decreased with increasing water deficits (e.g., *Alnus rugosa*, *A. oblongifolia*, *Fraxinus pennsylvanica*, *F. pennsylvanica* var. *velutina*; Bunce and Miller, 1976). By comparison, Bunce and Miller (1976) found that photorespiration increased with water deficits in xerophytic species (e.g., *Larrea divaricata*, *Acacia greggii*). They suggested that high rates of photorespiration under drought conditions would remove NADPH, which might either prevent direct injury or stimulate adenosine triphosphatase (ATP) synthesis. When critical (and perhaps irreversible) ψ^{leaf} values are reached, both light and dark respiration increase dramatically (Levitt, 1972).

Edwards and McLaughlin (1978) found that stem respiration in 15- to

20-m tall *Liriodendron tulipifera* trees decreased during the daylight hours, even though temperature conditions were more favorable for respiration than during the night. Changes in stem water potential and concentration of reducing sugars (substrate level) in the phloem were offered as possible reasons for these observations. It might be expected that growth would reflect changes in stem respiration. Higher rates of cambial growth of the stem occurred during the night than during the day when leaves were present (Fritts, 1958; Hinckley *et al.*, 1976). As illustrated by Fig. 11B, greater turgor pressures were encountered at night than during the day. In contrast, when *Quercus alba* leaves were not present in early spring, cambial growth was not limited by water deficits and occurred in short surges during the warmer, daylight hours (Hinckley *et al.*, 1976). These surges may have been related to mobilization of reserve foods. In *Quercus alba*, the rate of cambial growth in the spring was constantly greater when no leaves were present (Hinckley *et al.*, 1976, 1979).

C. Protein Synthesis

Brandle *et al.* (1977) studied the effect of drought on protein synthesis in leaves of 2-month-old *Robinia pseudoacacia* seedlings. As predawn ψ^{leaf} decreased from -0.48 to -1.72 MPa and midday ψ^{leaf} dropped from -0.82 to -2.54 MPa, there was a 92% decrease in g^{leaf}, 81% decrease in Ph_n, 302% increase in RNase, 21% decrease in the percentage of the ribosomes that were in the polysome form, and a 48% decrease in protein synthesis. The results suggested that protein synthesis was not as drought-sensitive as originally suspected (Hsiao, 1973).

D. Translocation of Photosynthates

Most studies indicate that carbohydrate translocation in hardwoods and conifers is affected little by water deficits, whereas vein loading, photosynthesis (source strength), and growth (sink strength) are all markedly affected. Flow of photosynthetic products is generally restricted to sieve tubes produced in the current growing season and is less than 100 cm hr^{-1} (e.g., *Fraxinus americana*, Zimmermann, 1969). Sieve tubes are very long cells, laterally semipermeable, and permeable longitudinally through the sieve plates. Flow theoretically occurs along a gradient of decreasing osmotic and turgor pressure (source to sink). Most investigators have found osmotic gradients which range between 0.02 and 0.04 MPa m^{-1}, for example in *Quercus rubra* (Huber *et al.*, 1937; Hammel, 1968), *Fraxinus americana* (Zimmermann, 1957), and *Acer rubrum* (Kaufmann and Kramer, 1967). Although ψ^{phloem} follows the daily and seasonal pattern of ψ^{xylem} or ψ^{leaf}, the change of ψ^{phloem} remains small (Kaufmann and Kramer, 1967). Even when leaves are wilting or senescing, turgor pressures greater than zero occur in sieve tubes (Hammel, 1968; Zimmermann, 1971). In fact, turgor pressures remain positive in sieve tubes until the first freeze of autumn (Hammel, 1968).

The seasonal pattern of carbohydrate translocation in *Quercus alba* (Fig. 7G) can be compared to the course of net photosynthesis and growth in the same tree (Fig. 7F and E). In *Quercus alba,* leaves did not begin to export photosynthate until they had achieved 90 to 100% of full expansion (McLaughlin et al., 1979). During the period when leaves were importing food and using currently produced photosynthate, root and stem growth were reduced dramatically (Dougherty et al., 1979). During this period, the amount of starch storage in root ray parenchyma and living fibers decreased rapidly (Wargo, 1979). Approximately a month later, after leaves were fully expanded, starch began to accumulate again in roots (Wargo, 1979). Conifers and evergreen hardwoods appear to have similar patterns of carbohydrate translocation; but older foliage plays an important role as a continuous source of photosynthate (Webb, 1977).

E. MEMBRANE FUNCTION

Studies of freezing injury in hardwoods (Siminovitch et al., 1975; Yoshida, 1979) and drought injury and stress in herbaceous plants (Fellows and Boyer, 1978) suggest that the structure or state of plant membranes is affected by stress and also plays an important role in stress adaptation and recovery. The activity of enzymes and hormones (e.g., ABA) is drastically altered by cell dehydration (Keck and Boyer, 1974; Mansfield et al., 1978), possibly as a result of alterations in membrane structure (e.g., phospholipid content) and permeability (Simon, 1978). In spite of the role of membranes in metabolism and regulation of exchange of nutrients and water, very little work has been done on the effect of drought on membranes in deciduous hardwood species.

F. HORMONES

Almost all of the research on growth regulating hormones and their relationships to changes in plant water status caused by drought or flooding has been done with crop species. However, some observations may be applied to trees.

Too much or too little water appears to cause disruption in hormone balance in trees. It is well known that flooding of soils leads to reduction of root growth, stem elongation, epinastic movement of leaves and petioles, and wilting and chlorosis of leaves (Burrows and Carr, 1969; Hook and Scholtens, 1978). Since these responses occur rapidly, they are not the result of nutrient deficiency and cannot be completely explained by shoot water deficits. Hence, they have been attributed to hormone imbalance between the root and shoot (Kramer, 1951; Kramer and Jackson, 1954). Reduced petiole elongation has been attributed to supraoptimal amounts of auxin in the stem (Phillips, 1964). Decreased gibberellic acid appears to be responsible for reduced stem elongation (Burrows and Carr, 1969; Reid et al., 1969). Increases in ABA in the stem may be responsible for early leaf senescence and stomatal closure. The accumulation of auxin at the base of the stem due to decreased stem-root transport appears to initiate adventitious roots at that point (Phillips, 1964). However, most of the effects of flooding are probably due to an imbalance in several hormones rather than to any individual hormone. High amounts of auxin and ethylene cause epinastic leaf movements (Jackson and Campbell, 1975). Wilting and chlorosis of lower leaves, which can appear very rapidly after flooding, appear to be responses to an imbalance in ethylene and cytokinins (Burrows and Carr, 1969; Jackson and Campbell, 1975). Abscisic acid (ABA) levels increased dramatically in *Zea mays* as drought intensified (Giles et al., 1976) and high ABA levels induce stomatal closure (Mansfield et al., 1978). Ethylene production by plants increases during drought, and the increase in ethylene and ABA probably reflects an adaptive mechanism for mesophytes under water deficits. Ethylene increases both leaf and fruit abscission, and ethylene in combination with ABA enhances leaf senescence and subsequent abscission (El-Beltagy and Hall, 1974).

VII. DROUGHT TOLERANCE

Every year, major droughts occur somewhere and usually have large scale social and economic impacts (Rosenberg, 1978). A drought typically begins when the available soil water supplies or atmospheric evaporative demand cause prolonged periods of stomatal closure or reductions in growth. Most hardwood trees are periodically subject to drought, which may occur regularly or infrequently. Highly predictable or regular droughts probably are important in restricting the range of a species and reducing the likelihood of habitat invasion by less drought-tolerant species. On the other hand, infrequent droughts are important in allowing drought tolerant species to compete successfully in more mesic environments and in restricting the ranges of mesophytic species.

During extreme or record droughts, observations on tree survival in nurseries, forests, and plantations often provide clues to adaptive mechanisms of trees for resisting droughts. The drought of 1925 in the southern Appalachians (Hursh and Haasis, 1931), the droughts of the 1930's (Albertson and Weaver, 1945) and the mid-1950's (Rosenberg, 1978) of the United States, the 1911 (Hübner, 1912) and the 1947 (Schmithüsen, 1948) droughts in Germany and the 1976 drought in Great Britain (Coultherd, 1978) are examples of such records.

Drought tolerance reflects drought avoidance and desiccation tolerance (Kramer and Kozlowski, 1979). Drought avoidance involves morphological features and physiological functions that minimize the stress situation or reduce the internal impact (strain) of the stress. Desiccation tolerance is the passive endurance of stress. Certain drought tolerance mechanisms represent both avoidance and tolerance. For example, osmotic adjustment maintains both positive turgor (avoidance) and higher water contents for a given water potential (tolerance). Drought recovery is an important component of drought tolerance and represents the ability to regain physiological functions after either reversible strain (e.g., leaf wilting) or irreversible strain (e.g., leaf death).

Drought damage may be caused by (1) low ψ^{tissue}; (2) low water content and shrinkage of tissues; (3) low osmotic potentials; (4) high leaf temperatures; and (5) the amount of water loss or the water potential decrease between the point of turgor loss and the point of injury or a combination of these factors. (Fig. 10 illustrates how some of these factors are interrelated.) Drought tolerance incorporates a combination of avoidance and tolerance mechanisms, which have differing significance depending on which factors cause drought damage.

Poikilohydric lichens, mosses, and liverworts are examples of plants that in the absence of a constant water supply rely on desiccation tolerance. These plants tolerate decreases in ψ^{tissue} to well below -140.0 MPa (Larcher, 1975). In comparison, a forest tree, such as *Quercus rubra*, is to some extent a drought-avoider but is still able to tolerate ψ^{tissue} to -5.5 MPa (Fig. 14) (Parker, 1968a,b; Seidel, 1972). At the other extreme, *Opuntia basilaris* can almost completely avoid droughts for years, but can tolerate ψ^{tissue} down to only -2.5 or -3.0 MPa (Szarek and Ting 1974). Although these three examples represent a wide range of drought tolerance, most plants, especially trees, show only small differences in desiccation tolerance, but wide differences in drought avoidance.

A. DESICCATION TOLERANCE

As in all higher plants, cells or tissues from most hardwood tree species are relatively intolerant to desiccation (Fig. 14). Readers are referred to Parker (1968a,b), Levitt (1972), Larcher (1975), Bewley (1979),

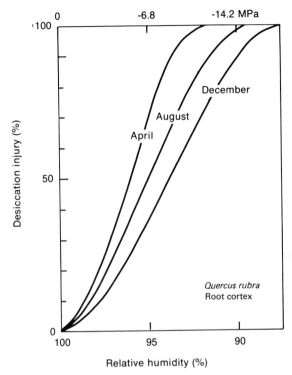

Fig. 14. Cellular injury as a result of desiccation of the cortical parenchyma of oak roots at different seasons. Desiccation injury was determined by exposing roots to different relative humidities. The relative humidity (or the corresponding absolute value of ψ^{cell}) at which the desiccation injury results in 50% or greater cell death is the measure of desiccation tolerance. (Redrawn from Parker, 1968b.)

and Kramer and Kozlowski (1979) for thorough discussions of desiccation tolerance.

B. DROUGHT AVOIDANCE

As mentioned, avoidance plays a major role in drought tolerance of forest trees. Drought avoidance may involve:

1. Maintenance of root water absorption during periods of low or decreasing soil moisture.
2. Reduction of the energy load on leaves by means of (a) change in leaf angle; (b) change in leaf reflectivity; (c) increase in leaf transmissivity; and (d) increase in convective cooling.
3. Reduction of foliar water loss by, (a) closure of stomata; (b) de-

layed stomatal opening following prolonged or severe droughts (see Section VI,D); (c) reduction of cuticular water loss; and (d) abscission of leaves and branches.

4. Maintenance of water status in transpiring organs by (a) internal water storage (i.e., capacitance); (b) osmotic adjustment; and (c) reduction in cell size.

1. Water Absorption

Adequate water absorption is possible in dry soils through the capacity of roots to decrease rhizosphere resistance by (1) growth; (2) maintenance of a large percentage of the root tip in an unsuberized state; and (3) development of a mycorrhizal association (Caldwell, 1976; Pereira and Kozlowski, 1976; Landsberg and Fowles, 1978; Teskey *et al.*, 1978). Water uptake is more closely related to soil water content than to root density (Luxmoore *et al.*, 1978); therefore, the capacity of some species to grow through zones of low ψ^{soil} into zones of higher ψ^{soil} gives them a competitive advantage on xeric sites (Fig. 6). Extensive root proliferation and increased rooting depth also result in higher predawn ψ^{leaf} (Fig. 6) and longer periods of transpiration of xeric sites, providing a further advantage to these species for growth and development (Pereira and Kozlowski, 1978; Teskey *et al.*, 1978). Root distribution appears to be a key factor in determining the extent and intensity of water deficits on a particular site in mediterranean environments (Griffin, 1973) and in temperate environments (Huzulák, 1976; Eliáš, 1978; Teskey *et al.*, 1978; Hinckley *et al.*, 1979). As a site becomes drier, a greater proportion of the biomass of a tree is in the root system (Yen *et al.*, 1978).

Mycorrhizal associations increase drought tolerance as both avoidance (more soil volume is exploited) and desiccation tolerance (roots are thicker and more resistant to desiccation damage) are enhanced (Marks and Kozlowski, 1973). Under conditions of extremely low ψ^{soil}, mycorrhizal associations either do not form or are greatly reduced (Kessler, 1966).

2. Reduction in Energy Load

Requirements for prolonged transpiration include a continuous input of energy, a vapor-permeable plant–air interface, and a vapor pressure gradient from the leaf to air. Because of the continuous input of energy during the daily light period, leaves sometimes are subjected to excessive heat loads. In order to avoid excessive leaf temperatures, radiative, convective, or transpirational energy exchange must be altered by decreasing input or increasing output of energy. For example, leaves in bright sunlight may be positioned in such a way as to reduce the radiation load during the midday period. Leaves at the top of a canopy tend to be more vertically oriented than those at the bottom (Kinerson, 1979) (Fig. 8B). Other examples include *Populus tremuloides*, which turns the leaf edge toward the sun, and the leaflets of the pinnate leaf of *Robinia pseudoacacia*, which rapidly fold down under conditions of high radiation.

Eller (1979) reported that both waxes and hairs on the upper side of a leaf may lead to appreciable reduction in absorbed radiation, mainly in the visible region of the spectrum. In *Kalanchoe pumila*, removal of the surface wax resulted in 40–70% increase in absorption of radiation. Reduction in photosynthetic capacity because of the presence of wax and hairs is compensated for by lower leaf temperatures, reduced transpiration, and higher ψ^{leaf}.

Chloroplast density seems to be reduced in sun leaves (Wallace and Dunn, 1980), thereby increasing leaf transmissivity. In addition, radiation affects the positioning of chloroplasts within cells of a number of species (Biebl, 1955). In the dark or in low-light intensities, they are positioned parallel to the leaf surface; at high light intensity they tend to accumulate along the anticlinal wall. This adaptation is particularly advantageous in sun grown leaves.

Convective heat exchange is enhanced by small, finely dissected leaves (Gates, 1976). Leaves of desert and mediterranean hardwoods tend to be more finely dissected and smaller than leaves of temperate hardwood trees. Temperate forest trees from a mesic habitat have a greater variation in leaf size than those from a xeric habitat (Zangerl, 1978). In addition, leaves at the top of a canopy tend to be smaller and more dissected than those at the bottom (Aubuchon *et al.*, 1978).

Leaves produced under drought conditions (including sun foliage) tend to be xeromorphic (Huzulák, 1976; Huzulák and Eliáš, 1976). They are small, strongly subdivided, have a high weight to area ratio, small cells, a thick cuticle, and high stomatal frequency. Such factors as crown position (both aspect and depth into the crown), leaf age, and even position on a branch affect these characters.

3. Reduction in Foliar Water Loss

Stomatal closure is probably the most important means of drought avoidance at the plant level. The severity of the plant water deficit is strongly correlated to the magnitude of transpiration and the timing and duration of stomatal closure. The postponement of stomatal closure by various pioneer *Quercus* species was suggested as a mechanism of reducing leaf temperature which can be appreciable on exposed xeric sites (Phelps *et al.*, 1976). Bunce *et al.* (1977) speculated that pioneer species maintain high rates of transpiration (stomata open) under drought conditions. The resulting reduction in both leaf temperature and soil moisture supplies exposes their competitors to more severe soil water deficits. Under these conditions, prolonging stomatal opening means that leaves will need to be more desiccation tolerant.

Stomatal regulation represents a trade-off between carbon exchange and water loss. Frequently, either the ratio of photosynthesis to transpiration or the ratio of $g_{H_2O}^{total}$ to $g_{CO_2}^{total}$ are used as a measure of plant capacity to reduce transpiration more rapidly than photosynthesis (Wuenscher and Kozlowski, 1970; Larcher, 1975). Species that trigger stomatal closure earlier than their competitors have an ecological advantage only if this strategy aids their survival, since their competitive advantage for photosynthate production is reduced by this strategy. In addition, if stomatal closure results in excessively high leaf temperatures, it may lead to mortality. It has been suggested that predawn ψ^{leaf} triggers stomatal closure in xerophytic trees in a cool environment (avoid low ψ^{soil}), ψ^{leaf} is important in warm environments (avoid critical ψ^{tissue} levels), and vapor density gradient triggers stomatal closure in mesophytic trees growing in a humid environment (Hinckley, 1971). However, this suggestion appears to be an oversimplification since it does not account for such factors as leaf shape, size, and degree of succulence (Larcher, 1975) and root distribution (Fig. 6).

Tree survival during prolonged droughts once stomata are closed depends on the magnitude of cuticular water loss, the plant's capacity to reduce the transpiring surface through complete or partial shedding of shoot tissues, and the amount of water loss possible between the turgor loss point and the point of first damage (Larcher, 1975). Cuticular conductance varies with the thickness of the cuticle and the arrangement, density, and number of the cutin and wax lamellae embedded in the outer wall of the epidermis. During prolonged droughts, the cutin and wax lamellae may further thicken as more waxes are deposited on the leaf and the leaf shrinks, thereby bringing the wax plates into closer proximity (Larcher, 1975). Since water loss can continue even with effective cuticular control, many trees shed leaves and branches (e.g., *Betula, Celtis, Fagus, Quercus, Salix, Ulmus*) in response to severe drought (Chaney, 1979). The capacity to resprout after severe droughts is also an important resistance mechanism.

Finally, water can be lost through the bark (peridermal transpiration), and trees without bark fissures and lenticels (e.g., *Fagus, Betula*) lose less water than trees having such openings (e.g., *Populus, Quercus, Malus, Acer*) (Larcher, 1975). This adaptation varies even within the range of a species. For example, *Ulmus minor* has strongly suberized twigs and branches in southeastern Europe, whereas it does not in Britain. Marchand and Chabot (1978) have observed appreciable decreases in stem water content and increases in water loss due to ice abrasion in *Betula papyrifera* located near the tree-line during the winter in New Hampshire. In addition, Grace (1976) observed that strong winds damage the structure of leaf cuticules.

4. Osmotic Adjustment

In addition to the use of internal reserves of water (capacitance), osmotic adjustment is also involved in the maintenance of water balance within the tree. Osmotic adjustments may be passive, apparently active, or active (i.e., metabolic) (Cutler *et al.*, 1977; Lakso, 1979; Hinckley *et al.*, 1980). Active osmotic adjustment is important in drought tolerance because it increases the water-absorbing power of the foliage, delays wilting and stomatal closure, and protects protoplasm from desiccation and coagulation (Levitt, 1972). The straight-line portion of a pressure–volume curve (Fig. 10) shows how passive osmotic adjustment occurs. As a cell or tissue dehydrates, the solute concentration must increase. The slope of this straight-line portion varies considerably due to the age of the leaf, the species and the severity of previous leaf water deficits. Passive changes in osmotic potential associated with decreases in relative water content from 100 to 80% can vary from 0.12 MPa [7% change from $\psi_{o(sat)}$] in *Cornus mas* to 1.95 MPa [248% change from $\psi_{o(sat)}$] in *Sorbus aria* from the same site (Hinckley *et al.*, 1980).

Active osmotic adjustments occur when the osmotic potential becomes more negative at a given relative water content. At a relative water content above the turgor loss point, active osmotic adjustment results in maintenance of turgor pressure sufficient for growth or stomatal opening. Large changes in osmotic potential are ubiquitous with leaf development (Tyree *et al.*, 1978) and in guard cells with photoactive stomatal opening (Raschke, 1975). In fully expanded leaves, active adjustment in osmotic potential is more common in herbaceous than in woody plants (Fereres *et al.*, 1978; Poole and Miller, 1978). However, Davies and Lakso (1979) observed 1.28 MPa of passive and 0.56 MPa of active osmotic adjustment in *Malus domestica* as relative water content decreased from 100 to 62%. In addition active osmotic adjustments were inferred from shifts in the ψ^{leaf} threshold for stomatal closure (see Fig. 9E,10) and were as large as 1.8 MPa in *Quercus coccifera* (Duhme, 1974), 0.77 MPa in *Q. alba* (Aubuchon *et al.*, 1978), 0.64 MPa in *Carpinus betulus,* and 0.41 MPa in *Q. cerris* (Huzulák, 1976).

Osonubi and Davies (1978) observed active osmotic adjustment of *Quercus robur* seedlings, which, coupled with changes in root morphology, resulted in increased ψ^{leaf} and no alteration in shoot growth during a drying cycle. In contrast, osmotic adjustments in roots of *Betula verrucosa* seedlings were small, ψ^{leaf} was more negative, and shoot growth was decreased under the same drying conditions.

C. Drought Recovery

In many ways, drought avoidance and desiccation tolerance are important components of plant capacity to recover fully from drought and associated strain or injury. Species differ considerably in their sensitivity to drought; for example, Davies and Kozlowski (1974) observed rapid recovery following drought in *Acer rubrum* and *Cornus amomum* and only incomplete recovery in *Fraxinus americana, Ulmus americana,* and *Acer saccharum.* The intensity and duration of the drought (i.e., the level of injury or strain) are also critical factors in the rate of recovery. For example, leaf conductance in *Eucalyptus globulus* and *E. camaldulensis* seedlings recovered completely if the seedlings were irrigated after only 24 hr at the turgor loss point (Pereira and Kozlowski, 1976). However, when *E. camaldulensis* remained for 4 days at or below the turgor loss point, leaf conductance did not fully recover (Quraishi and Kramer, 1970).

Full recovery from drought can be viewed from several aspects. The exposure of trees to short or mild droughts preconditions trees for future droughts. The longer and more intense the drought, the greater the likelihood or reversible and even irreversible strain either due to water deficits or to excessive leaf temperatures. If these types of droughts are infrequent and rather random in occurrence, a tree species or forest ecosystem may not be able to adjust (with the possible exception of resprouting) especially if its tolerance to water deficits is exceeded. However, if droughts occur frequently or are predictable, trees (through leaf abscission or resprouting) or the forest community (through seed production) may respond and recover.

D. Drought Tolerance at the Community Level

By comparing measures of hardwood tree growth (stem area, leaf area) or abundance (importance value) with a measure of the vailability of water on the site, it is possible to stratify species and to infer the relative tolerance of the species to drought (Figs. 2, 15; Grier and Running, 1977; Fralish *et al.,* 1978; Rogers and Hinckley, 1979). If these relationships are examined for mature, successionally stable communities (Fralish *et al.,* 1978), some generalizations can be made: (1) increasing available water content (AWC) increases stand basal area (i.e., productivity); (2) AWC determines which species can dominate and occupy a particular site; (3) disturbed stands composed of pioneer species do not fully utilize site potential; (4) as site stocking is reduced, the soil water available to an

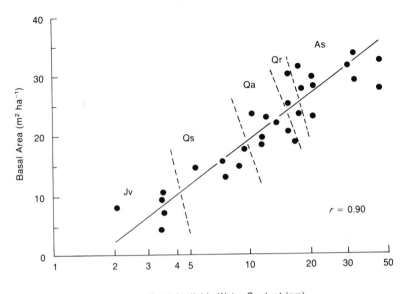

Fig. 15. Relationship between basal area and total available water content (AWC). Dashed lines separate stands dominated by *Juniperus virginiana* (Jv), *Quercus stellata* (Qs), *Q. alba* (Qa), *Q. rubra* (Qr), and *Acer saccharum* (As). AWC (cm/100) soil horizon = AWC (%) × bulk density × depth of horizon (cm/100) where AWC (%) = 6.5 + 0.26 (%silt). (Redrawn from Fralish *et al.*, 1978.)

individual tree increases; (5) as a corollary to (4), more mesophytic species occur on sites with low total AWC if the site is understocked, and (6) periods of dryness are more severe and frequent on sites with low AWC (Fralish *et al.*, 1978; Teskey *et al.*, 1978; Hinckley *et al.*, 1979). Similar observations have been made in coniferous communities (Jarvis, 1975; Grier and Running, 1977).

Coupled with these community-level observations (Figs. 3, 15) has been the detailed study of the ecophysiology of some of the same species (Table II). When such factors as depth of rooting (Fig. 6), evergreenness, threshold values for stomatal closure, the ability to shed plant parts, and tolerance of low ψ^{tissue} are studied, the relative drought tolerances of these species can be estimated and ranked. Nutrient deficiencies or toxicities can also be important considerations on droughty sites. For example, Reich and Hinckley (1980) found that although *Quercus stellata* (Figs. 2, 15) was extremely drought tolerant, it was not found on nutrient deficient sites while *Quercus rubra* was present on such sites. Ranking species only on the basis of one factor such as ψ^{leaf} threshold for stomatal closure or rooting depth often results in errors, because drought tolerance is a

TABLE II

Some Physiological and Morphological Adaptations Involved in Drought Tolerance for 13 Species[a]

Species	Threshold for stomatal closure — Ψ^{leaf} (−MPa)	Threshold for stomatal closure — Predawn Ψ^{leaf} (−MPa)	Number days stomata closed	% of time stomata closed[c]	Minimum B Ψ^{leaf} (−MPa)	Adaptation[d] — Stomatal closure	Adaptation[d] — Deep rooting	Adaptation[d] — Dieback aboveground	Adaptation[d] — Desiccation tolerance	Adaptation[d] — Water use efficiency	Species rank[e]
Acer saccharum	1.70	—	69	48	4.21	+	−	−	+	+	6( )
Andropogon scoparius	*[b]	*	*	*	*	+	+	++	++	*	1( )
Cladonia spp.	*	*	*	*	*	−	−	−	+++++	*	(1)
Cornus florida	2.10	—	83	59	4.52	+	−	−	+	*	5( )
Helianthus strumosus	1.82	—	51	70	4.20	+	−	++	+	*	5( )
Juglans nigra	—	7.5	78	58	2.78	++	++	−	+	*	5( )
Juniperus virginiana	1.65	9.0	71	21	4.33	++	+	−	+++	*	2( )
Opuntia compressa	*	*	*	*	*	++++	−	−	−	*	
Quercus alba	2.61	—	32	23	3.30	+	++	−	+	++	4(4)
Q. marilandica	3.70	—	8	6	3.61	+	++	−	++	*	3(2)
Q. rubra	2.25	—	42	33	3.49	+	++	−	+	++	5(3)
Q. stellata	*	*	*	*	3.55	+	++	−	++	*	3( )
Q. velutina	2.75	—	26	20	3.94	+	++	−	+	+++	4(4)

[a] This example comes from Missouri (*Quercus-Carya* forest, available water content = 6; see Fig. 2) and summarizes the work of Holch (1931), Hursh and Haas (1931), Albertson and Weaver (1945), Wuenscher and Kozlowski (1971), Phelps *et al.*, (1976), Teskey *et al.* (1978), Hinckley *et al.* (1979), and Reich and Hinckley (1980).

[b] *, Not measured.

[c] This category refers to the percentage of time during the period between complete maturation and leaf scenescence that stomata were closed during the intense drought of 1976 in Missouri.

[d] +++, high; +, low; −, absent or unimportant.

[e] Rank—1( ) most drought-resistant species under conditions of adequate soil nutrient status, not found where soil nutrient status was not adequate; 3(2) third level of drought resistance where soil nutrient status was adequate, second level where inadequate.

result of numerous factors (Bunce *et al.*, 1977, 1979; Pereira and Koz-lowski, 1977; Eliáš, 1978; Hinckley *et al.*, 1979). It is important to consider the whole-tree response to drying soil conditions, because growth and physiological activity both below and above ground contributes to a species drought resistance capabilities. Adjustments in leaf area are probably the most important means of drought avoidance at the plant community level. Community leaf area index decreased from 21.5 in an *Abies–Tsuga* forest receiving 200 cm of rainfall per year to 2.4 in an *Artemisia tridentata* stand with 20 cm of rain per year (Grier and Running, 1977). There is also a strong relationship between site water balance or rainfall and leaf area index in hardwood forests (Waring *et al.*, 1977; Fralish *et al.*, 1978; Rogers and Hinckley, 1979).

ACKNOWLEDGMENTS

The suggestions, counsel, and review of the manuscript by R. Borchert, G. Henderson, S. G. Pallardy, and J. P. Lassoie are gratefully acknowledged. Financial support was provided by Fonds zur Förderung der Wissenschaftlichen Forschung, Vienna (Project 3765) and Research Council of the Graduate School of the University of Missouri. Contribution of Missouri Agriculture Experimental Station, Journal Series #8447. This chapter was prepared while T. M. Hinckley and R. O. Teskey were members of the faculty of the School of Forestry, Fisheries and Wildlife, University of Missouri, Columbia, Missouri.

REFERENCES

Albertson, F. W., and Weaver, J. E. (1945). Injury and death or recovery of trees in prairie climates. *Ecol. Monogr.* **15**, 393–433.

Anderson, J. E., and McNaughton, S. J. (1973). Effects of low soil temperature on transpiration, photosynthesis, leaf relative water content, and growth among elevationally diverse plant populations. *Ecology* **54**, 1220–1233.

Aubuchon, R. R., Thompson, D. R., and Hinckley, T. M. (1978). Environmental influences on photosynthesis within the crown of a white oak. *Oecologia* **35**, 295–306.

Aussenac, G., Clement, A., and Valette, J. C. (1978). "Potential de sevé etude de quelques espèces forestières due Mont-Ventaux campagne 1977," Doc. No. 78/07. Station de Sylviculture et de Production, Champenoux, 54280 Seichamps.

Barbour, M. G., and Major, J., Jr. (1977). "Terrestrial Vegetation of California." Wiley, New York.

Bartholomew, E. T. (1926). Internal decline of lemons. III. Water deficit in lemon fruit caused by excessive leaf evaporation. *Am. J. Bot.* **13**, 102–117.

Bauer, H. (1979). Photosynthesis of ivy leaves (*Hedera helix* L.) after heat stress. III. Stomatal behavior. *Z. Pflanzenphysiol.* **92**, 277–284.

Baxter, P., and West, C. (1977). The flow of water into fruit trees. I. Resistances to water flow through roots and stems. *Ann. Appl. Biol.* **87**, 95–101.

Benecke, P. (1976). Soil water relations and water exchange of forest ecosystems. *Ecol. Stud.* **19**, 101–131.

Benecke, P., and Mayer, R. (1971). Aspects of soil water behavior as related to beech and spruce stands—some results of the water balance investigations. *Ecol. Stud.* **2**, 153–163.

Bewley, J. D. (1979). Physiological aspects of desiccation tolerance. *Annu. Rev. Plant Physiol.* **30**, 195–238.

Biebl, R. (1955). Tagesgänge der Lichttransmission verschiedener Blätter. *Flora (Jena)* **142**, 280–294.

Biswell, H. H. (1935). Effects of environment upon the root habits of certain deciduous forest trees. *Bot. Gaz. (Chicago)* **96**, 676–707.

Blake, J., and Ferrell, W. K. (1977). The association between soil and xylem water potential, leaf resistance and abscisic acid content in droughted seedlings of Douglas-fir (*Pseudotsuga menziesii*). *Physiol. Plant.* **39**, 106–109.

Boggess, W. R. (1956). Weekly diameter growth of shortleaf pine and white oak as related to soil moisture. *Proc. Soc. Am. For.* pp. 83–89.

Borchert, R. (1975). Endogenous shoot growth rhythms and indeterminate shoot growth in oak. *Physiol. Plant.* **35**, 152–157.

Borchert, R. (1978). Feedback control and age-related changes of shoot growth in seasonal and nonseasonal climates. *In* "Tropical Trees as Living Systems" (P. B. Tomlinson and M. H. Zimmermann, eds.), pp. 497–515. Cambridge Univ. Press, London and New York.

Boyer, J. S. (1976). Photosynthesis at low water potentials. *Philos. Trans. R. Soc. London, Ser. B* **273**, 501–512.

Braekke, F. H., and Kozlowski, T. T. (1977). Distribution and growth of roots in *Pinus resinosa* and *Betula papyrifera* stands. *Medd. Nor. Inst. Skogforsk.* **33**, 437–451.

Brandle, J. R., Hinckley, T. M., and Brown, G. N. (1977). The effects of dehydration-rehydration cycles on protein synthesis of black locust seedlings. *Physiol. Plant.* **40**, 1–5.

Bunce, J. A. (1977). Nonstomatal inhibition of photosynthesis at low water potentials in intact leaves of species from a variety of habitats. *Plant Physiol.* **59**, 348–350.

Bunce, J. A. (1978). Effects of shoot environment on apparent root resistance to water flow in whole soybean and cotton plants. *J. Exp. Bot.* **29**, 595–601.

Bunce, J. A., and Miller, L. N. (1976). Differential effects of water stress on respiration in the light in woody plants from wet and dry habitats. *Can. J. Bot.* **54**, 2457–2464.

Bunce, J. A., Miller, L. N., and Chabot, B. F. (1977). Competitive exploration of soil water by five eastern North American tree species. *Bot. Gaz. (Chicago)* **138**, 168–173.

Bunce, J. A., Chabot, B. F., and Miller, L. N. (1979). Role of annual leaf carbon balance in the distribution of plant species along an elevational gradient. *Bot. Gaz. (Chicago)* **140**, 288–294.

Burke, M. J., Gusta, L. V., Quamme, H. A., Weiser, C. J., and Li, P. H. (1976). Freezing and injury in plants. *Annu. Rev. Plant Physiol.* **27**, 507–528.

Burrows, W. J., and Carr, D. J. (1969). Effects of flooding the root system on sunflower plants on the cytokinin content in the xylem sap. *Physiol. Plant.* **22**, 1105–1112.

Caldwell, M. M. (1976). Root extension and water absorption. *Ecol. Stud.* **19**, 63–85.

Carlquist, S. (1975). "Ecological Strategies in Xylem Evolution." Univ. of California Press, Berkeley.

Chabot, B. F., and Bunce, J. A. (1979). Drought stress effects on leaf carbon balance. *In* "Topics in Plant Population Biology" (O. T. Solbrig, S. K. Jain, G. B. Johnson, and P. H. Raven, eds.), pp. 338–355. Columbia Univ. Press, New York.

Chaney, W. R. (1979). Leaf and twig abscission relationships in a mature white oak. *Can. J. For. Res.* **9**, 345–348.

Chaney, W. R., and Kozlowski, T. T. (1977). Patterns of water movement in intact and excised stems of *Fraxinus americana* and *Acer saccharum* seedlings. *Ann. Bot. (London)* [N. S.] **41**, 1093–1100.

Chung, H. H., and Kramer, P. J. (1975). Absorption of water and ^{32}P through suberized and unsuberized roots of loblolly pine. *Can. J. For. Res.* **5**, 229–235.

Clemens, J., Kirk, A. M., and Mills, P. D. (1978). The resistance to water-logging of three *Eucalyptus* species. *Oecologia* **34**, 125–131.

Cohen, D. B., Dumbroff, E. B., and Webb, D. P. (1978). Seasonal patterns of abscisic acid in roots of *Acer saccharum. Plant Sci. Lett.* **11**, 35–39.

Coultherd, P. (1978). Observations on the effects of drought on tree species (with particular reference to the summer of 1976). *Q. J. For.* **72**(2), 67–80.

Cutler, J. R., Rains, D. W, and Loomis, R. S. (1977). The importance of cell size in the water relations of plants. *Physiol. Plant.* **40**, 255–260.

Davies, F. S., and Lakso, A. N. (1978). Water relations in apple seedlings: Changes in water potential components, abscisic acid levels and stomatal conductances under irrigated and non-irrigated conditions. *J. Am. Soc. Hortic. Sci.* **103**, 310–313.

Davies, F. S., and Lakso, A. N. (1979). Diurnal and seasonal changes in leaf water potential components and elastic properties in response to water stress in apple leaves. *Physiol. Plant.* **46**, 109–114.

Davies, W. J., and Kozlowski, T. T. (1974). Stomatal responses of five woody angiosperms to light intensity and humidity. *Can. J. Bot.* **52**, 1525–1534.

Davies, W. J., and Kozlowski, T. T. (1975). Effects of applied abscisic acid and plant water stress on transpiration of woody angiosperms. *For. Sci.* **21**, 191–195.

Davies, W. J., Kozlowski, T. T., and Pereira, J. (1974). Effect of wind on transpiration and stomatal aperture of woody plants. *Bull.-R. Soc. N.Z.* **12**, 433–438.

Day, W., and Parkinson, K. J. (1979). Importance to gas exchange of mass flow of air through leaves. *Plant Physiol.* **64**, 345–346.

Dougherty, P. M. (1977). Net carbon exchange characteristics of a dominant white oak tree (*Quercus alba* L.). Ph.D. Dissertation, University of Missouri, Columbia (unpublished); *Diss. Abstr. Int. B* **38**, 3487 (1978).

Dougherty, P. M., Hinckley, T. M., and Lassoie, J. P. (1976). Leaf conductance to water vapor transfer from leaves of a black walnut (*Juglans nigra* L.) sapling. *In* "First Central Hardwood Forest Conference Proceedings" (J. S. Fralish, G. T. Weaver, and R. C. Schlesinger, eds.), pp. 259–268. Southern Illinois University, Carbondale.

Dougherty, P. M., Teskey, R. O., Phelps, J. E., and Hinckley, T. M. (1979). Net photosynthesis and early growth trends of a dominant white oak (*Quercus alba* L.). *Plant Physiol.* **64**, 930–935.

Duhme, F. (1974). Die Kennzeichnung der ökologischen Konstitution von Gehölzen im Hinblick auf den Wasserhaushalt. *Diss. Bot.* **28**, 1–115.

Duvigneaud, P., and Kestemont P. (1977). "Productivité biologique en Belgique." SCOPE, Travaux de la Section Belge du Programme Biologique International, Gemblaux, Belgium.

Edwards, M., and Meidner, H. (1978). Stomatal responses to humidity and the water potentials of epidermal and mesophyll tissue. *J. Exp. Bot.* **29**, 771–780.

Edwards, N. T., and McLaughlin, S. B. (1978). Temperature-independent diel variations of respiration rates in *Quercus alba* and *Liriodendron tulipifera. Oikos* **31**, 200–206.

Edwards, N. T., and Ross-Todd, B. M. (1979). The effects of stem girdling on biochemical cycles within a mixed deciduous forest in eastern Tennessee. I. Soil solution chemistry, soil respiration, litterfall and root biomass studies. *Oecologia* **40**, 247–257.

Ehret, D. L., and Boyer, J. S. (1979). Potassium loss from stomatal guard cells at low water potentials. *J. Exp. Bot.* **30**, 225–234.

El-Beltagy, A. S., and Hall, M. A. (1974). Effect of water stress upon endogenous ethylene levels in *Vicia faba. New Phytol.* **73**, 47–60.

Eliáš, P. (1978). Water deficits of plants in an oak-hornbeam forest. *Preslia* 50, 173–188.

Eliáš, P. (1979a). Leaf diffusion resistance pattern in an oak-hornbeam forest. *Biol. Plant.* 21, 1–8.

Eliáš, P. (1979b). Stomatal oscillations in adult forest trees in natural environment. *Biol. Plant.* 21, 71–74.

Eliáš, P. (1979c). Contribution to the ecophysiological study of the water relations of forest shrubs. *Preslia* 51, 77–90.

Eliasson, L. (1968). Dependence of root growth on photosynthesis in *Populus tremula*. *Physiol. Plant.* 21, 806–811.

Eliasson, L. (1978). Effects of nutrients and light on growth and root formation in *Pisum sativum* cuttings. *Physiol. Plant.* 43, 13–18.

Eller, B. M. (1979). Die strahlungsökologische Bedeutung von Epidemisauflagen. *Flora (Jena)* 168, 146–192.

Federer, C. A. (1977). Leaf resistance and xylem potential differ among broad-leafed species. *For. Sci.* 23, 411–419.

Federer, C. A. (1979). A soil-plant atmosphere model for transpiration and availability of soil water. *Water Resour. Res.* 15, 555–562.

Federer, C. A., and Gee, G. W. (1976). Diffusion resistance and xylem potential in stressed and unstressed northern hardwood trees. *Ecology* 57, 975–984.

Fellows, R. J., and Boyer, J. S. (1978). Altered ultrastructure of cells of sunflower leaves having low water potentials. *Protoplasma* 93, 381–395.

Fereres, E., Acevedo, E., Henderson, D. W., and Hsiao, T. C. (1978). Seasonal changes in water potential and turgor maintenance in sorghum and maize under water stress. *Physiol. Plant.* 44, 261–267.

Förster, M. (1979). Gesellschaften der xerothermen Eichenmischwälder des deutschen Mittelgebirgsraumes. *Phytocoenologia* 5, 367–446.

Fralish, J. S., Jones, S. M. O'Dell, R. K., and Chambers, J. L. (1978). The effect of soil moisture on site productivity and forest composition in the Shawnee hills of southern Illinois. *In* "Proceedings Soil Site Productivity Symposium" (W. E. Balmer, ed.), pp. 263–285. U.S. Dep. Agric. For. Serv., Washington, D.C.

Frenzel, B. (1968). The pleistocene vegetation of northern Europe. *Science* 161, 637–649.

Fritts, H. C. (1958). An analysis of radial growth of beech in a central Ohio forest during 1954–1955. *Ecology* 39, 705–720.

Garrett, H. E., and Cox, G. S. (1973). Carbon dioxide evolution from the floor of an oak-hickory forest. *Soil Sci. Soc. Am., Proc.* 37, 641–644.

Garrett, H. E., Cox, G. S., and Roberts, J. E. (1978). Spatial and temporal variations in carbon dioxide in an oak-hickory forest ravine. *For. Sci.* 24, 180–190.

Gates, D. M. (1976). Energy exchange and transpiration. *Ecol. Stud.* 19, 137–147.

Gee, G. S., and Federer, C. A. (1972). Stomatal resistance during senescence of hardwood leaves. *Water Resour. Res.* 8, 1456–1460.

Gholz, H. L., Fitz, F. K., and Waring, R. H. (1976). Leaf area differences associated with old-growth forest communities in the western Oregon Cascades. *Can. J. For. Res.* 6, 49–57.

Giles, K. L., Cohen, D., and Beardsell, M. F. (1976). Effects of water stress on the ultra-structure of leaf cells of *Sorghum bicolor*. *Plant Physiol.* 57, 11–14.

Goode, J. E., Higgs, K. H., and Hyrycz, K. J. (1978). Abscisic acid applied to orchard trees of golden delicious apple to control water stress. *J. Hortic. Sci.* 53, 99–103.

Grace, J. (1976). Wind damage to vegetation. *Comment. Plant Sci.* 17, 209–220.

Greenidge, K. N. H. (1958). Rules and patterns of moisture movement in trees. *In* "The Physiology of Forest Trees" (K. V. Thimann, ed.), pp. 19–41. Ronald Press, New York.

Grier, C. C., and Running, S. W. (1977). Leaf area of mature northwestern coniferous forest: Relation to site water balance. *Ecology* **58,** 893–899.

Griffin, J. R. (1973). Xylem sap tension in three woodland oaks of central California. *Ecology* **54,** 152–159.

Grosser, D. (1977). "Die Hölzer Mitteleuropas." Springer-Verlag, Berlin and New York.

Halbwachs, G. (1970). Vergleichende Untersuchungen über die Wasserbewegung in gesunden und fluorgeschädigten Holzgewächsen. *Centralbl. Gesamte Forstwes.* **87,** 1–20.

Hales, S. (1727). "Vegetable Staticks." W. and J. Innys and T. Woodward, London.

Hammel, H. T. (1967). Freezing of xylem sap without cavitation. *Plant Physiol.* **42,** 55–66.

Hammel, H. T. (1968). Measurement of turgor pressure and its gradient in the phloem of oak. *Plant Physiol.* **43,** 1042–1048.

Head, G. C. (1965). Studies of diurnal changes in cherry root growth and nutational movements of apple root tips by time-lapse cinematography. *Ann. Bot. (London)* [N.S.] **29,** 219–224.

Hellkvist, J., Richards, G. P., and Jarvis, P. G. (1974). Vertical gradients of water potential and tissue water relations in Sitka spruce trees measured with the pressure chamber. *J. Appl. Ecol.* **11,** 637–668.

Helvey, J. D., and Patric, J. H. (1965). Canopy and litter interception of rainfall by hardwoods of eastern United States. *Water Resour. Res.* **1,** 193–206.

Henderson, G. S., Hugg, D. D., and Grizzard, T. (1977). Hydrologic characteristic of Walker branch watershed. *In* "Watershed Research in Eastern North America" (D. L. Correll, ed.), pp. 195–209. Cheasapeake Bay Center for Environmental Studies, Smithsonian Institution, Edgewater, Maryland.

Henderson, G. S., Swank, W. T., Waide, J. B., and Grier, C. C. (1978). Nutrient budgets of Appalachian and Cascade region watersheds: A comparison. *For. Sci.* **24,** 385–397.

Hermann, R. (1977). Growth and production of tree roots: A review. *In* "The Below Ground Ecosystem: A Synthesis of Plant Associated Processes" (J. K. Marshall, ed.), Range Sci. Dep. Sci. Ser. No. 26, pp. 7–28. Colorado State University, Fort Collins.

Hinckley, T. M. (1971). Estimate of water flow in Douglas-fir seedlings. *Ecology* **52,** 525–528.

Hinckley, T. M., and Bruckerhoff, D. M. (1975). The effects of drought on water relations and stem shrinkage of *Quercus alba*. *Can. J. Bot.* **53,** 62–72.

Hinckley, T. M., Thompson, D. R., McGinness, M. P., and Hinckley, A. R. (1976). Stem growth and phenology of a dominant white oak. *In* "First Central Hardwood Forest Conference Proceedings" (J. S. Fralish, G. T. Weaver, and R. C. Schlesinger, eds.), pp. 187–202. Southern Illinois University, Carbondale.

Hinckley, T. M., Aslin, R. G., Aubuchon, R. R., Metcalf, C. L., and Roberts, J. E. (1978a). Leaf conductance and photosynthesis in four species of the oak-hickory forest type. *For. Sci.* **24,** 73–84.

Hinckley, T. M., Lassoie, J. P., and Running, S. W. (1978b). Temporal and spatial variations in the water status of forest trees. *For. Sci. Monogr.* **20,** 1–72.

Hinckley, T. M., Dougherty, P. M., Lassoie, J. P., Roberts, J. E., and Teskey, R. O. (1979). A severe drought: Impact on tree growth, phenology, net photosynthetic rate and water relations. *Am. Midl. Nat.* **102,** 307–316.

Hinckley, T. M., Duhme, F., Hinckley, A. R., and Richter, H. (1980). Water relations of drought hardy shrubs: Osmotic potential and stomatal reactivity. *Plant, Cell, & Environ.* **3,** 131–140.

Hogdkinson, K. C., and Becking, H. G. (1977). Effect of defoliation on root growth of some arid zone perennial plants. *Aust. J. Agric. Res.* **29,** 31–42.

Hoffmann, G. (1972). Wachstumsrhythmik der Wurzeln und Spropachsen von Forstgehölzen. *Fora (Jena)* **161**, 303–319.

Holch, A. E. (1931). Development of roots and shoots of certain deciduous tree seedlings in different forest sites. *Ecology* **12**, 259–298.

Hook, D. D., and Scholtens, J. R. (1978). Adaptations and flood tolerance of tree species. *In* "Plant Life in Anaerobic Environments" (D. D. Hook and R. M. M. Crawford, eds.), pp. 299–331. Ann Arbor Sci., Publ., Ann Arbor, Michigan.

Hook, D. D., Langdon, O. G., Stubbs, J., and Brown, C. L. (1970). Effect of water regimes on the survival, growth, and morphology of tupelo seedlings. *For. Sci.* **16**, 304–311.

Horvat, I., Glavac, V., and Ellenberg, H. (1974). "Vegetation Südosteuropas." Fisher, Stuttgart.

Hsaio, T. C. (1973). Plant responses to water stress. *Annu. Rev. Plant Physiol.* **24**, 519–570.

Huber, B., Schmidt, E., and Jahnel, H. (1937). Untersuchungen über den Assimilationsstrom. I. *Tharandter Forstl. Jahrb.* **88**, 1017–1049.

Hübner, O. (1912). Beobachtungen über die Einwirkung und Nachwirkung der Dürre des Sommers 1911 an den Baumpflanzungen, Parkanlagen und in den Forsten des Kreises Teltow. *Mitt. Dtsch. Dendrol. Ges.* **21**, 77–85.

Hursh, C. R., and Haasis, F. W. (1931). Effects of 1925 summer drought on southern Appalachian hardwoods. *Ecology* **12**, 380–386.

Huzulák, J. (1976). Relationship between water saturation deficit and water potential in the leaves of three forest tree species. *Biologia (Bratislava)* **31**, 25–32.

Huzulák, J., and Eliáš, P. (1976). The intensity of the transpiration flow in the trunk of *Quercus cerris. Biologia (Bratislava)* **31**, 537–543.

Jackson, M. B., and Campbell, D. J. (1975). Movement of ethylene from roots to shoots, a factor in the responses of tomato plants to waterlogged soil conditions. *New Phytol.* **74**, 397–406.

Jarvis, P. G. (1975). Water transfer in plants. *In* "Heat and Mass Transfer in the Environment of Vegetation" (D. A. deVries and N. K. van Alfen, eds.), pp. 369–394. Scripta Book Co., Washington, D. C.

Jarvis, P. G. (1976). The interpretation of the variations in leaf water potential and stomatal conductance found in canopies in the field. *Philos. Trans. R. Soc. London, Ser. B* **273**, 593–610.

Jarvis, P. G., and Jarvis, M. S. (1965). The water relations of tree seedlings. V. Growth and root respiration in relation to osmotic potential of the root medium. *In* "Water Stress in Plants" (B. Slavik, ed.), pp. 167–183. The Hague.

Kaufmann, M. R., and Kramer, P. J. (1967). Phloem water relations and translocation. *Plant Physiol.* **42**, 191–194.

Kazaryan, V. O., and Khurshudyan, P. A. (1966). Seasonal and daily growth rhythm of roots and stems in one-year-old oak and oleaster seedlings. *Sov. Plant Physiol. (Engl. Transl.)* **13**, 647–649; *Fiziol. Rast.* **13**, 725–728 (1965).

Keck, R. W., and Boyer, J. S. (1974). Chloroplast response to low leaf water potentials. I. Differing infiltration of electron transport and photophosphorylation. *Plant Physiol.* **53**, 474–479.

Kessler, K. J. (1966). Growth and development of mycorrhiza of *Acer saccharum* Marsh. *Can. J. Bot.* **44**, 1413–1425.

Kinerson, R. S. (1979). Studies of photosynthesis and diffusion resistance in paper birch (*Betula papyrifera* Marsh.) with synthesis through computer simulation. *Oecologia* **39**, 37–49.

King, R. W. (1976). Implications for plant growth of the transport of regulatory compounds in phloem and xylem. *In* "Transport and Transfer Processes in Plants" (I. F. Wardlaw and J. B. Passioura, eds.), pp. 415–432. Academic Press, New York.

Körner, C., Scheel, J. A., and Bauer, H. (1979). Maximum leaf diffusive conductance in vascular plants. *Photosynthetica* **13**, 45–82.

Kozlowski, T. T. (1961). The movement of water in trees. *For. Sci.* **7**, 177–192.

Kozlowski, T. T., ed. (1968). "Water Deficits and Plant Growth," Vols. 1 and 2. Academic Press, New York.

Kozlowski, T. T., ed. (1972). "Water Deficits and Plant Growth," Vol. 3, Academic Press, New York.

Kozlowski, T. T., ed. (1976). "Water Deficits and Plant Growth," Vol. 4. Academic Press, New York.

Kozlowski, T. T., ed. (1978). "Water Deficits and Plant Growth," Vol. 5. Academic Press, New York.

Kozlowski, T. T. (1979). "Tree Growth and Environmental Stresses." Univ. of Washington Press, Seattle.

Kozlowski, T. T., and Pallardy, S. G. (1979). Stomatal responses of *Fraxinus pennsylvanica* seedlings during and after flooding. *Physiol. Plant.* **46**, 155–158.

Kozlowski, T. T., and Winget, C. H. (1964). Diurnal and seasonal variation in radii of tree stems. *Ecology* **45**, 149–155.

Kramer, P. J. (1946). Absorption of water through suberized roots of trees. *Plant Physiol.* **21**, 37–41.

Kramer, P. J. (1951). Causes of injury to plants resulting from flooding of the soil. *Plant Physiol.* **26**, 772–736.

Kramer, P. J. (1969). "Plant and Soil Water Relationships: A Modern Synthesis." McGraw-Hill, New York.

Kramer, P. J., and Bullock, H. C. (1966). Seasonal variations in the proportions of suberized and unsuberized roots of trees in relation to the absorption of water. *Am. J. Bot.* **53**, 200–204.

Kramer, P. J., and Jackson, W. T. (1954). Causes of injury to flooded tobacco plants. *Plant Physiol.* **29**, 241–245.

Kramer, P. J., and Kozlowski, T. T. (1979). "Physiology of Woody Plants." Academic Press, New York.

Krygier, J. T. (1971). Comparative water loss of Douglas-fir and Oregon white oak. Ph.D. Dissertation, Oreg. State University, Corvallis (unpublished); *Diss. Abstr. Int. B* **32**, 2835 (1972).

Kuiper, D., and Kuiper, P. J. C. (1978). Temperature effects on water transport in plant roots. *Physiol. Plant.* **43**, 121–125.

Kuntz, J. E., and Riker, A. J. (1955). The use of radioactive isotopes to ascertain the role of root grafting in the translocation of water, nutrients and disease-inducing organisms. *Proc. Int. Conf. Peaceful Uses At. Energy,* Vol. 12, pp. 144–148.

Ladefoged, K. (1963). Transpiration of forest trees in closed stands. *Physiol. Plant.* **10**, 378–414.

Lakso, A. N. (1979). Seasonal changes in stomatal response to leaf water potential in apple. *J. Am. Soc. Hortic. Sci.* **104**, 58–60.

Landsberg, J. J., and Fowles, N. D. (1978). Water movement through plant roots. *Ann. Bot.* (*London*) [N. S.] **42**, 493–508.

Landsberg, J. J., Beadle, C. L., Biscoe, P. V., Butler, D. R., Davidson, B., Incoll, L. D., James, G. B., Jarvis, P. G., Martin, P. J., Neklson, R. E., Powell, D. B. B., Slack, E. M., Thorpe, M. R., Turner, N. C., Arrit, B., and Watts, W. R. (1975). Diurnal energy, water and CO_2 exchanges in an apple (*Malus pumila*) orchard. *J. Appl. Ecol.* **12**, 659–684.

Lange, O. L., Lösch, R., Schulze, E.-D., and Kappen, L. (1971). Responses of stomata to changes in humidity. *Planta* **100**, 76–86.

Larcher, W. (1975). "Physiological Plant Ecology." Springer-Verlag, Berlin and New York.

Larson, M. M. (1974). Effects of soil moisture on early growth of oak seedlings. *Ohio Agric. Res. Dev. Cent., For. Res. Rev.* **74**, 10–12.

Larson, M. M. and Palashev, I. (1973). Effects of osmotic water stress and gibberellic acid on initial growth of oak seedlings. *Can. J. For. Res.* **3**, 75–82.

Larson, M. M., and Whitmore, F. W. (1970). Moisture stress affects root regeneration and early growth of red oak seedlings. *For. Sci.* **16**, 495–498.

Larson, P. R. and Isebrands, J. G. (1978). Functional significance of the nodal constricted zone in *Populus deltoides. Can. J. Bot.* **56**, 801–804.

Lassoie, J. P. (1979). Stem dimensional fluctuation in Douglas-fir of different crown classes. *For. Sci.* **25**, 132–144.

Lassoie, J. P. (1981). Physiological activity in Douglas-fir. *In* "Coniferous Biome/IBP Synthesis" (R. L. Edmonds, ed.), Chapter 6. Dowden, Hutchinson & Ross, Inc., Stroudsburg, Pennsylvania (in press).

Lassoie, J. P., and Chambers, J. L. (1976). The effects of an extreme drought on tree water status and net assimilation rates of northern red oak under greenhouse conditions. *In* "First Central Hardwood Forest Conference Proceedings" (J. S. Fralish, G. T. Weaver, and R. C. Schlesinger, eds.), pp. 269–283. Southern Illinois University, Carbondale.

Lassoie, J. P., Scott, D. R. M., and Fritschen, L. J. (1977). Transpiration studies in Douglas-fir using the heat pulse technique. *For. Sci.* **23**, 377–390.

Levitt, J. (1972). "Responses of Plants to Environmental Stresses." Academic Press, New York.

Lieth, H. (1975). Modeling the primary productivity of the world. *In* "Primary Productivity of the Biosphere" (H. Lieth and R. H. Whittacker, eds.), pp. 237–263. Springer-Verlag, Berlin and New York.

Likens, G. E., Bormann, F. H. Pierce, R. S., and Reiners, W. A. (1978). Recovery of a deforested ecosystem. *Science* **199**, 492–496.

Longman, K. A., and Coutts, M. P. (1974). Physiology of the oak tree. *In* "The British Oak" (M. G. Morris and F. H. Perring, eds.), pp. 194–221. E. W. Classey Ltd., Farington, England.

Lopushinsky. W. (1975). Water relations and photosynthesis in lodgepole pine. *In* "Management of Lodgepole Pine Ecosystem" (D. M. Baumgartner, ed.), pp. 135–153. Coop. Ext. Serv., Washington State University, Pullman.

Lösch, R., and Schenk, B. (1978). Humidity responses of stomata and the potassium content of guard cells. *J. Exp. Bot.* **29**, 781–787.

Lucier, A. A., Teskey, R. O., and Hinckley, T. M. (1978). Forest tree responses on irrigated and non-irrigated sites. *Bull. Ecol. Soc. Am.* **59**, 84.

Luxmoore, R. J., Huff, D. D., McConathy, R. K., and Dinger, D. E., (1978). Some measured and simulated plant water relations of yellow-poplar. *For. Sci.* **24**, 327–341.

Lyr, H., and Hoffmann, G. (1967). Growth rates and growth periodicity of tree roots. *Int. Rev. For. Res.* **2**, 181–236.

Lyr, H., Erdmann, A., Hoffmann, G., and Kohler, S. (1968). Über den diurnalen Wachstumrhythmus von Gehölzen. *Flora (Jena)* **157**, 615–624.

McLaughlin, S. B., McConathy, R. K., and Beste, B. (1979). Seasonal changes in within-canopy allocation of ^{14}C-photosynthate by white oak. *For. Sci.* **25**, 361–370.

Mansfield, T. A., Wellburn, A. R., and Moreira, T. J. S. (1978). The role of abscisic acid and farnesol in the alleviation of water stress. *Philos. Trans. R. Soc. London, Ser. B* **284**, 471–482.

Marchand, P. J., and Chabot, B. F. (1978). Winter water relations of tree-line plant species on Mt. Washington, New Hampshire. *Arct. Alp. Res.* **10**, 105–116.

Marks, G. C., and Kozlowski, T. T., eds. (1973). "Ectomycorrhizae: Ecology and Physiology." Academic Press, New York.

Monteith, J. L., ed. (1975). "Vegetation and the Atmosphere," Vol. 1. Academic Press, New York.

Morrow, R. R. (1950). Periodicity and growth of sugar maple surface layer roots. *J. For.* **48,** 875–877.

Mudd, J. B., and Kozlowski, T. T., eds. (1975). "Response of Plants to Air Pollution." Academic Press, New York.

Newman, E. I. (1976). Water movement through root systems. *Philos. Trans. R. Soc. London, Ser. B* **273,** 463–478.

Noland, T. L., and Kozlowski, T. T. (1979). Effect of SO_2 on stomatal aperture and sulfur uptake of woody angiosperm seedlings. *Can. J. For. Res.* **9,** 57–62.

Nuritdinov, N., and Vartapetyan, B. B. (1976). Transport of oxygen from the overground parts into roots of cotton. *Sov. Plant Physiol. (Engl. Transl.)* **23,** 527–529; *Fiziol. Rast.* **23,** 622–624 (1975).

Osonubi, O., and Davies, W. J. (1978). Solute accumulation in leaves and roots of woody plants subjected to water stress. *Oecologia* **32,** 323–332.

Ovington, J. D., and Murray, G. (1968). Seasonal periodicity of root growth of birch trees. *In* "Methods of Productivity Studies in Root Systems and Rhizosphere Organisms" (M. S. Ghilarov, V. A. Kovda, L. N. Novichkova-Ivanova, L. E. Rodin, and V. M. Sveshnikova, eds.), pp. 146–154. Nauka, Leningrad.

Pallardy, S. G., and Kozlowski, T. T. (1979). Relationships of leaf diffusion resistance of *Populus* clones to leaf water potential and environment. *Oecologia* **40,** 371–380.

Parker, J. (1968a). Drought-resistance mechanisms. *In:* "Water Deficits and Plant Growth" (T. T. Kozlowski, ed.), Vol. 1, pp. 195–234. Academic Press, New York.

Parker, J. (1968b). Drought resistance of roots of white ash, sugar maple and red oak. *U.S., For. Serv., Res. Pap.* NE **NE-95,** 1–9.

Parker, J. (1970). Effects of defoliation and drought on root food reserves in sugar maple seedlings. *U.S., For. Serv., Res. Pap.* NE **NE-169.**

Pereira, J. S., and Kozlowski, T. T. (1976). Leaf anatomy and water relations of *Eucalyptus camaldulensis* and *E. globulus* seedlings. *Can. J. Bot.* **54,** 2868–2880.

Pereira, J. S., and Kozlowski, T. T. (1977). Influence of light intensity, temperature, and leaf area on stomatal aperture and water potential of woody plants. *Can. J. For. Res.* **7,** 145–153.

Pereira, J. S., and Kozlowski, T. T. (1978). Diurnal and seasonal changes in water balance of *Acer saccharum* and *Betula papyrifera*. *Physiol. Plant.* **43,** 19–30.

Phelps, J. E., Chambers, J. L., and Hinckley, T. M. (1976). Some morphological, ecological and physiological traits of forest species. *In:* "First Central Hardwood Forest Conference Proceedings" (J. S. Fralish, G. T. Weaver, and R. C. Schlesinger, eds.), pp. 231–243. Southern Illinois University, Carbondale.

Phillips, I. D. J. (1964). Root-shoot hormone relations. II. Change in endogenous auxin concentration produced by flooding of the root system in *Helianthus annuus*. *Ann. Bot. (London)* [N. S.] **28,** 37–45.

Phung, H. T., and Knipling, E. B. (1976). Photosynthesis and transpiration of citrus seedlings under flooded conditions. *HortScience* **11,** 131–133.

Ponnamperuma, F. N. (1972). The chemistry of submerged soils. *Adv. Agron.* **24,** 29–96.

Poole, D. K., and Miller, P. C. (1978). Water related characteristics of some evergreen sclerophyll shrubs in central Chile. *Oecol. Plant.* **13,** 289–299.

Powell, D. B. B., and Thorpe, M. R. (1977). Dynamic aspects of plant-water relations. *In*

"Environmental Effects on Crop Physiology" (J. J. Landsberg and C. V. Cutting, eds.), pp. 53–57. Academic Press, New York.

Proctor, J., and Woodell, W. (1971). The ecology of serpentine soils. *Adv. Ecol. Res.* **5**, 255–366.

Quraishi, M. A., and Kramer, P. J. (1970). Water stress in three species of *Eycalyptus*. *For. Sci.* **16**, 74–78.

Raschke, K. (1975). Stomatal action. *Annu. Rev. Plant Physiol.* **26**, 309–340.

Regehr, D. L., Bazzaz, F. A., and Boggess, W. R. (1975). Photosynthesis, transpiration, and leaf conductance of *Populus deltoides* in relation to flooding and drought. *Photosynthetica* **9**, 52–61.

Reich, P. B., and Hinckley, T. M. (1980). Water-relations, soil fertility and plant nutrient composition of a pygmy oak ecosystem. *Ecology* **61**, 400–416.

Reid, D. M., Crozier, A., and Harvey, B. M. R. (1969). The effects of flooding on the export of gibberellins from the root to the shoot. *Planta* **89**, 376–379.

Richardson, S. D. (1953). Studies of root growth in *Acer saccharinum* L. I. The relation between root growth and photosynthesis. *Proc. K. Ned. Akad. Wet., Ser. C* **56**, 185–193.

Richter, H. (1976). The water status in the plant—experimental evidence. *Ecol. Stud.* **19**, 42–58.

Richter, H., Duhme, F., Glatzel, G., Hinckley, T. M., and Karlic, H. (1981). Some limitations and applications of the pressure-volume curve technique in ecophysiological research. *In* "Plants and Their Atmospheric Environment" (J. Grace, P. J. Jarvis, and D. Ford, eds.), Chapter 15, pp. 263–272. Brit. Ecol. Soc.

Riedacker, A. (1976). Rythmes de croissance et de régénération des racines des végétaux ligneau. *Ann. Sci. For.* **33**, 109–138.

Roberts, S. W., Knoerr, K. R., and Strain, B. R. (1979). Comparative field water relations of four co-occurring forest tree species. *Can. J. Bot.* **57**, 1876–1882.

Rogers, R., and Hinckley, T. M. (1979). Foliar weight and area related to current sapwood area in oak. *For. Sci.* **25**, 298–303.

Rosenberg, N. J., ed. (1978). "North American Droughts." Westview Press, Boulder, Colorado.

Running, S. W. (1976). Environmental control of leaf water conductance in conifers. *Can. J. For. Res.* **6**, 104–112.

Rutter, A. J. (1975). The hydrological cycle in vegetation. *In* "Vegetation and the Atmosphere" (J. L. Monteith, ed.), pp. 111–154. Academic Press, New York.

Schmithüsen, J. (1948). Pflanzenschäden im Dürrejahr 1947. *Kosmos* **44**, 241–244.

Schulze, E.-D., Fuchs, M., and Fuchs, M. I. (1977). Spatial distribution of photosynthetic capacity and performance in a mountain spruce forest of northern Germany. III. The significance of the evergreen habit. *Oecologia* **30**, 239–248.

Scrivner, C. L., Baker, J. C., and Brees, D. R. (1973). Combined daily climatic data and dilute solution chemistry in studies of soil profile formation. *Soil Sci.* **115**, 213–223.

Seidel, K. W. (1972). Drought resistance and internal water balance of oak seedlings. *For. Sci.* **18**, 34–40.

Shaner, D. L., and Boyer, J. S. (1976). Nitrate reductase activity in maize (*Zea mays* L.) leaves. I. Regulation by nitrate flux. *Plant Physiol.* **58**, 499–504.

Sheriff, D. W. (1979). Stomatal aperture and the sensing of the environment by guard cells. *Plant, Cell, & Environ.* **2**, 15–22.

Shive, J. B., Jr., and Brown, K. W. (1978). Quaking and gas exchange in leaves of cottonwood (*Populus deltoides*, Marsh.). *Plant Physiol.* **61**, 331–333.

Siminovitch, D., Singh, J., and de la Roche, I. A. (1975). Studies on membranes in plant cells resistant to extreme freezing. I. Augumentation of phospholipids and membrane substance without changes in unsaturation of fatty acids during hardening of black locust bark. *Cryobiology* **12**, 144–153.

Simon, E. W. (1978). Plant membranes under dry conditions. *Pestic. Sci.* **9**, 169–172.

Sprackling, J. A., and Read, R. A. (1979). Tree root systems in eastern Nebraska. *Nebr. Conserv. Bull.* **37**, 1–73.

Stone, E. C. (1967). A nursery-conditioned root growth response to the field environment. *Proc. Kongr., Int. Union For. Res. Organ., 14th, 1967* Vol. III, pp. 248–264.

Stone, E. C., and Jenkinson, J. L. (1970). Influence of soil water on root growth capacity of ponderosa pine transplants. *For. Sci.* **16**, 230–239.

Swank, W. T., and Douglass, J. E. (1974). Streamflow greatly reduced by converting deciduous hardwood stands to pine. *Science* **185**, 857–859.

Swanson, R. H. (1967). Seasonal course of transpiration of lodgepole pine and Englemann spruce. *Proc. Int. Symp. For. Hydrol. 1965* pp. 417–432.

Szarek, S. R., and Ting, I. P. (1974). Seasonal patterns of acid metabolism and gas exchange in *Opuntia basilaris*. *Plant Physiol.* **54**, 76–81.

Teskey, R. O., and Hinckley, T. M. (1977). "Impact of Water Level Changes on Woody Riparian and Wetland Communities," Vol. I, FWS/OBS-77/58, U.S. Government Printing Office, Washington, D. C.

Teskey, R. O., Chambers, J. L., Cox, G. S., Hinckley, T. M., and Roberts, J. E. (1978). A severe drought. I. Soil-site relationships in an oak-hickory forest. *In* "Proceedings Soil Site Productivity Symposium" (W. E. Balmer, ed.), pp. 316–326. U.S. Dep. Agric. For. Serv., Washington, D.C.

Thompson, D. R., and Hinckley, T. M. (1977a). Effect of vertical and temporal variations in stand microclimate and soil moisture on water status of several species in an oak-hickory forest. *Am. Midl. Nat.* **97**, 373–380.

Thompson, D. R., and Hinckley, T. M. (1977b). A simulation of water relations of white oak based on soil moisture and atmospheric evaporative demand. *Can. J. For. Res.* **7**, 400–409.

Tinker, P. B. (1976). Transport of water to plant roots in soil. *Philos. Trans. R. Soc. London, Ser. B* **273**, 445–461.

Toumey, J. W. (1929). Initial root habit in American trees and its bearing on regeneration. *Proc. Int. Congr. Plant Sci., 1926* Vol. 1, pp. 713–728.

Turner, N. C., and Heichel, G. H., (1977). Stomatal development and seasonal changes in diffusive resistance of primary and regrowth foliage of red oak (*Quercus rubra* L.) and red maple (*Acer rubrum* L.). *New Phytol.* **78**, 71–81.

Tyree, M. T., Cheung, Y. N. S., MacGregor, M. E., and Talbot, A. J. B. (1978). The characteristics of seasonal and ontogentic changes in the tissue-water relations of *Acer, Populus, Tsuga* and *Picea*. *Can. J. Bot.* **56**, 635–647.

Wallace, L. L., and Dunn, E. L. (1980). Comparative photosynthesis three gap phase successional tree species. *Oeclogia* (Berl.) **45**, 331–340.

Walter, H. (1968). "Die Vegetation der Erde," Vol. II. Fischer, Stuttgart.

Wargo, P. M. (1979). Starch storage and radial growth in woody roots of sugar maple. *Can. J. For. Res.* **9**, 49–56.

Waring, R. H., and Franklin, J. F. (1979). Evergreen coniferous forests of the Pacific Northwest. *Science* **204**, 1380–1386.

Waring, R. H., and Running, S. W. (1978). Sapwood water storage: Its contribution to transpiration and effect upon water conductance through the stems of old growth Douglas-fir. *Plant, Cell, & Environ.* **1**, 131–140.

Waring, R. H., Gholz, H. L., Grier, C. C., and Plummer, M. L. (1977). Evaluating stem conducting tissue as an estimator of leaf area in four woody angiosperms. *Can. J. Bot.* **55**, 1474–1477.

Weaver, J. E., and Kramer, J. (1932). Root system of *Quercus macrocarpa* in relation to the invasion of prairie. *Bot. Gaz. (Chicago)* **94**, 51–85.

Webb, D. P. (1976). Root growth in *Acer saccharum* Marsh. seedlings: Effects of light intensity and photoperiod on root elongation rates. *Bot. Gaz. (Chicago)* **137**, 211–217.

Webb, W. L. (1977). Seasonal allocation of photoassimilated carbon in Douglas-fir seedlings. *Plant Physiol.* **60**, 320–322.

Weiser, C. J. (1970). Cold resistance and injury in woody plants. *Science* **169**, 1269–1278.

Went, F. W., and Babu, V. R. (1978). The effect of dew on plant water balance in *Citrullus vulgaris* and *Cucumis mebo*. *Physiol. Plant.* **44**, 307–311.

Wilcox, H. E., and Ganmore-Newman, R. (1975). Effects of temperature on root morphology and ectendomycorrhizal development in *Pinus resinosa* Ait. *Can. J. For. Res.* **5**, 171–175.

Woods, D. B., and Turner, N. C., (1971). Stomatal response to changing light by four tree species of varying shade tolerance. *New Phytol.* **70**, 77–84.

Wuenscher, J. E., and Kozlowski, T. T. (1970). Carbon dioxide transfer resistance as a factor in shade tolerance of tree seedlings. *Can. J. Bot.* **48**, 453–456.

Wuenscher, J. E., and Kozlowski, T. T. (1971). Relationship of gas-exchange resistance to tree-seedling ecology. *Ecology* **52**, 1016–1023.

Yen, C. P. (1973). The major patterns of root growth of the useful plants for soil conservation in Taiwan. *J. Chin. Soil & Water Conserv.* **4**, 65–85.

Yen, C. P., Pham, C. H., Cox, G. S., and Garrett, H. E. (1978). Soil depth and root development patterns of Missouri black walnut and certain Taiwan hardwoods. *In* "Root Form of Planted Trees" (E. V. Eerden and J. M. Kinghorn, eds.), Can. For. Serv. Joint Rep. No. 8, pp. 36–43. British Columbia Ministry of Forests.

Yoshida, S. (1979). Freezing injury and phospholipid degradation *in vivo* in woody plant cells. III. Effects of freezing on activity of membrane-ground phospholipase D in microsome-enriched membranes. *Plant Physiol.* **64**, 252–256.

Zangerl, A. R. (1978). Energy exchange phenomena, physiological rates and leaf size variation. *Oecologia* **34**, 107–112.

Zimmermann, M. H. (1957). Translocation of organic substances in trees. II. On the translocation mechanism in the phloem of white ash (*Fraxinus americana* L.). *Plant Physiol.* **32**, 399–404.

Zimmermann, M. H. (1969). Translocation velocity and specific mass transfer in the sieve tubes of *Fraxinus americana* L. *Planta* **84**, 272–278.

Zimmermann, M. H. (1971). Transport in the phloem. *In* "Tree Structure and Function" (M. H. Zimmermann and C. L. Brown, eds.), pp. 221–280. Springer-Verlag, Berlin and New York.

Zimmermann, M. H. (1978a). Hydraulic architecture of some diffuse-porous trees. *Can. J. Bot.* **56**, 2286–2295.

Zimmermann, M. H. (1978b). Structural requirements for optimal water conduction in tree stems. *In* "Tropical Trees as Living Systems" (P. B. Tomlinson and M. H. Zimmermann, eds.), pp. 517–532. Cambridge Univ. Press, London and New York.

Zimmermann, M. H., and Brown, C. L., eds. (1971). "Tree Structure and Function." Springer-Verlag, Berlin and New York.

CHAPTER 4

TROPICAL AND SUBTROPICAL FORESTS AND WOODLANDS

D. Doley

BOTANY DEPARTMENT, UNIVERSITY OF QUEENSLAND, ST. LUCIA,
QUEENSLAND, AUSTRALIA

Water Deficits and Plant Growth, Vol. VI
Copyright © 1981 by Academic Press, Inc.
All rights of reproduction in any form reserved.
ISBN 0-12-424156-5

I. INTRODUCTION

The water relations of woody plants in the tropics and subtropics have attracted interest from the time of the earliest scientific expeditions. In the deserts, interest was focussed on the means by which plants survived extended periods with little or no precipitation and extremely high temperatures. At the other extreme, in the rain forests, more popular concerns were adaptations to wet soils and disposal of the very large amounts of water intercepted by the forest canopy.

Quantitative studies of water consumption by tropical plants have generally followed similar developments in the temperate latitudes, but an increasing awareness of the importance of vegetation cover to the maintenance of stable landscapes and water supplies has led to a greater interest in the contribution of plants in tropical locations to regional water balance.

Many investigators have searched for unifying principles of structure or function among the species occupying a particular site. However, close study of individual plants in relation to their environment often reveals a great array of structural and physiological attributes that combine to permit the survival of many different species in one habitat. Because of this variation, it is presumptuous to judge the fitness of a species for a particular site on the basis of one or even a few structural or physiological attributes.

Limitations to the abilities of single researchers or even small teams to examine fully a species or a site have resulted in partial elucidation of most questions relating to the functioning of plant species in their environments. These limitations have been particularly apparent in the tropics, where a considerable proportion of the research has been of a short-term nature, organized and executed by investigators based in the temperate latitudes.

This chapter will discuss some of the diversity of function with respect to the water relations of woody plants in the wide variety of environments encountered in the tropics and subtropics.

II. HYDROLOGY

A. Classification of Vegetation Types

Vegetation classification in this survey is of secondary importance to the discussion of water relations, but some consideration must be given to methods of classification since they reflect the water balance of a region or plant community. A detailed comparison of several indices of climatic/vegetational classification has been made by Schulze and McGee (1978), and only brief mention of some pertinent aspects will be made here.

Strictly climatic indices, such as those of Köppen and Geiger (1930), Thornthwaite (1948), the xerothermic index and its derivatives (Legris, 1972), or the climatic diagram of Walter (1964) do not incorporate any vegetational component in their derivation or final description. They simply depict the climatic heat and water balances of a site, although Walter's climatic diagram can include a considerable amount of information that is useful directly in assessing plant behavior.

Another basically climatic index is the Holdridge life zone system (Holdridge *et al.*, 1971), which evaluates the seasonal courses of temperature, precipitation and evaporation, and predicts a typical climax vegetation. Adjustments to the climax type may then be made according to altitude and soil type. Field application of this method also involves a physiognomic assessment of the vegetation.

Classifications based solely on stand structure, or physiognomy, have been popular in tropical regions. Rain forest classifications according to structural criteria have been proposed by Richards (1952), Webb (1959), and Grubb *et al.* (1963), and classifications of different vegetation types have been devised by Kuchler (1949), Beard (1955), Cochrane (1963), Specht (1970), and Goodland (1971) for use in tropical and subtropical situations. These classifications commonly depend on the height of the predominant canopy stratum, the proportion of the ground covered by this canopy, and the presence and nature of lower vegetation strata. All these features of a plant community can convey a great deal of information about its water balance. Disturbed vegetation may have quite different characteristics (both physiognomic and physiological) from those that may develop in the absence of human interference (Puri, 1957), so the factual physiognomic classifications may well be at variance with the hypothetical climatic classifications.

In describing different vegetation types, the classifications used by researchers will be adopted unless a more general description is considered preferable.

B. General Hydrological Characteristics

Comprehensive reviews of the hydrology of vegetated land surfaces have been prepared by Penman (1963) and Pereira (1973), with an earlier contribution to this series (Rutter, 1968) dealing in particular with forests. The purpose of this section is to compare the hydrological characteristics of different forest and woodland types that may influence other aspects of the water relations of the constituent species.

One problem of hydrological studies in the tropics is the torrential nature of much of the rainfall (Jackson, 1971; Gilmour and Bonell, 1979). In order to measure the heaviest storms, high capacity and therefore rather coarse instruments are required, so that the precision of measurement of components of the hydrological balance may be low as compared with that attained in some studies at higher latitudes. Nevertheless, important work has been carried out in the tropics, and the shortcomings of some investigations cannot be attributed to such practical difficulties as those of coping with intense storms.

The components of the hydrological balance for an area can be described by the relationship

$$P = E + T + G + R + \Delta S \tag{1}$$

where P is precipitation, E is evaporation from the soil surface, T is transpiration from the vegetation, R and G are the net gains or losses of water from the site by surface or subsurface drainage, respectively, and ΔS is the change in soil water storage during the period under observation. This equation permits the yield of water as runoff to be assessed in relation to precipitation and also enables comparisons to be made of the rates of water use by vegetation of different types.

Another important hydrological relationship of vegetated areas is the partitioning of precipitation, since this affects the amount of water available for disposition at the soil surface. This is described by

$$P = I + SF + TF \tag{2}$$

where I is water intercepted by the vegetation and evaporated directly, SF is intercepted water that subsequently reaches the ground by stem flow, and TF is water that either falls directly to the ground through gaps in the canopy, or drips from the foliage.

Water reaching the soil surface (net precipitation, P') may either flow off a site as surface drainage or enter the soil water store. The properties of the soil surface are critical at this stage, as will be seen subsequently.

$$P' = SF + TF = R + \Delta S + E' \tag{3}$$

describes this relationship where E' is the component of total evaporation from the ground surface occurring during a precipitation event.

Where surface and subsurface drainage can be ignored, as during rainless periods in areas where the tree roots do not penetrate to the groundwater, consumption of water by vegetation T and evaporative loss E can be determined simply by

$$E + T = \Delta S \tag{4}$$

Separation of the soil evaporation and transpiration components is not easy, as will be shown in Section IV,A,2.

Rain forests are traditionally associated with perpetual abundance of water (Fosberg *et al.*, 1961; Baynton, 1969; Grubb, 1977), but even in very uniform climates such as lowland Sarawak, dry periods of up to 30 days duration may occur, with important consequences for the water balance of vegetation on shallow soils (Brunig, 1969; Baillie, 1976). In montane rain forests, where clouds and fogs are common, dry periods of up to 20 days occur and result in visible signs of water stress, particularly in epiphytic and moss vegetation (Richards, 1936; Grubb and Whitmore, 1966). As the seasonality of rainfall becomes more pronounced, the impact of variability of rainfall becomes more evident (Fitzpatrick *et al.*, 1967), even in regions with a high annual precipitation (Nieuwolt, 1968, 1972).

In regions of lower rainfall, the dominant feature of the water balance of plants is clearly the length and severity of the dry season, and the variability of occurrence of wet seasons of sufficient duration to secure the establishment of regeneration. Sarlin (1970) stressed the importance in dry forest areas of the excess of potential evaporation over rainfall and the disparity in time between the peaks of the two. Actual evaporation from dry forests and deserts may be almost zero for a considerable portion of the year (Turner, 1965; Birot and Galabert, 1972; Specht, 1972) and the existence of perennial vegetation may depend upon uneven distribution of water over the ground surface due to drainage (Koller, 1969; Pressland, 1976b). This is effected in some cases by development of groves, or stripes of vegetation that may alter the pattern of distribution of surface runoff (Turner, 1965; Catinot, 1967; White, 1971; Goodspeed and Winkworth, 1973) or the pattern of infiltration (Slatyer, 1961a; Stocker, 1970).

C. INTERCEPTION AND THROUGHFALL

In the hydrological sense, interception of rainfall is often considered as a loss from the water cycle (Penman, 1963), since much of this water is evaporated from the site without contributing to the soil water store.

1. Interception of Fog and Cloud

In particular cases, interception may provide an important source of water for plant survival. Coastal regions where temperature differences between marine and continental surfaces are sufficient or where hills are sufficiently high (Ekern, 1964; Grubb and Whitmore, 1966; Burgess, 1969a; Ellis, 1971) enable the formation of fogs or clouds that are intercepted by vegetation (Kerfoot, 1967; Shuttleworth, 1977).

This intercepted water may be of nuisance in carrying leachates from the leaves, which then suppress the growth of understory grasses (Del Moral and Muller, 1970), or it may be beneficial in that it can then pass into the soil and be used for survival and growth of plants (Stone, 1957; Gindel, 1966; Stark and Love, 1969), growth of tree crops, or irrigation (Carlson, 1961). Such fog drip must not be confused with the precipitation reported by de Rosayro (1960) to originate from sap-sucking insects in trees. This accession of water is generally confined to the wetter months of the year, and at least in the dry season, the dew point is not often reached in tropical dry forest areas (Slatyer, 1961a) or deserts (Migahid, 1961). At high-altitude locations, the benefit of intercepted water may be partly offset by higher evaporation rates that may be experienced in clear weather due to strong winds (Jenik and Hall, 1966; Burgess, 1969a; Ellis, 1971).

The amount of cloud water intercepted by an individual tree will clearly be related to the overall dimensions of the crown and to the surface area of foliage carried. Increasing these dimensions provides an opportunity for either greater subsequent evaporative loss of intercepted water or drainage as stem flow or drip. Carlson (1961) considered coniferous trees, because of their large surface area of foliage, to be more efficient interceptors of fog in the cloud belt on Lanai, Hawaii, than were dicotyledonous trees. *Araucaria heterophylla* trees about 10 m high could intercept approximately 750 mm of water per year, or an amount almost equal to the mean annual rainfall for the lower altitudes of Lanai. Supplementation of rainfall by up to 550 mm, or 40%, as a result of fog drip was reported for a high-altitude eucalypt forest by Ellis (1971). Even forests of low stature may intercept considerable quantities of water from clouds. Baynton (1969) estimated interception by the canopy of a 4-m high montane rain forest in Puerto Rico to be approximately 390 mm per year, but this was only 8.6% of the annual rainfall.

The ability of species to absorb directly intercepted water varies. Although some temperate zone species such as *Pinus ponderosa* (Stone, 1958) and *P. radiata* (Leyton and Armitage, 1968) are capable of absorbing water through the bases of their growing needles, species such as *Acacia aneura* are considered not to absorb intercepted rainfall to a significant degree (Slatyer, 1961a). A contrary opinion was expressed by Gindel (1966, 1973), who claimed that direct absorption of intercepted water could provide the transpirational needs of arid zone xerophytes. However, although the possibility of water uptake by submerged leaves and its transfer to neighboring plants was established, no proof was given of the adequacy of this phenomenon under natural conditions. Grubb and Whitmore (1966) considered that the absorption of water vapour by terrestrial plants in montane rain forest in Ecuador was an insignificant contribution to their total water balance. On the other hand, the epiphytes that are prominent in humid forests have arrangements of leaves and root structures that may facilitate collection and absorption of liquid water resulting from intercepted fog and rain (Gessner, 1956).

2. The Proportion of Rainfall Intercepted

For most forest and plantation types where precipitation arrives as rain, the amount of precipitation intercepted by the canopy is linearly related to storm size. However, in two studies in the Usambara Mountains, Tanzania, Jackson (1971, 1975) established semilogarithmic relationships, and Schulze et al. (1978) applied the same form of equation to the relationship between interception by Pinus patula plantation and duration of storms. The proportion of rain intercepted varies considerably between species and even within species growing under different conditions, as may be seen in Table I. For example, for rain events between 12.5 and 25 mm, interception may vary from 2.2% for Acacia aneura in central Australia (Slatyer, 1965) to 66.3% for Tectona grandis forest in Thailand (Chunkao et al., 1971). Even within taxonomic groups, interception varies widely, from 2.2% quoted above for Acacia aneura to 13% for the same species in semiarid Queensland, Australia (Pressland, 1973), and to 39.1% for Acacia catechu plantation in India (Dabral and Subba Rao, 1968). Among conifers, interception for rain events between 12.5 and 25 mm varies from 6.6% for Pinus caribaea plantation in Sao Paulo, Brazil (Lima, 1976), to 10% for P. roxburghii forest in West Pakistan (Raeder-Roitzsch and Masrur, 1969b) and 29.3% for P. roxburghii plantation in India (Dabral and Subba Rao, 1968). Hopkins (1960) considered that most of the interception by a tropical forest in Uganda was due to a Dracaena fragrans herb layer and not to the tree canopy.

a. Canopy Storage. It may be expected that the interception percentages would be influenced by the minimum rainfall resulting in throughfall or stemflow. This expectation is borne out in comparisons between species of Acacia, with A. aneura yielding throughfall after 0.5 or 0.8 mm rain and stemflow after 1.5 or 2.5 mm rain (Slatyer, 1965; Pressland, 1973). A. harpophylla showed no minimum rainfall for throughfall, but required at least 5 mm in a storm for stemflow to occur (Tunstall, 1973), whereas A. catechu yielded stemflow only after more than 12.5 mm of rain (Dabral and Subba Rao, 1969). The teak forest studied by Chunkao et al. (1971) showed no minimum rain event for throughfall, and stemflow was re-

TABLE I

PARTITIONING OF PRECIPITATION IN SOME TROPICAL AND SUBTROPICAL FORESTS AND WOODLANDS[a]

Vegetation type	Location	Basis of calculation	Trees (ha⁻¹)	P (mm)	SC (mm)	I (%)	SF (%)	TF (%)	Reference
Humid Forest									
Cloud forest	Puerto Rico	Total area			All	—	21.4	78.6	Baynton (1969)
Rain forest	El Verde, Puerto Rico	Crown area			2.5		0	8.5	Sollins and Drewry (1970)
					15.0		0.9	52.5	
Manilkara bidentata									
					All	27		73	
Dacryodes excelsa									
Tabanuco forest	Puerto Rico	Crown area		3300	All	38		62	Clegg (1963)
					2.8	64	—	(36)	
					22.9	52	—	(48)	
					All	57	—	(43)	
Upland forest	Macabé, Mauritius	Total area		3094	All	30.4	—	69.6	Vaughan and Wiehe (1947)
Lowland dipterocarp forest	Jelebu, Malaysia	Total area		2381	All	21.8	0.6	77.6	Manokaran (1979)
Dipterocarp forest	Selangor, Malaysia	Total area		2500	All	18	0	82	Kenworthy (1971)
Hill evergreen forest	Chiengmai, Thailand			1955	0–12.5	16.1	0.4	83.5	Chunkao et al. (1971)
					25–50	6.9	0.5	92.6	
					All	13.8	0.4	85.8	
Hill evergreen forest	Chiengmai, Thailand	Total area	593	1955	All	6.2	1.3	92.5	Tangtham (1970)
Teak forest	Lampang, Thailand				0–12.5	44.7	0	55.3	Chunkao et al. (1971)
					>50	60.2	0	39.8	
					All	63.3	0	36.7	
Dry dipterocarp forest	Lampang, Thailand				0–12.5	46.1	0.1	53.8	Chunkao et al. (1971)
					>50	60.7	0	39.3	
					All	60.8	0	39.2	
Dry dipterocarp forest	Nakhourajasima, Thailand				0–12.5	31.3	0.7	68.0	Chunkao et al. (1971)
					>25	5.5	0.9	93.6	
					All	13.8	0.8	95.4	

Forest type	Location	Basis		P	SC	I	SF	TF	Reference
Tectona grandis plantation	Dehra Dun, India	Crown area	472	2160	0–12.7	36.7	8.8	54.5	Dabral and Subba Rao (1968)
Shorea robusta plantation	Dehra Dun, India	Crown area	668	2160	0–12.7	56.6	1.9	41.5	Dabral and Subba Rao (1969)
					>50.8	27.9	9.5	62.6	
Acacia catechu plantation	Dehra Dun, India	Crown area	574	2160	0–12.7	52.8	0	47.2	Dabral and Subba Rao (1969)
					>50.8	9.0	6.1	84.9	
Pinus roxburghii plantation	Dehra Dun, India	Crown area	1156	2160	0–12.7	33.5	3.0	63.5	Dabral and Subba Rao (1968)
					>50.8	1.7	6.6	91.7	
Seasonal Forest									
Shorea robusta plantation	Arabari, India	Total area	1678		1–10	35.4	5.3	59.3	Ray (1970)
					61–70	16.5	9.3	74.2	
Alstonia scholaris plantation	Arabari, India	Total area	1675		1–10	30.4	20.4	48.7	Ray (1970)
					41–50	36.3	13.4	50.3	
Pinus roxburghii forest	Pakistan	Total area	550	1462	0–2.5	35.6	1.8	62.5	Raeder-Roitzsch and Masrur (1969b)
					>50.8	3.1	1.9	94.9	
Pinus caribaea plantation	Piracicaba, Brazil	Total area	1670	1280	All	6.6	3.0	90.4	Lima (1976)
Eucalyptus saligna plantation	Piracicaba, Brazil	Total area	1670	1280	All	12.2	4.2	83.6	Lima (1976)
Dry Forest									
Eucalyptus camaldulensis plantation	Ilanot, Israel	Total area	1100	762	All	14.3	3.3	82.4	Heth and Karschon (1963)
Eucalyptus camaldulensis plantation	Ilanot, Israel	Total area	1100	517		14.6	4.5	80.8	Karschon and Heth (1967)
Eucalyptus camaldulensis plantation	Ilanot, Israel	Total area	870	441	All	5.3	1.8	92.9	Karschon (1971)
Eucalyptus camaldulensis coppice 1–2 yr. old		Total area	870	563	All	7.1	2.0	90.0	
3–4 yr. old		Total area		618	0–12.5	6.2	0.3	93.5	
Acacia harpophylla forest	Meandarra, Australia	Total area			>50	16.1	2.0	81.9	Tunstall (1973)
					All	15.2	1.8	83.0	
Acacia aneura woodland	Charleville, Australia	Total area			All	13	18	69	Pressland (1973)
Acacia aneura woodland	Alice Springs, Australia	Total area	250		0–2.5	59.6	0	40.4	Slatyer (1965)
					>25	0.4	38.9	60.7	

[a] **Key to abbreviations:** P, precipitation; SC, storm class; I, interception; SF, stem flow; TF, throughfall.

corded after a storm of less than 6 mm, despite the very small percentage of any storm which appeared as stem flow. Canopy saturation in a mixed forest in Tanzania required 0.89 mm of precipitation (Jackson, 1975), compared with 3 mm for *Pinus patula* (Schulze et al., 1978). Interception by the canopy of a montane cloud forest in Puerto Rico was 0.89 mm, regardless of storm size (Baynton, 1969). This was associated with the low stature of the forest, which was only 3 to 4 m in height. If the canopy was already wet, there was no interception loss at the commencement of a storm due to low rates of evaporation in this forest. The minimum storm resulting in throughfall will be related to the degree of canopy cover, leaf orientation, water retaining capacity of the crown, and shaking of leaves by wind (Jackson, 1975; Manokaran, 1979). These attributes are not easy to measure, and quantitative studies of the duration of leaf wetness have not been applied to tropical species.

b. Rainfall Intensity. Forest types in Table I may be divided into three groups, depending on whether with increasing precipitation per event the percentage intercepted increases, remains constant, or decreases. An increasing interception percentage was recorded in *Acacia harpophylla* woodland (Tunstall, 1973) and *Pinus patula* plantation (Schulze et al., 1978). The second group, with a constant interception percentage, includes *Tectona grandis*, dry dipterocarp, and hill evergreen forest types (Chunkao et al., 1971), *Alstonia scholaris* plantation (Ray, 1970), and *Acacia aneura* woodland (Pressland, 1973). Forest types in which a decreasing interception percentage was recorded with increasing storm class included tropical upland rain forest (Vaughan and Wiehe, 1947), lowland dipterocarp forest (Manokaran, 1979), *Pinus roxburghii* plantation and forest (Dabral and Subba Rao, 1968; Raeder-Roitzsch and Masrur, 1969b), plantations of *Tectona grandis* (Dabral and Subba Rao, 1968), *Shorea robusta* (Dabral and Subba Rao, 1968; Ray, 1970), *A. catechu* (Dabral and Subba Rao, 1969), *Acacia aneura* woodland (Slatyer, 1965), and dry evergreen forest (Chunkao et al., 1971).

Factors contributing to these differences in behavior can be considered in connection with the two studies on *Acacia aneura* by Slatyer (1965) and Pressland (1973), since they show different responses. Pressland reported that *A. aneura* intercepted approximately 12% of the rainfall, irrespective of intensity, whereas Slatyer concluded that interception accounted for a constant 1.5 to 2.5 mm for all rain events above 2.5 mm. In this species, tree architecture varies noticeably with provenance (Everist, 1949), central Australian individuals having upward-slopiing branches and vertically held phyllodes, whereas those in the semiarid areas of eastern

Australia have more horizontal branches, without a definite orientation of phyllodes.

c. Stem Basal Area. Dabral and Subba Rao (1968) found that in a plantation of *Pinus roxburghii* with a relatively high crown cover there was a slight increase in interception with increased stem basal area of individual trees between 45 and 615 cm² particularly at rainfall intensities of less than about 50 mm. For *Tectona grandis,* there was no consistent trend in interception loss with stem basal area with rainfall groups of more than 25 mm (Dabral and Subba Rao, 1968). Pressland (1973) showed that the percentage of rainfall intercepted by individual *Acacia aneura* trees increased with stem basal area, probably due to an increased contribution of branches to the canopy, since leaf area index decreased with increasing stem basal area.

Comparisons between the data on interception, stemflow, and throughfall in Table I are made somewhat difficult by the fact that stem basal area or crown projection figures are not available for most studies. Hence, the differences in interception on a land area basis may be due to variation in stand density or to other characteristics of the trees. Pressland (1976b) established linear regressions for throughfall in *Acacia aneura* woodland for four densities of trees. Throughfall percentages for stands with 40, 160, 640, and 4000 trees ha⁻¹ were 94, 93, 88, and 86, respectively. These values were all significantly different, but it is interesting to note the relatively small absolute effect of increasing stand density on throughfall.

Jackson (1971) suggested that considerable variation in interception and throughfall in mixed forest in Tanzania, even within rainfall intensity classes, was associated with seasonal differences in leaf cover. This difference is obvious in deciduous woods (Reynolds and Henderson, 1967), but less so in evergreen forests, and it would probably be difficult to establish in *Acacia aneura* woodlands. However, Rutter *et al.* (1971) and Pressland (1973) emphasized the importance of the duration of rainfall events and the occurrence of short rainless periods during an individual event, when evaporation of intercepted water could occur. A similar conclusion was reached by Jackson (1975) from studies of mixed forest in Tanzania, and this phenomenon was quantified by Schulze *et al.* (1978) for a *Pinus patula* plantation in South Africa. The latter investigators showed that interception was a linear function of the logarithm of rainfall duration up to 12 hr, and that interception for a given duration of rainfall increased with rainfall intensity. The complexity of these effects is such that the only satisfactory solutions are likely to emerge from very detailed studies, such as those of

Rutter (1963), Rutter *et al.* (1971, 1975), Rutter and Morton (1977), and Calder (1978).

D. STEMFLOW AND INFILTRATION

1. Stemflow

Drainage of water from foliage to branches and trunks of trees is influenced by the amount of precipitation intercepted, the disposition of leaves and branches, and the roughness and water-holding capacity of the bark. Environmental conditions that influence the evaporation of intercepted water, such as air temperature and humidity, and the interval between successive storms also influence the proportion of precipitation that is recorded as stemflow.

Table I shows that, over a range of storm classes, the proportion of rainfall appearing as stemflow varies widely, from <0.1% in dry dipterocarp forest (Chunkao *et al.*, 1971) to 41% in *Acacia aneura* (Slatyer, 1965). The minimum storm resulting in stemflow varies from slightly more than 1 mm in *Acacia aneura* (Slatyer, 1965; Pressland, 1973) and *Pentacme contorta* (Fellizar, 1976) to 12.7 mm in *Acacia catechu* (Dabral and Subba Rao, 1969). In most studies stemflow increases in proportion as storm class increases, but it remains a constant proportion in some species, including *Alstonia scholaris* (Ray, 1970), *Pentacme contorta* and *Parashorea plicata* (Fellizar, 1976), *Tectona grandis* and dry dipterocarp forests (Chunkao *et al.*, 1971), *Acacia aneura* (Pressland, 1976b) and *Acacia harpophylla* (Tunstall, 1973), and in montane rain forest (Baynton, 1969). A few studies indicating a declining stem flow proportion with increasing precipitation include one in a plantation of *Tectona grandis* (Dabral and Subba Rao, 1968).

Where more than one study has dealt with a particular species, it is common for the stemflow estimates to differ considerably. For example, stemflow in *Tectona grandis* forest in Thailand showed a mean value of 0.02% (Tangtham, 1970; Chunkao *et al.*, 1971), whereas in an Indian plantation it varied between 3.6 and 8.8% with storm class (Dabral and Subba Rao, 1968). The latter investigators found variations in branch habit of *Tectona grandis* and *Pinus roxburghii* but did not associate them with differences in stemflow characteristics of individual trees.

Similarly, stemflow in *Acacia aneura* varied from a mean of 13% (Pressland, 1976b) to 14 to 41% depending on storm class (Slatyer, 1965). Differences in crown architecture can account for some, though probably not all, of these differences (Pressland, 1976b). It is not possible without further information to attribute any discrepancy to differences in methodology.

In a lowland dipterocarp forest in Peninsular Malaysia, Manokaran (1979) found general correspondence between stem basal area and the volume of stemflow, both in absolute terms for individual trees and as a percentage of the total stemflow for the 100-m² study area. However, the greatest stemflow was contributed by one of the smallest understory trees examined. *Knema malayana*, with a stem basal area of 0.013 m² contributed 52.4% of the total annual stemflow, whereas an overstory specimen of *Dipterocarpus cornutus*, with a stem basal area of 0.156 m², contributed 18.2%. The branches of *K. malayana* were ascending in habit and may have come into contact with canopies of other trees, collecting water that may otherwise have appeared as throughfall or stemflow on other boles.

The concentration of water at the base of the stem through stemflow of intercepted precipitation has been described as an important factor in the water relations of arid or semiarid zone species such as *Balanites aegyptica* (Glover et al., 1962) and *Acacia aneura* (Slatyer, 1961a, 1965; Pressland, 1976b).

2. Interception by Leaf Litter

Rainfall reaching the leaf litter may be evaporated, drain slowly to the soil beneath, or may be extracted by surface roots of trees. A widely recognized benefit of litter is the reduction, or at least retardation, of surface drainage (Soerjono, 1964; Bailly et al., 1974), particularly in areas of lower rainfall which have been subjected to heavy grazing pressures and other abuses, with a consequent decline in the hydrological quality of the site (Pereira, 1967; Raeder-Roitzsch and Masrur, 1969a). Losses to the soil water store of rainfall intercepted by the litter may be regarded as a small price to pay for soil stability and enhanced total infiltration.

In tropical lowland rain forests, the rate of decay of litter may be so great that there is practically no accumulation of water at the surface of the soil (e.g., Gilmour, 1975). In other situations, where litter fall may be appreciable but the rate of decay is slower, actual interception by litter may be significant and may be expected to vary with storm class and the degree of wetness of the litter. Dabral et al. (1963) used "undisturbed" litter samples of 0.6 to 2.2 kg m⁻² from four forest types and showed that interception was independent of storm class between 1 and 50 mm and varied from 1.2 mm for *Dendrocalamus strictus* to 2.0 mm for *Shorea robusta*. A similar experiment by Pradhan (1973) used packed litter samples of about 17.3 kg m⁻² in storms of 6.8 to 154.4 mm. Two species, *Tectona grandis* and *Dendrocalamus strictus*, were common to the two studies. When allowance is made for the fact that the litter samples may not have been saturated in storms less than about 50 mm in Pradhan's (1973) work, the interception in heavier storms is more or less independent of rainfall, and the amount of interception per unit weight of litter is quite similar for the two studies (Table II).

3. Infiltration

Infiltration of rainfall into soil depends primarily on the volume of large pore spaces in the soil and therefore on its bulk density. Undisturbed forest soils commonly have relatively low bulk densities at the surface (Wilde, 1958), and in comparison with cultivated or grazed lands, they also have high rates of infiltration of water in both dry and saturated conditions. An example of this is shown by Inthasothi and Chunkao (1976), who compared natural forests, *Pinus kesiya* plantation, tea plantation, and an old cultivated area in Thailand (Table III). The developed sites had initial infiltration rates that were comparable with those for forest, but under artificial rain conditions, the constant rates of infiltration into the forest soils were 7 to 145% greater than that for the old agricultural soil. Chunkao and Naksiri (1976) observed that infiltration of water into soils beneath hill evergreen forest in the same locality did not vary greatly with increasing slope of the ground surface (Table III). This capacity of undisturbed forest soils to convey water into the profile is very important for reduction of the surface drainage peaks (Pereira, 1973). In the present context, its importance also lies in the recharge of the soil water store for the maintenance of vegetation on hill slopes during rainless periods.

TABLE II

INTERCEPTION OF RAINFALL BY LEAF LITTER OF MONSOON FOREST, INDIA

		Interception		
Species	Litter weight (kg m^{-2})	mm	mm kg^{-1}m^{-2}	Reference[a]
Tectona grandis	1.238	1.961	1.584	1
	17.361	24.76	1.426	2
Dendrocalamus strictus	0.708	1.242	1.754	1
	17.361	21.72	1.251	2
Shorea robusta	1.263	2.019	1.598	1
Pinus roxburghii	1.964	1.689	0.860	1
Acacia arabica	17.361	22.12	1.274	2
Dalbergia sissoo	17.361	19.78	1.139	2
Albizia lebbek	17.361	14.68	0.845	2

[a]References: (1) Dabral *et al.* (1963); (2) Pradham (1973).

TABLE III

RATES OF INFILTRATION OF WATER INTO FOREST AND CULTIVATED TROPICAL SOILS

| Forest type | Infiltration rate (μm sec^{-1}) | | Reference |
	Initial	Constant	
Hill evergreen			
Slope 0–25%	239.4	85.5	Chunkao and Naksiri (1976)
Slope 25–50%	198.9	81.9	Chunkao and Naksiri (1976)
Slope 50–75%	244.2	76.4	Chunkao and Naksiri (1976)
Slope 75–100%	225.5	66.7	Chunkao and Naksiri (1976)
Dry dipterocarp forest	139.4	58.5	Inthasothi and Chunkao (1976)
Mixed deciduous forest	154.3	89.0	Inthasothi and Chunkao (1976)
Evergreen forest	210.0	133.3	Inthasothi and Chunkao (1976)
Pinus kesiya plantation	234.2	133.9	Inthasothi and Chunkao (1976)
Tea plantation	196.7	73.3	Inthasothi and Chunkao (1976)
Old cultivation	170.8	54.7	Inthasothi and Chunkao (1976)

In drier regions, infiltration, particularly of stemflow water, may be critical to the water balance of perennial vegetation. The entry of most of the stemflow into the soil within about 50 cm from the base of *Acacia aneura* trees has been attributed to a higher rate of infiltration in this zone than in treeless intergrove areas (Slatyer, 1961a) or near the margins of canopies in uniformly stocked woodland (Pressland, 1976b). The water-holding capacity of soil close to the stem base is also about 50% greater than that between groves (Slatyer, 1961a), and the amount of water added to the soil within the zone of infiltration may be up to 200% greater than that due to precipitation outside the canopy (Pressland, 1976b). A similar phenomenon was described for alternating "ripples" of tree and shrub and vegetation in Nigeria, where infiltration into the soil beneath the trees was faster than into soil under shrubs (Clayton, 1966).

III. ABSORPTION OF WATER

A. GENERAL CHARACTERISTICS

Perennial vegetation requires a continuous supply of water for survival, whereas annual species can become independent of the environment during the phase of seed dormancy (Koller, 1969). The occurrence of trees and shrubs is therefore associated with areas of higher precipitation, or in semiarid and arid regions with locations where soil depth and texture permit accumulation of water in the profile and its availability to plant roots (Morison *et al.,* 1948; Slatyer, 1965; Eiten, 1972; Walter, 1972).

These distribution patterns are emphasized in tropical and subtropical latitudes, where potential evapotranspiration is greater than at temperate latitudes. A further limitation to the occurrence of woody plants is that they do not normally persist in areas subjected to alternate waterlogging and severe drought (van Donselaar, 1965). Finally, growth forms may be considerably modified by the nutritional status of soils (Grubb, 1977), with high nutrient contents tending to be associated with mesic plant forms (Webb, 1969) and low nutrient contents with sclerophylly (Arens, 1958; Beadle, 1962; Goodland and Pollard, 1973) and, in extreme cases, the development of geoxylic suffrutices (Rawitscher, 1948; van Donselaar-ten Bokkel Huinink, 1966; White, 1977).

The roles of soil and root system characteristics in uptake of water by agricultural crop species have been reviewed in detail by Newman (1976), Tinker (1976), and Taylor and Klepper (1978). It is useful to examine two important characteristics of soils and root systems of trees in particular that may influence their water relations before surveying the combinations of these characteristics that have been observed in the field.

1. Root Density and Permeability

It is accepted that, at least for grasses and cereals, the supply of water to the plant is determined principally by root density, the length of roots per unit volume of soil (Greacen et al., 1976). This assumption is tenable for annuals, where at any stage of development the greater part of the root system may be composed of newly elongated members (Tinker, 1969; Hackett, 1973). On the other hand, Kramer and Bullock (1966) and Caldwell (1976) indicated that a relatively small proportion of the root surface in perennial species may be composed of actively growing roots, and that absorption of water through suberized root members, particularly at points of damage or rupture due to branching, may be quantitatively most important. Under these circumstances, the density of the enduring, though not necessarily permanent, fine root system becomes critical to the water balance of a tree.

One of the costs to the metabolism of a perennial plant is the maintenance of these roots or their regeneration under favorable conditions. Even though suberization of the root surface may extend to the apical meristem under drought conditions in some species (Leshem, 1965), minimizing the loss of water from roots to dry soil, the enduring root members must have available a continuing supply of water. Some of this may become available to roots near the soil surface through vapor transport at night from lower soil layers (Migahid, 1961; Stark and Love, 1969) or by redistribution from other more deeply penetrating parts of the root system.

2. Diffusivity of Water in the Soil

Under normal conditions of plant growth, the rate of supply of water from the soil to the root surface will depend on the water diffusivity of the soil, that is, its hydraulic conductivity in the unsaturated state. This varies considerably with soil water content and texture and also depends on whether the soil is drying or wetting (Nye and Tinker, 1977). Common values of water diffusivity in relatively dry soils range from about 10^{-5} to 10^{-4} cm^{2}sec^{-1} (Hillel, 1971), with higher values associated with coarse textured soils. These greater conductivities only occur at higher water contents; dry soils of any texture have similar low conductivities (Gardner, 1965). The flux of water toward unit length root (I_w) is a function of root radius (r), diffusivity (D), and the gradient in soil water potential around the root ($d\psi/dr$) (Nye and Tinker, 1977)

$$I_w = 2\,rD\,\frac{d\psi}{dr} \tag{5}$$

Diffusivity is itself a function of soil water content (θ) or water potential (ψ) (Caldwell, 1976).

$$D = m \exp (g\theta) \tag{6}$$

where m and g are constants.

It follows that an increase in the value of D by a factor of 10 will compensate for a lessening of the ψ gradient of the same order without any reduction in the water flux. Therefore, sparse rooting or roots of low surface permeability do not constitute a serious disadvantage in coarse-textured soils, whereas they may be inimical to satisfaction of the transpirational demand of plants in fine textured soils. Hillel (1971) and Nye and Tinker (1977) emphasized that nonuniformity of many factors, including soil texture, structure, and temperature, complicate the process of water movement in unsaturated soils to the extent that detailed quantitative relationships between soil water content and uptake rates are difficult to establish. A further complication is the growth of new roots in the soil volume already explored by the perennial root system (Caldwell, 1976).

3. The Volume of Available Water

Sandy soils may not necessarily be more favorable for tree growth than clay soils because of the low volume of soil water that remains available to plants after free drainage. Sarlin (1964) developed an empirical formula relating available water (W_A) to the fractions of sand (S) and clay (C) in the profile, bulk density (ρ_B), and soil depth (Z).

$$W_A = (S + C)\,\rho_B Z \tag{7}$$

Using the assumption that tropical evergreen vegetation withdrew 100 mm of water per month from the soil, Sarlin suggested the following associations between soils and forest types: (a) No dry season or dry season of less than 1.5 months permits evergreen forest; (b) dry season between 1.5 and 3 months permits evergreen forests on clay soils, and deciduous forests on sandy soils; and (c) dry season longer than 3 months affects vegetation principally on sandy soils.

4. Aeration

Trees with special adaptations in their root systems will survive under conditions of regular and sometimes extended flooding (Jenik, 1967), but for most species, root development is limited to the more or less well-aerated soil horizons. In some situations, the demarcation between forest types is very clear, and can be attributed to factors such as aeration (Tracey, 1969), although the same demarcations are also associated with differences in soil chemistry (Webb, 1969). These two soil attributes may themselves be associated with, and due to, differences in parent material and the processes of soil formation.

Root relations of trees may be influenced by the formation of grafts. This phenomenon has been closely studied for temperate latitude species (Graham and Bormann, 1966), but little has been published on the existence, not to mention the function, of root grafts in tropical species. Corner (1940) and Ng (1975) comment on the development of extensive grafting between the roots of strangler figs (*Ficus* spp.), which begin as aerial roots and eventually fuse to perform the function of a stem. Ng (1975) recorded root grafts in 16 species, of which 9 belonged to the Dipterocarpaceae, with 5 of these being species of *Shorea*. It is unlikely that the frequency of root grafts in most species would be sufficient to contribute substantially to the water balance of a large tree, although it could prolong the life of a damaged or disadvantaged individual. Close contact between adjacent roots of different plants may transfer sufficient water to enable a severely droughted small plant to survive (Gindel, 1973).

B. HUMID FORESTS

1. Roots in the Litter and Air

In extremely humid situations, the whole soil profile may be wet for almost the whole year (Lyford, 1969) and mucky soils may develop, with most of the roots concentrated close to the surface or even being aerial in nature (Gill, 1969). Lyford (1969) reported that the amount of root material in the mucky forest-floor horizons of a montane rain forest in Puerto

Rico was 577 gm m^{-2}. Many of the roots were free from adhering soil particles, presumably because they had developed between the freshly fallen leaves and the decomposing organic muck beneath. Few roots appeared to extend to soil depths greater than about 10 cm. Soil of a type also occurred in the bases of epiphytes growing on tree stems, and the amount of root material here was estimated to be 125 gm m^{-2}. Air-layering in the peaty bases of the epiphytic fern, *Platycerium grande,* was observed by Herbert (1959) in *Ficus watkinsiana* and *Macadamia tenuifolia.*

Gill (1969) reported the formation of aerial roots in 21 species of trees and shrubs in a montane rain forest. They generally grew without branching for up to 3 m in the absence of damage, although on reaching the soil surface there was profuse branching. In some species, particularly *Hedyosmum arborescens,* the root tip is covered with a gelatinous material that was presumed to prevent desiccation of the root tip. The capacity of these roots to absorb water was not determined, although van den Honert (1941) showed that aerial roots of the climber *Cissus sicyoides* were effective in water uptake.

2. Roots in the Soil

The root mass of tropical forests is important in nutrient cycling (Grubb, 1977) and in maintenance of stability in wet soils (Henwood, 1973). It may be anticipated, then, that in either wet or dry forest situations, the biomass of roots will be commensurate with that of shoots. In fact, considerable variation in root biomass has been reported for tropical forests, as may be seen in Table IV. Part of this variation is due to the methods of sampling used. For example, auger and monolith samples are usually taken at some distance from the base of trees in order to avoid the very large skeleton roots or root buttresses (Mensah and Jenik, 1968; Klinge, 1973). This automatically excludes most of the mass of the root system, but concentrates attention on the fine roots, which are of greater interest from the point of view of water and nutrient uptake.

a. Root Density. For assessing the water relations of plants, the data of greatest value are the lengths of roots per unit volume of soil. Very few studies of this attribute have been made for tropical forests, the most detailed being that by Klinge (1973) at two sites in the Amazon basin. Figure 1 shows the relationship between root size, length per unit volume, and soil depth for a latosol and humus podzol. A much greater proportion of the roots occurred in the 1–5 mm diameter class in the podzol than in the latosol profile. The lengths of roots per unit volume in these soils are small in comparisons with grasses, but more extensive root development may be expected in other forest types where greater weights of fine roots have been observed but lengths not measured.

TABLE IV

DISTRIBUTION OF ROOT MASS (MG CM^{-3}) WITH DEPTH IN HUMID FOREST SOILS[a]

Rainforest, depth (cm)	Latosol (diam. in mm)[b]			
	<2	2–10	10–50	Total <50
0–2	2.50	1.25	0.10	3.85
2–4	4.26	6.77	1.9	13.02
4–6	6.29	10.56	15.38	32.23
6–16	1.50	1.87	3.69	7.06
16–27	1.10	1.81	2.08	4.99
27–47	0.51	0.54	2.01	3.06
47–69	0.42	0.48	1.24	2.04
69–89	0.32	0.50	1.24	2.06
89–107	2.07	0.13	0.86	3.06

Rainforest, depth (cm)	Esukawkaw (diam. in mm)[c]			
	<2	2–10	>10	Total
0–5	5.49	5.00	0	10.49
5–10	2.53	0.43	2.07	5.03
10–20	0.64	0.29	8.17	9.10
20–30	0.02	0	0	0.02
30–40	0.02	0	0	0.02
40–50	0.02	0	0	0.02

Rainforest, depth (cm)	Humus podzol (diam. in mm)[b]			
	<2	2–10	10–50	Total <50
0–2	0.05	0	0	0.05
2–3	2.60	0	0	2.60
3–6	3.32	0	0	3.32
6–9	3.82	8.99	0	12.81
9–11	5.72	9.08	73.12	87.92
11–13	1.24	0.08	0	1.32
13–19	0.46	0.64	0	1.10
19–29	0.90	1.55	0	2.45
29–49	0.15	0.40	0	0.55
49–89	0.07	0.16	0	0.23

Rainforest, depth (cm)	Kade No. 1 (diam. in mm)[c]			
	<2	2–10	>10	Total
0–5	5.33	2.27	0	7.60
5–10	1.18	0.70	0	1.88
10–20	0.80	0.23	0	1.03
20–30	0.43	0.56	1.60	2.59
30–40	0.27	0.32	0	0.59
40–50	0.22	0.32	0	0.54

Evergreen forest, depth (cm)	Plateau (diam. in mm)[d]			
	<2	2–5	>5	Total
0–10	0.258	0.077	0.183	0.518
10–20	0.091	0.058	0.319	0.468
20–30	0.055	0.052	0.145	0.252
30–40	0.055	0.031	0.180	0.266
40–50	0.040	0.022	0.080	0.142
50–70	0.028	0.011	0.081	0.120
70–90	0.038	0.005	0.090	0.133
90–110	0.028	0.007	0.071	0.098
110–130	0.014	0.006	0.040	0.060

Evergreen forest, depth (cm)	Valley (diam. in mm)[d]			
	<2	2–10	>10	Total
0–10	0.223	0.313	0.530	1.066
10–20	0.079	0.158	0.325	0.562
20–30	0.061	0.070	0.090	0.221
30–40	0.050	0.102	0.025	0.177
40–50	0.039	0.037	0	0.076
50–70	0.017	0.012	0	0.029
70–90	0.012	0.008	0	0.020
90–110	0.010	0.008	0	0.018
110–130	0.012	0.005	0	0.017

[a] Calculated from author's data.
[b] Brazil; from Klinge (1973).
[c] Ghana, *Chlorophora excelsa* rainforest; from Mensah and Jenik (1968).
[d] Ivory Coast; from Huttel (1975).

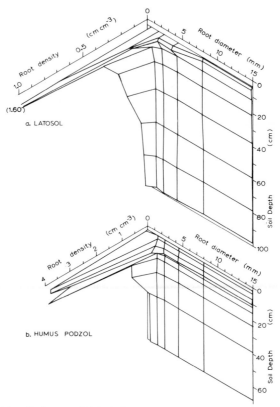

Fig. 1. Root density (cm cm⁻³) in relation to root diameter and soil depth in (a) a latosol, and (b) a humus podzol under rain forest in Brazil. (Drawn from data of Klinge, 1973.)

b. Root Mass. A more common dimension of root systems is the dry weight of roots per unit volume of soil. In these estimations, the roots may be separated into size classes, but the very large ratio of root length to weight of fine roots, and the frequent loss of much of the fine root fraction during extraction, means that little precise information is available on the extent of root surface active in water and nutrient uptake in these forests.

One of the simplest techniques for estimating root development of a species is to sample a pure stand, such as a plantation. Hosegood and Howland (1966) studied root distribution beneath 8-yr-old plantations of *Eucalyptus saligna, Pinus patula, Cupressus benthamii,* and *Cupressus lusitanica* at Muguga, Kenya. Auger samples were taken at the midpoint between four trees planted at a spacing of 2.7 × 2.7 m. In all species, the greatest density of roots occurred in the zone 0 to 30 cm from the surface and the root systems penetrated to 4.86 m. At the end of the dry season, all portions of the soil profile were dried close to -1.5 MPa despite the sparse root distribution at greater depth.

Individual trees of different ages were excavated in a teak plantation in Thailand in order to assess distribution and total dry weight of roots (Ngampongsai, 1973). Penetration of the taproot was limited to less than 1 m for all trees studied up to the age of 20 years, and in three of the five trees excavated, all major lateral roots occurred at depths less than 30 cm. The vertical growth of the root system appeared to cease at about 0.8 m, before the trees reached 5 years of age, and lateral extension to about 5.5 m was completed between 5 and 10 years of age. However, the biomass of the root system continued to increase up to age 20, presumably by secondary development of the skeleton roots, since their total length did not increase after the trees were more than about 5 years old. The volume of soil available for water supply to these trees is, therefore, rather limited, and it must be expected that the water supplies of the root zone would be exhausted after a brief drought unless they were supplemented by subsurface drainage.

This conclusion is supported by observations in the teak-growing areas of India, Burma, and Indonesia (Seth and Yadav, 1959) that the most favored sites are well-drained alluvial soils near river banks. Poor drainage or hardpan formations in the soil lead to deterioration of the root systems in older trees.

Mensah and Jenik (1968) showed a strong taproot development in seedlings of *Chlorophora excelsa*, there being linear relationships between the thickness of the taproot at the root collar and both plant height and total length of the taproot and primary laterals (the skeleton roots). The lateral extent of these roots in mature trees is up to 7-m radius, and near their ends sinker roots develop that branch profusely in the more compact soil layers at depths of 50 to 70 cm.

The depth at which the greatest root development occurs is influenced considerably by soil type and topographic situation. Lawson et al. (1970) studied a moist semideciduous forest in Ghana, and found at upper and lower slope situations a concentration of roots, particularly of larger diameter, in the upper 10 cm of the profile. In a midslope location, there was the usual concentration of fine roots at the surface, but appreciable development of larger roots at 20–30 cm depth. These patterns are presumably related to a combination of conditions that may result in low water supply at depth in the upper slope and deficient aeration in the lower slope profile, with a permanent supply of groundwater moving through the midslope soil. More detailed analysis of root development and function in these situations would require information on physical characteristics of the soil with regard to water transport, seasonal water regime, and the inherent growth patterns of the species concerned.

C. Dry Forests and Shrublands

It has been recognized for many years (e.g., Troup, 1921), that trees growing in dry areas usually require a relatively large volume of fairly coarse-textured soil in which their root systems may develop. This is associated with the generally sparse nature of the root systems of trees (Tables IV and V) and the importance of hydraulic conductivity of the soil in the supply of water to the roots. The possibilities of establishing trees in semiarid and arid regions depend on the soil characteristics (Kaushik *et al.*, 1969) and, under natural conditions, upon a relatively infrequent sequence of seasons favorable for production of seed and establishment of the resulting seedlings (Preece, 1971a,b). Where these conditions do not obtain, some manipulation of the environment is necessary, and a knowledge of the relationships between soil and root characteristics, and water balance of the shoot, is essential.

Soil and root relationships in semiarid and arid regions are dominated by the fact that rates of direct evaporation from the soil are high (Pressland, 1976a), and there may be sufficient water from the growth of perennial woody plants only if they are located in areas receiving surface or subsurface drainage. This accession of drainage water often brings with it the problem of salt, which is part of the aerosol content of the atmosphere and may be deposited as particles or in solution in raindrops (Peck and Hurle, 1972). In general, deep coarse-textured soils do not have serious salinity problems because vertical drainage tends to remove salts from the rooting zone of plants. Shallow, or fine-textured soils, on the other hand, commonly have water tables close to or at the surface for part of the year, and the salts are held in the normal rooting zones of plants.

It is possible to see a variety of responses to water stress in these environments, but the behavior of different species appears to differ more in degree than in principle as compared with the patterns of behavior in species from more humid regions. One important difference between humid and arid areas is that the latter more frequently have been the targets for afforestation projects designed to protect the soil or to provide a local source of timber or fuel. Consequently, exotic species feature more prominently in studies of water relations of woody plants, with understandable ambivalence of the results.

1. Rate of Root Growth and Regeneration

The speed of development of root systems may be crucial to the survival of seedlings in arid areas, but this must be considered in association with the speed of development of the shoot system. Bhimaya and Kaul (1966) studied four arid zone species, and found considerable differ-

TABLE V
DISTRIBUTION OF ROOT MASS[a] WITH DEPTH BENEATH SAVANNA AND THICKET, GHANA[b]

Savanna and thicket, depth (cm)[c]	Grassland (diam. in mm)				Thicket edge (diam. in mm)				Thicket center (diam. in mm)			
	<2	2–5	>5	Total	<2	2–5	>5	Total	<2	2–5	>5	Total
0–20	2.29	0	1.94	4.23	4.00	1.57	9.78	15.35	1.91	1.86	20.58	23.31
20–40	3.02	0.05	0.10	3.35	2.62	1.37	1.58	5.57	0.44	0.44	7.38	8.26
40–60	1.58	0	0	1.58	0.87	0.15	0.10	1.12	0.60	0.36	1.61	2.57
60–80	1.15	0	0	1.15	0.65	0.04	1.63	2.32	0.38	0.22	0	0.60
80–100	1.12	0	0	1.12	0.75	0.12	0	0.87	0.25	0.10	0.10	0.45
100–120	0.14	0	0	0.14	0.31	0.06	0.64	1.01	0.26	0.09	3.02	3.37

Guinean savanna, colluvial soil, depth (cm)[d]	Diam (mm)			
	<2	2–10	>10	Total
0–10	8.19	0	0	8.19
10–20	6.81	0.33	0	7.14
20–30	2.20	1.18	6.05	9.93
30–40	0.81	0	0	0.81
40–50	0.90	0.35	0	1.25
50–60	0.26	0.36	0	0.62
60–70	0.25	0.05	0	0.30

[a] In mg cm^{-3}.
[b] Calculated from authors' data.
[c] From Okali et al. (1973).
[d] From Lawson et al. (1968).

ences in the extent of development of the taproot, degree of lateral branch formation, and the depth at which lateral root development commenced. Greatest percentage of survival of seedlings in *Albizia lebbek* was associated with a short taproot but also with the greatest development of lateral branches and commencement of branching close to the soil surface. The lowest percentage of survival, in *Prosopis spicigera*, was associated with a taproot of intermediate length, a long unbranched region on the taproot, and few laterals. Evidently, exploitation of surface soil water soon after seed germination is more beneficial to the further development of the whole plant, especially if this is associated with relatively slow development of the shoot.

Another study of root development in young seedlings of *Acacia spirocarpa* var. *tortilis* and *Prosopis juliflora* showed a great deal of variation within species and insignificant differences between species in depths of penetration of roots up to the age of 10 weeks (Gupta and Balara, 1972). The variability within species would be associated with the percentage of seedlings surviving an unfavorable period, but this connection has not been established experimentally for these species.

The development of new fine roots after rain can be rapid. Hodgkinson *et al.* (1978) observed a significant increase in root density, from about 8.9 to 12.0 cm cm^{-3} beneath bushes of *Atriplex confertifolia* within 10 days of application of 48 cm of water. However, within 60 days about one-half of this increment in root density had been lost, although about 15% of the roots remained turgid throughout the summer in dry soil.

2. Distribution of Roots in the Soil

In most species, trees are not able to grow successfully where soils are waterlogged and then become seasonally very dry unless they occur as small groves or clumps located on low mounds (Okali *et al.*, 1973; Eiten, 1978). Shrubs, especially those characterized by geoxylic suffrutices (White, 1977), replace trees on these gently sloping, usually oligotrophic sites. Where waterlogging is prolonged, woody species are totally excluded. White suggested that the suffrutex growth habit is a response primarily to extremely poor nutritional conditions in sandy soils where a seasonally fluctuating water table leads to development of an impervious layer close to the surface. This restricts the soil volume available to root systems and therefore the potential growth of plants. In the Zambezian region of Southern Africa, suffrutices produce shoot growth earlier in the growing season than do grasses, and thereby escape competition for light and the diminishing water supplies. Where soils are deeper, suffrutices are replaced by trees.

a. West African Savanna and Woody Thicket. Okali *et al.* (1973) described the distribution of roots beneath a thicket on the Accra Plains, Ghana, in relation to soil properties and water supply. The thickets develop on slight mounds, commonly occupied by termitaria, and consist of dark brown sandy loam to sandy clay loams up to 1.2 m deep, whereas the grassland top soils are loamy sands less than 0.2 m deep.

A detailed study was made of the distribution of roots greater than 5 mm diameter within the soil profile according to species. More than 40% of the large roots of *Azadirachta indica, Clausena anisata, Elaeophorbia drupifera,* and *Grewia carpinifolia* were located in the upper 10 cm of soil, and no large roots were found at depths exceeding 40 cm. In five other species, *Allophylus spicatus, Ehretia cymosa, Fagara zanthoxyloides, Flacourtia flavescens,* and *Securinega virosa,* the majority of large roots were located at 10–30 cm depth with the maximum depth of large roots being 30–50 cm. In contrast to these rooting habits, *Capparis* spp. showed a more uniform distribution of large roots up to 120 cm depth, and a maximum penetration of 160 cm. This much greater rooting depth was not associated with any greater shoot development than in other species. Considerable differences in the patterns of distribution of fine roots with depth were observed between grassland and thicket (Okali *et al.,* 1973).

Under the grassland, distribution was fairly uniform in the top 40 cm of soil, whereas under the thicket, the greatest weight of fine roots was recorded in the upper 20 cm of the profile (Table V). Okali *et al.* obtained a total of 93.69 gm of roots less than 2 mm diameter per m^3 of soil from the grassland profile and an inexplicably small mass of 44.8 gm m^{-3} in soil from the thicket profile. These roots are the most important in water absorption, although their effectiveness depends upon being connected to the shoot by pathways of high hydraulic conductivity, which implies a well developed permanent root system with relatively wide vessels. However, the most outstanding differences are seen in the mass of roots greater than 5 mm diameter, these being relatively uncommon and shallow under grassland, and abundant and deeply penetrating under the thicket. It is interesting that the species with the deepest root system, *Capparis erythrocarpos,* was reported by Okali (1971) to be the most drought-resistant, and that other thicket species appeared to survive through deciduousness or extensive lateral development of their roots.

Lawson *et al.* (1968) examined several aspects of the Guinean savanna catena vegetation in northern Ghana, including the soils and development of the root system. The middle slope soils of the catena were about 70 cm deep, with a water-holding capacity decreasing from 30% at the surface to about 25% below a 50-cm depth. Many of the woody species develop a lignotuberous habit, which sprouts coppice shoots after burning, and some species produce adventitious shoots from shallow roots. Thick horizontal lateral roots are produced at a depth of 20–30 cm, below the zone of greatest concentration of grass roots (Table V).

Similar effects of small changes in depth and/or texture of soils, to which are related the volume and availability of soil water, have been noted in other vegetation types. For example, Clayton (1966) described vegetation ripples in northern Nigeria where groves of trees 100 m wide and dominated by 10-m-high *Combretum nigricans* var. *elliottii* were separated by 300–400 m wide bands of *C. nigricans* and *Guiera senegalensis* about 3 m high. The only difference between the grove and intergrove area was a greater rate of water infiltration into the soil of the former. Alternatively, brousse tigree (groves) of shrubby vegetation in the semiarid zone of the Niger Republic may be separated by more or less bare ground (Catinot, 1967). A very similar difference between grove and intergrove areas in *Acacia aneura* woodland has been described by Slatyer (1961a). The terrain in the *Acacia* woodlands is gently sloping, with the *Acacia* groves occurring on flatter terraces, and the bare ground or grassland occupying slightly steeper slopes. In contrast, the *Combretum–Guiera* ripples occur on quite flat terrain.

b. East African Savannas and Woodlands. A generally similar situation in the Sudan was described by Adams (1967). On clay soils, penetration of water was limited, as was the depth of rooting of *Acacia mellifera* in short grass scrubs with an associated and presumably compensating great lateral extent. In tall grass forests, with somewhat better soil water supply, depth of rooting increased and the lateral extent of roots was reduced. *Acacia seyal* and *Balanites aegyptica* growing together on clay soils had quite different root systems, although both had lateral roots which spread 7–8 m from the stem at a depth of 0.15–0.18 m. In *A. seyal,* a tapering taproot penetrated to about 1.2 m with relatively few deep laterals, whereas in *B. aegyptica* a complex system of strong diagonal radiating roots formed a hemispherical mass in the soil. The normal depth of wetting in these clay soils is about 35 cm (Bunting and Lea, 1962), and Adams (1967) suggested that the infiltration of stemflow in *B. aegyptica,* reported by Glover *et al.* (1962) may be crucial in extending the zone of moist soil immediately below the base of the tree. Naturally, this is only important when a large aerial shoot can develop, a condition that appears sensitive to the nutrient as well as the water status of the soil.

c. South American Savannas. Soils of the orchard savanna of Venezuela are relatively shallow with a concretionary hardpan usually between 0.5

and 1.0 m from the surface. Tree species growing in these savannas are distributed according to depth of the hardpan, duration of flooding, and texture of the soil (Foldats and Rutkis, 1968). In general, they are characteristic of the caatinga. *Bowdichia virgilioides* is the most abundant species, being found on shallow, poorly drained soils with *Curatella americana.* The lateral extent of these root systems may be up to 18 m. *Byrsonima crassifolia* occurs on deeper, better drained soils, and its roots have a much more limited lateral spread.

In the savannas of Surinam, van Donselaar (1965), and van Donselaar-ten Bokkel Huinink (1966) described vegetation types and their root systems in relation to soil conditions. Dicotyledons, including woody species, developed root system structures which could be related to the position of the water table: (1) Where the water table is at the soil surface for some time, root development is superficial only. (2) Where the water table occurs at depth, some roots penetrate to this level but no further. In this second situation, the greater part of the root system is situated in the surface soil, the distribution depending upon hydraulic conductivity of the soil. Soils with low conductivity are moistened only very close to the surface, and plants have extensive superficial root systems. Soils with high conductivity and a more or less permanent water table at depth have deeply descending root systems. Where hydraulic conductivity is high and there is no water table within several meters of the surface, the immediate interception of precipitation is critical, and root systems tend to be very extensive and superficial.

The shapes of root systems are shown in Fig. 2 and the variation in one plant species, *Tibouchina aspera,* growing on different sites, is shown in Fig. 3. This shows clearly the association between soil texture, water table height during the rainy season, and patterns of root development. Van Donselaar-ten Bokkel Huinink (1966) noted that in the *Mesoseto–Trachypogonetum* association, the sandy soil dried quickly to great depth when rainfall ceased. This was attributed to the large soil pores and high evaporation rates, but is more likely due to rapid vertical drainage and extraction of water by a few sparse, but deeply penetrating roots or to very deep subsurface drainage. The fact that the depth of root penetration of all species in the sand of *Mesoseto–Trachypogonetum* did not extend below about 1.5 m suggests the very deep subsurface drainage as the most likely fate of percolating water, since Rawitscher (1948) reported root depths of up to 18 m in the deep drained soils of the Brazilian cerrado.

Although no quantitative studies were carried out, van Donselaar-ten Bokkel Huinink (1966) emphasized the ephemeral nature of the fine roots of tree and shrub species of the savanna. Large lateral roots extended for many meters from the stem base, and only branched close to

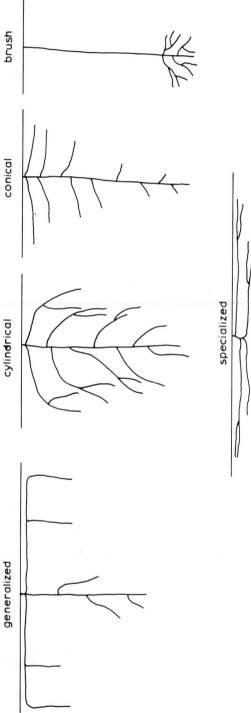

Fig. 2. Shapes of root systems of perennial dicotyledon species in Surinam savannas. (Redrawn from van Donselaar-ten Bokkel Huinink, 1966.)

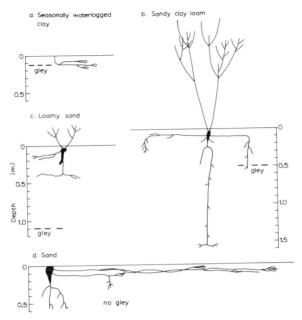

Fig. 3. Shapes of root systems of *Tibouchina aspera* growing on (a) seasonally waterlogged clay, (b) sandy clay loam, (c) loamy sand, and (d) deep sand in Surinam. (Redrawn from van Donselaar-ten Bokkel Huinink, 1966.)

the tip, producing both vertically and horizontally growing fine roots. A single root, not necessarily the original taproot, may penetrate through very dry soil to the vicinity of the water table. Van Donselaar-ten Bokkel Huinink (1966) suggested that infiltration of rainwater was greater at the stem base than elsewhere (cf. Section II,C), and that this percolated down the root surface and permitted continued vertical growth. Another explanation may be the successful pursuit of a slowly retreating water table during the onset of the dry season, the roots following a course through the soil that provides sufficient water for growth and subsequently aeration for its persistence in later rainy seasons.

The severity of seasonal drought and low nutrient status in all these soils, the apparently limited opportunities for dry matter production during the rainy season, and the additional hazard of frequent fires are reflected in the habits of the plants. A geoxylic suffrutex (xylopodium) is common, with relatively short-lived aerial shoots. The allocation of assimilates to root growth and to shoot regeneration is not known for these plants, but van Donselaar-ten Bokkel Huinink (1966) reported that a specimen of *Tibouchina aspera* that had not been burned for at least two years had a larger and deeper root system than would be anticipated for the site concerned (cf. Fig. 3). This suggested that under normal conditions regeneration of the shoot after a fire depletes reserves that may otherwise lead to root growth and an improved water balance.

d. *Australian Semiarid Woodland.* Pressland (1975) examined root distribution beneath *Acacia aneura* woodland by intensive core sampling on a concentric circular pattern around the stems of isolated mature trees. Table VI shows that the weight of roots less than 5 mm in diameter per unit volume of soil was maximal at a 15- to 30-cm depth, particularly at points beyond the spread of the canopy. The relatively high root density in the surface soil beneath the canopy may be due to shading and moderation of soil temperatures, and penetration of more roots to depths up to 1.2 m reflects the greater infiltration rates and storage of water which occurs in this zone (Pressland, 1976b). Lateral roots extended up to 14 m from the stem base of an isolated tree, and Pressland (1975) concluded that these extensive, rather shallow roots were able to absorb water effectively from light falls of rain.

e. *Effects of Irrigation.* Root development of young irrigated trees at Khartoum was investigated by Bosshard and Wunder (1966). With flood or furrow irrigation, depth of rooting was limited, generally to less than 0.5 m, particularly if the site received excessive irrigation or was relatively infertile. For planted species, these investigators emphasized the importance of treatment of trees in the nursery in determining the characteris-

TABLE VI

DISTRIBUTION OF ROOTS AROUND AN ISOLATED *ACACIA ANEURA* TREE IN SEMIARID WOODLAND AUSTRALIA[a]

	Root mass (mg cm^{-3})			
Depth (cm)	1 m from tree	2 m from tree	3 m from tree	4 m from tree
0–15	3.97	1.95	1.04	0.30
15–30	4.04	2.73	1.81	1.74
30–45	1.31	0.78	0.56	0.28
45–60	0.74	0.38	0.21	0.20
60–75	0.42	0.20	0.15	0.21
75–90	0.22	0.21	0.21	0.21
90–105	0.14	0.11	0.08	0.20
105–120	0.07	0.11	0.06	0.21

[a]Data calculated from Pressland (1975).

tics of root systems in the field. Similar conclusions have led others to root prune seedlings in the nursery to stimulate lateral root development (e.g., Bacon and Hawkins, 1977). In both *Azadirachta indica* and *Eucalyptus microtheca,* Bosshard and Wunder (1966) found that overirrigation reduced the depth of rooting significantly ($p < 0.05$) and increased the density of rooting. This was achieved in *A. indica* by a decrease in the number and length of roots, but a very great decrease in root depth. On the other hand, *E. microtheca* showed an increase in both number and length of roots under excessive irrigation, even though there was a decrease in the exploited soil volume.

Khan (1965) examined root development in *Dalbergia sissoo* planted in an arid area of Pakistan and liberally provided with irrigation water. All trees varying in age between 9 months and 9.67 years had root systems that penetrated to the water table. The lateral extension of the root system was almost complete within 5 years after planting. Khan commented that one problem if irrigation in these sandy soils was elevation of the water table and restriction of the soil volume available for exploitation by plant roots.

f. Experimental Studies. In an attempt to establish the effect of soil texture on root growth of *Shorea robusta* seedlings, Seth and Srivastava (1972) grew seedlings in artificial soils containing alternate layers of sandy loam soil and pure sand, each either 7.5 or 15 cm deep. Plants were watered as required for 6 months before harvest. The weight of roots in each layer was determined, as were height increment and total weights of shoot and root. Great variability in development of individual plants was observed, but in the treatment with 15 cm of soil in the surface layer, the mean root density was 2.12 mg cm^{-3}, whereas in the treatment with 7.5 cm of soil in the surface layer, the value was 1.49 mg cm^{-3}. On the other hand, depth of rooting was greater in the alternating 7.5-cm layers than in the 15-cm deep layers of soil and sand. No details were given concerning soil water status during the experiment, but it seems that addition of water "whenever required" may have allowed the 15-cm sand layers to dry to the point at which root growth was not possible. In order to provide sufficient available water in the sand stratum to support growth, the sandy loam strata could be undergoing impaired aeration. In the 7.5-cm sand layer, capillary movement of water into the sand was apparently just sufficient to enable root growth to proceed. Therefore, where textural changes occur in the profile, aeration is not impaired, and nutrient and water supplies are adequate, it is reasonable to expect greater root development in fine than in coarse textured soils.

D. SOIL WATER ABSORPTION PATTERNS

1. Patterns with Time and Soil Depth

The patterns of extraction of soil moisture by different vegetation types reflect the distribution of roots, which in turn reflect the conditions of water supply, nutrition, aeration, and physical characteristics of the soils. Where dense forest cover develops, the water content of the soil is often high throughout the year, and reduction of water content is more uniform with depth (Hill, 1972; Chaicharus and Chunkao, 1973; Gilmour, 1975). This may be influenced to some extent by the fact that periodic falls of rain rewet the surface soil, even though drainage to deeper soil layers may not occur. In many forests of lower rainfall regions, changes in water content near the soil surface are very rapid after the cessation of rain, particularly where direct solar radiation may reach the ground (Slatyer, 1961a; Birot and Galabert, 1972), and the amplitude of water content in the surface soil may be several times greater than at depths in excess of about 2 m (Slatyer, 1961a; Bailly *et al.*, 1974). It is in these drier forests and woodlands that most interest lies in water absorption patterns by trees, since under suitable conditions these can indicate directly the rate of evapotranspiration from the site.

Babalola and Samie (1972) followed seasonal changes in the volumetric water content of soils in the northern Guinea savanna zone of Nigeria, using the neutron scattering technique (Holmes, 1956). Natural vegetation, dominated by *Isoberlinia doka* and a 10-yr-old plantation of *Eucalyptus citriodora* were compared.

At the beginning of the wet season, soil moisture contents beneath *Isoberlinia* woodland were relatively uniform at depths between about 100 and 250 cm, varying from 0.31 to 0.26 cm^3 cm^{-3} (Fig. 4). Percolating rainwater produced a clear wetting front which, by the end of the wet season, had penetrated to 406 cm. Drying of the profile was relatively uniform at depths between about 50 and 300 cm and declined progressively at greater depths. In contrast, water contents beneath *Eucalyptus citriodora* plantation at the beginning of the wet season varied from 0.16 cm^3 cm^{-3} at 140 cm to 0.37 cm^3 cm^{-3} at about 550 cm.

It is apparent from the data in Fig. 4 that the volume of water added to the soil profile beneath *Isoberlinia* woodland during the wet season is greater than that beneath eucalypt plantation. Consequently, in the succeeding dry season, the woodland soil returned to essentially the same soil water content as pertained 12 months previously, while the soil beneath the plantation was drier throughout the profile by about 0.03 cm^3 cm^{-3}. Also, the plantation trees increased in diameter by 0.127 to 1.270 mm/

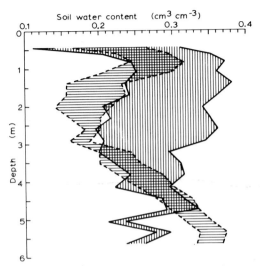

Fig. 4. Maximum and minimum soil water contents beneath natural *Isoberlinia doka* woodland (vertical hatching) and *Eucalyptus citriodora* plantation (horizontal hatching) in Nigeria. (Drawn from data of Babalola and Samie, 1972.)

week and in height by 12.7 to 20.3 mm/week during the dry season, whereas *I. doka* trees showed no growth. Babalola and Samie (1972) suggest that *E. citriodora,* in consuming 34.7 cm of water during the dry season, was more efficient in water use than the natural *Isoberlinia* woodland, which consumed 44.4 cm of water. It may also be concluded that the eucalypt plantation was depleting soil water reserves to a considerable depth, certainly beyond 5.5 m (Fig. 4), and that this depletion was possibly an accelerating process as the depth and density of rooting of the trees increased. Added to this is the greater interception of rainfall by the eucalypt plantation, and therefore a continually diminishing replenishment. It would be interesting to know what was the change in condition of the plantation, and the soil beneath it as tree age increased.

2. Effects of Soil Salinity

In dry soils, salinity is likely to affect the development of roots. Tunstall (1973) examined the salt and water relations of a clay soil in Queensland, Australia, supporting *Acacia harpophylla* forest. He showed that the relationship between gravimetric soil water content and water potential depended quite sensitively upon the salinity, but that the matric and osmotic components of soil water potential were not entirely independent of one another. The relationship between volumetric water content and the dawn value of plant water potential (ψ) differed between the levels

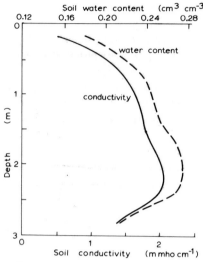

Fig. 5. Variation in soil water content and salt content, indicated by conductivity, with depth in a clay soil beneath *Acacia harpophylla* woodland. (Redrawn from Tunstall, 1973.)

in the soil profile under consideration, and could be related to the changing salt concentration with depth (Fig. 5).

The greatest differences in water content of soil beneath *Acacia harpophylla* forest varied between approximately 3.5% at a depth of 25 cm to less than 0.5% at 250 cm (Tunstall, 1973). Therefore, although high volumetric water contents were maintained at depth in the soil, and although roots of *A. harpophylla* penetrate to more than 3 m, the effective soil volume for the supply of water to the trees may be relatively small.

Gupta *et al.* (1975) suggested that establishment of deep-rooted species in arid and semiarid areas conserved surface soil water that could then be used for growth of shallow rooted crops and grasses. They compared patterns of soil water depletion beneath 3-yr-old eucalypt plantation, 11-yr-old *Acacia* plantation, 3-yr-old *Panicum antidotale* grassland, and fallow at Jodhpur, India, during a seasonal drying cycle. Soil water storage after rain was greatest beneath the young eucalypt plantation, and relatively low under the other two cover types and fallow, suggesting differences in interception and surface runoff between the treatments. Water penetrated beyond 60 cm in the soil only beneath the eucalypt plantation. Gupta *et al.* (1975) inferred that the eucalypts did not use water from the upper soil layers, but their data show depletion from the surface soil which, although less complete than that from the other cover types, was no less rapid. It would be unfortunate if such ill conceived and short-term experiments encouraged management practices that were almost certainly doomed to failure.

IV. WATER USE AND TRANSPIRATION

Water consumption by vegetation may be considered from the point of view of evapotranspiration from a community, or as a physiological process of an individual plant or plant part. The first approach tends to minimize differences between species and individual plants in their response to the environment, whereas these differences are emphasized in the second approach. It is necessary to emphasize that transpiration from a plant community is the result of the contributions of all plants and plant parts, so that the sum of individual plant behavior should be identical with water use measured on a large scale. Unfortunately, direct comparison of evapotranspiration or transpiration at different levels of organization is extremely difficult because it requires precise description of the various types of plant material with respect to water loss, quantitative estimation of their abundance in a plant community, and estimation of the component of water loss due to direct evaporation from the soil.

The structural complexity of forests makes studies at the ecosystem or regional level much easier to conduct than comprehensive investigation at the individual plant level (McGinnis *et al.,* 1969).

A. REGIONAL AND COMMUNITY WATER BALANCE

Studies of water balance of vegetation on a regional basis rely on climatic indices (Penman, 1948, 1956; Thornthwaite, 1948), with or without modification according to topographic and soil characteristics and the nature of vegetation occurring on a particular site. Recent advances in computational techniques have enabled more realistic models of water consumption to be developed, so that account can be taken of changes between seasons in the amount of transpiring foliage and available soil water storage (e.g., Specht, 1972).

1. Forest Structure and Water Balance

It is widely recognized (e.g., Walter, 1971) that the structures of plant communities reflect to a great extent the water balance of the site, and that water and energy balances are closely associated (Gates and Papian, 1971). Also, since the size and shape of a single leaf are important in determining energy and water balances (Parkhurst and Loucks, 1972), it is not difficult to accept that the structure of a whole plant canopy is similarly important to these energy and water balances.

Brunig (1970) observed that forest types in the humid environment of Sarawak could be differentiated according to stand and crown structural characteristics, which corresponded closely to moisture-related gradients in upland forest types and to a seral gradient in peat swamp types. The general roughness of the canopy decreased progressively from the riparian and lowland mixed dipterocarp forest to kerangas (heath forest) on sloping sites and to kerangas on level sites with podzolic soils (Fig. 6).

Greater canopy roughness, as in mixed dipterocarp forest, would increase the amount of radiation intercepted by an emergent crown as compared with a horizontal surface when the sun is not overhead. One consequence is that emergent trees would be subjected to a much greater daily heat load than trees in lower canopy strata or in more uniform forests.

Irregularity of the canopy also increases the aerodynamic roughness of the surface, with the result that both heat and mass transfer from the canopy are increased in relation to the amount of radiation received.

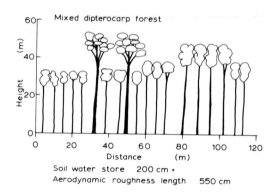

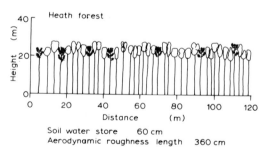

Fig. 6. Simplified canopy structure model of mixed dipterocarp forest and heath forest on shallow humus podzol in Sarawak, showing the relationships to soil water storage capacity. (Redrawn from Brunig, 1970.)

Brunig considered that the relationships customarily used to estimate aerodynamic roughness of vegetated surfaces were not appropriate for tropical forest conditions, and suggested a formulation which included the height range (h), diameter (r), and the distance between the tops (t) of roughness elements, their irregularities (i), and parameters describing stiffness of crown elements (s) and leaf size and arrangement (l). Not all these data could be described, and the following formula was used to determine roughness length, z_0:

$$\log z_0 = \log (h - d) + \log (\Delta r) + \log (\Delta t) - 2.94 \qquad (8)$$

where $h - d$ is the difference in height between emergent and codominant crowns, Δr is the range of diameters of emergent trees and groups, and Δt is the range of distances between roughness elements. These data can be determined readily from aerial photographs and ground surveys.

As the level of net radiation to the leaf increases, the value of the Bowen ratio exceeds unity, and a greater proportion of energy is dissipated from the tree crown by convection. Wind increases this convective transfer, lowers temperature, and tends to reduce transpiration. However, the thickness of the boundary layer is reduced at the same time, and this tends to increase transpiration. Brunig (1970) suggested that there should be a trend of canopy pattern and structure along a water regimen gradient which would result in minimizing factors increasing potential transpiration from progressively drier sites, and maximizing factors controlling leaf surface temperature by sensible heat transfer. In the progression from mixed dipterocarp forest to peatswamp forest, there was a general reduction in leaf size, and a tendency toward erect disposition of smaller and needle-like leaves, factors that both reduce heat absorption and increase dissipation. Also, reduction of the canopy roughness would lessen forced convection in the canopy and evapotranspiration rate. Brunig (1970, 1976) concluded that by these means the water economy of the stand would be improved while water was available, and the tendency to overheating of leaves in a drought would be minimized. This supports the novel but convincing view that stand morphology is primarily related to the water regime of the sites. It clearly demonstrates that more detailed structural analyses of forests of all types and application of simply physical principles can add greatly to our understanding of the relationships between forest type and site conditions.

2. Watershed Studies

A considerable refinement in community water balance studies is the use of watersheds in which various components of the hydrological balance can be measured or estimated. In such cases, total water use or actual evapotranspiration, E_a, is computed from Equation (1). The ratio of actual to potential evapotranspiration (E_a/E_0) is commonly used to describe the degree to which water consumption by a plant community is limited by soil water availability. Penman (1963), Sopper and Lull (1967), and Pereira (1973) summarized many such estimates for a wide range of vegetation types, and the present survey will be concerned with some of those relating to tropical and subtropical forests.

In watershed studies measurement of surface and subsurface drainage is important. Large watershed experiments rely on the existence of watertight basement strata, and studies on small plots must estimate percolation through the soil by means of a lysimeter, or must assume that there is no subsurface drainage below the zone of soil moisture measurement and that the soil moisture sampling includes the whole of the root zone. In stony soils this clearly does not apply, and reliable estimates of water consumption cannot be made. A further requirement is that of careful calibration over a period of several years, although often for practical reasons this is not met.

The most comprehensive series of watershed studies in the tropics is that in East Africa (Pereira, 1962, 1967, 1973), where the effects of different forms of land use on the water balances of watersheds in several contrasting environments were determined. More recently, experimental watersheds have been established in Madagascar (Bailly et al., 1974) and Thailand (Chunkao et al., 1971) with the same objectives in mind.

a. Estimates of Evapotranspiration. In the present context, the behavior of control watersheds is of greatest interest. Table VII shows that for humid forests (receiving more than 2000 mm precipitation per annum), the ratio E_a/E_0 is almost always greater than 0.8, and that it may exceed 0.9. Studies of several years' duration show considerable variation between years, so that the very high E_a/E_0 ratio of 0.98 reported in a short-term study by Low and Goh (1972) does not necessarily indicate a permanent condition.

The mean annual rates of evapotranspiration referred to above conceal widely fluctuating short-term or seasonal rates (Penman, 1963; Rutter, 1968). Even in locations where the annual precipitation is almost 4000 mm and rainless periods are few and short, evapotranspiration from forests may be limited by soil water content. Gilmour (1975) showed that in one disturbed watershed in Queensland, E_a/E_0 declined from 1.0 to 0.6 as volumetric soil water content decreased from about 0.35 to 0.20. In an adjacent watershed, E_a/E_0 fell from 1.0 to 0.2 as soil water content decreased from 0.45 to 0.37. This reinforces Brunig's (1970) contention concerning the severity of drought that may develop in generally humid cli-

TABLE VII

EVAPOTRANSPIRATION FROM HUMID FORESTS USING RAINFALL AND RUNOFF DATA FROM SMALL CATCHMENTS[a]

Forest type	Location	Period of observation	P (mm)	R (mm)	ΔS (mm)	E_a (mm)	E_o (mm)	E_a/E_o	Reference
Rainforest, undisturbed	Babinda, Queensland	1969–1973	3899	2372	+23	1502	1725	0.87	Gilmour (1975)
Rainforest, logged	Babinda, Queensland	1969–1973	3899	2344	+53	1501	1725	0.87	
Rainforest	Kericho, Kenya	1958–1964	2198.4	698.2	+3.3	1503.6	1665.7	0.90	Pereira (1967)
Bamboo forest	Kimakia, Kenya	1957–1964	2506.3	1428.1	−1.4	1079.6	1434.2	0.75	Pereira (1967)
Pinus patula plantation	Kimakia, Kenya	1962–1964	2598.3	1540.2	20.9	1037.1	1374.0	0.75	Pereira (1967)
Dipterocarp forest	Ulu Gombak, Malaysia	1968	2500	750	—	1750			Kenworthy (1971)
Evergreen forest	Selangor, Malaysia								Low and Goh (1972)
Sungei Lui No. 1		1968–1969	2155.7	1076.7		1079	1247	0.86	
Sungei Lui No. 2		1968–1969	2108.7	1100.1		1008.9	1241.6	0.81	
Sungei Lui No. 3		1968–1969	2162	1100		1061.7	1236.5	0.86	
Ulu Langat		1969	2482	1219		1263	1291	0.98	
Evergreen forest	Ampangamatsary, Madagascar	1964–1972	2103	634		1469			Bailly et al. (1974)
Shrubland	Marolaona, Madagascar	1964–1972	1882	844		1038			
Eucalyptus robusta plantation	Betsakosako, Madagascar	1964–1972	1595	216		1379			
Natural grassland	Manankazo, Madagascar	1967–1973	1857	432		1425			Bailly et al. (1974)
Pinus patula plantation	Manankazo, Madagascar	1967–1973	1884	272		1612			Bailly et al. (1974)
Evergreen forest	Abidjan, Ivory Coast	1969–1971	1449	(374)[b]	—	(1075)[b]			Huttel (1971)

[a]Key to symbols: P, precipitation; R, runoff; ΔS, change in soil moisture storage; E_a, actual evapotranspiration; E_o, poential evapotranspiration (from pan evaporation or Penman equation).
[b]Parentheses indicate values estimated.

mates, and has a bearing on later discussion of morphological characteristics of leaves.

The effect of changes in land use on water consumption by vegetation are considered by Penman (1963) and Pereira (1967, 1973). The examples in Table VII that indicate changed land use in humid regions show relatively little change in water balance over a period of a few years, either where *Pinus patula* plantation replaced bamboo forest (Pereira, 1967) or where rainforest was logged and the area left untreated (Gilmour, 1975). In both these cases, surface compaction due to grazing animals was not a factor in the changed land use, and the mechanical disturbance involved in the change of vegetation type was of only short duration. Also, the *P. patula* plantation at Kimakia, Kenya, was only 5 to 7 years old when the figures reported by Pereira (1967) were obtained. Further development of the trees could change the magnitude of water use, although the pre-existing bamboo forest had reached a steady rate of water loss at the same E_a/E_o ratio.

In a slightly lower rainfall zone, Bailly *et al.* (1974) observed that *Eucalyptus robusta* plantation in Madagascar lost a greater proportion of rainfall through evapotranspiration than did native forest or secondary shrubland, both of which received higher rainfalls. The increasing tendency to greater water loss from forest than from bare or grassed land under lower rainfall regimes is clearly associated with the greater depth of rooting of trees than of grasses. An extreme example of low water yield from a forest is that of *Eucalyptus marginata* growing on deep latosols in Western Australia. Runoff into streams may be only 2% of the total annual rainfall of approximately 1275 mm (Shea *et al.*, 1975), and water extraction is presumed to occur through roots which may penetrate the soil profile to a depth of 15 m (Kimber, 1972).

b. Estimates of Water Storage and Flow. Measurement of the water contents of components of a forest community permits estimation of rates of flow through components in relation to their storage capacity. This can be achieved by hydrological analysis plus estimation of biomass and the water content of the various components of the system. Such an approach was used by McGinnis *et al.* (1969) in a rain forest at Darien, Panama during a wet season lasting 221 days. The water contents of the components and the rates of transfer between selected components are shown in Fig. 7. This analysis assumed that the water balance of the forest was in a steady-state condition, that there was no net growth of the forest, and that the transfer of water could be described by a series of ordinary linear differential equations of the form

$$\frac{dW_i}{dt} = \sum_{j=1}^{n} \lambda_{ji} W_j - \sum_{j=1}^{n} \lambda_{ij} W_i \qquad i = 1, \ldots n$$

where W_i is the water content of component i, λ is the function describing the transfer of water between components i and j, and t is time.

It can be seen that, even in a wet season condition, the rate of movement of water through the leaves is greater than the leaf water content, and that evaporation of intercepted water may add 44% to this rate of water loss from the canopy. At the mean rate of water uptake from the soil for the period, the soil water store would be totally exhausted in 37 days, assuming no supplement from rain or drainage. Obviously not all the soil water is available to plants, and the above analysis is grossly simplified, but it suggests that water stress can be expected to develop in rain forests after about 30 days without rain (cf. Brunig, 1969).

A similar but more detailed analysis was carried out in the rain forest of El Verde, Puerto Rico, by Odum *et al.* (1970c) using hydrological analysis, direct estimates of evapotranspiration by forest enclosed in a giant plastic cylinder, and application of tritiated water to the soil. The concen-

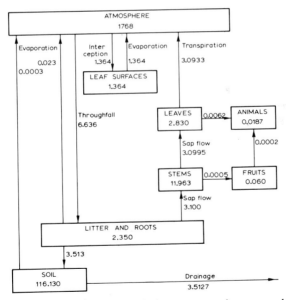

Fig. 7. Compartments and flow of water during wet season in a seasonal forest, Darien, Panama. Figures within rectangles indicate water storage (kg m^{-2} land surface); figures on arrows indicate water flux per day (kg m^{-2} land surface). (Redrawn from McGinnis *et al.*, 1969.)

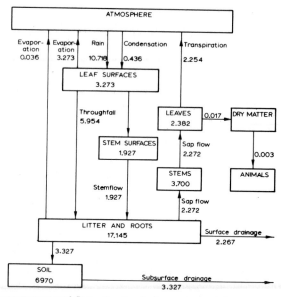

Fig. 8. Compartments and flow of water during one month in a rain forest at El Verde, Puerto Rico. Figures within rectangles indicate water storage (kg m⁻² land surface); figures on arrows indicate water flux per day (kg m⁻² land surface). (Redrawn from Odum et al., 1970c.)

tration of tritium in a component of the ecosystem is a function of the quantity applied, the rates of transfer between components, and their storage capacities (Zimmermann et al., 1966; Kline and Jordan, 1968; Bloom and Raines, 1970). The flow rates of water through the El Verde site were greater than those for Darien, Panama, due to the higher rainfall (Fig. 8). However, the quantities of water stored in the various components of the ecosystem were generally similar. The considerably greater total soil water store at El Verde would sustain transpiration for about 82 days using the simplifying assumptions outlined above.

It is clear that there must be a close connection between the quantity of available water stored in a plant community and the rate of evapotranspiration that can be sustained in the absence of rainfall. Where this stored water can be exhausted in a short time, the vegetation can survive by having either a relatively small transpiring surface, or structural and physiological attributes that reduce the rate of water loss per unit area of soil and reduce the heat load on the shoot. It seems somewhat less important that the more common condition of the plant community is one of an abundant water supply.

B. SINGLE TREE STUDIES

Transpirational loss from a forested area is essentially the sum of losses from all the trees, particularly those in the overstory. Determination of transpiration can thus be made on single trees, and the results extrapolated to an area of forest in a more or less uniform environment. Three methods have been applied to this problem: (1) whole tree gas exchange; (2) tritiated water injection; and (3) heat pulse velocity. ˙

1. Whole Tree Gas Exchange

Gas exchange techniques are widely used for determination of transpiration rates from single leaves and whole small plants, but their application to trees has been limited by the large size of the subjects and difficulties of maintaining ambient environmental conditions within the enclosures (Odum and Jordan, 1970).

Greenwood and Beresford (1979) employed whole tree enclosures with open tops for estimation of transpirational losses from juvenile trees of 14 *Eucalyptus* species established in plantations on agricultural land in Western Australia suffering from secondary salinity (Peck, 1978). Air was blown into the bottom of the enclosure at a measured rate and the increase in water vapor concentration between this point and the open top was determined by infrared gas analysis, giving a direct estimate of water vapor flux from the enclosed plant.

The transpiration rates from each species varied between topographic position at one location and also between locations. There was no consistent pattern of response of the various species at different locations. A relatively high rate of water loss per tree per day in 24-month-old *Eucalyptus globulus* tree (37 liters day^{-1}) on a site with a mean annual rainfall of 850 mm was associated with high leaf area index (1.40) and a low rate of transpiration per unit of leaf area (2.20 kg m^{-2} day^{-1}). On the other hand, 27-month-old *E. wandoo* at a different site with a mean annual rainfall of 420 mm had a total daily water consumption of 21.3 liters per tree from a plantation with a leaf area index of 0.10 and a very high rate of transpiration per unit of leaf area (17.80 kg m^{-2} day^{-1}). It was concluded that the roots of *E. wandoo* had reached an aquifer before the reported measurements were made at the end of the summer dry season. Comparison of the high rates of water consumption from these young trees with the estimates made for larger trees (Tables VIII and IX) emphasize the importance of the gradient in water vapor pressure between leaf and air in the water balance of plants in different environments.

2. Tritiated Water Velocity

Tritiated water was injected into stems of rain forest trees at El Verde, Puerto Rico (Kline *et al.*, 1970), and at San Carlos, Venezuela (Jordan and Kline, 1977). Table VIII shows the results for individuals of *Dacryodes excelsa* and *Sloanea berteriana*.

Differences in residence time, calculated as the water content of the tree divided by the rate of transpiration per day, are due to differences in both water capacity and transpiration rate. Clearly, *D. excelsa* had a lower rate of transpiration per unit volume of stem (indicated by the cylindrical product of basal area and height). This may be related to factors such as wood density and moisture content, the relative areas of sapwood and heartwood, the relationship between leaf surface and sapwood cross-sectional area, and leaf characters which might influence their rate of transpiration. It is interesting that in *D. excelsa*, Kline *et al.* (1970) found rather similar residence times for water in emergent and understory specimens. This may be interpreted as adaptability to different conditions of exposure, whereby the development of leaves and stems is consistently related to transpiration rate (Kline *et al.*, 1976; Jordan and Kline, 1977; Whitehead, 1978), and presumably to photosynthetic production.

In their Venezuelan study, Jordan and Kline (1977) established a good correlation ($r = 0.96$) between sapwood sectional area of a tree and its total daily water consumption. Nevertheless, as may be seen in Table VIII, an overstory tree of *Micranda spruceana* had a very much higher transpirational flux per unit area of sapwood than did an understory individual of the same species. Even where overstory *Eperua leucantha* trees of different size were compared, the larger individuals appeared to have the greater transpirational flux per unit area of sapwood. Jordan and Kline stressed the variability in the relationship beteen daily transpiration and sapwood area that was observed in small trees that had only part of their canopy exposed to full sunlight. This condition obviously affects development of the crown as well as the rate of water loss from it.

Kline *et al.* (1970) also showed a decrease in tritium concentration in leaves following rainfall, with an increase after rain ceased. This suggests that direct absorption of intercepted rainfall occurred in leaves of *D. excelsa*. Indirect absorption of intercepted water also may occur through roots arising on branches and penetrating the peaty bases of epiphytic ferns (Lyford, 1969).

3. Heat Pulse Velocity

An alternative method of estimating water flux through trees is that of Kenworthy (1969), who compared hydrological observations in a hill dipterocarp forest at Ulu Gombak, Malaysia, with estimates of heat pulse

TABLE VIII
DAILY WATER CONSUMPTION OF RAIN FOREST TREES ESTIMATED BY TRITIATED WATER APPLICATION[a]

Species	Position	H (m)	BA (m²)	BA · H (m³)	T (liters day⁻¹)	$\frac{T}{BA \cdot H}$	RT (days)	Reference[a]
Dacryodes excelsa	Emergent	19.8	0.2367	4.687	372	79.37	11	1
	Understory	7.3	0.0022	0.0199	1.75	87.54	9.6	1
Eperua leucantha	Overstory	—	0.0131[b]	—	91.4	6977[c]	—	2
	Overstory	—	0.0114[b]	—	36.1	3166[c]	—	2
Eperua purpurea	Overstory	—	0.1521[b]	—	1180.0	7758[c]	—	2
Micranda spruceana	Overstory	—	0.0421[b]	—	301.0	7149[c]	—	2
	Understory	—	0.0017[b]	—	2.7	1588[c]	—	2
Sloanea berteriana	Emergent	18.3	0.0564	1.0323	140	136.52	3.9	1

[a] Key to abbreviations: H, height; BA, basal area; T, transpiration; RT, residence time of water in stem.
[b] Sectional area of sapwood including sapwood–heartwood transition.
[c] Ratio of transpiration to sapwood sectional area.
[d] References: (1) Kline *et al.* (1970); (2) Jordan and Kline (1977).

velocity in individual trees. Hydrological data for the period August 1968–January 1969 were used, so that the change in soil water content was zero and evapotranspiration ($E_a = E + T$) could be derived from the equation (1). The estimated evapotranspiration for the period of 1300 mm was equivalent to 3.56 mm day^{-1}. Dye injection studies indicated that the cross-sectional area of functioning vessels in trees growing in a 400-m^2 plot was 180 cm^2, contained in a stem basal area of 7000 cm^2.

Assuming that the xylem sap flows at constant speed for 12 hr/day, Kenworthy (1969) estimated a mean sap speed of 6.6 m/hr. This estimate was confirmed by the heat pulse technique (Huber, 1956), mean heat pulse velocities for *Shorea leprosula* being 6.97 and 6.85 m/hr on two different days. This is a rather crude calibration as heat pulse and sap velocities are not identical (Huber, 1956; Marshall, 1958).

Other data from the Ulu Gombak Study (Kenworthy, 1971) indicate the effect of disruption of conducting elements resulting from the placement of sensors. The heat pulse velocity at 2000 hr on the first day was about twice the midday maximum and about 10 times the value recorded at about 1830 hr on the second day. Kenworthy attributed this to the release of tension after implacement of the sap flow sensors. Doley and Grieve (1966) considered that disruption of the conducting system of eucalypts could occur as a result of increasing sap tension some time after implacement of the sensors, and could spread progressively, retarding the transfer of heat from the moving sap to the heat sensors. On the other hand, Swanson (1974) concluded that disruption of the conducting system as a result of drilling holes in conifer stems was not serious. The difference between tree types may lie in the fact that aspiration of the bordered pits of conifers leads to very effective sealing of the elements against the passage of an air–water interface (Petty, 1972), whereas the anastomosing vessels of dicotyledons (Zimmermann and Brown, 1971) and the rather large pit pores at the ends of vessels (Jane, 1970) would permit eventual spread of gas through many adjacent vessels.

Where measurements are made on individual trees, it is important to sample sufficiently in time and within a population in order to derive useful data on a land area basis. Doley and Grieve (1966) studied a *Eucalyptus marginata* forest over a period of 4 months from the end of the wet season (winter) until midsummer, measuring heat pulse velocities on 13 trees with a projected crown cover of 45% in a 0.04-ha plot. On any day, the time courses of heat pulse velocity varied for different trees, and between days the relative totals of sap flux changed considerably, as is apparent in Table IX. The rates of transpiration per day from the plot varied from 0.34 to 1.10 mm, depending chiefly on air temperature and the degree of cloud cover on the day of measurement. These values are rather

TABLE IX

DAILY WATER CONSUMPTION OF *Eucalyptus marginata* TREES ESTIMATED BY HEAT PULSE VELOCITY[a]

	Tree No.					Plot total (mm/day)
	1	5	8	9	13	
Height (m)	23.5	19.0	18.4	22.0	20.7	
Stem basal area (cm²)	817	240	712	630	905	
Sapwood area (cm²)	182	63	177	105	201	
Water consumption (liters day⁻¹)						
Oct. 16, 1963	24.0	8.3	17.6	15.1	21.7	0.34
Nov. 14, 1963	65.4	8.5	62.5	33.7	103	1.10
Nov. 28, 1963	47.1	6.9	56.5	38.4	100	1.00
Dec. 12, 1963	27.2	9.6	49.0	23.8	56.6	0.65
Jan. 1, 1964	44.0	8.3	54.4	25.5	210	1.10

[a]Data of Doley and Grieve (1966).

small compared with the annual rainfall of approximately 1300 mm, taking into account the fact that the water yield of watersheds in this region is low (Shea *et al.*, 1975).

C. TRANSPIRATION FROM INDIVIDUAL LEAVES

1. Rain Forest

Transpiration from individual leaves of montane rain forest trees in Puerto Rico was estimated by Gates (1969), using an energy budget approach. In the cool, wet, windy, low illumination conditions experienced, transpiration rates from *Tabebuia rigida* and *Octoea spathulata* leaves ranged between 2.5 and 10.5 mg m⁻² sec⁻¹. These values are very low, but are attained under low absorbed radiant fluxes of 0.57 to 0.78 cal cm⁻² min⁻¹. It can be assumed that, at atmospheric relative humidities around 85%, water stress was insignificant.

Odum *et al.* (1970b,c) used a leaf chamber to measure carbon dioxide and water vapor exchange in leaves of rain forest tree species *in situ*. The rates of transpiration were influenced greatly by insolation and atmospheric water vapor pressure and to some extent by the speed of air movement through the leaf chamber, but maximum and total daily rates are of considerable interest. The highest rate of water loss for leaves of species in upper crown positions ranged from 346 gm m⁻² day⁻¹ for *Manilkara bidentata* to 1360 gm m⁻² day⁻¹ for *Dacryodes excelsa*. Water loss from shade leaves ranged from 134 gm m⁻² day⁻¹ for *Euterpe globosa* to 1130 gm m⁻² day⁻¹ for *Sloanea berteriana*. The maximum rates of transpiration are generally lower than those recorded by Coutinho (1962) for humid forest species in Brazil (Table X), but Odum *et al.* (1970b,c) computed transpiration on an hourly basis whereas Coutinho estimated transpiration by weight loss over intervals of no more than a few minutes.

It is interesting that transpiration from shade leaves of one species may considerably exceed that from sun leaves of another species. Another rather unusual aspect of this study is the relatively large proportion of the total daily water loss that occurred during the 12 hr of darkness, particularly from shade leaves. In *Sloanea berteriana*, transpiration during the night was often greater than during the day.

Stocker (1935) applied his rapid weighing technique for transpiration measurement (Stocker, 1929) to rain forest species at Bogor, Indonesia. On a dry, generally sunny day, three tree species showed relatively high rates of transpiration as compared with those in other species (Table X). Transpiration from leaves of understory species was generally slow (Table X), except that sun-flecks falling on leaves during measurements resulted in up to sixfold increases in transpiration rate.

More extensive studies of the water relations of humid forest species at Paranapiacaba, Brazil, were reported by Coutinho (1962). Transpiration was estimated by Stocker's (1929) rapid weighing technique, and evaporation by Piche evaporimeter. Cuticular transpiration was taken as the steady weight loss of leaves 30 min. or longer after excision from the plant. The mean annual rainfall for the site was 3145.7 mm.

Variation in behavior within life forms was in most cases greater than that between life forms, except that epiphytes had low rates of transpiration and lower relative leaf water contents (Table X). Both lianes and epiphytes were able to effectively control transpiration through stomatal closure, although two of the tree species, *Inga sessilis* and *Miconia willdenowii*, showed cuticular transpiration rates exceeding 8% of the rate of water loss from the Piche evaporimeter. The daily course of transpiration for most species followed that of free water evaporation, except for the epiphytes.

Although this study used detached leaves and Piche evaporimeters, comparisons between species are likely to be valid, provided that care is taken to make weight loss measurements for transpiration before there is any substantial change in the behavior of leaves. Comparisons between the behavior of species from different regions are most conveniently carried out under controlled conditions. The transpiration rates of potted seedlings of *Khaya ivorensis*, a high forest zone species, and *K. senegalensis*, a savanna species from West Africa, were compared in a glasshouse study by Okali and Dodoo (1973).

When soil water was freely available, both maximum and mean daily

TABLE X

TRANSPIRATIONAL CHARACTERISTICS OF RAIN FOREST SPECIES IN
BRAZIL AND PUERTO RICO

Species	T_{max} (mg m^{-2}sec^{-1})	$\Sigma T/\Sigma E$ (%)	T_{max}/E (%)	T_{cut}/E (%)	RLWC
Trees					
Affonsea edwallii[b]	7.5	14.4	28.4	1.4	—
Alchornea triplinervia[b]	9.1	33.6	40.0	1.0	0.965
Euterpe edulis[b]	2.6	9.2	12.2	3.4	0.985
Inga sessilis[b]	5.6	27.7	30.4	8.4	—
Miconia willdenowii[b]	4.8	11.4	13.7	8.6	0.908
Weinmannia hirta[b]	6.1	—	—	—	—
Bathysa stipulata[b]	4.3	29.2	24.5	2.0	—
Cocoloba martii[b]	3.5	24.1	24.4	2.0	—
Dacryodes excelsa[c]	2.4				
Manilkara bidentata[c]	1.2				
Anthocephalus cadamba[c]	3.5				
Cecropia peltata[c]	0.5				
Calophyllum inophyllum[d]	28.0	25	41.0	—	—
Cassia fistula[d]	21.7	23	21.3'	—	—
Elaeodendron glaucum[d]	28.3	28	64.1	—	—
Shrubs					
Geonoma gamiova[b]	5.1	14.1	15.1	2.4	0.981
Geonoma schottiana[b]	6.4	14.6	18.6	3.4	0.999
Begonia isoptera[d]	2.5	17	25.0	—	—
Cyrtandra perdula[d]	5.0	29	35.3	—	—
Lianes					
Abuta selloana[b]	5.3	12.1	17.8	1.0	0.987
Mikania trinervis[b]	2.1	10.2	23.3	1.4	—
Nematanthus fritschii[b]	3.5	12.9	20.0	1.1	—
Epiphytes					
Codonanthe gracilis[b]	1.9	4.1	11.1	1.0	0.867
Hypocyrta radicans[b]	3.5	6.8	11.5	—	0.578
Maxillaria picta[b]	0.3	0.7	1.1	0.6	—
Vriesia altodaserrae[b]	2.1	7.6	12.4	1.7	—
Vriesia inflata[b]	1.1	2.5	4.4	1.0	0.923
Hymenophyllum polyanthes[b]	—	—	—	—	0.062

[a] Key to symbols: T_{max}, maximum transpiration for the day; ΣT, total daily transpiration; E, Piché evaporation at the time of T_{max}; ΣE, total daily Piché evaporation; T_{cut}, cuticular transpiration; RLWC, relative leaf water content at 1100–1300 hr.

[b] Data calculated from Coutinho (1962).

[c] Data calculated from Odum et al. (1970b).

[d] Calculated from Stocker (1935).

transpiration were greatest in *K. senegalensis*. As soil matric potential decreased from -0.03 to -0.45 MPa, transpiration from *K. senegalensis* was reduced by 72.5% and from *K. ivorensis* by 54.5%.

The greater maximum transpiration in *K. senegalensis* was attributed to wider stomatal apertures, as estimated using a kerosene/liquid paraffin infiltration series (Alvim and Havis, 1954). Unfortunately, the experiment did not discriminate between the behavior of the two species under lower soil matric potential conditions.

These differences in the behavior of soil-grown plants of the two species were confirmed by studies on seedlings grown in nutrient solutions, the osmotic potentials of which were regulated by polyethylene glycol. Again, Okali and Dodoo (1973) found distinct differences in behavior between high forest and savanna species.

The measuring technique resulted in the nutrient solution being without aeration for 4 hr at a time. Nevertheless, *K. senegalensis* showed a greater response to reduction of root medium osmotic potential from -0.03 to -1.03 MPa, and at the end of the experiment had more completely restricted transpiration than had *K. ivorensis*.

2. Cerrado and Llaneros

The generally evergreen, xeromorphic vegetation of the cerrado of Brazil has been of great interest to ecologists and plant physiologists for many years, particularly with regard to plant water relations (Rawitscher, 1948; Ferri, 1955, 1961; Eiten, 1972, 1978). Coutinho and Ferri (1956) studied transpiration rates of three species of perennial cerrado plants at Emas, Sao Paolo, Brazil during the wet season. The diurnal patterns of transpiration more or less paralleled the patterns for evaporation from leaf atmometers on clear days. Atmospheric relative humidities fell from 80% or more at 0800 hr to 22–40% by early afternoon, when both transpiration and evaporation were maximal. The rate of transpiration as a percentage of evaporation (relative transpiration) shows interesting differences between the three species (Table XI). *Stryphnodendron barbaditemimam* showed a continuous decline in the ratio during the day and *Bowdichia virgilioides* showed a decline in the ratio during the morning and no change after the maximum transpiration was reached. On the other hand, *Platypodium elegans* exhibited a relatively high early morning rate of relative transpiration, a low minimum in the early afternoon, and complete recovery by late afternoon.

Stomata of these species were judged to be open all day (Coutinho and Ferri, 1956) and to be sluggish in their closing reactions. Certainly, there seems to be no stomatal control over transpiration within 20 min. of detachment in *S. barbaditemimam* and *B. virgiliodes,* although *P. elegans*

TABLE XI
Transpiration from Cerrado Species at Different Seasons[a,b]

Species	Season	Time T_{max} (hr)	T_{max} (mg m^{-2} sec^{-1})	T_{max}/E	$(T/E)_{min}$	Reference
Anacardium occidentale	Late dry	0845	16.3	0.210	0.114	Ferri and Lamberti (1960)
Andira humilis	Early dry	1426	24.1	0.195	0.195	Ferri (1955)
	Late dry	1330	32.0	0.107	0.107	Ferri (1955)
Annona coriacea	Early dry	1300	20.0	0.255	0.255	Ferri (1955)
	Early dry	1440	18.6	0.330	0.256	Coutinho and Ferri (1960)
	Late dry	1200	25.9	0.216	0.076	Ferri (1955)
Bowdichia virgilioides	Wet	1330	28.0	0.219	0.219	Coutinho and Ferri (1956)
Byrsonima coccolobifolia	Early dry	1430	22.1	0.226	0.226	Ferri (1955)
	Late dry	1045	28.8	0.277	0.070	Ferri (1955)
Curatella americana	Late dry	0920	15.5	0.196	0.098	Ferri and Lamberti (1960)
Duguetia furfuracea	Early dry	1545	12.9	0.241	0.208	Coutinho and Ferri (1960)
Erythroxylum suberosum	Early dry	1450	21.2	0.354	0.190	Coutinho and Ferri (1960)
Eucalyptus saligna	Early dry	1420	28.8	—	—	Franco and Inforzato (1967)
	Late dry	0850	17.7	—	—	
Eucalyptus tereticornis	—	1100	16.0	0.445	0.445	Villaca and Ferri (1954)
Ouratea sp.	Late dry	1050	26.2	0.277	0.113	Ferri and Lamberti (1960)
Platypodium elegans	Wet	1330	36.0	0.154	0.154	Coutinho and Ferri (1956)
Stryphnodendron barbadetimam	Wet	1410	30.4	0.240	0.240	Coutinho and Ferri (1956)

[a] Calculated from authors data.

[b] Key to symbols: T_{max}, maximum daily rate of transpiration; T_{max}/E, ratio of transpiration/evaporation at time of T_{max}; $(T/E)_{min}$, daily minimum of ratio of transpiration/evaporation.

did show complete and effective closure within 15 min. It is suggested that the xylol infiltration method used by Coutinho and Ferri was not sufficiently sensitive to detect changes in stomatal aperture during the day, particularly in *P. elegans,* which exhibited quite sensitive control over water loss, considering the possibility that the soil water content was relatively high.

Rawitscher (1948) reviewed extensive measurements of the water relations of cerrado plant species at Emas, Sao Paulo, at different times of the year. The climate is characterized by relatively little variation in mean monthly maximum temperature, but much greater variation in mean monthly minimum temperature, with occasional frosts. The months of May–August are dry, but atmospheric relative humidity is very high until 0900 or 1000 hr except toward the end of the dry season when humidities are consistently lower. An important aspect of the cerrado is the very deep soils (about 18 m) and the permanent and relatively stable nature of the water table which makes water available to the deep-rooting perennial plants. Long-established burning and grazing practices have resulted in many species having much reduced or even ephemeral aerial shoot systems and very large underground stem and/or root systems (the geoxylic suffrutix of White, 1977). Rawitscher (1948) noted 29 (possibly 32) deep-rooted evergreen perennials, 25 shallow-rooted, summer-green species, 8 with intermediate root habits, and 13 grasses.

At the beginning of the dry season, transpiration from deep rooted evergreen species followed closely the daily pattern of evaporation, indicating little or no stomatal control over water loss (Ferri, 1955). At the end of the dry season, *Andira humilis* still showed more or less parallel trends of evaporation and transpiration, but the ratio of transpiration/evaporation at the time of maximum transpiration was reduced (Table XI), and recovered more noticeably toward late afternoon.

Ferri (1955) presented data for another deep-rooted species, *Byrsonima coccolobifolia,* from the same area. At the beginning of the dry season, transpiration and evaporation followed parallel courses. However, late in the dry season, maximum daily transpiration was attained at about 1045 hr and declined to a minimum at 1430 hr, recovering briefly late in the afternoon. This indicates a considerable degree of stomatal control, the relative rates of transpiration falling from 0.277 in the early morning to 0.070 in the early afternoon.

A third type of transpirational response was cited by Ferri (1955) for *Anona coriacea,* another deep-rooted species. As with the other species studied, transpiration at the beginning of the dry period followed closely upon evaporation, but at the end of the dry season maximum transpiration was achieved earlier in the day and slowly declined toward late afternoon.

Again, there was evidence of substantial stomatal control, but without the classical midday depression of transpiration.

Coutinho and Ferri (1960) examined a subtropical example of cerrado vegetation at Campo de Mourao, Parana, with an annual rainfall between 1500 and 1700 mm. At the beginning of the dry season, transpiration followed the course of daily evaporation. At Goiana, in the north of Brazil, the total annual rainfall was 1981 mm, with relative drought in October and November and 3 months of heavy rainfall in April, May, and June. Measurements of water loss from leaves of perennial plant species in January showed a general correlation between evaporation and transpiration, with restriction of water loss during the afternoon (Table XI). Under the moderate conditions of temperature and relative humidity, the rates of transpiration were not high. Cuticular transpiration varied from 5% in *Anacardium occidentale* and *Ouratea* sp. to 10% total transpiration in *Curatella americana*. Root penetration by some species was restricted in these soils by impermeable horizons, although some roots penetrated to more than 2.5 m.

The introduction of exotic species to an area may substantially change the water balance of a site, and it is of interest and importance to determine the magnitude of such changes. Villaca and Ferri (1954) examined the pattern of transpiration of *Eucalyptus tereticornis* at Sao Paulo, Brazil. Maximum transpiration varied from 14.8 mg m^{-2} m^{-1} for very old leaves to 16.0 mg m^{-2} sec^{-1} for mature leaves, and occurred at 1100 hr, which corresponded with maximum evaporation for the site. The diurnal trends in relative transpiration were inconsistent between leaves of different age, but were related to the patterns of stomatal opening. A clear distinction existed between the tendency for stomata of young leaves to close in the afternoon and for stomata of mature and old leaves to remain open until about sunset. Cuticular transpiration in leaves of all ages was estimated to vary from 4.6 to 5.9% of evaporation from wet filter paper, and mean relative transpiration on a daily basis was 50–60%, which was equivalent to a daily water consumption of 150 liters per tree. Assuming that the trees retained their leaves for 250 days per year, Villaca and Ferri (1954) estimated annual water consumption to be 37,5000 liters per tree. The trees examined were grown in a botanic garden rather than in a commercial plantation, so it may be anticipated that water consumption would be greater than in the field due to the possibility of greater crown development and better soil water supply.

Transpiration rates of several perennial species of the llaneros of Venezuela were studied during the dry season by Vareschi (1960). Maximum rates varied widely, from about 78 mg m^{-2} sec^{-1} in *Hibiscus rosasinensis* to 69 mg m^{-2} sec^{-1} in *Byrsonima crassifolia,* 32 mg m^{-2} sec^{-1} in *Curatella americana* and 11 mg m^{-2} sec^{-1} in *Cecropia* sp.

Transpiration in *Byrsonima crassifolia* was more closely related to the daily course of air temperature than that of evaporation. *Curatella americana,* another evergreen tree with coriaceous leaves, showed a distinct midday depression of transpiration. The roots of *B. crassifolia* penetrated to a depth of at least 2.6 m, and probably to the water table, thereby supporting almost free transpiration. Vareschi (1960) estimated the total leaf surface area of an individual *B. crassifolia* tree to be 15 m^2, so that between 1130 and 1430 hr, when a relatively constant transpiration rate of about 62 mg m^{-2} sec^{-1} was maintained, the water consumption of the tree would be approximately 10 liters.

3. Caatinga

The caatinga vegetation type of Brazil is found in a hotter, drier climate than the cerrado (Ferri, 1955). Two dry periods occur, in September–October and in January, but in no month of the year is there an excess of rainfall over evaporation. Therefore, the vegetation is likely to experience water stress at any time of the year and the patterns of water use by plants can be expected to differ from those of the cerrado. Some transpiration data for species from this vegetation type are summarized in Table XII.

Spondias tuberosa is notable for maximum transpiration early in the day, and for low rates during the late afternoon (Ferri and Labouriaou, 1952; Ferri, 1955). This behavior is associated with rapid and complete closure of stomata in detached leaves, and with very low maximum rates of transpiration during both January and July. The sensitivity of the stomatal response is indicated by the very high relative leaf water contents recorded in this species.

An interesting feature of Ferri's (1955) data is the occurrence of rain during the measurements of July 1953. The first recorded rates of transpiration for the day are somewhat higher than those for April 1953, and in *Spondias tuberosa,* which was measured on a day free from rain, this first measured rate was the highest for the day. For the other three species studied, transpiration increased after a brief shower, and in *Maytenus rigida* and *Caesalpinia pyramidalis,* relative transpiration was higher at about 1130 than 0900 hr. This suggests that at least some of the rainfall was utilized directly for replenishment of leaf water supplies and transpiration. Unfortunately relative leaf water content data are not available for this occasion.

Later in the year (October–November), early morning transpiration rates from species that retained their leaves in the dry season were similar

TABLE XII

TRANSPIRATION RATES OF SINGLE LEAVES OF CAATINGA SPECIES[a,b]

Species	Season	Time T_{max} (hr)	T_{max} (mg m^{-2}sec^{-1})	T_{max}/E	$(T/E)_{min}$	Reference[c]
Aspidosperma pyrifolium	Dry	1140	20.0	0.231	0.117	1
Caesalpinia pyramidalis	Wet	1030	13.3	0.148	0.043	1
	Early dry	1400	17.1	0.139	0.016	2
	Dry	1130	8.6	0.225	0.225	1
Jatropha phyllacantha	Wet	0900	3.2	0.100	0.001	1
	Early dry	1200	25.0	0.198	0.057	2
	Dry	1400	19.2	0.135	0.135	1
	Late dry	0630	13.3	0.268	0.014	1
Maytenus rigida	Wet	1410	10.2	0.050	0.050	1
	Early dry	1020	21.9	0.285	0.046	2
	Dry	1400	14.4	0.409	0.409	1
	Late dry	0730	11.4	0.209	0.011	1
Spondias tuberosa	Wet	0740	1.9	0.062	0.003	1
	Early dry	0930	17.6	0.220	0.018	2
	Dry	0930	3.1	0.084	0.015	1
Tabebuia caraiba	Late dry	1030	13.3	0.092	0.050	1

[a] Calculated from author's data.

[b] Key to symbols: T_{max}, maximum daily rate of transpiration; T_{max}/E, ratio of transpiration/evaporation at time of T_{max}; $(T/E)_{min}$, daily minimum of ratio transpiration/evaporation.

[c] References: (1) Ferri (1955); (2) Ferri and Labouriaou (1952).

to those observed in July (Ferri, 1955). However, high evaporation rates and rapidly rising air temperatures in the morning were associated with an early decline in transpiration rates, although water loss was not restricted as completely as in January, when the new season's leaves had developed. An exception to this pattern is *Tabebuia caraiba*, which occurs along water courses in the caatinga. Transpiration was first reduced after about 1200 hr, but there was no evidence of a drastic decline in relative transpiration. Obviously, the supply of water to the roots of *T. caraiba* is abundant as compared with the situation of other tree species in the caatinga.

Cuticular transpiration rates for most of the species studied by Ferri and Labouriaou (1952) were surprisingly high. In *Spondias tuberosa*, cuticular transpiration was estimated to be 5% of total transpiration, in *Maytenus rigida* 10%, in *Caesalpina pyramidalis* 15%, and in *Jatropha phyllacantha* 20%. Since *S. tuberosa* and *J. phyllacantha* were able to reduce transpiration to very low values (Table XII), it follows that closure of stomata must be very effective indeed, particularly in younger leaves.

4. Semiarid Woodlands

Semiarid regions are characterized by relatively short periods of water availability and extended periods of water stress. Perennial species must undergo considerable change in water consumption and be able to resume physiological activity quickly when water becomes available. Hellmuth (1971) compared the rates of transpiration from leaves of *Acacia cuminata* and *Anthocercis odgersii* under conditions of optimal water supply and stress in Western Australia. These two species exhibit different patterns of behavior, particularly under stress. *Acacia acuminata* maintained a low but uniform rate of water loss throughout the day, whereas *Anthocercis odgersii* showed a rapid early morning rise in transpiration, followed by an equally rapid decline to a low value during the middle of the day. These differences were associated with variations in leaf structure. *Acacia acuminata* is a phyllodinous sclerophyll, and *Anthocercis odgersii* is a mesomorph which sheds most of its leaves during the dry season.

Transpiration from several Australian semiarid zone species was studied in the field by Kreeb (1966), using the torsion-balance technique. The maximum rates for *Eucalyptus camaldulensis* were between 5.0 and 7.9 mg gm^{-1} min.$^{-1}$, these being attained up to 3 hr before maximum rates in *E. papuana* growing in the same site, and up to 5 hr before maximum air temperatures were reached.

Where soil water was freely available, water loss from detached leaves of *E. camaldulensis, E. campanulata* and *Acacia kempiana* continued at undiminished rates for up to 20 min. However, after a week of warm dry weather, transpiration of *E. camaldulensis* and *E. papuana* was negligible within 16 to 18 min. of detachment (Kreeb, 1966). Presumably these differences in behavior at the two times of sampling indicate a sensitive stomatal response to relative leaf water content, which would be presumed to be lower under dry conditions.

Transpiration from *Eucalyptus crebra* at Ougadougou, Upper Volta, was rapid during the wet season, reaching a maximum of 43.5 mg m^{-2} sec^{-1} during the afternoon (Birot and Galabert, 1972). At this time an almost constant ratio of transpiration to evaporation (0.31 to 0.36) was maintained throughout the day. In the dry season, maximum absolute transpiration was 18.8 mg m^{-2} sec^{-1}, and relative transpiration rates varied between 0.06 and 0.15, the maxima occurring early in the day. There was no marked midday reduction in absolute transpiration, but a steady and low rate was maintained during the day. Consequently, relative transpiration showed a pronounced decline from the early morning peak, indicating that stomatal regulation of water loss was very effective.

Comparisons among three *Eucalyptus* species with different site requirements in their native range were made by Quraishi and Kramer (1970). *E. camaldulensis* (syn. *E. rostrata*) is normally found on alluvial sands along inland water courses, *E. polyanthemos* on loamy clay soils, and *E. sideroxylon* on drier ridges in semiarid areas of Australia. Potted seedlings were allowed to deplete their soil water supply over periods of 6–18 days.

The rate of desiccation in *E. camaldulensis* seedlings was much greater than in the other species because of the greater size of the plants. However, by the time moderate water stress had developed, transpiration rates were similar for all species, and under severe stress, *E. sideroxylon* and *E. polyanthemos* maintained transpiration rates almost twice those of *E. camaldulensis* (Table XIII). Recovery of transpiration after rewatering of *E. camuldulensis* was much less complete than in the other two species.

Another comparison of transpiration rates from potted seedlings was made by Pereira and Kozlowski (1976), using *Eucalyptus camaldulensis* and *E. globulus* subjected to declining soil water content. *E. camaldulensis* was able to maintain transpiration at lower plant water potential than *E. globulus*. When rates were expressed on the basis of stomatal pore area or leaf dry weight, there was no difference between the two species, although *E. camaldulensis* is amphistomatous and *E. globulus* is hypostomatous.

5. Arid Shrublands

Rates of water loss from shrubs in arid regions, like those of perennial plants in semiarid woodlands, must vary greatly between seasons. Part of this control of water loss is achieved by seasonal shedding of leaves, or

268   D. Doley

TABLE XIII
Transpiration from Tree Seedlings (*Eucalyptus* Species) at Different Leaf Water Potentials (ψ)[a]

Species	Origin	ψ_{leaf}[b] (MPa)	RLWC[b]	Transpiration (kg m^{-2}day^{-1})
E. camaldulensis		−0.8	0.952	39.9
		−2.4	0.478	24.2
		−2.8	0.246	13.7
E. polyanthemos		−1.2	0.971	37.2
		−3.0	0.524	21.4
		−3.2	0.407	24.9
E. sideroxylon		−0.4	0.978	32.6
		−3.2	0.618	22.0
		−3.6	0.512	25.6

[a] From Quaraishi and Kramer (1970).
[b] Key to symbols: ψ, leaf water potential (MPa); RLWC, relative leaf water content.

development of structural modifications of stems to perform functions of photosynthesis (Stocker, 1970, 1971). In addition, there can be considerable control of transpiration from leaves, providing that at the same time they are not damaged by the increased temperatures which accompany the cessation of water vapor exchange (Gates, 1968).

Table XIV shows the range of transpiration rates during wet and dry seasons for a variety of arid zone species. In general the rates are very high when compared with those for species from other vegetation types, even during the dry season. These high maximum daily rates conceal the fact that transpiration may be restricted almost completely later in the day (Migahid, 1961; Stocker, 1970). Total daily transpiration as a fraction of evaporation may be very low in some species during the dry season, e.g. 0.07 in *Adenium hongel* (Stocker, 1971), but in others may be rather high, e.g., 0.34 in *Nitraria retusa* (Stocker, 1971).

Moore *et al.* (1972) related plant water potential to transpiration of desert shrubs. At the relatively high potentials of −2.8 or −5.4 MPa, transpiration from *Eurotia lanata* and *Atriplex confertifolia* in the Curlew Valley, California, reached maximum rates of 110 to 126.7 mg m^{-2} sec^{-1}, although mean rates were much lower. Under severe water stress (−9.5 to −12.0 MPa), transpiration rates were reduced to 2.7 to 18.6 mg m^{-2} sec^{-1}.

In all cases, these perennial plants occur in situations where the roots have access to permanent water supplies (Stocker, 1970, 1971), even though these supplies may be reduced during the dry season. It must also be recognized that the rates of transpiration in relation to evaporation quoted in Table XIV refer to leaf area equivalents, and not to land surface area, as is the case in estimates of E_a/E_o that have been made for some

TABLE XIV
Transpiration from Arid Zone Woody Plants[a]

Species	Location	Condition or season	ΣT (kg m^{-2}day^{-1})	T_{max} (mg m^{-2}sec^{-1})	$\Sigma T/\Sigma E$	Reference
Acacia raddiana	Israel	Wet	1.06	23.7	—	Zohary and Orshan (1959)
		Dry	2.04	45.9		
Acacia raddiana	Mauretania	Wet	2.92	133.6	0.70	Stocker (1970)
		Dry	0.98	38.2	0.21	
Adenium hongel	Mauretania	Dry	0.47	26.4	0.07	Stocker (1971)
Atriplex confertifolia	California	−5.4MPa	—	110.0	—	Moore et al. (1972)
		−9.6, −11.4 MPa	—	2.7–18.6	—	
Boscia senegalensis	Mauretania	Dry	3.21	98.4	0.28	Stocker (1970)
Calotropis procera	Mauretania	Wet	1.59	59.8	0.37	Stocker (1971)
		Dry	2.30	81.4	0.24	
Capparis decidua	Mauretania	Dry	2.50	89.8	0.17	Stocker (1971)
Colocynthis vulgaris	Egypt	Dry	—	120.8	—	Migahid (1961)
Eurotia lanata	California	−2.8MPa	—	126.7	—	Moore et al. (1972)
		−9.6, −12.0 MPa	—	4.3–6.7	—	
Nitraria retusa	Mauretania	Wet	1.91	81.3	0.55	Stocker (1971)
		Dry	1.50	46.6	0.34	
Salvadora persica	Mauretania	Dry	0.96	52.2	0.10	Stocker (1970)

[a]Calculated from author's data.

vegetation types in more humid regions. The sparse nature of vegetation distribution in arid areas means that the total water loss from the land surface is very small in relation to potential evaportion, and that the concentration of this loss into a few plants enables them to maintain rather high rates (Migahid, 1961; Stocker, 1970, 1971).

Transpiration rates of different species in an area may vary considerably from year to year, depending on seasonal conditions and characteristics of the individual plants studied. Stark and Love (1969) found that on a dry, gravelly site in Death Valley, California, *Peucephyllum schottii* showed a higher noon rate of transpiration than did *Larrea divaricata* or *Atriplex hymenelytra*. In the following year, plants of the same species but growing on a more moist site had substantially higher transpiration rates, with those of *L. divaricata* highest, followed by those of *P. schottii* and *A. hymenelytra*.

V. STOMATA AND WATER POTENTIAL

A. STOMATA

Stomata are the points of exit for water vapor from leaves, so it follows that their characteristics will influence both transpiration and the status of water in the leaf. Some aspects of stomatal function have been referred to already in Section III,C.

Basic aspects of stomatal behavior have been reviewed by Allaway and Milthorpe (1976), Burrows and Milthorpe (1976), Cowan (1977), Sheriff (1979), and Jarvis (1976) considered the variation in stomatal conductances of plants in the field. The present survey is concerned with examples of stomatal arrangement and behavior in woody plants from lower latitudes, since general reviews of the water relationships of trees (Hinckley *et al.*, 1978) tend to concentrate on species native to temperate regions.

1. Stomatal Conductance and Leaf Structure

Under conditions of adequate water supply, stomata open in the morning with increasing radiant flux and maintain a relatively constant aperture until radiant flux declines in the late afternoon (Schulze *et al.*, 1972). This pattern may be modified by shading due to adjacent tree crowns or cloud cover, with consequent changes in leaf temperature, water vapor pressure gradient between leaf and air, and, therefore, transpiration rate (Schulze *et al.*, 1973, 1975).

The maximum stomatal conductance will be characteristic for a species, but may depend upon factors influencing leaf development. Table XV shows that there may be a wide range of conductances in species

Vegetation type	Species	ψ (MPa)	g (cm sec^{-1})	Data source[d]	Reference
Rain forest	Tabebuia rigida		(0.2)[c]	F	Gates (1969)
	Ocotea spathulata		(0.2)[c]	F	Gates (1969)
Evergreen forest	Alectryon tomentosa	-3.5	0.23	F	D. Doley (unpublished data)
	Excoecaria dallachyana	-1.0	0.20	F	D. Doley (unpublished data)
	Flindersia collina	-2.1	0.50	F	D. Doley (unpublished data)
	Mallotus philippensis	-1.2	0.16	F	D. Doley (unpublished data)
	Araucaria cunninghamii	-0.5	0.10–0.41	L	D. Doley (unpublished data)
	Eucalyptus grandis	-0.5	0.59	L	D. Doley (unpublished data)
	Eucalyptus globulus	-0.5	0.50	L	D. Doley (unpublished data)
		-1.8	0.036	L	Pereira and Kozlowski (1976)
Semiarid woodland	Eucalyptus camaldulensis	-0.5	0.67	L	Pereira and Kozlowski (1976)
		-1.8	0.033	L	Collatz et al. (1976)
	Eucalyptus socialis		(0.58)[c]	L	van den Driessche et al. (1971)
	Acacia harpophylla	-1.47	0.22	L	Tunstall (1973)
	Acacia harpophylla	-6.63	0.34	F	
		-6.63	0.017	F	
Arid shrubland	Pinus halepensis		0.24	F	Whiteman and Koller (1967)
	Prunus armeniaca[b]		0.83	F	Schulze et al. (1972)
	Prunus armeniaca		0.40	F	Schulze et al. (1975)
	Zygophyllum dumosum[b]		0.40	F	Schulze et al. (1973)
	Zygophyllum dumosum		0.36	F	Whiteman and Koller (1967)
	Artemesia herba-alba[b]	-2.7	0.25	F	Schulze et al. (1973)
	Reaumuria negevensis[b]	-3.9	0.77	F	Schulze et al. (1973)
	Reaumuria negevensis	-8.7	0.22	F	Schulze et al. (1973)
	Reaumuria hirtella		0.29	F	Whiteman and Koller (1967)
	Atriplex halimus		0.18	F	Whiteman and Koller (1967)
	Atriplex lentiformis		1.02	L	Pearcy (1977)
	Atriplex hymenelytra		0.19	F	Pearcy et al. (1974)
	Larrea divaricata		1.10	F	Mooney et al. (1978)

[a] Calculated from authors data.   [c] Estimated.
[b] Plants irrigated.   [d] F, field; L, laboratory.

occupying a particular habitat. The relationship between stomatal conductance and structure is complex (Allaway and Milthorpe, 1976), and within any community of plants there is a great variety of leaf structures (Beiguelman, 1962a–d; Vasiliev, 1972; Grubb, 1977), and frequency and size of stomata (Merida and Medina, 1967; Gentry, 1969; Howard, 1969; Vasiliev, 1972; Connor and Doley, 1980).

Bearing in mind this structural variety, attempts have been made to establish characteristic leaf structures and stomatal patterns for vegetation types. Four classes of leaf structures were recognized by Vasiliev (1972) as being characteristic of different forest types:

1. Loose, few layered mesophyll in medium sized dorsiventral leaves from temperate forests;

2. Closely arranged, few layered mesophyll in thin, compound, isobilateral leaves from tropical deciduous forests;

3. Loose, multiseriate mesophyll in small, thick, entire, dorsiventral leaves from tropical rain and semi-evergreen forests;

4. Closely arranged multiseriate mesophyll in entire, thick, simple, isobilateral leaves from dry forest and woodland.

Grubb (1977) compared the leaf structure of species from different altitudinal rain forest zones with those of mediterranean climate sclerophylls. He observed that rain forest leaves became more xeromorphic in structure with increasing altitude, having greater thickness, more common occurrence of hypodermis, thicker outer epidermal walls, and more frequent occurrence of multiseriate palisade. Stomatal frequency did not vary consistently with forest type.

A similar gradation in structural characteristics of leaves between the base and upper levels of the crown of a single tree was noted by Kenworthy (1971), who suggested that the lower, more mesomorphic leaves exercised more effective control over water loss than did the upper more xeromorphic leaves. Unfortunately, difficulty of access to different parts of the large crowns of such trees has so far prevented *in situ* experimental verification of these notions.

Pyykkö (1979) made a detailed study of the morphology and anatomy of 27 woody species from a humid tropical forest in Venezuela. There was very much more variation in leaf thickness between species in the upper and middle strata of the forest than there was between these two strata, but the species occupying the lowest stratum possessed generally thinner leaves. The internal leaf structures, venation patterns, and stomatal densities also varied more within than between forest strata. However, all species had stomata on the lower surface only, with the exception of *Solanum salviifolium,* in which stomata occasionally occurred on the upper leaf surface. In most species, the upper leaf surface was shiny, and the leaf apex extended to form a drip tip which Pyykkö considered to be important in rapidly draining intercepted rain water from the leaf. The study was conducted from dried herbarium material so it is not possible to establish whether the variation in leaf characteristics within species was associated with their position in the canopy during development. However, it reinforces the conclusion that a wide range of individual leaf characteristics may combine to permit successful development of plants in a humid forest environment.

Leaf anatomy of cerrado species has been described by Ferri (1955) and Beiguelman (1962a,b). In general the leaves are thick, with several layers of palisade cells and relatively little tissue which could be regarded as spongy mesophyll. Even species with a marked dorsiventral structure (*Duguetia furfuracea, Aspidosperma pyrifolium, Spondias tuberosa*) have a relatively densely packed palisade parenchyma on the ventral surface. *Anona coriacea* and *Erythroxylum suberosum* combine a somewhat open structure within the leaf with extensive development of long, thick-walled fibers and a thick epidermis supported by a hypodermis or collenchyma (Beiguelman, 1962a,b). Stomata in these species are not usually sunken but may be embellished by processes on the guard cells or ridges on the subsidiary cells (Ferri, 1955).

In the majority of species there are two, and sometimes three, points of contact between the occluding surfaces of a guard cell pair, with the result that one or two more-or-less isolated regions of intermediate water vapor pressure will be established between the interior of the leaf and the atmosphere when the stomata are closed. In a few species, such as *Duguetia furfuracea*, the lower leaf surface is covered by an imbricate array of multicellular stellate hairs that further restrict the exchange of water vapor between the leaf and air. The fact that these features, which must also restrict carbon dioxide exchange, are retained in deep-rooted species that, in many cases, have access to freely available soil water (Rawitscher, 1948) indicates that they are regularly exposed to water stress conditions at least for short periods.

Howard (1969) and Cintron (1970) observed a wide range of both stomatal frequencies and size within forest types in Puerto Rico (Table XVI). Stomata in species of a high altitude mossy forest were generally more numerous and larger than in a lower montane forest, and of 30 species recorded, only one had stomata on the upper leaf surface. The larger stomatal pore dimensions in most mossy forest species were associated with the low water vapor pressure gradients between leaf and air which prevail in this zone.

Because of the sensitive and seasonally changing relationships between internal leaf structures, stomatal characteristics, and rates of carbon dioxide exchange in leaves (Allaway and Milthorpe, 1976; Nobel *et*

TABLE XVI

STOMATAL FREQUENCIES AND SIZES FROM CONTRASTING VEGETATION TYPES

Vegetation type	Species	Stomatal frequency (mm^{-2})	Stomatal length (μm)	Leaf area (cm^2)	Reference[b]
Rain forest, montane	Ardisia luquillensis	230	25	9.4	2
	Cecropia peltata	500	18	1250	2
	Clusia grisebachiana	177	36	8.0	2
	Clusia krugiana	240	24		1
	Eugenia borinquensis	884–920	22.5–23	41.0	1, 2
	Micropholis garciniaefolia	170–230	24–27	9.1	1, 2
	Mikania pachyphylla	196	35	5.9	2
	Ocotea spathulata	306–480	30–32	24.1	1, 2
Rain forest, lower montane	Alchornea latifolia	120	21		1
	Dacryodes excelsa	450	19.5		1
	Manilkara bidentata	320	30		1
	Miconia tetranda	400	15		1
	Sloanea berteriana	400	16.5		1
Savanna	Bowdichia virgilioides	370	20		3
	Byrsonima crassifolia	145	31		3
	Casearia sylvestris	642	13		3
	Curatella americana	162	17		3
Desert	Acacia raddiana	177	27		4
	Boscia senegalensis (L)[a]	177	36		4
	Capparis decidua	177	36		4
	Nitraria retusa (U)[a]	204	19–22		4
	Salvadora persica	69	22		4

[a] Key to abbreviations: U, upper leaf surface; L, lower leaf surface.
[b] References: (1) Cintron (1970); (2) Howard (1969); (3) Merida and Medina (1967); (4) Stocker (1971).

al., 1975), it is not possible to generalize on the most effective arrangement of tissues or stomatal structures for species in a particular situation. The water balance of plant depends not only upon leaf characteristics that may influence the loss of water, but also on attributes of the stem and roots which influence water uptake, transport, and storage. A combination of factors that results in competitive success for the plant may be achieved in many ways.

2. Stomatal Response to Desiccation

A stomatal characteristic that has been used extensively to indicate the capacity of leaves to withstand water stress is the speed of stomatal closure in detached leaves and the steady rate of water loss from leaves with closed stomata (Stålfelt, 1929; Pisek and Winkler, 1953). Detached leaves of rain forest species lost water at a faster rate over a period of 24 hr than did European sclerophylls, and were less tolerant of desiccation (Grubb, 1977). From this and structural studies, Grubb concluded that xeromorphism did not represent an adaptation to drought stress, particularly as higher altitude rain forests were the least likely to be subjected to drought.

However, forests growing on shallow soils may experience quite rapid and severe droughting (Brunig, 1970, 1971), so that higher-altitude stands on skeletal soils may in fact be exposed to drought stresses in rainless periods, particularly when high wind speeds occur (Wadsworth, 1948). It is important to know, in analyzing Grubb's hypothesis, whether the soil water store was completely exhausted or whether a small amount was available from subsurface flow. If the latter were the case, then the ability to totally restrict water loss would not necessarily be an advantage to survival of a plant. Extensive studies of species from the humid forests, cerrado, and caatinga of Brazil, some of which are summarized in Table XVII, show that within each vegetation type there is a very wide range of rates at which stomata close in detached leaves.

In general, Ferri and Coutinho (1958) indicated that species which do not show rapid closure of stomata after detachment are either drought deciduous or have access to permanent soil water supplies. Therefore, we may not be entitled to draw definite conclusions on the total water balance of a site or the adaptations of a plant community on the basis of one or two foliar attributes.

A common characteristic of species native to humid regions is their comparative inability to restrict stomatal transpiration under conditions of water stress. Okali and Dodoo (1973) compared the diurnal patterns of stomatal opening in *Khaya ivorensis,* a forest species, and *K. senegalensis,* a savanna species from Ghana. At a soil water potential of -0.03 MPa, *K. ivorensis* maintained a steady stomatal aperture for about 3 hr during the morning, with slight closure during the afternoon. *K. senegalensis* reached a wider stomatal aperture early in the morning and this was followed by progressive closure during the rest of the day. At a soil ψ of -0.45 MPa, both species opened their stomata slightly in the early morning, but *K. senegalensis* showed complete closure soon after while *K. ivorensis* did not achieve this state for almost 4 hr. This stomatal closure occurred at higher leaf relative water contents in *K. senegalensis* than in *K. ivorensis,* a drought-avoiding pattern of behavior (Levitt, 1972). The lack of a drought-avoiding behavior in *K. ivorensis* was paralleled by allocation of a larger proportion of plant dry weight to the leaves than in *K. senegalensis.*

A similar comparison between humid (*Eucalyptus globulus*) and arid zone (*E. camaldulensis*) species was conducted by Pereira and Kozlowski (1976). Maximum stomatal conductances for the two species were similar,

TABLE XVII

RATE OF STOMATAL CLOSURE IN DETACHED LEAVES OF TREES AND SHRUBS[a,b]

Vegetation type	Species	E	t_E	t_o	Reference[c]
Rain forest Trees	Affonsea edwallii	50	10	<60	1
	Alchornea triplinerva	50	5	10	1
	Bathysa stipulata	16	20	>60	1
	Coccoloba martii	50	15	>60	1
	Euterpe edulis	60	5	25	1
	Inga sessilis	55	15	20	1
	Miconia willdenowii	50	25	>60	1
Shrubs	Abuta selloana	40	5	20	1
	Alsophila longipetiolata	50	10	25	1
	Geonoma gamiova	25	5	25	1
	Geonoma schottiana	20	10	15	1
	Nematanthus fritschii	10	5	10	1
Cerrado	Anacardium occidentale	50	3	30	4
	Andira humilis	60	20	—	3
	Anona coriacea	40	30	—	3
	Byrsonima coccolobifolia	50	30	—	3
	Curatella americana	50	2	30	4
	Ouratea sp.	50	5	9	4
	Platypodium elegans	50	5	—	2
	Stryphnodendron adstringens	50	15	>60	2
	Stryphnodendron barbatimao	50	20	—	3
Caatinga	Bumelia sartorum	50	3	—	3
	Caesalpinia pyramidalis	50	10	—	3
	Jatropha phyllacantha	50	2	—	3
	Maytenus rigida	50	10	—	3
	Spondias tuberosa	50	2	—	3
	Zizyphus joazeiro	50	5	—	3
Woodland	Acacia kempeana (morning)	90	>15	—	5
	(afternoon)	50	6	12	5
	Eucalyptus camaldulensis (morning)	50	10	>20	5
	(afternoon)	50	6	12	5
	Eucalyptus tereticornis (young)	50	5	20	6
	(mature)	50	8	20	6

[a] Calculated from authors' data.

[b] Key to symbols: E, percentage of initial transpiration rate; t_E, time from detachment to attainment of E (minutes); t_o, time from detachment to complete stomal closure (minutes).

[c] References: (1) Coutinho (1962); (2) Coutinho and Ferri (1956); (3) Ferri (1955); (4) Ferri and Lamberti (1960); (5) Kreeb (1966); (6) Villaca and Ferri (1954).

and about 0.57 cm sec^{-1}, both before and after a droughting–rewatering cycle. The stomata of *E. camaldulensis* began to close at about −1.0 MPa, and at −1.5 MPa stomatal conductance had declined to about 0.05 cm sec^{-1}. In *E. globulus,* stomatal closure began at about −0.8 MPa and at −1.5 MPa conductance was approximately 0.06 cm sec^{-1}.

Sensitive stomatal movements in response to changes in atmospheric humidity have been reported by Schulze *et al.* (1972, 1975), who concluded that the water status and rate of transpiration from the leaf were functions of stomatal conductance. In some species it appears that the linkage between internal water status of the leaf and stomatal function is not very direct, since oscillations of stomatal aperture may occur as a result of manipulation (Cowan, 1972) or naturally in the field (Teoh and Palmer, 1971). In the latter case, stomata of *Eucalyptus umbra* oscillated with a period of about 60 to 95 min. when the soil ψ was approximately −0.2 MPa, but not at −0.02 MPa when there was a slow decrease in conductance throughout the day, or at −0.5 MPa when there was only a brief early morning opening. When oscillations developed, they were not in phase between different leaves of the same plant, suggesting that the vascular connections between the roots and individual leaves of seedlings are relatively independent. It could be expected, then, that appreciable differences in leaf water status could develop between leaves of even a small woody plant.

B. WATER POTENTIAL

Patterns of variation in leaf water potential (ψ) of forest trees have been examined by Hinckley *et al.* (1978), and it is immediately clear from this work that relatively little is known of the behavior of tropical and subtropical woody plants. Of the information that is available, most concerns desert species.

1. Diurnal Variation

It is well established that the diurnal range of leaf ψ or the associated leaf parameter, relative water content (RWC), is a useful index of the water status of plants. Slatyer (1961b) showed that in *Acacia aneura,* high soil moisture content was associated with high dawn values and a large diurnal range of RWC. Low soil water content was associated with low dawn RWC and a smaller diurnal amplitude. This finding has been substantiated by Tunstall and Connor (1975) for *A. harpophylla,* using water potentials (ψ) determined by pressure chamber (Fig. 9).

The data in Fig. 9 suggest that there is little stomatal control over water loss at ψ values above −3 MPa, and progressively greater control as

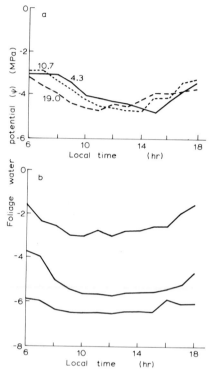

Fig. 9. Patterns of daily variation in foliage water potential of *Acacia harpophylla*. (a) Days with similar dawn water potential but differing radiant flux totals: 4.3 MJ m⁻², 10.7 MJ m⁻², and 19.0 MJ m⁻². (b) Days with similar total radiant flux (13.8 MJ m⁻²) but different dawn water potentials. (Redrawn from Tunstall and Connor, 1975.)

dawn ψ declines towards -7 MPa. However, detailed curves of the diurnal variation in shoot ψ shown in Fig. 9 do indicate considerable stomatal control of water loss, even on days when dawn ψ is high. The days represented in Fig. 9a had a wide range of total radiant fluxes, from 4.3 to 19.0 MJ m⁻², but the amplitudes of water potential were very similar. Differences did occur in the time at which ψ began to decline from the dawn value, being earlier in summer than in spring or winter, and in the time at which the minimum ψ was reached, again being earlier in summer than in spring and winter.

On the other occasions, when days of equal radiation load (13.8 MJm⁻²) were associated with very different dawn shoot ψ values, marked differences in the amplitude of water potential during the day were observed (Fig. 9b). It is interesting that the greatest amplitude was recorded in the drying soil, and that depressed ψ's were sustained for a longer period during the day. A similar phenomenon was reported by Slatyer (1961b) with respect to the variation in RWC in *Acacia aneura* leaves during a soil-drying cycle.

Schulze *et al.* (1972, 1975) indicated that stomatal opening in desert species was extremely sensitive to atmospheric humidity. This observation is supported by the results of Tunstall and Connor (1975), who found that early morning stomatal conductances in *A. harpophylla* varied from 0.17 to 0.48 cm sec^{-1}, the lowest value being associated with a dawn shoot ψ of about -1.5 MPa and 87% relative humidity, and the highest conductance a dawn ψ of -2.1 MPa and a maximum relative humidity of 100%.

Not only do species in arid and semiarid regions undergo considerable diurnal variation in shoot or leaf ψ values (Table XVIII). For example, in a semi-evergreen forest on Barro Colorado Island, Panama, Fetcher (1979) observed considerable change in ψ of *Trichilia cepo* and *Cordia alliodora* leaves before and after the commencement of the wet season. The general relationship between dawn ψ and amplitude discussed previously also applied to these species. As may be expected, there is considerable variation in the dawn values, ranges, and patterns of variation in ψ for different species growing on the same site.

Although there are difficulties in correlating osmotic and xylem pressure potentials (Tyree, 1976a), estimates of the former have been made for many vegetation types of the world (Walter, 1964). These also show the same diurnal and seasonal variations in a qualitative fashion as the patterns of xylem pressure potential discussed above.

2. Maintenance of Water Supply to the Leaf

One aspect of the water relations of trees that deserves further attention is the quantitative relationship between transpirational loss and flow through the plant. This has been studied in conifers (Waring and Running, 1976), and attempts to relate some parameters of water flow in hardwoods were made by Doley (1967) and Kenworthy (1971). In these two situations, soil water was freely available, but in many tropical and subtropical conditions it is not so, and the integrity of the conducting system may be threatened. Huber (1956) suggested that greater security of the conducting system is provided by having shorter, narrower vessels, which limit the extent of disruption to the vascular system in the event of failure of one element. Briggs (1967) and Zimmermann and Brown (1971) emphasized the relative importance of end-wall and lumen resistances to water flow, so that one may expect, teleologically, that narrow vascular elements will also be short.

TABLE XVIII

DIURNAL AND SEASONAL VARIATION IN FOLIAGE WATER POTENTIAL (ψ) IN DIFFERENT VEGETATION TYPES

Species	Location	Wet conditions (MPa)[a]			Dry conditions (MPa)[a]			Reference[c]
		ψ_{max}	ψ_{min}	$\Delta\psi$	ψ_{max}	ψ_{min}	$\Delta\psi$	
Humid forest								
Trichilia cipo	Barro Colorado Islands, Panama	−0.2	−2.7	2.5	−2.6	−3.9	1.3	1
Cordia alliodora	Barro Colorado Islands, Panama	−0.1	−1.3	1.2	−0.7	−1.9	1.2	1
Araucaria cunninghamii	Brisbane, Australia	−0.2	−2.0	1.8	−1.2	−1.5	0.3	4
Flindersia collina	Brisbane, Australia	−0.8	−3.0	2.2	−2.3	−3.0	0.7	4
Flindersia australis	Brisbane, Australia	—	—	—	−3.4	−5.5	2.1	4
Mallotus philippensis	Brisbane, Australia	−1.0	−2.2	1.2	−0.7	−3.4	2.7	4
Semiarid woodland								
Acacia harpophylla	Meandarra, Australia	−1.6	−3.1	1.5	−5.8	−6.5	0.7	3
Desert								
Cercidium microphyllum	Cave Creek (ht)[b], Arizona	−0.96	−2.08	1.12	−4.83	−7.41	3.58	2
	Cave Creek (ls)[b], Arizona	−0.67	−1.45	0.78	−4.50	−8.45	3.95	2
Franseria deltoidea	Cave Creek (ht)[b], Arizona	−1.33	−1.66	0.33	−3.16	−3.23	0.07	2
	Cave Creek (ls)[b], Arizona	−1.33	−1.50	0.17	−3.25	−3.58	0.33	2

[a] Key to symbols: ψ_{max}, daily maximum foliage water potential; ψ_{min}, daily minimum water potential; $\Delta\psi$, daily amplitude of water potential.

[b] ht, hill top; ls, lower slope.

[c] References: (1) Fetcher (1979); (2) Halvorson and Patten (1974); (3) Tunstall and Connor (1975); (4) D. Doley (unpublished data).

The ease with which water columns in these elements collapse is an important factor in maintaining continuity of water supply to leaves (Huber, 1956). Milburn (1973) described a method of acoustic detection of the collapse of these columns and used it in studies of water relations of *Ricinus communis* leaves, palms (Milburn and Davis, 1973), and *Plantago major* (Milburn and McLaughlin, 1974). Application of the technique to detached leaves of *Eucalyptus maculata* (D. Doley and D. S. Crombie, unpublished) indicated cavitation at xylem sap tensions of 3.5–4.5 MPa, which were reached at RWC values of 46–49%. Other studies (D. Doley, unpublished) suggest that wilting of intact seedlings of this species occurs at leaf ψ values of about -3.0 MPa. Cavitation of water columns in immature leaves was much less common than in mature leaves with lignified vascular tissues and may partly explain the greater ability of immature leaves to recover following water stress.

Application of the acoustic detection technique to trees in the field is difficult because of the susceptibility of the acoustic system to interference from moving stems or flapping leaves. It may be argued that the use of detached leaves is likely to lead to cavitation of water columns at lower tensions than in intact plants (West and Gaff, 1976), so that the tensions recorded here may be a minimum for the species.

VI. EFFECTS OF WATER DEFICITS

A. Physiological Processes

The level of physiological activity in plants will be determined to a large extent by the photosynthetic capacity of leaves. This in turn depends on development of the photosynthetic mechanism and on availability of light, carbon dioxide, and water.

1. Humid Forests

Water deficits usually are not associated with the limitation of photosynthesis in tropical rain forests (Gessner, 1960). However, the different conditions experienced by emergent and understory crowns is reflected in differences in their development and chlorophyll contents of leaves. Odum *et al.* (1970a) compared winter and summer chlorophyll *a* concentrations in leaves of *Dacryodes excelsa* from lower montane rain forest in Puerto Rico, using leaves from two levels in the crown. In winter, upper and lower canopy concentrations were 0.403 and 0.434 gm m^{-2}, respectively; in summer they declined to 0.334 and 0.369 gm m^{-2}, respectively. Leaf sizes were very different, being 34 cm^2 in the upper canopy and 105 cm^2 in the lower canopy. This seasonal change in chlorophyll concentration may indicate that chlorophyll concentration is sensitive to leaf age, temperature, or water balance of the whole tree, since light and humidity conditions at the lower canopy level were relatively stable.

The relative rates of photosynthesis of pioneer forest species adapted to high levels of solar radiation and of shade-adapted seedlings in rain forest at El Verde, Puerto Rico, were compared by Lugo (1970). Sun plants (*Cecropia peltata* and *Anthocephalus cadamba*) had relatively high maximum rates of photosynthesis (about 0.3 gm $C/m^2/hr$) and high light-compensation points (about 0.3 cal/cm^2/min.). The shade-adapted species, *Sloanea berteriana* and *Dacryodes excelsa,* had maximum rates of photosynthesis of 0.06 to 0.14 gm $C/m^2/hr$ and light-compensation points of 0.007 to 0.03 gm cal/cm^2/min.

No direct measurements of water stress or stomatal condition were made by Lugo (1970), but *Dacryodes excelsa* commenced CO_2 uptake very early in the morning and ceased early in the afternoon. *Sloanea berteriana,* another shade species, also showed early reduction of CO_2 uptake, particularly on a sunny day. On the other hand, *Cecropia peltata,* a sun species, showed a close correspondence between solar radiation and photosynthetic rate. It may be concluded that the shade species were more sensitive to diurnal water stress than the sun species.

2. Dry Forests and Savanna

Studies of the effects of water stress on physiological activity are much more common for drier-zone species. Medina (1967) studied gas exchange of isolated trees in the *Trachypogon* savanna of Venezuela, using a colorometric method for estimation of CO_2 exchange (Lange, 1956). Under dry season conditions in March, early morning rates of photosynthesis for four tree species ranged from 2.84 ± 0.28 mg dm^{-2} hr^{-1} for *Casearia sylvestris* to 4.34 ± 0.69 mg dm^{-2} hr^{-1} for *Curatella americana.* From calculations of the rate of nocturnal respiration, Medina estimated that compensation for respiratory losses would be achieved between 1.6 and 2.1 hr after commencement of daily photosynthesis. Even if stomatal closure occurred later in the day, these trees, which appeared to have access to groundwater supplies, were able to maintain physiological activity throughout the dry season.

The chlorophyll concentration in leaves of savanna species from Venezuela increased from 0.123–0.214 gm m^{-2} at the time of their appearance in the late dry season to 0.279–0.406 gm m^{-2} in the middle portion of the wet season (Medina *et al.,* 1969). By the end of the wet season leaves had senesced and chlorophyll contents were very low (0.014–0.026 gm m^{-2}). Respiration decreased from the time of appearance of the leaves until the senescence, so that activity was greatest during the dry season. A

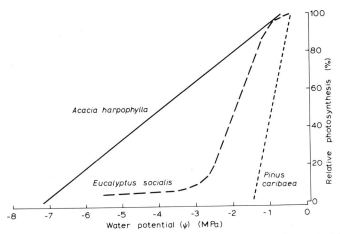

Fig. 10. Relationship between the rate of apparent photosynthesis expressed as a percentage of the maximum and foliage water potential for *Acacia harpophylla* (van den Driessche *et al.*, 1971), *Eucalyptus socialis* (drought-hardened and control, Collatz *et al.*, 1976), and *Pinus caribaea* var. *hondurensis* (drought-hardened and control, Bacon, 1978). (Redrawn from author's data.)

midday decrease in respiration rate on one occasion of observation was attributed to stomatal closure (Medina, 1967).

Semiarid tree species have been studied under glasshouse or controlled environmental conditions. A detailed study of the photosynthetic response of *Acacia harpophylla* was made by van den Driessche *et al.* (1971). Significant linear relationships were established between net photosynthesis and stomatal water vapor conductance under all experimental conditions of phyllode temperature and ψ (Fig. 10).

Photosynthesis ceased at a leaf ψ between -6 and -8 MPa, declining from a mean maximum value of 13.6 mg dm^{-2} hr^{-1}. The limitations to dry matter production in such an environment were not determined solely by water supply, although at certain times of the year it was dominant (Connor *et al.*, 1971).

Figure 11 shows an interesting condition of a midday depression in photosynthesis under conditions of high plant ψ (-1.4 MPa) and high solar radiation. The almost vertical leaf orientation (83° to the horizontal) meant that energy incident on the phyllode was less around noon than when the sun was below the zenith. Therefore, a decline in photosynthesis occurred which was independent of water-induced stomatal closure. In fact, diurnal variation in shoot water potential played a relatively small part in control of photosynthetic CO_2 exchange in this species.

On a seasonal basis, the restriction of photosynthesis in *Acacia harpophylla* by water stress occurred irregularly in response to distribution of

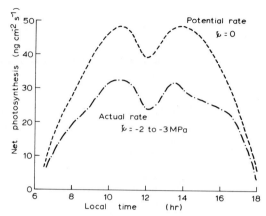

Fig. 11. Diurnal pattern of photosynthesis in *Acacia harpophylla* showing midday depression resulting from near-vertical orientation of phyllodes. (Redrawn from Connor *et al.*, 1971.)

rainfall. This variability of environmental conditions is disguised in many climatic indices, and it emphasizes the sporadic nature of physiological activity, establishment, and survival that must be expected in semiarid and arid regions.

The physiological behavior of perennial plants may change on exposure to water stress in such a way that the effects of this stress are reduced (Levitt, 1972). This has been shown many times for water exchange patterns, and by inference also for carbon dioxide exchange. Direct measurements of photosynthesis in response to seasonal water stress are less common, but some detailed studies (Hellmuth, 1971) indicate a variety of responses by different species.

In order to assess the mechanisms by which such adaptation may occur, Collatz *et al.* (1976) grew *Eucalyptus socialis* seedlings under controlled environmental conditions and observed similar responses of net photosynthesis to leaf ψ in plants which had been well watered or exposed to mild water stress (hardened) for 2.5 months (Fig. 10). The response of stomatal conductance to ψ was also similar in hardened and nonhardened plants during the second of two droughting and rewatering cycles. These responses were obtained in plants that had developed quite different leaf structures and led Collatz *et al.* (1976) to conclude that the assimilatory apparatus was more stable than was leaf structure. Nevertheless, damage due to water stress was visible in nonhardened plants desiccated to -3.6 MPa but not in hardened plants.

This work also showed various aspects of the photosynthetic process were affected differently by drought stress. There was a sharp decrease in intracellular conductance to CO_2 transfer as leaf ψ fell below -2.0 MPa, although there was no significant change in the CO_2 compensation concentration. This latter point suggests that the level of photorespiration declined in proportion to photosynthesis.

The difference between a species native to semiarid conditions and ones native to more humid temperate regions may be seen by comparing *E. socialis* (Collatz *et al.*, 1976) with six species studied by Pereira and Kozlowski (1977) (Fig. 10). Unhardened *E. socialis* behaved in a manner similar to *Juglans nigra,* the temperate species exhibiting least stomatal control. Hardened *E. socialis* restricted water loss even less and exhibited relative rates of transpiration intermediate between the extremes observed by Pereira and Kozlowski at ψ values of about -2.5 MPa.

3. Deserts

The seasonal photosynthetic behavior of desert plants has been studied extensively in recent years. Seasonal adaptation of the photosynthetic apparatus to changing temperature has been described for *Hammada scoparia* and *Prunus armeniaca* (Lange *et al.*, 1974, 1975, 1978; Schulze *et al.*, 1976), *Larrea divaricata* (Oechel *et al.*, 1972; Odening *et al.*, 1974; Mooney *et al.*, 1978; Armond *et al.*, 1978), *Encelia farinosa* (Nobel *et al.*, 1978; Odening *et al.*, 1974; Ehleringer and Mooney, 1978), *Atriplex lentiformis* (Pearcy, 1977), and *Ceratoides lanata* and *Atriplex confertifolia* (Caldwell *et al.*, 1977).

Under controlled environmental conditions, there was a decrease of 7°C in the optimum temperature for photosynthesis of *Encelia farinosa* as leaf ψ decreased from -0.02 to -1.8 MPa (Nobel *et al.*, 1978). Lange *et al.* (1974) observed that the optimum temperatures for photosynthesis in *Hammada scoparia* and *Prunus armenica* were about 3°C lower in unirrigated than in irrigated plants, although the degree of water stress was not defined. In *Encelia farinosa,* the decrease in optimum temperature for photosynthesis was closely matched by that for the residual CO_2 conductance (Nobel *et al.*, 1978). At the same time, there was a decrease in water vapor conductance, the relative magnitude of these changes increasing with temperature as water stress increased.

Depression of the temperature optimum for photosynthesis in *E. farinosa* would account for its brief periods of photosynthetic activity during the day in summer (Strain, 1975), although it has been shown by Ehleringer and Björkman (1978) that the pubescent leaf surface acts as an effective reflector of visible radiation, particularly during the dry periods of early summer and later autumn when more densely pubescent leaves are produced (Cunningham and Strain, 1969). This effect was considered by Ehleringer and Mooney (1978) to benefit the plant by reducing leaf temperature. The extent of the reduction would depend on both absorptance and water vapor conductance. At an air temperature of 39.3°C, leaf temperatures were about 34°C, this difference in temperature being consistent with predictions from calculations of absorption (40%) and water vapor conductance (0.7 cm sec^{-1}). This reduction of leaf temperature does not obviate the effects of reduced ψ, but may prevent the attainment of lethal temperatures.

The response to water stress of desert species with different adaptations was studied by Odening et al. (1974). *Larrea divaricata,* a drought-tolerant evergreen shrub, *Encelia farinosa,* a drought-deciduous shrub, and *Chilopsis linearis,* a winter deciduous phreatophytic shrub, were studied in the field and under controlled conditions. The three species showed similar rates of photosynthesis at high ψ, and the ψ associated with the cessation of apparent photosynthesis decreased from about -3.6 MPa in *C. linearis* to -7.8 MPa in *L. divaricata* (Fig. 10).

Stocker (1970, 1971) studied photosynthesis in the field of several desert species in Mauretania, concentrating on the differences in activity of species with various structural and phenological adaptations. Two evergreen trees, *Salvadora persica* and *Boscia senegalensis,* showed very different gas exchange characteristics during the dry season (Stocker, 1970). Apparent photosynthesis was eliminated for about 4 hr during the middle of the day in *S. persica,* whereas *B. senegalensis* maintained photosynthesis, albeit at a reduced rate, throughout the day. These differences were attributed to the permanent water supplies in sites occupied by *B. senegalensis* but not *S. persica.*

Acacia raddiana, a rain-green species, showed a depression of photosynthesis during the early afternoon in both the wet and dry seasons, and in the dry season photosynthesis was eliminated altogether. The existence of an early morning maximum for photosynthesis is common in desert species (Stocker, 1970; Hellmuth, 1971; Lange et al., 1975; Strain, 1975), but even relatively brief periods of photosynthetic activity may be sufficient to preserve a positive carbon balance in the plant during the dry season (Lange et al., 1975).

Species from the Mauretanian desert with alternating leaf forms in the wet and dry seasons (*Nitraria retusa* and *Calotropis procera*) and a species with green stems (*Tamarix aphylla*) showed lower rates of photosynthesis than did the evergreen and rain-green species, and less distinct restriction of gas exchange during the middle of the day.

A similar variety of behavior was observed in arid zone species of Western Australia by Hellmuth (1971), who classified species according to their relative changes in transpiration and photosynthesis between optimal and water stress conditions. The rankings of species according to rates of photosynthesis changed substantially between wet and dry seasons, and a wide variety of daily patterns of gas exchange was recorded. It is unfortunate that gas conductance values are not available from these studies, since they would help to clarify some of the interesting field responses observed. Finally, the importance of repeated measurements of photosynthetic activity during any season is shown by the reviews of Lange et al. (1975) and Strain (1975).

B. Vegetative Growth

1. Humid Forests

The vegetative growth of trees of the humid tropics gives an impression of being extremely rapid, although this may be associated more with the extended period of the year during which growth occurs than with the intrinsic growth rate (Whitmore, 1975). A considerable literature exists on rates of dry matter production of forest communities (Müller and Nielsen, 1965; Ogawa et al., 1965; Duvigneaud, 1971; Kira and Ogawa, 1971) and of annual rates of increase in stem volume or weight of commercial timber species (Dawkins, 1964; Whitmore, 1975). On the other hand, there is little information concerning the patterns of growth on a seasonal basis, particularly as these may be affected by water deficits.

Rapid growth of individual plants is usually associated with a plentiful water supply, and species with the potential for rapid growth may be particularly sensitive to water stress (Kramer and Kozlowski, 1960, 1979). Fernando and De la Cruz (1976) recorded height growth in 3- or 4-month old seedlings of Anthocephalus chinensis and Albizia falcataria at soil water contents which were maintained at four levels by the regular addition of calculated amounts of water.

In Anthocephalus chinensis, maximum stem elongation of 0.79 mm day^{-1} was attained in slightly dry soil (70% of water-holding capacity), and the relative performance of this species in drier soils was better than that of Albizia falcataria, a species having a much higher growth rate (2.65 mm day^{-1}) under optimum conditions of wet soil. This suggests an inverse relationship between potential growth rate and the capacity to withstand water stress.

An interesting attempt to relate the rate of xylem growth in Cordia alliodora to rainfall was made by Blake et al. (1976). In this species, stem diameter growth varied widely between individuals of the same age and growing in the same situations, and individuals that flushed at the beginning of the wet season were much smaller than those which flushed up to 4 months later. This difference in growth was related successfully to an index of the probability of daily rainfall being less than potential daily evapotranspiration of 4 mm. Blake *et al.* (1976) concluded that the higher probability of water stress occurring during the formation and early growth of shoot primordia would limit the amount of extension growth in early flushing plants as compared with later flushing individuals. Cambial activity appears to be regulated closely by activity terminal buds, a situation typical for temperate zone species with diffuse porous woods (Wareing and Roberts, 1956; Digby and Wareing, 1966; Doley, 1970; Kramer and Kozlowski, 1979). Davis (1970) showed that, for *Cecropia peltata* trees growing in a semi-deciduous forest in Costa Rica, there was a significant correlation between mean internode length and the weekly sum of precipitation at the time of internode development. Mean internode length during the dry season was approximately 3 cm, and during the wet season it was approximately 8 cm. The total number of internodes produced per year varied from 52 to 84 in a sample of nine trees, but total height growth varied from 1.7 to 6.2 m. This suggests that the process of shoot initiation is more regular between seasons than is the process of internode expansion, at least in *C. peltata,* which possesses the capacity for indeterminate growth under favorable conditions (Davis, 1970). The periods of slower axial growth appeared to coincide with the development of secondary xylem with thicker fiber walls, a phenomenon that is consistent with the incidence of moderate water stress in conifers (Zahner, 1963) but not necessarily in dicotyledons (Doley, 1970).

The relationship between climate and wood density was examined by Chudnoff (1976), who used data from Central and South America and the Holdridge World Life Zone Classification system (Holdridge *et al.,* 1971). In the Tropical Moist Forest zone, with annual rainfall between 2000 and 4000 mm, a large number of species had wood densities exceeding 0.69 gm cm^{-3}, the proportion increasing from about 30% at latitude 10°N (Costa Rica) to about 50% at latitude 3°S (Amazon Basin, Brazil).

Under conditions of lower or higher rainfall (<2000 or >4000 mm) there was a more uniform distribution of species between wood density classes from <0.30 to >0.69, or a preponderance of species in the 0.40 to 0.59 gm cm^{-3} density range. Therefore, although the processes of xylem formation and, consequently, wood density, are influenced by water supply, many other factors are also involved. It is regrettable that this question has received virtually no attention.

2. Dry Forests and Woodlands

Growth of trees and shrubs in tropical dry forests and woodlands can be associated much more readily with water stress where a distinct, albeit short, dry season develops than in regions with more or less sufficient water supply throughout the year. It is not surprising that many studies have been made of this association between environment and plant response, although the detail in such studies is commonly less than is reported for many investigations on temperate zone species.

a. *Seedling Growth.* Seed germination and early growth of seedlings are critical in areas of deficient rainfall. This means, first, that water supplies sufficient to ensure establishment will be available with decreasing frequency in progressively more arid areas. Second, the time available for establishment of seedlings before surface soil water supplies are exhausted by direct evaporation and transpiration may be short. Access of roots to lower, possibly wetter, soil levels is critical, so that a high rate of seedling root growth is essential. Deep penetration of the soil by roots is especially important.

Species native to arid environments often have high root/shoot ratios (Jacobs, 1955), and development of this characteristic in provenances of *Eucalyptus camaldulensis* was studied by Awe et al. (1976). Seeds collected from six more-or-less-continental locations in Australia, at latitudes ranging from 14°25′ to 36°00′ were germinated and grown on columns of a sand and loam mixture at a temperature of about 30°C. In three experiments, watering was discontinued at different times after germination. Comparisons were made with *E. saligna* and *E. pilularis,* two species from humid coastal eastern Australia.

The data of Awe et al. cover variations in weight of seedlings of *E. camaldulensis* from 0.916 to 12.06 gm, and in root/shoot ratios from 0.31 to 1.21, but the estimate of the allometric constant for the relationship between root weight (W_r) and shoot weight (W_s) was fairly precise. The equation of best fit was

$$\log_{10}W_r = -0.787 + 1.164 \log_{10}W_s \qquad r^2 = 0.893$$

This relationship is very similar to that determined by Ledig et al. (1970) for *Pinus taeda* in a similar experiment, namely

$$\log_{10}W_r = -0.612 + 1.158 \log_{10}W_s$$

These latter investigators did not establish any significant difference in the patterns of allocation of dry matter in treatments combining three levels of light intensity and two levels of moisture. It may be coincidental that two quite different species have essentially the same patterns of shoot and root growth, or it may simply indicate that allometry is not a sufficiently sensitive tool for studying the development of plants under these experimental conditions. Mechanistic models of plant development may be more appropriate to this task than the conventional growth analysis approach.

Strang (1966) observed that seedlings of *Brachystegia spiciformis* and *Julbernardia globiflora* that germinated at the beginning of the wet season in the highveld of Zimbabwe all perished, presumably due to alternation of very hot dry periods and short periods of cooler wet weather. Later, under more humid overcast conditions, survival of young seedlings was greater in experimental seed trays placed in the open or under shade, but parallel experiments in the field at this time were an almost complete failure. It is noteworthy that these two species normally regenerate only by root suckers, so that seedling establishment may be an extremely rare event without continuation of the population being placed at risk.

Direct comparisons of root development in seedlings of *Prosopis juliflora* and *Acacia tortilis* spp. *tortilis* were made by Gupta and Balara (1972). Both these species are relatively fast-growing and drought-resistant in Rajasthan, with *P. juliflora* able to tolerate rather saline soils. Early root extension in *P. juliflora* was faster than in *A. tortilis,* but 10 weeks after germination *A. tortilis* had a slightly deeper root system (55.8 cm) than did *P. juliflora* (52.3 cm). These differences are small when compared with variation in the rate of development of the plants sampled from week to week, but the importance of this study may be the demonstration that, at one week of age, *A. tortilis* and *P. juliflora* had produced root systems penetrating to 11.83 and 20.73 cm, though having dry weights of only 3.00 and 3.20 mg, respectively.

In a monsoonal environment, establishment of *Shorea robusta* is considered to depend upon water availability, which in turn influences nutrient uptake. Seth and Srivastava (1971) grew seedlings in sloping tubes designed to maintain a gradient in soil water content from 0.17 to 0.46 gm gm^{-1}. Maximum height and dry weight increments of the seedlings were attained at 0.37 gm gm^{-1} water content. No height growth occurred at 0.17 gm gm^{-1}, and at 0.46 gm gm^{-1} increment was about one-fourth of the maximum value, due to waterlogging. The sensitivity of growth rate in *Shorea robusta* to soil water content suggests that soil depth is an important determinant of the natural distribution.

The establishment of four species indigenous to the arid zone of Rajasthan, *Prosopis cineraria, Albizia lebbek, Acacia senegal,* and *Tecomella undulata* was studied by Muthana *et al.* (1976). Direct sowing of seeds resulted in failure of every species, regardless of soil preparation treatments, whereas acceptable establishment and growth were achieved in transplanted one-yr-old seedlings. No significant difference was recorded between soil preparation treatments which included cultivated trenches of 30 × 30 cm or 60 × 60 cm cross-section or pits 60 × 60× 60 cm. Height increment during the first 6 years of growth was greater in *Albizia lebbek* than in the other species, but in the second 6 years of the study growth of *A. lebbek* slowed, except where the trees were planted in pits. The reason for this was not indicated, but may be due to limited soil water storage.

The limitation of growth by shallow soil may set in from the time of establishment of seedlings. Muthana and Arora (1976) observed marked differences in height and stem diameter growth of *Eucalyptus camaldulensis* planted in loamy soils of 75 and 180 cm depth, particularly during the first few years of growth when annual height increments were approximately 75 and 230 cm, respectively.

b. Shoot Growth of Mature Plants. Development of foliage of woody plants in arid and semiarid areas may be very important as a source of fodder for livestock (Everist, 1949; Poupon, 1976; Skerman, 1977). The rate of growth of foliage of *Eremophila longifolia* and *Acacia* species in relation to rainfall, and its modification by simulated grazing treatments, were studied by Maconachie (1973). Distinct differences occurred between species with respect to the time of maximum foliage production in Central Australia. Leaf growth of *E. longifolia* coincided with the incidence of spring and summer rainfall and appeared to be independent of temperature. The *Acacia* species, on the other hand, showed some effect of temperature on their response to rainfall. *Acacia ligulata* and *A. murrayana* produced new foliage only when water was available during periods of increasing daily temperature. Foliage growth of *A. kempeana* occurred only in response to summer rains, and growth of *A. aneura* occurred after either summer or winter rains. The loss of foliage occurred under two conditions—during new shoot development, and during periods of water stress. A coincidence of leaf growth and shedding has been noted for several Australian forest and woodland species (Hatch, 1955; Specht and Brouwer, 1975; Tunstall and Connor, 1975), and leaf fall has been used as an index of the timing of new leaf growth. However, in the more humid forest conditions, it has been suggested that temperature rather than water supply determines the development and shedding of leaves.

Growth abnormalities of trees, such as the foxtails of some conifers, may be attributed to climatic differences between the source and place of cultivation of a species (Kozlowski and Greathouse, 1970). However, Ferreira (1970) could not establish a relationship between the water balance of *Pinus kesiya* (*P. khasya*) and the incidence of foxtail development in Brazil. In this and many other cases, the effect of water supply may be involved, but not independently of thermoperiod and photoperiod.

The death of trees due to brief or prolonged water stress is a common feature of arid woodlands and may even occur occasionally in more humid areas (Pook *et al.*, 1966). Guy (1970) attributed the widespread death of *Adansonia digitata* trees in south central Africa to a secular decline in rainfall over several decades. Over a timespan of approximately 14.5 years, Cunningham and Walker (1973) observed considerable fluctuations in the populations of *Acacia aneura* in semiarid eastern Australia. Trees originating from different regeneration events were identified, and their numbers decreased almost linearly with time. At the termination of observations, the original population of 2050 trees ha^{-1} had declined to 100 trees ha^{-1}, and the contribution of three regeneration events was 475 trees ha^{-1}. The frequency of successful regeneration in this species may be once in 9–10 years (Preece, 1971b).

c. Secondary Growth of Stems. Seasonal cambial growth of different species occurring in one locality may vary greatly, as shown by Daubenmire (1972) in a study of 25 species from a tropical dry forest in Guanacaste, Costa Rica. In 12 species, secondary growth was confined to the duration of the wet season, and in the other 13, growth continued for a month or more into the dry season. As compared with temperate zone species (Daubenmire and Deters, 1947), rates of cambial growth of tropical species were slow but continued for a longer period.

Shrinkage of stems was observed in species with both growth habits and also in *Calycophyllum candidissimum* and *Swietenia macrophylla* growing in riparian situations. Daubenmire (1972) also found that complete leaf shedding did not prevent marked stem shrinkage during the dry season in *Chomelia spinosa, Guettarda macrosperma, Sapranthus palanga,* and *Tabebuia neochrysantha.* In some species, shrinkage was greater than total annual increment, taking into account the possible effects of drying of the bark. This occurred in *Swietenia macrophylla,* despite constant access of the roots to water.

The patterns of wood formation in tropical trees are usually indistinct, except where ring porous structures develop. In many species there may be regular and close banding, as in the Moraceae (Metcalfe and Chalk, 1950), or growth bands may form that do not correspond with annual growth layers. Daubenmire (1972) concluded that it was not possible to assess annual growth accurately from increment cores from trees of the dry forest of Costa Rica.

Eucalyptus citriodora planted in the Northern Guinea zone of Nigeria maintained cambial growth, although at a reduced rate, throughout the dry season, whereas cambial growth of *Isoberlinia doka,* a native tree of the same area, promptly ceased at the end of the wet season and stems exhibited some shrinkage during the dry season (McComb and Ogigirigi, 1970). No explanation for the difference in behavior of the two species could be found on the basis of soil water content measurements, and Babalola and Samie (1972) showed that, in the same area, soil water content beneath the eucalypt plantation was being gradually depleted, whereas beneath the *Isoberlinia* woodland it was relatively constant between successive years (Fig. 4). A lack of concurrence between seasonal conditions and height growth of *Eucalyptus camdulensis* and *E. gomphocephala* was observed by Karschon (1964) in Israel, although diameter growth in these two species ceased for 2 months during summer.

Another example of the variability in the relationship between water stress and cambial activity was provided by Fahn (1963), who examined 21 species native to the semiarid region of Israel and two introduced species, and found four types of behavior with respect to cambial activity: (1) activity commencing in early winter, dormant for 4–8 months; (2) activity commencing in spring, dormant for 0–8 months; (3) active more or less throughout the year, but with activity increasing at the end of summer; (4) uniformly active throughout the year.

It is interesting that the first category contained only shrubs, and that categories 2, 3, and 4 were almost all tree species. The only shrub species that maintained cambial activity throughout the year, *Thymelaea hirsuta,* was collected from a wadi bed (Fahn and Sarnat, 1963), where the roots presumably had access to ground water, since new leaves were produced throughout the year. In other species, cambial activity was generally associated with extension growth in species of tropical or subtropical origin (Fahn, 1958; Waisel *et al.*, 1966; Liphschitz and Waisel, 1970b), but not in *Robinia pseudoacacia,* a species native to temperate latitudes (Waisel and Fahn, 1965). An important qualification is that species with extensive ranges such as *Eucalyptus camaldulensis* showed considerable genotypic variability with respect to wood structural characteristics (Rudman, 1970), and to distribution of extension and radial growth within the tree (Waisel *et al.*, 1966).

This general picture may be modified by local variations in environment, as shown by Liphschitz and Waisel (1970a,b,c). Cambial activity of *Ziziphus spina-christi* growing on a dry hillside was closely associated with the single occasion of active shoot extension, whereas on a lower slope where plant roots had access to groundwater there were two bursts of cambial activity, one during the main period of shoot extension and another during fruit development and a late flush of shoot extension

(Liphschitz and Waisel, 1970a). The restricted period of growth in the drier situations was associated with the production of fewer cambial derivatives. Liphschitz and Waisel (1970a) concluded that there was a direct effect of water availability on cambial activity, rather than the effect being mediated through the activity of the terminal buds (Larson, 1963). This behavior apparently differs from that in some shrubs from the region, in which cambial activity was reported to commence independently of the onset of the wet season (Fahn and Sarnat, 1963).

Irrigation promoted both active shoot extension and radial growth of *Populus euphratica* (Liphschitz and Waisel, 1970b), but there were differences in xylem increment and structure within well-watered trees, depending upon the nature of shoot development. In all shoots, a certain amount of extension growth occurred as a result of expansion of the preformed bud. This was associated with a narrow ring of xylem containing large vessels and relatively thin-walled fibers. Where water supply was limited, there was no shoot extension and subsequent xylem development was limited to thick-walled fibers and very small vessels, producing a ring porous appearance. Where water was more readily available, shoot extension of an indeterminate nature occurred, and the cambium continued to produce relatively thin-walled fibers and moderately large vessels, resulting in xylem with a diffuse porous appearance. The response of wood structure to water stress is not uniform between species of indeterminate shoot growth potential, since Doley (1970) found a decrease rather than an increase in the proportion of fibers produced in *Liriodendron tulipifera* stems during simulated drought.

The effects of water stress may be developed by an increase in soil salinity as well as by low soil water content (Liphschitz and Waisel, 1970c). A soil osmotic potential of -0.6 MPa almost completely prevented extension and radial growth in *Populus euphratica,* and resulted in production of only a few flattened fibers.

An effect on wood formation that is indirectly related to drought is that of fire, which is so common in many Australian forests that it is regarded as factor of evolutionary importance (Gardner, 1942). Nicholls (1974) observed that in *Eucalyptus marginata,* leaf production during midsummer preceded initiation of cambial activity, which then varied in proportion to rainfall. Removal of leaves by burning in spring did not interfere with the xylem increment in the following autumn, since a new canopy was regenerated during the intervening summer. On the other hand, an autumn fire of sufficient intensity prevented replacement of a full canopy for about 9 months, and also eliminated xylem increment during the wet season immediately following the fire.

C. Phenology

1. Humid Forests

The phenology of tropical tree species is complex and variable (Holttum, 1931; Koriba, 1958; Pinto, 1970; Daubenmire, 1972; Huxley and van Eck, 1974; Lieth, 1974; Tomlinson and Zimmermann, 1978; Putz, 1979). In some conditions water deficits appear to be involved in the timing of both vegetative and reproductive growth, and this aspect is the concern of the present chapter.

A common feature of growth in tropical tree species is its intermittent nature, with multiple flushes of leaf expansion and stem elongation interspersed with periods of quiescence (Richards, 1952; Njoku, 1964). Amobi (1972) found that seedlings of *Monodora tenuifolia* from different parent trees underwent an average of 5.0 to 6.1 bud flushes in their first 23 months of growth if watered only with natural rainfall at Nsukka, Nigeria, and an average of 6.0 to 6.8 flushes if watered twice daily in addition to rainfall. Within progeny from a single female parent, a significant increase in the number of flushes, from 6.1 to 7.6, was obtained by supplementary watering. There was no clear association between the number of bud flushes and the number of growth rings visible in the xylem of these seedling stems. Clearly, the rather obscure connection between environmental factors such as rainfall, day length and temperature, and the growth of shoots in *M. tenuifolia* (Njoko, 1964; Amobi, 1972) is even more obscure when secondary xylem development is being considered. The inability to relate the incidence of growth to environmental variables precludes the establishment of any quantitative relationship between the two.

Unusually widespread, heavy flowering and fruiting of rain forest trees in Malaya during 1968 were associated by Burgess (1969b), with a period of very low rainfall during the early months of that year. However, flowering within the genus *Shorea* was quite irregular, with individuals of some species flowering while other members of the genus were not. Even in years of gregarious flowering, only about 50% of the individuals of any one species may be in flower (Burgess, 1972). In general, however, heavy flowering occurs at intervals of 2 to 3 and occasionally 5 years.

Burgess (1972) showed that most abundant flowering of different *Shorea* species occurred consistently during May in the southern Malay peninsula, but flowering behavior was sometimes different in northern Malaya. *Shorea* species endemic to northern Malaya showed much less regular flowering patterns. A connection between leaf flushing and flowering was established by Medway (1972), with the interval between the two events varying from about 2 weeks to 8 months, but the causal factors for either event are not yet understood. Burgess (1972) concluded that flowering in *Shorea* was "not unrelated to periodic water stress," even though "many apparently adequate droughts are not followed by flowering," and emphasized the lack of and need for detailed studies of this problem.

In humid forests of Central America, Croat (1969) recorded a peak of flowering of tree species about 1 month after the commencement of the dry season, but considerable flowering could occur in the early wet season. Fruit development was completed somewhat later, with the peak of maturity occurring in the late dry season and early wet season.

Fournier (1976) observed two peaks of flowering and fruiting in the premontane wet forest of Costa Rica, the main peak of flowering coinciding with the long dry season, from January to May, and a secondary peak in September–October following a short dry season in July–August. Nevertheless, some flowering and fruit maturation could be observed at any time of the year. Leaf fall occurred in the dry seasons and leaf growth followed with the commencement of wet seasons.

In a humid subtropical forest of Brazil, Jackson (1978) found that the abundance of flowering was almost inversely proportional to the mean monthly rainfall, with a peak of flowering at the end of the dry season. On the other hand, rates of leaf fall coincided with mean monthly rainfall. Jackson (1978) proposed a scheme describing the incidence of leaf fall and replacement under different conditions of temperature and water supply seasonality (Fig. 12). The various behavior patterns were seen as adaptations which avoided temperature and/or drought stress, with drought the dominant factor at lower latitudes.

2. Dry Forests

As the aridity of a site increases, the coincidence of most aspects of growth, both vegetative and reproductive, increases with the season of water availability. However, there remain many anomalies of behavior that defy correlation with easily measured or deduced environmental factors, and on any site a wide variety of phenological patterns may be evident. This is shown by Daubenmire's (1972) study of development of tree species from upland forest and savanna and riparian forest in Guanacaste, Costa Rica. In the upland forest, leaf appearance (flushing) was most common at the end of the dry season, with another minor peak late in the wet season. Leaf shedding occurred mainly during the middle of the dry season, and up to 40% of the species were leafless in the latter dry season, coincident with the peak of flowering. Considerable emphasis is placed by population biologists upon the occurrence of flowering during

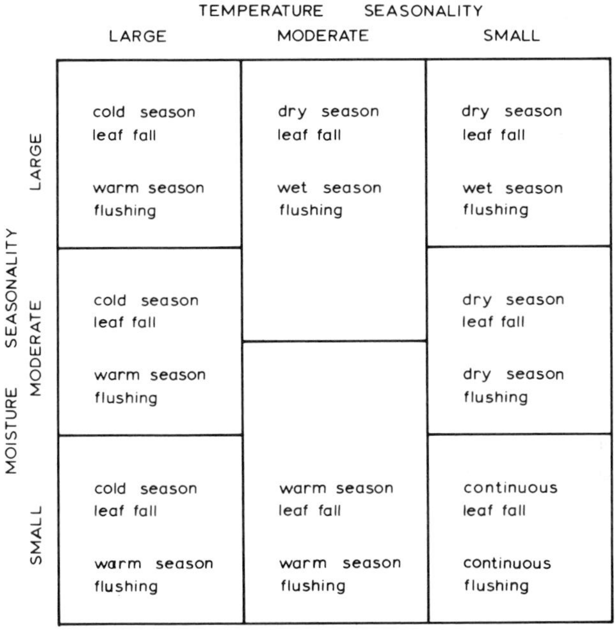

TEMPERATURE   SEASONALITY
LARGE   MODERATE   SMALL

cold season leaf fall warm season flushing	dry season leaf fall wet season flushing	dry season leaf fall wet season flushing
cold season leaf fall warm season flushing		dry season leaf fall dry season flushing
cold season leaf fall warm season flushing	warm season leaf fall warm season flushing	continuous leaf fall continuous flushing

MOISTURE SEASONALITY — LARGE / MODERATE / SMALL

Fig. 12. Model of seasonality of development in tropical and subtropical trees showing effects of temperature and moisture variation. (Redrawn from Jackson, 1978.)

periods of leaflessness (Janzen, 1967; Daubenmire, 1972), but the mechanisms by which these processes are regulated are almost totally unknown.

Contrary to the normal pattern of leaf-shed during dry weather, Janzen and Wilson (1974) described the converse pattern in *Jacquinia pungens,* an understory species of deciduous forests in Costa Rica. This species produced its canopy early in the dry season, at the time when other species were beginning to shed their leaves. Leaf fall in *J. pungens* occurred quite rapidly approximately 1 month after the beginning of the wet season. Janzen and Wilson observed a decline of about 50% in total available carbohydrates in this species during its period of leaflessness, and concluded that dormancy was an expensive means of avoiding competition for light from overstory species. This observation also begs the question of the comparative water relations of *J. pungens* and its associated species during the dry season.

Zapata and Arroyo (1978) observed that three species in tropical dry premontane forest in Venezuela flowered between the end of the dry season and the middle of the wet season. In tropical dry forests of Ceylon, Koelmeyer (1960) found that leaf fall commenced when stored soil water depletion began, and that high rainfall and humidity were associated with the initiation of leaf flushing, which may actually take place during dry weather. Flowering was also associated with depletion of soil water reserves, and this association became more regular as one progressed from wetter to drier sites.

The periodicity of vegetative and reproductive growth in tropical vegetation, and the lack of uniformity of behavior both between and within species, has been of interest for many years (Schimper, 1903). A great deal of painstaking observation has led to the general conclusion that no single environmental factor is responsible for the type or timing of growth in tropical species. Water stress is often implicated in initiation of vegetative and reproductive growth (Koelmeyer, 1960; Janzen, 1967; Alvim and Alvim, 1978), but other factors such as day length and plant age (Longman, 1978) or internal growth correlations (Borchert, 1978) may be critical. The literature reveals a close association between water supply and developmental events in dry situations, and a progressively more obscure association between these parameters as the supply becomes more abundant and more uniformly distributed throughout the year (Richards, 1952; Medway, 1972; Wycherley, 1973; Whitmore, 1975; Jackson, 1978; Putz, 1979). These generalizations do not account for the very considerable variation between individuals of one species, and the extent to which this variation is environmentally or genetically determined is only now, perhaps belatedly, being studied (Burley and Styles, 1976; Longman, 1978).

VII. DROUGHT TOLERANCE

Levitt (1972) described the responses of plants in general to drought stress and suggested categories within which species might be placed in this regard. The responses of woody plants in particular have been reviewed by Parker (1969), Kozlowski (1976), Tyree (1976b), and Kramer and Kozlowski (1979) but most of the information considered was obtained from temperate zone species. The present survey extends this coverage to tropical and subtropical species.

Tree species native to even humid areas may be subjected to periodic and possibly considerable drought (Brunig, 1969, 1970). The result of this exposure may ultimately be evolutionary selection for small or acicular leaves with an erect orientation and relatively open branching patterns, all characteristics that reduce the heat load on the canopy and, therefore, the tendency for rapid transpiration (Taylor, 1975; Brunig, 1976; Hallé et al., 1978; Oldeman, 1978).

In addition to architectural modifications that reduce the tendency for water loss, there are histological and physiological responses of leaves that may make important contributions to the survival of leaves during periods of water stress.

Using Levitt's (1972) classification of plant responses to drought stress, Kozlowski (1976) showed that most woody plants were drought avoiders rather than being tolerant of the direct effects of drought. Tyree (1976b) elaborated on aspects of the tissue water relations that may be associated with drought resistance in trees. Some leaf and plant characteristics of tropical and subtropical trees are pertinent to the present survey.

A. CONTROL OF TRANSPIRATION

The rate of closure of stomata, and the rate of water loss through the cuticle, have been used by many investigators as measures of the capacity of leaves to restrict water loss under drought conditions (Ferri, 1955; Polster and Reichenbach, 1957; Parker, 1969), and some examples are discussed in Section V,A,2. This measurement is based on the assumption that the availability of water for evaporation from cell walls within the leaf is not materially affected by the process of desiccation (Slatyer, 1967; Jarvis and Slatyer, 1970). No substantial reduction in water vapor pressure in the leaf is likely to occur as a result of reduced ψ above values of about -3.0 MPa, and at this level of water stress the stomata of intact leaves would be almost closed, even in relatively drought tolerant species. Indeed, closure of stomata as a result of water stress would lead in most species to an increase in leaf temperature (Gates, 1976), and the associated increase in water vapor pressure would more than offset any reduction consequent upon a lowered leaf water potential.

The pattern of stomatal behavior can vary widely among species in a particular area. For example, Ferri (1955) observed rapid stomatal closure in some, but not all, cerrado species in Brazil. Those showing most rapid and effective closure, such as *Spondias tuberosa*, were notable for retaining their foliage longer in the dry season than most other species. This may also be a reflection of the patterns of root development, discussed in Section III,B,2.

B. MECHANICAL ATTRIBUTES OF LEAVES

Eiten (1972) indicated that leaves of all cerrado species had at least 13 xeromorphic features that could be associated with restriction of water transport from the vascular system to the mesophyll and to the atmosphere, and with maintenance of structural integrity in leaves. Such developments have often been inferred to enable the cells of leaves to withstand low ψ by development of negative turgor pressures (Grieve, 1961; Kappen et al., 1972). However, Tyree (1976a,b) argued that the bases on which negative turgors were predicted were not valid, and that examination of pressure–volume curves did not indicate development of negative turgor.

If, then, negative turgor is not a factor in the water balance of leaves, the mechanical stiffening evident in some species presumably has an adaptive advantage. Otherwise this investment of resources in structural components of leaves would be inimical for the competitive status of the plant. Tyree (1976a) suggested that the modulus of elasticity of cell walls is important in determining the drought resistance of the leaves, because a high bulk modulus of elasticity would increase the change in water potential for a given change in water content of tissues.

The most appropriate measure of modulus of elasticity is that made at maximum pressure potential (Cheung et al., 1976; Tyree, 1976b), where the bulk modulus is maximal and constant over a range of pressure potentials. Tyree (1976b) suggested that tree breeders should select for high bulk modulus in order to improve drought tolerance, but Bacon (1978) showed that the bulk modulus for Pinus caribaea var. hondurensis seedlings could be increased from 25 to 67.5 MPa by seven successive drought cycles, each of 3 weeks duration. If such conditioning treatment can cause almost threefold changes in bulk modulus, it is reasonable to expect that changes in seasonal water supply will induce short-term changes which may be important in adaptation to drought. Increase in stiffness of leaves could occur as a result of secondary wall thickening or lignification, developments that may be consequences of reduced translocation (Roberts, 1964) and accumulation of photosynthate in the leaves. To what extent the modulus of elasticity is determined by the chlorenchymatous cells which could respond to changes in substrate concentration and by the mechanical tissues of different types and arrangements is an interesting but unsolved problem.

C. Osmotic Characteristics and Water Content

Another aspect of tissue water relations that has been measured more frequently than bulk modulus of elasticity is the relationship between RWC and ψ of foliage. This relationship is conveniently shown as the reciprocal of ψ versus RWC (Fig. 13). Such a representation permits estimation of RWC and ψ at which turgor pressure reaches zero (π_p), the proportion of water in the apoplast (θ_a) and the osmotic potential at full turgor (π_o). The alternative method of determining these parameters, by

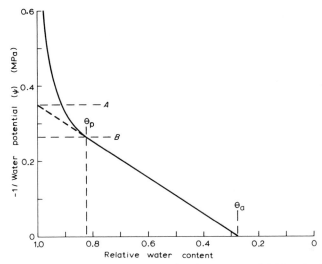

Fig. 13. Relationship between the inverse of leaf water potential (ψ) and relative water content (θ) of leaves. A, inverse of osmotic potential at full turgor; B, inverse of osmotic potential at incipient plasmolysis; θ_p, relative foliage water content at incipient plasmolysis; θ_a, apoplastic fraction of leaf water content.

comparing the inverse of xylem pressure potential against the volume of sap expressed from a shoot in a pressure chamber (Tyree and Hammel, 1972), does not give identical values for π_0 and θ_a, the greatest discrepancies being in the estimates of θ_a. Some comparisons between these parameters for tropical and subtropical species are made in Table XIX.

An interesting feature of Table XIX is the range of osmotic potentials associated with maximum and zero turgors in species that are native to relatively humid (*Pinus caribaea*) and arid (*Acacia* spp. and *Ceratonia siliqua*) areas. Connor and Tunstall (1968) emphasized the drought resistance of *A. harpophylla* as compared with *A. aneura*, and this method of analysis supports their conclusion. It also shows that turgor in *A. harpophylla* is lost at −5.4 MPa, a ψ close to that associated with cessation of apparent photosynthesis in this species (van den Driessche *et al.*, 1971).

Ladiges (1974) and Ladiges and Ashton (1974) showed that considerable variation in drought tolerance could not only occur between geographical provenances of a single species but also that this could be related to detailed site characteristics. *Eucalyptus viminalis* seedlings from low and high rainfall provenances possessed similar relationships between leaf ψ and RWC, but a low rainfall provanance from a podzolized soil showed higher transpiration rates during early drought stress and faster recovery after rewatering than did high rainfall provenances or a low

TABLE XIX
Critical Osmotic Potentials, Relative Water Contents, and Apoplastic Water Content of Tropical and Subtropical Woody Species[a]

Species	π_o (MPa)	π_p (MPa)	$RWC_{(\pi_p)}$	$RWC_{(-10\ MPa)}$	θ_a	Reference
Acacia aneura	−2.22	−3.67	0.69	0.36	0.17	Slatyer (1960)
Acacia aneura	−3.35	−5.46	0.68	0.46	0.19	Connor and Tunstall (1968)
Acacia harpophylla	−5.62	−10.41	0.75	0.76	0.45	Connor and Tunstall (1968)
Acacia harpophylla	−4.34	−5.40	0.85	0.60	0.27	Tunstall and Connor (1975)
Capparis erythrocarpus	−4.08	−5.68	0.73	0.44	0.07	Okali (1971)
Ceratonia siliqua	−2.38	−3.67	0.73	0.30	0.08	Noy-Meir and Ginzburg (1969)
Eucalyptus globulus	−1.49	−2.56	0.72	0.43	0.32	Carr and Gaff (1961)
Eucalyptus maculata	−1.90	−2.40	0.91	—	0.40	Unpublished
Eucalyptus viminalis						
Seedling, dry	−1.40	−1.70	0.89	—	0.36	Ladiges (1975)
Seedling, wet	—	−1.20	0.88	—	0.24	Ladiges (1975)
Tree, dry	−1.79	—	—	—	0.39	Ladiges (1975)
Fagara zanthoxyloides	−2.39	—	—	0.24	0.0	Okali (1971)
Flacourtia flavescens	−2.96	−4.08	0.73	0.34	0.06	Okali (1971)
Pinus caribaea var. *hondurensis*						
Seedling, wet	−1.18	−1.63	0.76	0.24	0.14	Bacon (1978)
Seedling, dry	−1.55	−1.95	0.85	0.31	0.19	Bacon (1978)
Securinega virosa	−2.78	—	—	0.30	0.03	Okali (1971)

[a] Key to symbols: π_o, osmotic potential at full turgor; π_p, osmotic potential at incipient plasmolysis; $RWC_{(\pi_p)}$, relative water content at incipient plasmolysis; $RWC_{(-10\ MPa)}$, relative water content at −10 MPa water potential; θ_a, apoplastic fraction of leaf water content.

rainfall provenance from basaltic loam soil. Further, the proportion of leaf material killed during drought stress in the low rainfall podzol provenance was lower than in three other provenances tested. This difference in drought tolerance was associated with significant variations in the fraction of water contained in the cell walls (Table XIX) and in the ratio of dry weight to turgid weight of leaves (Ladiges, 1975).

This indicates the importance of cell walls to water relations of leaves of *Eucalyptus* spp., a fact emphasized by Carr and Gaff (1961) and shown in Table XIX. Of all the species examined, eucalypts appear to maintain the highest proportions of water in cell walls, with even *E. viminalis* from a wet temperate climate having a cell-wall water capacity little less than *Acacia harpophylla* and greater than *A. aneura*. Even allowing for the fact that determinations of apoplastic fractions of leaf water content are relatively inaccurate, the values of Noy-Meir and Ginzburg (1969) and Okali (1971) are very low compared to those obtained by Australian workers. The methods employed in these determinations cannot account entirely for the differences, as Slatyer (1960) and Okali (1971) both used vapor equilibration methods. Apoplastic water content alone cannot account for differences in drought tolerance between species, but within species it may be associated with factors such as cell wall thickness that can be shown to impart some drought tolerance (Tyree, 1976b).

Differences in the characteristic osmotic properties of woody species from many water supply situations have been amply demonstrated by Walter (1972), and recent work has shown the importance of short-term variation in osmotic potential as an adaptive process (Turner and Begg, 1977). Few detailed studies of osmotic adjustment in woody species are available, but Bacon (1978) showed that, in seedlings of *Pinus caribaea* var. *hondurensis*, exposure to drought stress resulted in reduction of the osmotic potentials of leaves by about 0.32 to 0.37 MPa at zero and full turgor, respectively. As a consequence, RWC at zero turgor increased from 0.76 in well watered to 0.85 in repeatedly droughted seedlings. Such changes, if they occurred during the course of a season in field grown plants, would extend the range of soil water contents within which water could be available to maintain turgor, gas exchange, and growth (Tyree, 1976b).

The shape and orientation of leaves are both important in determining the amount of heat incident and the rate of loss of heat which can be independent of water loss (Parkhurst and Loucks, 1972; Taylor, 1975; Gates, 1976). In some species the shape and size of leaves may change when grown under different water supply conditions (Stocker, 1970) although the shapes formed under intermediate water stress may not confer an obvious advantage as far as energy exchange is concerned.

Natural variation in drought tolerance of provenances of *Pinus caribaea* was studied by Venator (1976), who found that *P. caribaea* var. *caribaea*, and var. *bahamensis*, which were subjected to long dry seasons in their insular environments underwent a transition from primary to secondary needles sooner than did *P. caribaea* var. *hondurensis* from mainland Central America. The insular varieties became semidormant 16 to 20 weeks after germination whereas var. *hondurensis* continued to grow vigorously for at least 20 weeks. Venator suggested that the primary needles were a feature inimical to drought tolerance, but it may also be argued that the restricted period of juvenile growth before the onset of semidormancy is important because seedlings subjected to water stress during active growth are more likely to succumb than those in which growth has previously ceased (Williams, 1974; Bacon and Hawkins, 1977; Bacon and Bachelard, 1978).

D. WATER USE EFFICIENCY

The rate of fixation of carbon dioxide per unit of water lost has been used as an index of water use efficiency in assessment of the suitability of crop species for dry environments (Fischer and Turner, 1978). In keeping with the few physiological studies of tropical tree species, there are few estimates of water use efficiency of photosynthesis. Schulze *et al.* (1975), compared the water use efficiencies of arid zone species from different parts of the world and concluded that the ratio of photosynthesis to transpiration was extremely sensitive to the nature of the stomatal response to temperature and atmospheric humidity. A distinct difference between the water use efficiencies of C_3 and C_4 species was shown by Ludlow (1976) to be independent of the life form or natural habitat of the species concerned. The fact that tree species do not normally possess the C_4 photosynthetic pathway (Schaedle, 1975; Kramer and Kozlowski, 1979) would suggest that their water use efficiencies are low. The observations summarized by Stocker (1976) indicate that variation in water use efficiency within woody species may be considerably greater than the differences between C_3 and C_4 species in a given location, as reported by Ludlow (1976). Further, water use efficiency appears to diminish as one progresses from the humid environment of Central Europe to the arid environment of the northern Sahara (Stocker, 1976), a situation of teleological paradox.

Water use efficiency can be seen to be most important where growth commences from a small perennating body, such as a seed or a small rhizome. Growth of the aerial shoot and root systems must occur rapidly if a plant is to withstand competition from neighbors, and if the water supply is limited, a high water use efficiency could be a competitive advantage. In perennial plants, on the other hand, the existence of comparatively large shoot and root systems, with considerable reserves, means that extensive structures for the uptake of water and carbon dioxide are in existence at the commencement of a growing season, and the chief requirement for survival of a plant is the maintenance of these systems, replenishment of reserves, and conversion of a relatively small amount of dry matter to reproductive structures.

One common characteristic of woody plants is their access to a permanent supply of water, even though this may be seasonally limited. In such a situation, the water use efficiency of dry matter production may be less critical to the survival of the plant than is the access to some water, at whatever distance from the base of the shoot this may be.

ACKNOWLEDGMENTS

The assistance of the Faculty of Forestry, Kasetsart University, Mr. B. G. Dabral and Dr. D. A. Gilmour in providing information, and Professor M. G. Ferri and M. C. Bailly for their comments, is gratefully acknowledged.

REFERENCES

Adams, M. E. (1967). A study of the ecology of *Acacia mellifera, A. seyal* and *Balanites aegyptica* in relation to land-clearing. *J. Appl. Ecol.* **4,** 221–237.

Allaway, W. G., and Milthorpe, F. L. (1976). Structure and functioning of stomata. *In* "Water Deficits and Plant Growth" (T. T. Kozlowski, ed.), Vol. 4, pp. 57–102. Academic Press, New York.

Alvim, P. de T., and Alvim, R. (1978). Relation of climate to growth periodicity in tropical trees. *In* "Tropical Trees as Living Systems" (P. B. Tomlinson and M. H. Zimmermann, eds.), pp. 445–464. Cambridge Univ. Press, London and New York.

Alvim, P. de T., and Havis, J. R. (1954). An improved infiltration series for studying stomatal opening as illustrated with coffee. *Plant Physiol.* **29,** 97–98.

Amobi, C. C. (1972). Multiple bud growth and multiple wood formation in the seedlings of *Monodora tenuifolia* Benth. *Ann. Bot. (London)* [N. S.] **36,** 199–205.

Arens, K. (1958). O cerrado como vegetação oligótrofica. *Univ. Sao paulo, Fac. Filos. Cienc. Let., Bol., Bot.* **15,** No. 224, 59–77.

Armond, P. A., Schreiber, U., and Björkman, O. (1978). Photosynthetic acclimation to temperature in the desert shrub, *Larrea divaricata*. II. Light-harvesting efficiency and electron transport. *Plant Physiol.* **61,** 411–415.

Awe, J. O., Shepherd, K. R., and Florence, R. G. (1976). Root development in provenances of *Eucalyptus camaldulensis* Dehn. *Aust. For.* **39,** 201–209.

Babalola, O., and Samie, A. G. (1972). The use of a neutron technique in studying soil moisture profiles under forest vegetation in the northern Guinea zone of Nigeria. *Trop. Sci.* **14,** 159–168.

Bacon, G. J. (1978). Physiological aspects of conditioning *Pinus caribaea* Mor. var. *hondurensis* B. & G. seedlings to withstand water stress. Ph.D. Thesis, Australian National University, Canberra.

Bacon, G. J., and Bachelard, E. P. (1978). The influence of nursery conditioning treatments on some physiological responses of recently transplanted seedlings of *Pinus caribaea* Mor. var. *hondurensis* B. & G. *Aust. For. Res.* **8**, 171–183.

Bacon, G. J., and Hawkins, P. J. (1977). Studies on the establishment of open root Caribbean pine planting stock in southern Queensland. *Aust. For.* **40**, 173–191.

Baillie, I. C. (1976). Further studies on drought in Sarawak, East Malaysia. *J. Trop. Geogr.* **43**, 20–29.

Bailly, I. C., de Coignac, G. B., Malvos, C., Ningre, J. M., and Sarrailh, J. M. (1974). Etude de l'influence du couvert naturel et ses modifications a Madagascar. *Cent. Tech. For. Trop. Cah. Sci.*, No. 4.

Baynton, H. W. (1969). The ecology of an elfin forest in Puerto Rico. 3. Hilltop and forest influences on the microclimate of Pico del Oeste. *J. Arnold Arbor., Harv. Univ.* **50**, 80–92.

Beadle, N. C. W. (1962). Soil phosphate and the delimitation of plant communities in eastern Australia. *Ecology* **43**, 281–288.

Beard, J. S. (1955). The classification of tropical American vegetation-types. *Ecology* **36**, 89–100.

Beiguelman, B. (1962a). Contribuição para o estudo anatómico de plantas do cerrado. I. Anatomia de fôlha e caule de *Erythroxylum suberosum* St. Hil. *Rev. Biol. (Lisbon)* **3**, 97–110.

Beiguelman, B. (1962b). Contribuição para o estudo anatómico de plantas do cerrado. II. Anatomia de fôlha e caule de *Byrsonima coccolobifolia* Kth. *Rev. Bio. (Lisbon)* **3**, 111–123.

Beiguelman, B. (1962c). Contribuição para o estudo anatómico de plantas do cerrado. III. Anatomia de fôlha e caule de *Anona coriacea* Mart. *Rev. Biol. (Lisbon)* **4**, 1–12.

Beiguelman, B. (1962d). Considerações sôbre a morfologia dos estomas de *Anona coriacea* Mart., *Byrsonima coccolobifolia* Kth., *Erythroxylum suberosum* St. Hil. e *Ouratea spectabilis* (Mart.) Engl. *Rev. Bras. Biol.* **22**, 115–124.

Bhimaya, C. P., and Kaul, R. N. (1966). Role of seedling root habit on initial seedling survival in some desert tree species. *Sci. Cult.* **32**, 204–206.

Birot, Y., and Galabert, J. (1972). Bioclimatologie et dynamique de l'eau dans une plantation d'*Eucalyptus*. *Cent. Tech. For. Trop. Cah. Sci.* No. 1.

Blake, J., Rosero, P., and Lojan, L. (1976). The interaction between phenology and rainfall in the growth of *Cordia alliodora* (R & P) Oken in a plantation at Turrialba, Costa Rica. *Commonw. For. Rev.* **55**, 37–40.

Bloom, S. G., and Raines, G. E. (1970). Kinetic study of hydrogen flow through the El Verde forest system. *In* "A Tropical Rain Forest" (H. T. Odum and R. F. Pigeon, eds.), pp. H123–H128. U. S. At. Energy Comm., Washington, D. C.

Borchert, R. (1978). Feedback control and age-related changes of shoot growth in seasonal and nonseasonal climates. *In* "Tropical Trees as Living Systems" (P. B. Tomlinson and M. H. Zimmermann, eds.), pp. 497–515. Cambridge Univ. Press, London and New York.

Bosshard, W. C., and Wunder, W. G. (1966). Root studies in irrigated tree plantations. *Pam. For. Res. Educ. Proj. For. Dep. Sudan* No. 29.

Briggs, G. E. (1967). "Movement of Water in Plants." Blackwell, Oxford.

Brunig, E. F. (1969). On the seasonality of droughts in the lowlands of Sarawak (Borneo). *Erdkunde* **23**, 127–133.

Brunig, E. F. (1970). Stand structure, physiognomy and environmental factors in some lowland forests in Sarawak. *Trop. Ecol.* **11**, 26–43.

Brunig, E. F. (1971). On the ecological significance of drought in the equatorial wet evergreen (rain) forest of Sarawak (Borneo). *Univ. Hull, Dep. Geogr., Misc. Publ.* No. 11, pp. 66–96.

Brunig, E. F. (1976). Tree forms in relation to environmental conditions: An ecological viewpoint. *In* "Tree Physiology and Yield Improvement" (M. G. R. Cannell and F. T. Last, eds.), pp. 139–156. Academic Press, New York.

Bunting, A. H., and Lea, J. D. (1962). The soils and vegetation of the Fung, East Central Sudan. *J. Ecol.* **50**, 529–558.

Burgess, P. F. (1969a). Ecological factors in hill and mountain forests of the States of Malaya *Malay. Nat. J.* **22**, 119–128.

Burgess, P. F. (1969b). Colour changes in the forest: 1968–69. *Malay. Nat. J.* **22**, 171–173.

Burgess, P. F. (1972). Studies on the regeneration of the hill forests of the Malay Peninsula: The phenology of dipterocarps. *Malay. For.* **35**, 102–123.

Burley, J., and Styles, B. T., eds. (1976). "Tropical Trees: Variation, Breeding and Conservation." Academic Press, New York.

Burrows, F. J., and Milthorpe, F. L. (1976). Stomatal conductance in the control of gas exchange. *In* "Water Deficits and Plant Growth" (T. T. Kozlowski, ed.), Vol. 4, pp. 103–152. Academic Press, New York.

Calder, I. R. (1978). A model of transpiration and interception loss from a spruce forest at Plynlimon, Central Wales. *J. Hydrol.* **33**, 247–265.

Caldwell, M. M. (1976). Root extension and water absorption. *In* "Water and Plant Life" (O. L. Lange, L. Kappen, and E.-D. Schulze, eds.), pp. 63–85. Springer-Verlag, Berlin and New York.

Caldwell, M. M., White, R. S., Moore, R. T., and Camp, L. B. (1977). Carbon balance, productivity, and water use of cold-winter desert shrub communities dominated by C_3 and C_4 species. *Oecologia* **29**, 275–300.

Carlson, N. K. (1961). Fog and lava rock, pines and pineapples. *Am. For.* **67**(2), 8–11.

Carr, D. J., and Gaff, D. F. (1961). The role of cell-wall water in the water relations of leaves. *In* "Plant-Water Relationships in Arid and Semi-Arid Conditions," pp. 117–125. UNESCO, Paris.

Catinot, R. (1967). Sylviculture tropicale dans les zones seches de l'Afrique. *Bois For. Trop.* **111**, 19–32.

Chaicharus, S., and Chunkao, K. (1973). Fluctuation of soil moisture in hill-evergreen forest of Doi Pui, Chiengmai. *Kasetsart Univ., Fac. For. Kog-Ma Watershed Res. Bull.* No. 15.

Cheung, Y. N. S., Tyree, M. T., and Dainty, J. (1976). Some possible sources of error in determining bulk elastic moduli and other parameters from pressure-volume curves of shoots and leaves. *Can. J. Bot.* **54**, 758–765.

Chudnoff, M. (1976). Density of tropical timbers as influenced by climatic life zones. *Commonw. For. Rev.* **55**, 203–217.

Chunkao, K., and Naksiri, C. (1976). Infiltration capacity of natural hill-evergreen forest at Doi Pui, Chiangmai. *Kasetsart Univ., Fac. For., Kog-Ma Watershed Res. Bull.* No. 25.

Chunkao, K., Tangtham, N., and Ungkulpakdikul, S. (1971). Measurements of rainfall in early wet season under hill- and dry-evergreen, natural teak, and dry-dipterocarp forests of Thailand. *Kasetsart Univ., Fac. For., Kog-Ma Watershed Res. Bull.* No. 10.

Cintron, G. (1970). Variation in size and frequency of stomata with altitude in the Luquillo Mountains. *In* "A Tropical Rain Forest" (H. T. Odum and R. F. Pigeon, eds.), pp. H133–H135. U. S. At. Energy Comm., Washington, D. C.

Clayton, W. D. (1966). Vegetation ripples near Gummi, Nigeria. *J. Ecol.* **54**, 415–417.

Clegg, A. G. (1963). Rainfall interception in a tropical forest. *Caribb. For.* **24**, 75–79.

Cochrane, G. R. (1963). A physiognomic vegetation map of Australia. *J. Ecol.* **51**, 639–655.

Collatz, J., Ferrar, P. J., and Slatyer, R. O. (1976). Effects of water stress and differential hardening treatments on photosynthetic characteristics of a xeromorphic shrub, *Eucalyptus socialis*, F. Muell. *Oecologia* **23**, 95–105.

Connor, D. J., and Doley, D. (1981). The water relations of heathlands: Physiological adaptation to drought. *In* "Ecosystems of the World: Heathlands and Related Shrublands: Experimental Studies" (R. L. Specht, ed.), Vol. 9B. Elsevier, Amsterdam and New York.

Connor, D. J., and Tunstall, B. R. (1968). Tissue water relations for brigalow and mulga. *Aust. J. Bot.* **16**, 487–490.

Connor, D. J., Tunstall, B. R., and van den Driessche, R. (1971). An analysis of photosynthetic response in a brigalow forest. *Photosynthetica* **5**, 218–225.

Corner, E. J. H. (1940). "Wayside Trees of Malaya." Government Printing Office, Singapore.

Coutinho, L. M. (1962). Contribuição ao conhecimento da ecologia da mata pluvial tropical. *Univ. Sao Paulo, Fac. Filos. Cienc. Let., Bol., Bot.* **18**, No. 257, 9–219.

Coutinho, L. M., and Ferri, M. G. (1956). Transpiração de plantas permanentes do cerrado na estação das chuvas. *Rev. Bras. Biol.* **16**, 501–518.

Coutinho, L. M., and Ferri, M. G. (1960). Transpiração e comportamento estomático de plantas permanentes de cerrado em Campo do Mourão (Est. do Paraná). *Univ. Sao Paulo, Fac. Filos. Cienc. Let., Bol. Bot.* **17**, No. 247, 119–130.

Cowan, I. R. (1972). Oscillations in stomatal conductance and plant functioning associated with stomatal conductance: Observations and a model. *Planta* **106**, 185–219.

Cowan, I. R. (1977). Stomatal behaviour and environment. *Adv. Bot. Res.* **4**, 117–228.

Croat, T. B. (1969). Seasonal flowering behaviour in central Panama. *Ann. Mo. Bot. Gard.* **56**, 295–307.

Cunningham, G. L., and Strain, B. R. (1969). An ecological significance of seasonal leaf variability in a desert shrub. *Ecology* **50**, 400–408.

Cunningham, G. M., and Walker, P. J. (1973). Growth and survival of mulga (*Acacia aneura* F. Muell. ex Benth.) in western New South Wales. *Trop. Grassl.* **7**, 69–77.

Dabral, B. G., and Subba Rao, B. K. (1968). Interception studies in chir and teak plantations - New Forest. *Indian For.* **94**, 541–551.

Dabral, B. G., and Subba Rao, B. K. (1969). Interception studies in sal (*Shorea robusta*) and khair (*Acacia catechu*) plantations - New Forest. *Indian For.* **95**, 314–323.

Dabral, B. G., Premnath, and Ramswarup (1963). Some preliminary investigations on the rainfall interception by leaf litter. *Indian For.* **89**, 112–116.

Daubenmire, R. (1972). Phenology and other characteristics of tropical semi-deciduous forest in north-western Costa Rica. *J. Ecol.* **60**, 147–170.

Daubenmire, R., and Deters, M. E. (1947). Comparative studies of growth in deciduous and evergreen trees. *Bot. Gaz. (Chicago)* **109**, 1–12.

Davis, R. B. (1970). Seasonal differences in internodal lengths in *Cecropia* trees; a suggested method for measurement of past growth in height. *Turrialba* **20**, 100–104.

Dawkins, H. C. (1964). The productivity of lowland tropical high forest and some comparisons with its competitors. *J. Oxford Univ. For. Soc.* [5] **12**, 15–18.

Del Moral, R., and Muller, C. H. (1970). The allelopathic effects of *Eucalyptus camaldulensis*. *Am. Midl. Nat.* **83**, 254–282.

de Rosayro, R. A. (1960). Rain from trees. *Ceylon For.* **4**, 381–382.

Digby, J., and Wareing, P. F. (1966). The relationship between endogenous hormone levels in the plant and seasonal aspects of cambial activity. *Ann. Bot. (London)* [N.S.] **30,** 607–622.

Doley, D. (1967). Water relations of *Eucalyptus marginata* Sm. under natural conditions. *J. Ecol.* **55,** 597–614.

Doley, D. (1970). Effects of simulated drought on shoot development in *Liriodendron* seedlings. *New Phytol.* **69,** 655–673.

Doley, D., and Grieve, B. J. (1966). Measurement of sap flow in a eucalypt by thermoelectric methods. *Aust. For. Res.* **2,** 3–27.

Duvigneaud, P., ed. (1971). "Productivity of Forest Ecosystems." UNESCO, Paris.

Ehleringer, J. R., and Björkman, O. (1978). Pubescence and leaf spectral characteristics in a desert shrub, *Encelia farinosa. Oecologia* **36,** 151–162.

Ehleringer, J. R., and Mooney, H. A. (1978). Leaf hairs: Effects on physiological activity and adaptive value to a desert shrub. *Oecologia* **37,** 183–200.

Eiten, G. (1972). The cerrado vegetation of Brazil. *Bot. Rev.* **38,** 201–341.

Eiten, G. (1978). Delimitation of the cerrado concept. *Vegetatio, Haag* **36,** 169–178.

Ekern, P. C. (1964). Direct interception of cloud water at Lanaihale, Hawaii. *Soil Sci. Soc. Am., Proc.* **28,** 419–421.

Ellis, R. C. (1971). Rainfall, fog drip and evaporation in a mountainous area of southern Australia. *Aust. For.* **35,** 99–106.

Everist, S. L. (1949). Mulga (*Acacia aneura* F. Muell.) in Queensland. *Queensl. J. Agric. Sci.* **6,** 87–131.

Fahn, A. (1958). Xylem structure and annual rhythm of development in trees and shrubs of the desert. I. *Tamarix aphylla, T. jordanis* var. *negevensis, T. gallica* var. *marismortui. Trop. Woods* **109,** 81–94.

Fahn, A. (1963). Xylem structure and the annual rhythm of cambial activity in woody species. *La-Yaaran* **13,** 5–9.

Fahn, A., and Sarnat, C. (1963). Xylem structure and annual rhythm of development in trees and shrubs of the desert. IV. Shrubs. *Bull. Res. Counc. Israel, Ser. D* **11,** 198–209.

Fellizar, F. P. (1976). Stemflow characteristics of *Parashorea plicata, Pentacme contorta* and *Arenga pinnata. Pterocarpus* **2,** 86–92.

Fernando, E. S., and De la Cruz, R. E. (1976). Effect of soil moisture stress on the shoot growth of *Anthocephalus chinensis* Rich. ex Walp. and *Albizia falcataria* (L.) Fosb. seedlings. *Pterocarpus* **2,** 65–67.

Ferreira, C. A. (1970). Estudo da economia de água de um povoamento de *Pinus khasya* (Royle) em Piracicaba (SP). *O Solo, Piracicaba* **62,** 35–39.

Ferri, M. G. (1955). Contribuição ao conhecimento da ecologia do cerrado e da caatinga. *Univ. Sao Paulo, Fac. Filos., Cienc. Let., Bol., Bot.* **12,** No. 195, 1–170.

Ferri, M. G. (1961). Problems of water relations of some Brazilian vegetation types, with special consideration of the concepts of xeromorphy and xerophytism. *In* "Plant-Water Relationships in Arid and Semi-Arid Conditions," pp. 191–197. UNESCO, Paris.

Ferri, M. G., and Coutinho, L. M. (1958). Contribuição ao conhecimento da ecologia do cerrado. Estudo comparativo da economia d'água de sua vegetação, em Emas (Est. de São Paulo), Campo Grande (Est. de Mato Grosso) e Goiâna (Est. de Goiás). *Univ. Sao Paulo, Fac. Filos., Cienc. Let., Bol., Bot.* **15,** No. 224, 103–150.

Ferri, M. G., and Labouriaou, L. G. (1952). Water balance of plants from the caatinga of Paulo alfonso (Bahia) in the rainy season. *Rev. Bras. Biol.* **12,** 301–312.

Ferri, M. G., and Lamberti, A. (1960). Informaçoes sôbre a economia d'água de plantas de um tabuleiro no município de Goiana (Pernambuco). *Univ. Sao Paulo, Fac. Filos., Cienc. Let., Bol., Bot.* **17,** No. 247, 133–145.

Fetcher, N. (1979). Water relations of five tropical tree species on Barro Colorado Island, Panama. *Oecologia* **40**, 229–233.

Fischer, R. A., and Turner, N. C. (1978). Plant productivity in the arid and semi-arid zones. *Annu. Rev. Plant Physiol.* **29**, 277–317.

Fitzpatrick, E. A., Slatyer, R. O., and Krishnan, A. I. (1967). Incidence and duration of periods of plant growth in central Australia as estimated from climatic data. *Agric. Meteorol.* **4**, 389–404.

Foldats, E., and Rutkis, E. (1969). Suelo y agua como factores determinantes en la selección de algunas de las especies de arboles que en form aislada acompañan nuestros pastizales. *Soc. Venez. Cienc. Nat., Bol.* **28**, 9–30.

Fosberg, F. R., Garnier, B. J., and Küchler, A. W. (1961). Delimitation of the humid tropics. *Geogr. Rev.* **51**, 333–347.

Fournier, L. A. (1976). Observaciones fenológicas en el bosque húmedo de premontano de San Pedro de Montes de Oca, Costa Rica. *Turrialba* **26**, 54–59.

Franco, C. M., and Inforzato, R. (1967). Transpiração de *Eucalyptus saligna* Sm. en condições de cultura. *Phyton* **24**, 35–41.

Gardner, C. A. (1942). The vegetation of Western Australia with special reference to the climate and soils. *J. R. Soc. West. Aust.* **28**, 9–87.

Gardner, W. R. (1965). Dynamic aspects of soil water availability to plants. *Annu. Rev. Plant Physiol.* **16**, 323–342.

Gates, D. M. (1968). Transpiration and leaf temperature. *Annu. Rev. Plant Physiol.* **19**, 211–238.

Gates, D. M. (1969). The ecology of an elfin forest in Puerto Rico. 4. Transpiration rates and temperatures of leaves in cool humid environment. *J. Arnold Arbor., Harv. Univ.* **50**, 93–98.

Gates, D. M. (1976). Energy exchange and transpiration. *In* "Water and Plant Life" (O. L. Lange, L. Kappen, and E.-D. Schulze, eds.), pp. 137–147. Springer-Verlag, Berlin and New York.

Gates, D. M., and Papian, L. E., eds. (1971). "Atlas of Energy Budgets of Plant Leaves." Academic Press, New York.

Gentry, A. H. (1969). A comparison of some leaf characteristics of tropical dry forest and tropical wet forest in Costa Rica. *Turrialba* **19**, 419–428.

Gessner, F. (1956). Der Wasserhaushalt der Epiphyten und Lianen. *Encycl. Plant Physiol. New Ser.* **3**, 915–960.

Gessner, F. (1960). Die Assimilationsbedingungen im tropischen Regenwald. *Encycl. Plant Physiol. New Ser.* **5**(2), 492–505.

Gill, A. M. (1969). The ecology of an elfin forest in Puerto Rico. 6. Aerial roots. *J. Arnold Arbor., Harv. Univ.* **50**, 197–209.

Gilmour, D. A. (1975). Catchment water balance studies on the wet tropical coast of North Queensland. Ph.D. Thesis, James Cook University of North Queensland, Townsville.

Gilmour, D. A., and Bonell, M. (1979). Six-minute rainfall intensity data for an exceptionally heavy tropical rainstorm. *Weather* **34**, 148–158.

Gindel, I. (1966). Attraction of atmospheric moisture by woody xerophytes in arid climates. *Commonw. For. Rev.* **45**, 297–321.

Gindel, I. (1973). "A New Eco-physiological Approach to Forest-Water Relationships in Arid Climates." Junk, The Hague.

Glover, P. E., Glover, J., and Gwynne, M. D. (1962). Light rainfall and plant survival in East Africa. *J. Ecol.* **50**, 199–206.

Goodland, R. (1971). A physiognomic analysis of the 'cerrado' vegetation of central Brazil. *J. Ecol.* **59**, 411–419.

Goodland, R., and Pollard, R. (1973). The Brazilian cerrado vegetation: A fertility gradient. *J. Ecol.* **61**, 219–224.

Goodspeed, M. J., and Winkworth, R. E. (1973). Fate and effect of runoff. *In* "Studies of the Australian Arid Zone. III. Water in Rangelands Symposium, Alice Springs," pp. 56–73. CSIRO, Canberra, Australia.

Graham, B. B., and Bormann, F. H. (1966). Natural root grafts. *Bot. Rev.* **32**, 255–292.

Greacen, E. L., Ponsana, P., and Barley, K. P. (1976). Resistance to water flow in the roots of cereals. *In* "Water and Plant Life" (O. L. Lange, L. Kappen, and E.-D. Schulze, eds.), pp. 86–100. Springer-Verlag, Berlin and New York.

Greenwood, E. A. N., and Berestord, J. D. (1979). Evaporation from vegetation in landscapes developing secondary salinity using the ventilated chamber technique. I. Comparative transpiration from juvenile *Eucalyptus* a bore saline groundwater seeps. *J. Hydrol.* **42**, 369–382.

Grieve, B. J. (1961). Negative turgor pressures in sclerophyll plants. *Aust. J. Sci.* **23**, 376–377.

Grubb, P. J. (1977). Control of forest growth and distribution on wet tropical mountains: With special reference to mineral nutrition. *Annu. Rev. Ecol. Syst.* **8**, 83–107.

Grubb, P. J., and Whitmore, T. C. (1966). A comparison of montane and lowland rain forest in Ecuador. II. The climate and its effects on the distribution and physiognomy of the forests. *J. Ecol.* **54**, 303–333.

Grubb, P. J., Lloyd, J. R., Pennington, T. D., and Whitmore, T. C. (1963). A comparison of montane and lowland rain forest in Ecuador. I. The forest structure, physiognomy, and floristics. *J. Ecol.* **51**, 567–601.

Gupta, J. P., Ullah, W., and Issac, V. C. (1975). A note on some soil moisture changes under permanent vegetative cover. *Indian For.* **101**, 523–526.

Gupta, R. K., and Balara, G. S. (1972). Comparative studies on the germination, growth and seedling biomass of two promising exotics in Rajasthan desert. *Indian For.* **98**, 280–285.

Guy, G. L. (1970). *Adansonia digitata* and its rate of growth in relation to rainfall in south central Africa. *Proc. Trans. Rhod. Sci. Assoc.* **54**, 68–84.

Hackett, C. (1973). A growth analysis of the young sorghum root system. *Aust. J. Biol. Sci.* **26**, 1211–1214.

Hallé, F., Oldeman, R. A. A., and Tomlinson, P. B. (1978). "Tropical Trees and Forests: An Architectural Analysis." Springer-Verlag, Berlin and New York.

Halvorson, W. L., and Patten, D. T. (1974). Seasonal water potential changes in Sonoran desert shrubs in relation to topography. *Ecology* **55**, 173–177.

Hatch, A. B. (1955). The influence of plant litter on the jarrah forest soils of the Dwellingup region - Western Australia. *Commonw. Aust., For. Timber Bur. Leafl.* No. 70.

Hellmuth, E. O. (1971). Eco-physiological studies on plants in arid and semi-arid regions of Western Australia. III. Comparative studies on photosynthesis, respiration and water relations of ten arid zone and two semi-arid zone plants under winter and late summer climatic conditions. *J. Ecol.* **59**, 225–259.

Henwood, K. (1973). A structural model of forces in buttressed rain forest trees. *Biotropica* **5**, 83–93.

Herbert, D. A. (1959). Natural air-layering in humus-collecting epiphytes. *Queensl. Nat.* **16**, 22–23.

Heth, D., and Karschon, R. (1963). Interception of rainfall by *Eucalyptus camaldulensis* Dehn. *Contrib. Eucalypts Isr.* **2**, 7–12.

Hill, R. D. (1972). Soil moisture under forest, Bukit Timah Nature Reserve, Singapore. *Gardens' Bull., Singapore* **26**, 85–93.

Hillel, D. (1971). "Soil and Water: Physical Principles and Processes." Academic Press, New York.

Hinckley, T. M., Lassoie, J. P., and Running, S. W. (1978). Temporal and spatial variations in the water status of forest trees. For. Sci. Monogr. 20.

Hodgkinson, K. C., Johnson, P. S., and Norton, B. E. (1978). Influence of summer rainfall on root and shoot growth of a cold-winter desert shrub, Atriplex confertifolia. Oecologia 34, 353–362.

Holdridge, L. R., Grenke, W. C., Hatheway, W. H., Liang, T., and Tosi, J. A. (1971). "Forest Environments in Tropical Life Zones: A Pilot Study." Pergamon, Oxford.

Holmes, J. W. (1956). Calibration and field use of the neutron scattering method of measuring soil water content. Aust. J. Appl. Sci. 7, 45–58.

Holttum, R. E. (1931). On periodic leaf-change and flowering of trees in Singapore. Gardens' Bull., Straits Settlements, Singapore 5, 173–206.

Hopkins, B. (1960). Rainfall interception by a tropical forest in Uganda. East Afr. Agric. For. J. 25, 255–258.

Hosegood, P. H., and Howland, P. (1966). A preliminary study of the root distribution of some exotic tree crops, evaluated by a rapid sampling method. East Afr. Agric. For. J. 32, 16–18.

Howard, R. A. (1969). The ecology of an elfin forest in Puerto Rico. 8. Studies of stem growth and form and of leaf structure. J. Arnold Arbor., Harv. Univ. 50, 225–261.

Huber, B. (1956). Die Gefässleitung. Encycl. Plant Physiol., New Ser. 3, 541–582.

Huttel, C. (1971). Estimation du bilan hydrique dans une forêt sempervivente de basse Côte d'Ivoire. In "Isotopes and Radiation in Soil-Plant Relationships Including Forestry," pp. 439–452. IAEA, Vienna.

Huttel, C. (1975). Root distribution and biomass in three Ivory Coast rain forest plots. In "Tropical Ecological Systems. Trends in Terrestrial and Aquatic Research" (F. B. Golley and E. Medina, eds.), pp. 123–130. Springer-Verlag, Berlin and New York.

Huxley, P. A., and van Eck, W. A. (1974). Seasonal changes in growth and development of some woody perennials near Kampala, Uganda. J. Ecol. 62, 579–592.

Inthasothi, S., and Chunkao, C. (1976). Infiltration capacity of soils in various types of land use at Mae Thalai watershed, Chiengdao. Kasetsart Univ., Fac. For., Kog-Ma Watershed Res. Bull. No. 26.

Jackson, I. J. (1971). Problems of throughfall and interception assessment under a tropical forest. J. Hydrol. 12, 234–254.

Jackson, I. J. (1975). Relationships between rainfall parameters and interception by tropical forest. J. Hydrol. 24, 215–238.

Jackson, J. F. (1978). Seasonality of flowering and leaf-fall in a Brazilian subtropical lower montane moist forest. Biotropica 10, 38–42.

Jacobs, M. R. (1955). "Growth Habits of the Eucalypts." Government Printer, Canberra, Australia.

Jane, F. W. (1970). "The Structure of Wood." Chapman & Hall, London.

Janzen, D. H. (1967). Synchronization of sexual reproduction of trees within the dry season in Central America. Evolution 21, 620–637.

Janzen, D. H., and Wilson, D. E. (1974). The cost of being dormant in the tropics. Biotropica 6, 260–262.

Jarvis, P. G. (1976). The interpretations of the variations in leaf water potential and stomatal conductance found in canopies in the field. Philos. Trans. R. Soc. London, Ser. B 273, 593–610.

Jarvis, P. G., and Slatyer, R. O. (1970). The role of the mesophyll cell wall in leaf transpiration. Planta 90, 303–322.

Jenik, J. (1967). Root adaptations in West African trees. *J. Linn. Soc. London, (Bot.)* **60**, 25–29.

Jenik, J., and Hall, J. B. (1966). The ecological effects of the Harmattan wind in the Djebobo massif (Togo Mountains, Ghana). *J. Ecol.* **54**, 767–779.

Jordan, C. F., and Kline, J. R. (1977). Transpiration of trees in a tropical rainforest. *J. Appl. Ecol.* **14**, 853–860.

Kappen, L., Lange, O. L., Schulze, E.-D., Evanari, M., and Buschbom, U. (1972). Extreme water stress and photosynthetic activity of the desert plant *Artemesia herba-alba* Asso. *Oecologia* **10**, 177–182.

Karschon, R. (1964). Periodicity of growth in *Eucalyptus camaldulensis* Dehn. and *E. gomphocephala* A. DC. *Natl. Univ. Inst. Agric., Ilanot, Leafl., For. Div.* No. 24.

Karschon, R. (1971). The effect of coppice cutting on the water balance of *Eucalyptus camaldulensis* Dehn. *Isr. J. Agric. Res.* **21**, 115–126.

Karschon, R., and Heth, D. (1967). The water balance of a plantation of *Eucalyptus camaldulensis* Dehn. *Contrib. Eucalypts Isr.* **3**, 7–34.

Kaushik, R. C., Qureshi, I. M. , Yadav, J. S. P., and Prakash, J. (1969). Suitability of soils for *Eucalyptus hybrid* (Mysore gum syn. *E. tereticornis*) in Haryana and Punjab. *Indian For.* **95**, 377–388.

Kenworthy, J. B. (1969). Water balance in the tropical rain forest: A preliminary study in the Ulu Gombak Forest Reserve. *Malays. Nat. J.* **22**, 129–135.

Kenworthy, J. B. (1971). Water and nutrient cycling in a tropical rain forest. *Univ. Hull, Dep. Geogr., Misc. Publ. Geogr.* No. 11, pp. 49–65.

Kerfoot, O. (1967). Mist precipitation on vegetation. *For. Abstr.* **29**, 8–20.

Khan, A. A. (1965). Study of root development in *Dalbergia sissoo* Roxb. raised from stumps in irrigated plantations of Leiah Forest Division. *Pak. J. For.* **15**, 172–187.

Kimber, P. C. (1972). The root system of jarrah (*Eucalyptus marginata*). *West. Aust., For. Dep., Res. Pap.* **10**.

Kira, T., and Ogawa, H. (1971). Assessment of primary production in tropical and equatorial forests. *In* "Productivity of Forest Ecosystems" (P. Duvigneaud, ed.), pp. 309–321. UNESCO, Paris.

Kline, J. R., and Jordan, C. F. (1968). Tritium movement in soil of tropical rain forest. *Science* **160**, 550–551.

Kline, J. R., Martin, J. R., Jordan, C. F., and Koranda, J. J. (1970). Measurement of transpiration in tropical trees with tritiated water. *Ecology* **51**, 1068–1073.

Kline, J. R., Reed, K. L., Waring, R. H., and Stewart, M. L. (1976). Field measurements of transpiration in Douglas Fir. *J. Appl. Ecol.* **13**, 272–283.

Klinge, H. (1973). Root mass estimation in lowland tropical rain forests of central Amazonia, Brazil. I. Fine root masses of a pale yellow latosol and a giant humus podzol. *Trop. Ecol.* **14**, 29–37.

Koelmeyer, K. O. (1960). The periodicity of leaf change and flowering in the principal forest communities of Ceylon. Part II. Phenology of the tropical dry mixed evergreen forest. *Ceylon For.* **4**, 308–364.

Koller, D. (1969). The physiology of dormancy and survival of plants in desert environments. *Symp. Soc. Exp. Biol.* **23**, 449–469.

Köppen, W., and Geiger, R. (1930). "Handbuch der Klimatologie," Vol. I." Borntraeger, Berlin.

Koriba, K. (1958). On the periodicity of tree growth in the tropics. *Gardens' Bull., Singapore* **17**, 11–81.

Kozlowski, T. T. (1976). Water relations and tree improvement. *In* "Tree Physiology and

Yield Improvement'' (M. G. R. Cannell and F. T. Last, eds.), pp. 307–327. Academic Press, New York.

Kozlowski, T. T., and Greathouse, T. E. (1970). Shoot growth characteristics of tropical pines. *Unasylva* **24**, 1–10.

Kramer, P. J., and Bullock, H. C. (1966). Seasonal variations in the proportion of suberized and unsuberized roots of trees in relation to the absorption of water. *Am. J. Bot.* **53**, 200–204.

Kramer, P. J., and Kozlowski, T. T. (1960). "Physiology of Trees." McGraw-Hill, New York.

Kramer, P. J., and Kozlowski, T. T. (1979). "Physiology of Woody Plants." Academic Press, New York.

Kreeb, K. (1966). Transpirationmessungen an einigen australiaschen Immergrünen in Bereich des feuchten Eucalyptuswaldes und der Mulgabuschsavanne. *Oecol. Plant.* **1**, 235–244.

Kuchler, A. W. (1949). A physiognomic classification of vegetation. *Ann. Assoc. Am. Geog.* **39**, 201–210.

Ladiges, P. Y. (1974). Variation in drought tolerance in *Eucalyptus viminalis* Labill. *Aust. J. Bot.* **22**, 489–500.

Ladiges, P. Y. (1975). Some aspects of tissue water relations in three populations of *Eucalyptus viminalis* Labill. *New Phytol.* **75**, 53–62.

Ladiges, P. Y., and Ashton, D. H. (1974). Variation in some central Victoria populations of *Eucalyptus viminalis* Labill. *Aust. J. Bot.* **22**, 81–102.

Lange, O. L. (1956). Zur Methodik der Kolorimetrischen CO_2-Bestimmung nach Alvik. *Ber. Dscht. Bot. Ges.* **69**, 49–60.

Lange, O. L., Schulze, E.-D., Evanari, M., Kappen, L., and Buschbom, U. (1974). The temperature-related photosynthetic capacity of plants under desert conditions. I. Seasonal changes of the photosynthetic response to temperature. *Oecologia* **17**, 97–110.

Lange, O. L., Schulze, E.-D., Evanari, M., Kappen, L., and Buschbom, U. (1975). The temperature-related photosynthetic capacity of plants under desert conditions. II. Possible controlling mechanisms for the seasonal changes of the photosynthetic response to temperature. *Oecologia* **18**, 45–53.

Lange, O. L., Schulze, E.-D., Evanari, M., Kappen, L., and Buschbom, U. (1978). The temperature-related photosynthetic capacity of plants under desert conditions. III. Ecological significance of the seasonal changes of the photosynthetic response to temperature. *Oecologia* **34**, 89–100.

Larson, P. R. (1963). The indirect effect of drought on tracheid diameter in red pine. *For. Sci.* **9**, 52–63.

Lawson, G. W., Jenik, J., and Armstrong-Mensah, K. O. (1968). A study of a vegetation catena in Guinea savanna at Mole Game Reserve (Ghana). *J. Ecol.* **56**, 505–522.

Lawson, G. W., Armstrong-Mensah, K. O., and Jenik, J. (1970). A catena in tropical moist semi-deciduous forest near Kade, Ghana. *J. Ecol.* **58**, 371–398.

Ledig, F. T., Bormann, F. H., and Wenger, K. F. (1970). The distribution of dry matter growth between shoot and roots in loblolly pine. *Bot. Gaz. (Chicago)* **131**, 349–359.

Legris, P. (1972). A new formula of hydric balance. *Trop. Ecol.* **13**, 12–26.

Leshem, B. (1965). The annual activity of intermediary roots of the aleppo pine. *For. Sci.* **11**, 291–298.

Levitt, J. (1972). "Responses of Plants to Environmental Stresses." Academic Press, New York.

Leyton, L., and Armitage, I. P. (1968). Cuticle structure and water relations of the needles of *Pinus radiata* (D. Don). *New Phytol.* **67**, 31–38.

Leith, H., ed. (1974). "Phenology and Seasonality Modelling." Springer-Verlag, Berlin and New York.

Lima, W. de P. (1976). Interceptação da chuva em povoamentos de eucalipto e de pinheiro. *IPEF, Piracicaba* **13**, 75–90.

Liphschitz, N., and Waisel, Y. (1970a). Environmental effects on wood production and cambial activity in *Ziziphus spina-christi* (L.) Willd. *Isr. J. Bot.* **19**, 592–598.

Liphschitz, N., and Waisel, Y. (1970b). Effects of environment on relations between extension and cambial growth of *Populus euphratica* Oliv. *New Phytol.* **69**, 1059–1064.

Liphschitz, N., and Waisel, Y. (1970c). The effect of water stresses on radial growth of *Populus euphratica* Oliv. *La-Yaaran* **20**, 53–61.

Longman, K. A. (1978). Control of shoot extension and dormancy: External and internal factors. In "Tropical Trees as Living Systems" (P. B. Tomlinson and M. H. Zimmermann, eds.), pp. 465–495. Cambridge Univ. Press, London and New York.

Low, K. S., and Goh, K. C. (1972). The water balance of five catchments in Selangor, West Malaysia. *J. Trop. Geogr.* **35**, 60–66.

Ludlow, M. M. (1976). Ecophysiology of C$_4$ grasses. In "Water and Plant Life" (O. L. Lange, L. Kappen, and E.-D. Schulze, eds.), pp. 364–386. Springer-Verlag, Berlin and New York.

Lugo, A. (1970). Photosynthetic studies of four species of rain forest seedlings. In "A Tropical Rain Forest" (H. T. Odum and R. F. Pigeon, eds.), pp. I81–I102. U.S. At. Energy Comm., Washington, D. C.

Lyford, W. H. (1969). The ecology of an elfin forest in Puerto Rico. 7. Soil, root, and earthworm relationships. *J. Arnold Arbor., Harv. Univ.* **50**, 210–224.

McComb, A. L., and Ogigirigi, M. (1970). "Features of the Growth of *Eucalyptus citriodora* and *Isoberlinia doka* in the Northern Guinea zon Savanna Zone of Nigeria," Savanna For. Res. Stn., Res. Pap. 3. Ministry of Agriculture and Nature Resources, Federal Republic of Nigeria.

Maconachie, J. R. (1973). Leaf and shoot growth on *Acacia kempeana* F. Muell. and selected other arid-zone species. *Trop. Grassl.* **7**, 49–55.

McGinnis, J. T., Golley, F. B., Clements, R. G., Child, G. I., and Duever, M. J. (1969). Elemental and hydrologic budgets of the Panamanian tropical moist forest. *BioScience* **19**, 697–700.

Manokaran, N. (1979). Stemflow, throughfall and rainfall interception in a lowland tropical rain forest in Peninsular Malaysia. *Malays. For.* **42**, 174–201.

Marshall, D. C. (1958). Measurement of sap flow in conifers by heat transport. *Plant Physiol.* **33**, 385–396.

Medina, E. (1967). Intercambio gaseoso de arboles de las sabanas de *Trachypogon* en Venezuela. *Soc. Venez. Cienc. Nat., Bol.* **27**(111), 56–69.

Medina, E., Silva, J., and Castellanos, E. (1969). Variaciones estacionales del crecimiento y la respiracion foliar de algunas plantas leñosas de las sabanas de *Trachypogon*. *Soc. Venez. Cienc. Nat., Bol.* **28**(115/116), 67–82.

Medway, Lord (1972). Phenology of tropical rain forest in Malaya. *Biol. J. Linn. Soc.* **4**, 117–146.

Mensah, K. O. A., and Jenik, J. (1968). Root system of tropical trees. 2. Features of the root system of iroko (*Chlorophora excelsa* Benth. et Hook.). *Preslia* **40**, 21–27.

Merida, T., and Medina, E. (1967). Anatomia y composicion foliar de arboles de las sabanas de *Trachypogon* en Venezuela. *Soc. Venez. Cienc. Nat., Bol.* **27**(111), 45–55.

Metcalfe, C. R., and Chalk, L. (1950). "The Anatomy of the Dicotyledons." Oxford Univ. Press (Clarendon), London and New York.

Migahid, A. M. (1961). The drought resistance of Egyptian desert plants. *In* "Plant-Water Relationships in Arid and Semi-Arid Conditions," pp. 213–233. UNESCO, Paris.

Milburn, J. A. (1973). Cavitation in *Ricinus* by acoustic detection: Induction in excised leaves by various factors. *Planta* 110, 253–265.

Milburn, J. A., and Davis, T. A. (1973). Role of pressure in xylem transport of coconut and other palms. *Physiol. Plant.* 29, 415–420.

Milburn, J. A., and McLaughlin, M. E. (1974). Studies of cavitation in isolated vascular bundles and whole leaves of *Plantago major* L. *New Phytol.* 73, 861–871.

Mooney, H. A., Björkman, O., and Collatz, G. J. (1978). Photosynthetic acclimation to temperature in the desert shrub, *Larrea divaricata*. I. Carbon dioxide exchange characteristics of intact leaves. *Plant Physiol.* 61, 406–410.

Moore, R. T., White, R. S., and Caldwell, M. M. (1972). Transpiration of *Atriplex confertifolia* and *Eurotia lanata* in relation to soil, plant, and atmospheric moisture stresses. *Can. J. Bot.* 50, 2411–2418.

Morison, C. G. T., Hoyle, A. C., and Hope-Simpson, J. F. (1948). Tropical soil-vegetation catenas and mosaics. A study in the south-western part of the Anglo-Egyptian Sudan. *J. Ecol.* 36, 1–84.

Müller, D., and Nielsen, J. (1965). Production brute, pertes par respiration et production nette dans la forêt ombrophile tropicale. *Forstl. Forsoegsvaes. Dan.* 29, 73–160.

Muthana, K. D., and Arora, G. D. (1976). Performance of *Eucalyptus camaldulensis* on shallow and deep sandy loam of Pali, (W. Rajasthan). *Ann. Arid Zone* 15, 297–300.

Muthana, K. D., Arora, G. D., and Gianchand (1976). Comparative performance of indigenous trees in arid zone under different soil working techniques. *Ann. Arid Zone* 15, 67–76.

Newman, E. I. (1976). Water movement through root systems. *Philos. Trans. R. Soc. London, Ser. B* 273, 463–478.

Ng, F. S. P. (1975). A note on natural root grafts in Malaysian trees. *Malays. For.* 38, 153–159.

Ngampongsai, C. (1973). The distribution and development of teakroot in different ages plantation. *Kasetsart Univ., Fac. For., For. Res. Bull.* No. 28.

Nicholls, J. W. P. (1974). Effect of prescribed burning in a forest on wood characteristics of jarrah. *Aust. For.* 36, 178–189.

Nieuwolt, S. (1968). Uniformity and variation in an equatorial climate. *J. Trop. Geogr.* 27, 23–39.

Nieuwolt, S. (1972). Rainfall variability in Zambia. *J. Trop. Geogr.* 34, 44–57.

Njoku, E. (1964). Seasonal periodicity in the growth and development of some forest trees in Nigeria. II. Observations on seedlings. *J. Ecol.* 52, 19–26.

Nobel, P. S., Zarazoga, L. J., and Smith, W. H. (1975). Relation between mesophyll surface area, photosynthetic rate, and illumination level during development for leaves of *Plectranthus parviflorus* Henckel. *Plant Physiol.* 55, 1067–1070.

Nobel, P. S., Longstreth, D. J., and Hartstock, T. L. (1978). Effect of water stress on the temperature optima of net CO_2 exchange for two desert species. *Plant Physiol.* 44, 97–101.

Noy-Meir, I., and Ginzburg, B. Z. (1969). An analysis of the water potential isotherm in plant tissue. II. Comparative studies on leaves of different types. *Aust. J. Biol. Sci.* 22, 35–52.

Nye, P. H., and Tinker, P. B. (1977). "Solute Movement in the Soil-Root System." Univ. of California Press, Berkeley and Los Angeles.

Odening, W. R., Strain, B. R., and Oechel, W. C. (1974). The effect of decreasing water potentiaal on net CO_2 exchange of intact desert shrubs. *Ecology* 55, 1086–1095.

Odum, H. T., and Jordan, C. F. (1970). Metabolism and evapotranspiration of the lower forest in a giant plastic cylinder. *In* "A Tropical Rain Forest" (H. T. Odum and R. F. Pigeon, eds.), pp. I 165–I 189. U. S. At. Energy Comm., Washington, D. C.

Odum, H. T., Abbott, W., Selander, R. K., Golley, F. B., and Wilson, R. F. (1970a). Estimates of chlorophyll and biomass of the tabonuco forest of Puerto Rico. *In* "A Tropical Rain Forest" (H. T. Odum and R. F. Pigeon, eds.), pp. I3–I19. U. S. At. Energy Comm. Washington, D. C.

Odum, H. T., Lugo, A., Cintron, G., and Jordan, C. F. (1970b). Metabolism and evapotranspiration of some rain forest plants and soil. *In* "A Tropical Rain Forest" (H. T. Odum and R. F. Pigeon, eds), pp. I103–I164. U. S. At. Energy Comm., Washington, D. C.

Odum, H. T., Moore, A. M., and Burns, L. A. (1970c). Hydrogen budget and compartments in the rain forest. *In* "A Tropical Rain Forest" (H. T. Odum and R. F. Pigeon, eds.), pp. H105–H122. U. S. At. Energy Comm., Washington, D. C.

Oechel, W. C., Strain, B. R., and Odening, W. R. (1972). Tissue water potential, photosynthesis, ^{14}C-labelled photosynthate utilization, and growth in the desert shrub *Larrea divaricata*. *Ecol. Monogr.* **42**, 127–141.

Ogawa, H., Yoda, K., Ogino, K., and Kira, T. (1965). Comparative ecological studies on three main types of forest vegetation of Thailand. II. Plant biomass. *Nat. Life South East Asia, Kyoto* **4**, 49–80.

Okali, D. U. U. (1971). Tissue water relations of some woody species of the Accra Plains, Ghana. *J. Ecol.* **59**, 89–101.

Okali, D. U. U., and Dodoo, G. (1973). Seedling growth and transpiration of two West African mahogany species in relation to water stress in the root medium. *J. Ecol.* **61**, 421–438.

Okali, D. U. U., Hall, J. B., and Lawson, G. W. (1973). Root distribution under a thicket clump on the Accra Plains, Ghana: Its relevance to clump localization and water relations. *J. Ecol.* **61**, 439–454.

Oldeman, R. A. A. (1978). Architecture and energy exchange of dicotyledonous trees in the forest. *In* "Tropical Trees as Living Systems" (P. B. Tomlinson and M. H. Zimmermann, eds.), pp. 535–560. Cambridge Univ. Press, London and New York.

Parker, J. (1969). Further studies of drought resistance in woody plants. *Bot. Rev.* **35**, 317–371.

Parkhurst, D. F., and Loucks, O. L. (1972). Optimal leaf size in relation to environment. *J. Ecol.* **60**, 505–537.

Pearcy, R. W. (1977). Acclimation of photosynthetic and respiratory carbon dioxide exchange to growth temperature in *Atriplex lentiformis* (Torr.) Wats. *Plant Physiol.* **59**, 795–799.

Pearcy, R. W., Harrison, A. T., Mooney, H. A., and Björkman, O. (1974). Seasonal changes in net photosynthesis of *Atriplex hymenelytra* shrubs growing in Death Valley, California. *Oecologia* **17**, 111–121.

Peck, A. J. (1978). Salinization of non-irrigated soils: A review. *Aust. J. Soil Res.* **16**, 157–168.

Peck, A. J., and Hurle, D. H. (1972). Chloride balance of some farmed and forested catchments in south-western Australia. *Water Resour. Res.* **9**, 648–657.

Penman, H. L. (1948). Natural evaporation from open water, bare soil and grass. *Proc. R. Soc. London, Ser. A* **193**, 120–145.

Penman, H. L. (1956). Evaporation—an introductory survey. *Neth. J. Agric. Sci.* **4**, 9–29.

Penman, H. L. (1963). "Vegetation and Hydrology," Commonw. Bur. Soils, Harpenden, Tech. Commun. No. 53. Commonwealth Agricultural Bureaus, Farnham Royal.

Pereira, H. C., compiler (1962). Hydrological effects of changes in land use in some East African catchment areas. *East Afr. Agric. For. J.* **27**, Spec. Issue, 1–130.

Pereira, H. C. (1967). Effects of land-use on the water and energy budgets of tropical watersheds. *In* "International Symposium on Forest Hydrology" (W. E. Sopper and H. W. Lull, eds.), pp. 435–450. Pergamon, Oxford.

Pereira, H. C. (1973). "Land Use and Water Resources in Temperate and Tropical Climates." Cambridge Univ. Press, London and New York.

Pereira, J. S., and Kozlowski, T. T. (1976). Leaf anatomy and water relations of *Eucalyptus camaldulensis* and *E. globulus* seedlings. *Can. J. Bot.* **54**, 2868–2880.

Pereira, J. S., and Kozlowski, T. T. (1977). Variations among woody angiosperms in response to flooding. *Physiol. Plant.* **41**, 184–192.

Petty, J. A. (1972). The aspiration of bordered pits in conifer wood. *Proc. R. Soc. London, Ser. B* **181**, 395–406.

Pinto, A. E. (1970). Phenological studies of trees at El Verde. *In* "A Tropical Rain Forest" (H. T. Odum and R. F. Pigeon, eds.), pp. D237–D269. U. S. At. Energy Comm., Washington, D. C.

Pisek, A., and Winkler, E. (1953). Die Schliessbewegung der Stomata bei ökologisch verschiedenen Pflanzentypen in Abhängigkeit von Wassersättigungszustand der Blätter und vom Licht. *Planta* **42**, 253–258.

Polster, H., and Reichenbach, H. (1957). Ein Verfahren zur Prognose der vitalen Dürreresistenz durch Ermittlung des Stomataregulationsvermögens abgeschnittener Pflanzensprosse. *Biol. Zentralbl.* **76**, 700–721.

Pook, E. W., Costin, A. B., and Moore, C. W. E. (1966). Water stress in native vegetation during the drought of 1965. *Aust. J. Bot.* **14**, 257–267.

Poupon, H. (1976). La biomasse et l'évolution de sa répartition au cours de la croissance d'*Acacia senegal* dans une savane Sahélienne (Sénégal). *Bois For. Trop.* **166**, 23–38.

Pradhan, I. P. (1973). Preliminary study on rainfall interception through leaflitter. *Indian For.* **99**, 440–445.

Preece, P. B. (1971a). Contributions to the biology of mulga. I. Flowering. *Aust. J. Bot.* **19**, 21–38.

Preece, P. B. (1971b). Contributions to the biology of mulga. II. Germination. *Aust. J. Bot.* **19**, 39–49.

Pressland, A. J. (1973). Rainfall partitioning by an arid woodland (*Acacia aneura* F. Muell.) in south-western Queensland. *Aust. J. Bot.* **21**, 235–245.

Pressland, A. J. (1975). Productivity and management of mulga in south-western Queensland in relation to tree structure and density. *Aust. J. Bot.* **23**, 965–976.

Pressland, A. J. (1976a). Effect of stand density on water use of mulga (*Acacia aneura* F. Muell.) woodlands in south-western Queensland. *Aust. J. Bot.* **24**, 177–191.

Pressland, A. J. (1976b). Soil moisture redistribution as affected by throughfall and stemflow in an arid zone shrub community. *Aust. J. Bot.* **24**, 641–649.

Puri, G. S. (1957). The relict vegetation of Sheo Bari, Sohan Valley in the Hoshiarpur Siwaliks. *Indian For.* **83**, 718–723.

Putz, F. E. (1979). Aseasonality in Malaysian tree phenology. *Malays For.* **42**, 1–24.

Pyykkö, M. (1979). Morphology and anatomy of leaves from some woody plants in a humid tropical forest of Venezuelan Guayana. *Act. Bot. Fenn.* **112**, 1–41.

Quraishi, M. A., and Kramer, P. J. (1970). Water stress in three species of *Eucalyptus. For. Sci.* **16**, 74–78.

Raeder-Roitzsch, J. E., and Masrur, A. (1969a). Some hydrologic relationships of natural vegetation in the chir pine belt of West Pakistan. *Pak. J. For.* **19**, 81–98.

Raeder-Roitzsch, J. E., and Masrur, A. (1969b). Observations on rainfall interception in chir pine (*Pinus roxburghii*) woodland. *Pak. J. For.* **19,** 99–111.

Rawitscher, F. (1948). The water economy of the vegetation of the 'campos cerrados' in southern Brazil. *J. Ecol.* **36,** 237–268.

Ray, M. P. (1970). Preliminary observations on stem-flow, etc., in *Alstonia scholaris* and *Shorea robusta* plantations at Arabari, West Bengal. *Indian For.* **96,** 482–493.

Reynolds, E. R. C., and Henderson, C. S. (1967). Rainfall interception by beech, larch and Norway spruce. *Forestry* **40,** 165–184.

Richards, P. W. (1936). Ecological observations on the rain forest of Mount Dulit, Sarawak. Part II. *J. Ecol.* **24,** 340–360.

Richards, P. W. (1952). "The Tropical Rain Forest." Cambridge Univ. Press, London and New York.

Roberts, B. R. (1964). Effects of water stress on the translocation of photosynthetically assimilated carbon-14 in yellow poplar. *In* "The Formation of Wood in Forest Trees" (M. H. Zimmermann, ed.), pp. 273–288. Academic Press, New York.

Rudman, P. (1970). The influence of genotype and environment on wood properties of juvenile *Eucalyptus camaldulensis* Dehn. *Silvae Genet.* **19,** 49–54.

Rutter, A. J. (1963). Studies on the water relations of *Pinus sylvestris* in plantation conditions. I. Measurement of rainfall and interception. *J. Ecol.* **51,** 191–203.

Rutter, A. J. (1968). Water consumption by forests. *In* "Water Deficits and Plant Growth" (T. T. Kozlowski, ed.), Vol. 2, pp. 23–84. Academic Press, New York.

Rutter, A. J., and Morton, A. J. (1977). A predictive model of rainfall interception in forests. III. Sensitivity of the model to stand parameters and meteorological variables. *J. Appl. Ecol.* **14,** 567–588.

Rutter, A. J., Kershaw, K. A., Robins, P. C., and Morton, A. J. (1971). A predictive model of rainfall interception in forests. I. Derivation of the model from observations in a plantation of corsican pine. *Agric. Meteorol.* **9,** 367–384.

Rutter, A. J., Morton, A. J., and Robins, P. C. (1975). A predictive model of rainfall inerception in forests. II. Generalization of the model and comparison with observations in some coniferous and hardwood stands. *J. Appl. Ecol.* **12,** 367–380.

Sarlin, P. (1964). L'eau et le sol(L'eau en forêt, en savane et dans les reboisements. *Bois For. Trop.* **89,** 11–29.

Sarlin, P. (1970). Evapotranspiration et végétation forestière tropicale. *Bois For. Trop.* **133,** 17–26.

Schaedle, M. (1975). Tree photosynthesis. *Annu. Rev. Plant Physiol.* **26,** 101–115.

Schimper, A. F. W. (1903). "Plant-Geography Upon a Physiological Basis." Oxford Univ. Press (Clarendon), London and New York.

Schulze, E.-D., Lange, O. L., Buschbom, U., Kappen, L., and Evanari, M. (1972). Stomatal responses to changes in humidity in plants growing in the desert. *Planta* **108,** 259–270.

Schulze, E.-D., Lange, O. L., Kappen, L., Buschbom, U., and Evanari, M. (1973). Stomatal responses to changes in temperature at increasing water stress. *Planta* **110,** 29–42.

Schulze, E.-D., Lange, O. L. Kappen, L., Evanari, M., and Buschbom, U. (1975). The role of air humidity and leaf temperature in controlling stomatal resistance of *Prunus armeniaca* L under desert conditions. II. The significance of leaf water status and internal carbon dioxide concentration. *Oecologia* **18,** 219–233.

Schulze, E.-D., Lange, O. L., Evanari, M., Kappen, L., and Buschbom, U. (1976). An empirical model of net photosynthesis for the desert plant *Hammada scoparia* (Pomel) Iljin. I. Description and test of the model. *Oecologia* **22,** 355–372.

Schulze, R. E., and McGee, O. S. (1978). Climatic indices and classifications in relation to the biogeography of southern Africa. *Monog. Biol.* **31**, 21–52.

Schulze, R. E., Scott-Shaw, C. R., and Nanni, U. W. (1978). Interception by *Pinus patula* in relation to rainfall parameters. *J. Hydrol.* **36**, 393–396.

Seth, S. K., and Srivastava, P. B. L. (1971). Effect of moisture gradient on the growth and nutrient uptake of sal (*Shorea robusta*) seedlings. *Indian For.* **97**, 615–625.

Seth, S. K., and Srivastava, P. B. L. (1972). Effect of profile morphology on root development of sal (*Shorea robusta*) seedlings. *Indian For.* **98**, 155–167.

Seth, S. K., and Yadav, J. S. P. (1959). Teak soils. *Indian For.* **85**, 2–16.

Shea, S. R., Hatch, A. B., Havel, J. J., and Ritson, P. (1975). The effect of changes in forest structure and composition on water quality and yield from the northern jarrah forest. *Proc. Ecol. Soc. Aust.* **9**, 58–73.

Sheriff, D. W. (1979). Stomatal aperture and the sensing of the environment by guard cells. *Plant, Cell, & Environ.* **2**, 15–22.

Shuttleworth, W. J. (1977). The exchange of wind-driven fog and mist between vegetation and the atmosphere. *Boundary-Layer Meteorol.* **12**, 463–489.

Skerman, P. J. (1977). "Tropical Forage Legumes," FAO Plant Prod. Prot. Ser. No. 2. FAO, Rome.

Slatyer, R. O. (1960). Aspects of the tissue water relationships of an important arid zone species (*Acacia aneura* F. Muell.) in comparison with two mesophytes. *Bull. Res. Counc. Isr., Ser. D* **8**, 159–168.

Slatyer, R. O. (1961a). Methodology of a water balance study conducted on a desert woodland (*Acacia aneura* F. Muell) community in central Australia. *In* "Plant-Water Relationships in Arid and Semi-Arid Conditions," pp. 15–26. UNESCO, Paris.

Slatyer, R. O. (1961b). Internal water balance of *Acacia aneura* F. Muell. in relation to environmental conditions. *In* "Plant-Water Relationships in Arid and Semi-Arid Conditions," pp. 137–146. UNESCO, Paris.

Slatyer, R. O. (1965). Measurements of precipitation interception by an arid zone plant community *Acacia aneura* F. Muell. *In* "Methodology of Plant Eco-physiology," pp. 181–192. UNESCO, Paris.

Slatyer, R. O. (1967). "Plant-Water Relationships." Academic Press, New York.

Soerjono, R. (1964). Hutan *Pinus merkusii* ditindjau dari segi pengawetan tanah dan air. *Rimba Indonesia* **9**, 311–318.

Sollins, P., and Drewry, G. (1970). Electrical conductivity and flow rate of water through the forest canopy. *In* "A Tropical Rain Forest" (H. T. Odum and R. F. Pigeon, eds.), pp. H137–H153. U. S. At. Energy Comm., Washington, D. C.

Sopper, W. E., and Lull, H. W., eds. (1967). "International Symposium on Forest Hydrology." Pergamon, Oxford.

Specht, R. L. (1970). Vegetation. *In* "The Australian Environment" (G. W. Leeper, eds.), pp. 44–67. CSIRO-Melbourne University Press, Melbourne.

Specht, R. L. (1972). Water use by perennial evergreen plant communities in Australia and Papua New Guinea. *Aust. J. Bot.* **20**, 273–299.

Specht, R. L., and Brouwer, Y. M. (1975). Seasonal shoot growth of *Eucalyptus* spp. in the Brisbane area of Queensland (with notes on shoot growth and litter fall in other areas of Australia). *Aust. J. Bot.* **23**, 459–474.

Stålfelt, M. G. (1929). Die Abhängigkeit der Spaltöffnungsreaktionen von der Wasserbilanz. *Planta* **8**, 287–340.

Stark, N., and Love, L. D. (1969). Water relations of three warm desert species. *Isr. J. Bot.* **18**, 175–190.

Stocker, O. (1929). Eine Feldmethode zur Bestimmung der Transpirations - und Evaporationsgross. *Ber. Dsch. Bot. Ges.* **47**, 126–136.

Stocker, O. (1935). Transpiration und Wasserhaushalt in verschiedenen Klimazonen. III. Ein Beitrag zur Transpirationsgrosse im javanischen Regenwald. *Jahrb. Wiss. Bot.* **81**, 464–496.

Stocker, O. (1970). Der Wasser-und Photosynthese-Haushalt von Wüstenpflanzen der mauretanischen Sahara. I. Regengrüne und immergrüne Bäume. *Flora (Jena)* **159**, 539–572.

Stocker, O. (1971). Der Wasser-und Photosynthese-Haushalt von Wüstenpflanzen der mauretanischen Sahara. II. Wechselgrüne, Rutenzweig-und stammsukulente Bäume. *Flora (Jena)* **160**, 445–494.

Stocker, O. (1976). The water-photosynthesis syndrome and the geographical plant distribution in the Saharan deserts. *In* "Water and Plant Life" (O. L. Lange, L. Kappen, and E.-D. Schulze, eds.), pp. 506–521. Springer-Verlag, Berlin and New York.

Stone, E. C. (1957). Dew as an ecological factor. I. A review of the literature. *Ecology* **38**, 407–413.

Stone, E. C. (1958). Dew absorption by conifers. *In* "Physiology of Forest Trees" (K. V. Thimann, ed.), pp. 125–151. Ronald Press, New York.

Strain, B. R. (1975). Field measurements of carbon dioxide exchange in some woody perennials. *In* "Perspectives of Biophysical Ecology" (D. M. Gates and R. B. Schmerl, eds.), pp. 145–158. Springer-Verlag, Berlin and New York.

Strang, R. M. (1966). The spread and establishment of *Brachystegia spiciformis* Benth. and *Julbernardia globiflora* (Benth.) Troupin in the Rhodesian highveld. *Commonw. For. Rev.* **45**, 253–256.

Swanson, R. H. (1974). A thermal flowmeter for estimating the rate of xylem sap ascent in trees. *In* "Flow: Its Measurement and Control in Science and Industry" (R. B. Dowdell, ed.), Vol. 1, pp. 647–652. Instrum. Soc. Am., Pittsburgh.

Tangtham, N. (1970). Preliminary study of rainfall and some nutrients under crown canopy in Doi Pui hill evergreen forest, Chiengmai. *Kasetsart Univ., Fac. For. Kog-Ma Watershed Res. Bull.* No. 5.

Taylor, H. M., and Klepper, B. (1978). The role of rooting characteristics in the supply of water to plants. *Adv. Agron.* **30**, 99–128.

Taylor, S. E. (1975). Optimal leaf form. *In* "Perspectives of Biophysical Ecology" (D. W. Gates and R. B. Schmerl, eds.), pp. 73–86. Springer-Verlag, Berlin and New York.

Teoh, C. T., and Palmer, J. H. (1971). Nonsynchronized oscillations in stomatal resistance among sclerophylls of *Eucalyptus umbra*. *Plant Physiol.* **47**, 409–411.

Thornthwaite, C. W. (1948). An approach toward a rational classification of climate. *Geogr. Rev.* **38**, 55–94.

Tinker, P. B. (1969). The transport of ions in the soil around plant roots. *Symp. Br. Ecol. Soc.* **9**, 135–147.

Tinker, P. B. (1976). Transport of water to plant roots in soil. *Philos. Trans. R. Soc. London, Ser. B* **273**, 445–461.

Tomlinson, P. B., and Zimmermann, M. H., eds. (1978). "Tropical Trees as Living Systems." Cambridge Univ. Press, London and New York.

Tracey, J. G. (1969). Edaphic differentiation of some forest types in Eastern Australia. I. Soil physical factors. *J. Ecol.* **57**, 805–816.

Troup, R. S. (1921). "The Silviculture of Indian Trees," 3 vols. Oxford Univ. Press, London and New York.

Tunstall, B. R. (1973). Water relations of a Brigalow community. Ph.D. Thesis, University of Queensland, Brisbane.

Tunstall, B. R., and Connor, D. J. (1975). Internal water balance of brigalow (*Acacia harpophylla* F. Muell.) under natural conditions. *Aust. J. Plant Physiol.* **2**, 489–499.

Turner, N. C. (1965). Some energy and microclimate measurements in a natural arid zone plant community. *In* "Methodology of Plant Eco-physiology," pp. 63–70. UNESCO, Paris.

Turner, N. C., and Begg, J. E. (1977). Responses of pasture plants to water deficits. *In* "Plant Relations in Pastures" (J. R. Wilson, ed.), pp. 50–66. CSIRO, Melbourne, Australia.

Tyree, M. T. (1976a). Negative turgor pressure in plant cells: Fact or fallacy? *Can. J. Bot.* **54**, 2738–2746.

Tyree, M. T. (1976b). Physical parameters of the soil-plant-atmosphere system; breeding for drought resistance characteristics that might improve wood yield. *In* "Tree Physiology and Yield Improvement" (M. G. R. Cannell and F. T. Last, eds.), pp. 327–348. Academic Press, New York.

Tyree, M. T., and Hammel, H. T. (1972). The measurement of the turgor pressure and the water relations of plants by the pressure bomb technique. *J. Exp. Bot.* **23**, 267–282.

van den Driessche, R., Connor, D. J., and Tunstall, B. R. (1971). Photosynthetic response of brigalow to irradiance, temperature and water potential. *Photosynthetica* **5**, 210–217.

van den Honert, T. H. (1941). Experiments on the water household of tropical plants. I. Water balance in *Cissus sicyoides* L. *Ann. Bot. Gard., Buitenzorg* **41**, 58–82.

van Donselaar, J. (1965). An ecological and phytogeographic study of northern Surinam savannas. *Wentia* **14**, 1–163.

van Donselaar-ten Bokkel Huinink, W. A. E. (1966). Structure, root systems and periodicity of savanna plants and vegetations in northern Surinam. *Wentia* **17**, 1–162.

Vareschi, V. (1960). Observaciones sobre la transpiracion de arboles llaneros, durante la epoca de sequia. *Soc. Venez. Cienc. Nat., Bol.* **21** (96), 128–134.

Vasiliev, B. R. (1972). Types of anatomical structure of leaves of arboreal and fruticose plants of West African dry savanna. *In* "Eco-physiological Foundation of Ecosystems Productivity in Arid Zone," pp. 93–95. USSR Acad. Sci., Leningrad.

Vaughan, R. F., and Wiehe, P. O. (1947). Studies on the vegetation of Mauritius. IV. Some notes on the internal climate of the upland climax forest. *J. Ecol.* **34**, 126–136.

Venator, C. R. (1976). Natural selection for drought resistance in *Pinus caribaea* Morelet. *Turrialba* **26**, 381–387.

Villaca, H., and Ferri, M. G. (1954). Transpiracão de *Eucalyptus tereticornis*. *Univ. Sao Paulo, Fac. Filos., Cienc. Letras, Bol., Bot.* **11**, No. 173, 3–30.

Wadsworth, F. H. (1948). The climate of the Luquillo Mountains and its significance to the people of Puerto Rico. *Caribb. For.* **9**, 321–344.

Waisel, Y., and Fahn, A. (1965). The effects of environment on wood formation and cambial activity in *Robinia pseudoacacia* L. *New Phytol.* **64**, 436–442.

Waisel, Y., Noah, I., and Fahn, A. (1966). Cambial activity in *Eucalyptus camaldulensis* Dehn. I. The relation to extension growth in young saplings. *La-Yaaran* **16**, 59–72, 103–108.

Walter, H. (1964). "Die Vegetation der Erde in öko-physiologischer Betrachtung," Vol. I. Fischer, Stuttgart.

Walter, H. (1971). "Ecoloty of Tropical and Subtropical Vegetation" (Engl. transl. by D. Mueller-Dombois). Oliver & Boyd, Edinburgh.

Walter, H. (1972). "Vegetation of the Earth in Relation to Climate and the Eco-Physiological Conditions" (transl. by J. Weiser), 2nd ed. Springer-Verlag, Berlin and New York.

Wareing, P. F., and Roberts, D. L. (1956). Photoperiodic control of cambial activity in *Robinia pseudoacacia*. *New Phytol.* **55**, 356–366.

Waring, R. H., and Running, S. W. (1976). Water uptake, storage and transpiration by conifers: A physiological model. *In* "Water and Plant Life" (O. L. Lange, L. Kappen, and E.-D. Schulze, eds.), pp. 189–202. Springer-Verlag, Berlin and New York.

Webb, L. J. (1959). A physiognomic classification of Australian rain forests. *J. Ecol.* **47**, 551–570.

Webb, L. J. (1969). Edaphic differentiation of some forest types in eastern Australia. II. Soil chemical factors. *J. Ecol.* **57**, 817–830.

West, D. W., and Gaff, D. F. (1976). Xylem cavitation in excised leaves of *Malus sylvestris* Mill. and measurements of leaf water status with the pressure chamber. *Planta* **129**, 15–18.

White, F. (1977). The underground forests of Africa: A preliminary review. *Gardens' Bull., Singapore* **29**, 57–71.

White, L. P. (1971). Vegetation stripes on sheet wash surfaces. *J. Ecol.* **59**, 615–622.

Whitehead, D. (1978). The estimation of foliage area from sapwood basal area in Scots pine. *Forestry* **51**, 137–149.

Whiteman, P. C., and Koller, D. (1967). Species characteristics in whole plant resistances to water vapour and CO_2 diffusion. *J. Appl. Ecol.* **4**, 363–377.

Whitmore, T. C. (1975). "The Tropical Rain Forests of the Far East." Oxford Univ. Press (Clarendon), London and New York.

Wilde, S. A. (1958). "Forest Soils." Ronald Press, New York.

Williams, J. (1974). Water relations of three planting stock types on *Pinus caribaea* following transplanting. *N. Z. J. For. Sci.* **5**, 87–104.

Wycherley, P. R. (1973). The phenology of plants in the humid tropics. *Micronesica* **9**, 75–96.

Zahner, R. (1963). Internal moisture stress and wood formation in conifers. *For. Prod. J.* **13**, 240–247.

Zapata, T. R., and Arroyo, M. T. K. (1978). Plant reproductive ecology of a secondary deciduous tropical forest in Venezuela. *Biotropica* **10**, 221–230.

Zimmermann, M. H., and Brown, C. L., eds. (1971). "Trees: Structure and Function." Springer-Verlag, Berlin and New York.

Zimmermann, U., Kruetz, W., Schubach, K., and Siegel, O. (1966). Tracers determine movement of soil moisture and evapotranspiration. *Science* **152**, 346–347.

Zohary, M., and Orshan, G. (1959). The maquis of *Ceratonia siliqua* in Israel. *Vegetatio, Haag* **8**, 285–297.

CHAPTER 5

CITRUS ORCHARDS

P. E. Kriedemann and H. D. Barrs

DIVISION OF IRRIGATION RESEARCH, C.S.I.R.O., GRIFFITH, NEW SOUTH WALES, AUSTRALIA

"a very thriving *Limon-tree,* which grew in a garden pot . . . perspires much less than the Sunflower, or than the Vine or the Apple-tree, whose leaves fall off in the winter . . . ."

(Stephen Hales, 1727)

I. INTRODUCTION

A crop of great antiquity, citrus fruit has become a major item of world trade. Second only to apples, production and export in the coming decade will probably exceed 60 million tons annually (Anonymous, 1979), with fruits as highly regarded now as was recorded in devotional texts of The Sanskrit Literature dating from 800 B.C. (Scora, 1975).

True citrus fruits of the genera *Fortunella, Poncirus,* and *Citrus* are native to a large Asiatic region extending from N. E. India to the Philippines and Indonesia (Chapot, 1975). Progenitors of these present day cultivars almost certainly evolved as substory trees in low latitude forests

325

Water Deficits and Plant Growth, Vol. VI

and even now contain many attributes that are still appropriate to that situation, e.g.,

1. Vegetative growth readily assumes dominance over reproductive development as manifest in growth flushes that can occur at the expense of fruit set. Such dominance appears to be important for tree survival in substory space but is less desirable horticulturally.

2. Foliar development is luxurious (up to 25% of tree fresh weight) (Jones and Steinacker, 1951) and leaves constitute a major carbohydrate resource due to high sugar and starch content.

3. High stomatal density predisposes citrus toward potentially fast transpiration, but the network of first- and second-order venation in leaves is poorly developed compared to the deciduous perennial *Vitis vinifera* (Possingham *et al.*, 1980). Moreover, low hydraulic conductivity in a shallow suberized root system which is equipped with only vestigial root hairs (Castle, 1980, and Fig. 1) exacerbates the problem. Consequently, transpirational losses under adverse conditions (e.g., strong in-

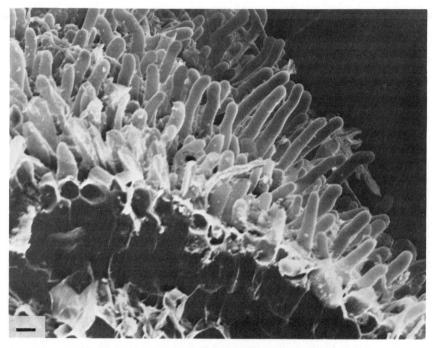

Fig. 1. Although root hairs can be numerous in citrus, as in this grapefruit seedling, they are abnormally short and so may be less effective absorbers than in other species (the bar represents 10 μm). Material was prepared by critical point drying followed by gold-palladium coating. (Unpublished SEM micrograph by N. Schaefer.)

solation and advection) readily exceed the tree's absorptive capacity, unless soil water is maintained at a high level by irrigation.

Despite such mesophytic characteristics, citrus orchards in semiarid environments are generally more productive than those in wet tropics (Reuther, 1973). However, water management then becomes especially critical because the crop grows best on lighter soils with lower water-holding capacity. Any margin for error in irrigation scheduling diminishes accordingly.

Such exacting cultural requirements, superimposed on the commercial significance of citrus, have called for fundamental and applied research that has burgeoned over the past 75 years (Bielorai, 1980). From tentative beginnings in the eighteenth century at the hands of Stephen Hales, citrus water relations have now been extensively documented. Research has ranged from basic physiology to crop management and irrigation practice. This chapter will cover that range.

II. TREES

A. ANATOMICAL FEATURES AND WATER RELATIONS

Citrus trees are perennial evergreen plants that evolved under humid tropic conditions but have become widely adapted to semiarid regions of the world. Luxurious foliar development and a massive canopy present an enormous evaporative surface, and yet root growth is shallow and water absorptive surface limited, resulting in a great imbalance in root:shoot biomass. Despite these constraints, mature 9-yr-old Valencia orange trees in California had a total dry weight of about 200 lbs., of which 17% was leaves (about 60,000 in all), 45% was trunk wood, and 20% roots (Wallace and Nauriyal, 1958). Uncrowded trees on favorable soils attain heights and widths of 30 feet and are more productive than their counterparts at low latitudes (Reuther, 1973). Clearly, some morphological and/or anatomical adjustment to the arid zone has been achieved.

1. Leaves

The morphology and anatomy of citrus leaves have been interpreted as either essentially mesophytic (Webber, 1943; Kalma and Fuchs, 1976) or xeromorphic (Turrell, 1936; Levitt and Ben Zaken, 1975). Before attempting to decide between or to reconcile these conflicting views it is necessary to summarize current information on those aspects of the morphology and anatomy of citrus leaves that bear on their water relations. Detailed illustrated accounts of citrus leaves, including their development, have been given by Scott et al. (1948) and Schneider (1968).

Leaf life may be from 9 months to 2 years or more, and growth in lamina length and width can occur for up to 130 days, being especially pronounced during spring and autumn shoot growth flushes. During this period of expansion, the leaf is light green and soft in texture; it wilts immediately on picking. The leaf subsequently matures to become dark green, firm in texture, and does not wilt readily when picked. Associated with this development are secondary thickening of the veins and cuticle and increased suberization of internal surfaces (Scott et al., 1948). Such changes may explain the observation (Albrigo, 1977) that leaf diffusion resistance is considerably lower in young (3.4 sec cm^{-1}) than in old leaves (7.2 sec cm^{-1}) and that the leaf diffusion error when determining leaf water potential (ψ) by the isopiestic psychrometric technique is 18% for old tissue but only 11.3% for young leaf tissue (Kaufmann, 1968). Levitt and Ben Zaken (1975) concluded that, because of these changes the mature orange leaf became a sclerophyll, thus decreasing the loss of intercellular space that normally accompanies loss in turgor of mesophytic leaves, since volume loss is brought about by decrease in leaf thickness rather than area. The resultant conservation of intercellular space was seen as important in minimizing the consequent secondary oxygen deficit stress.

Average life for Valencia orange leaves in southern California is 1.4 years, with the majority shed when 12 to 18 months old. Leaves are shed during all months of the year but most are lost in the months immediately following blossoming, or toward the end of the blossoming period (Wallace and Nauriyal, 1958).

According to Schneider (1968) and Spurling (1951), normal leaf senescence is terminated by abscission at an abscission zone at the base of the petiole, after labile constituents are withdrawn from the lamina. Spurling (1951) interprets a second abscission zone, between the lamina and the petiole, as an emergency abscission zone which permits abnormal and rapid excision of leaves when the tree is under severe moisture stress, enabling it to reduce quickly its transpiration rate. Observations that this occurs in response to desiccating winds (Schneider, 1968), temperature and soil moisture extremes, high wind velocity, low relative humidity, and excessive root injury from fungi and nematodes (Erickson, 1968) are consistent with Spurling's view. In this mode of abscission petioles are left behind on the tree for some time before they too are finally shed. However, Scott et al. (1948) consider this second method of leaf abscission to be normal in citrus orchards. As in other species, there is evidence (Aharoni, 1978; Young, 1971) for a direct causal link between stimulation of ethylene production by water stress and citrus leaf abscission. Concomitant stomatal closure associated with the water stress facilitates the rise in ethylene concentration in citrus leaves (Barmoew and Briggs, 1972).

The occurrence of two leaf abscission zones, one of which is apparently much more sensitive to water stress than the other, seems to be more consistent with capacity to adapt to drought than to be a typically mesophytic character. Begg and Turner (1970) suggested that leaf petioles may constitute a locally high hydraulic resistance in tobacco. It would be interesting to evaluate this hypothesis for citrus leaves, since Schneider (1968) described the petiole veins as passing through the distal abscission zone in particular in a zigzag fashion.

The outermost layer of the citrus leaf comprises an epicuticular wax, which according to Scott *et al.* (1948) is secreted through special plasmodesmatal wax canals on the outer wall of the epidermis, although Albrigo (1972b) considers this point not yet fully resolved. The wax builds up over a period of 4 months to an approximate plateau of between 20 and 30 μg cm^{-2} depending on species, being thicker in lemon and clementine leaves than in mandarin or orange leaves (Baker *et al.*, 1975). The ultrastructure of this surface wax has been described by Albrigo (1972b). An important feature is that, although some surface wax structure does develop, no platelet wax is formed. This permits the surface wax to remain in close contact with the leaf while it expands and so to remain a continuing effective barrier to water loss. The epicuticular wax consists predominantly of hydrocarbons, followed by primary alcohols, and then fatty acids, with some variation among species. Further details of the composition of these compounds are given by Baker *et al.* (1975), who concluded that these wax deposits are ideally suited to suppress cuticular transpiration, thereby decreasing sensitivity to environmental stress. An earlier determination of leaf wax composition in *C. aurantifolia* (Holloway and Baker, 1970) gave similar results to those above.

Beneath the epicuticular wax lies the well-developed cuticular membrane (Baker and Procopiou, 1975), weighing about 300 μg cm^{-2} for lemon, orange, clementine and mandarin leaves, with relatively little difference (280–316 μg cm^{-2}) among the four species. The membrane contains high proportions of cutin (77–87%) in which isomers of dehydrohexadecanoic acid (79–82%) are the most abundant monomers. Substantial amounts of cuticular wax (34–68μg cm^{-2}) are located within the membranes and consist largely of fatty acids, 68–97%, with hexadecanoic acid (58–78%) and octadecanoic acid (11–21%) as the main components.

Permeability of the cuticle to water in *C. aurantium* leaves has been investigated by Schönherr and Schmidt (1979), who found that the driving force for water movement across the membrane is the ψ gradient across the membrane. They interpreted this as implying that the membrane is porous, being transversed by polar water-filled pores that swell or shrink depending on the activity of external water vapor (a_{wv}). Hence the permeability coefficient of citrus leaf cuticle decreases with decreasing water vapor activity, by as much as a factor of two as (a_{wv}) decreases from 1.0 to 0.22. This is in contrast to the cuticular membrane of onion scale in which the permeability coefficient is independent of (a_{wv}), suggesting that the latter is a solubility membrane in which all molecules of the hydrophilic matrix are tightly packed, and discrete pores are therefore lacking.

Soluble cuticular lipids play a crucial role in maintaining a low permeability coefficient for citrus leaf cuticle, since although they comprise only 3% of the cuticular membrane, their removal increases the permeability coefficient of the membrane by a factor of about 500. The cuticular resistance of citrus is calculated to be as high as 400–1600 sec cm^{-1} by Schönherr and Schmidt (1979), who give reasons for such extraordinarily high values, which relate to problems and assumptions of conventional methods for determining cuticular resistance. Such resistance values are among the highest reported for any species, including xerophytes (Cowan and Milthorpe, 1971). Until data for a range of other species have been determined similarly it is difficult to compare cuticular resistance for citrus leaves with those for other species. Meanwhile, more conventional methods of measurement provide an estimate of cuticular resistance to water vapor of the lower surface of orange leaves of about 100 sec cm^{-1} (Elfving et al., 1972); this may be only the lower limit because the stomata may not have been completely closed.

The smooth waxy cuticle makes it difficult to wet the citrus leaf; consequently, it is resistant to nutrient leaching by rain (Tukey, 1970). Presumably this is an adaptation to its native tropical rain forests (Section I). The smooth waxy cuticle also confers on the citrus leaf very high reflectivity in the infrared, in fact the highest for 27 species and genera examined by Gates and Tantraporn (1952). Turrell et al. (1962) suggested that this high reflectivity may have aided survival of the genus in humid but highly insolated tropical rain forests by preventing lethal leaf temperatures. However, arguing from radiation balance considerations, Levitt (1972) doubts whether this property materially influences citrus leaf temperature.

Stomata are usually described as essentially confined to the lower surface of citrus leaves, only a few appearing along the midrib on the upper surface (Scott et al., 1948); Ehrler and Van Bavel (1968) measured 800 mm^{-2} on the lower and 40 mm^{-2} on the upper epidermis of lemon.

Beneath the guard cells lies an extremely large and deep stomatal chamber [7.8 μm deep in Valencia orange leaf according to Turrell (1947)], formed by loosely arranged spongy mesophyll cells with arms and large air spaces between them. The stomatal pore is frequently filled by a loosely fitting stomatal plug (Fig. 2), described by Turrell (1947) as having

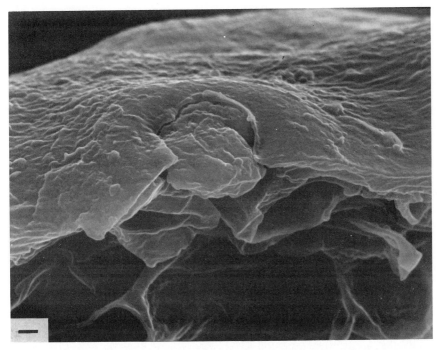

Fig. 2. Citrus stomata may often be obstructed by a plug, as in this transverse section of the lower surface of a mandarin leaf. Consequences for stomatal functioning and water-use efficiency require clarification. Sectioned fresh material was coated with gold-palladium. The bar represents 1 μm. (Unpublished SEM micrograph by N. Schaefer.)

a clearance of about 0.5 μm and possibly formed of citrus resin. This plug, like the epicuticular wax, is secreted by wax canals in the outer epidermis (Scott *et al.,* 1948).

Dimensions of guard cells and substomatal chambers vary appreciably among species and varieties of citrus (Abdalla *et al.,* 1978). Specific differences in stomatal sizes have also been reported by Kishore and Chand (1972) and by Turrell (1947). Regrettably, neither investigation reported associated stomatal densities. However, Kishore and Chand gave numbers of stomata per microscopic field, and it is apparent from their data that there is a positive correlation between these and the areas of guard cells. *Citrus aurantifolia* had the largest stomatal area of four species of citrus by a factor of more than two, and also had the most stomata. Such positive correlations are unusual; more commonly stomatal size and frequency are inversely correlated, (Ciha and Brun, 1975; Miskin and Rasmusson, 1970; Sapra *et al.,* 1975; Tan and Dunn, 1975). Turrell's data showed that average pore area differed by a factor of more than three between Valencia orange leaves (26.6 mm^{-2}) and Washington navel leaves (8.7 mm^{-2}). It would be interesting to compare transpiration rates of these varieties to see if transpiration is higher in the former, since data of Hirano (1931) suggested there was little difference in stomatal density of the two varieties and transpiration rate is proportional to stomatal pore area (Burrows and Milthorpe, 1976). An unknown complicating factor, however, is the frequency and effectiveness of the stomatal plug. Turrell (1947) considered plugged stomata as still functional and suggested that the plug prevents water infiltration of citrus leaves via the stomata by monsoonal rains in the area of origin of citrus. Erickson (1968), however, implied that plugged stomata are nonfunctional, although he considered that enough stomata remain unplugged to allow normal gas exchange. This point appears worthy of further study. If the stomata remain functional, the possibility that the plug may increase water use efficiency by increasing resistance to water vapor loss relatively more than overall resistance to inward diffusion and fixation of CO_2 [as Jeffree *et al.* (1971) reported for *Picea sitchensis* leaves, which also have plugged stomatal pores] should be investigated.

Stomatal pores of *Citrus* appear to be small compared to those of other genera. A comparison of pore sizes for four citrus species (Turrell, 1947) with those for a wide range of other genera (Meyer and Anderson, 1952) show that stomatal pore lengths and widths were smaller only in *Phaseolus vulgaris*.

Stomatal density has been studied in many species and cultivars of citrus by Hirano (1931). The extreme values he found were 326 mm^{-2} in *C. aurantifolia* and 873 mm^{-2} in Italian citron. Comparison with data for other species collated by Meyer and Anderson (1952) suggests that these densities are high, even among other tree species whose stomata are confined to the lower leaf surface.

Table I (Davies and Kozlowski, 1974) further illustrates the small size and relatively high frequency of citrus stomata, compared with four other woody angiosperm species. There may be considerable difficulty in relating citrus stomatal parameters to transpiration rate since among those five species transpiration rates were not consistently related to stomatal frequency, pore size, or stomatal area; transpiration was lowest in *C. mitis* even though it had the highest stomatal frequency and third highest relative pore area per unit leaf area.

Hirano's (1931) data also suggest that, except for lemon and grapefruit, a density greater than 500 stomata mm^{-2} is found on tropical trees and that hardier varieties and species, with some exceptions, have lower stomatal densities. Stomatal density in citrus is affected by a number of

TABLE I
VARIATIONS IN DIMENSIONS AND FREQUENCY OF STOMATA OF FIVE SPECIES OF WOODY ANGIOSPERMS[a]

	Acer saccharum	Fraxinua americana	Cercia canadensis	Quercus macrocarpa	Citrus mitis
Stomatal length (μm)	15.28 ± 0.25[c]	26.74 ± 0.76	21.47 ± 0.24	22.0 ± 0.34	19.19 ± 0.20
Stomatal frequency (per mm^2)	504 ± 13	118 ± 10	569 ± 14	587 ± 15	596 ± 12
Stomatal pore length (μm)	4.8 ± 0.10	10.2 ± 0.18	6.4 ± 0.12	6.1 ± 0.09	4.7 ± 0.14
Relative pore area per unit leaf area[b]	67	33	100	98	77

[a] From Davies and Kozlowski (1974). Reproduced by permission of the National Research Council of Canada from the *Canadian Journal of Botany*, Vol. 52, pp. 1525–1534.
[b] Assumes constant pore width.
[c] Standard error of mean.

factors. Dominant among these is the extent to which the leaves can expand. Hirano (1931) cited unpublished data of Bahgat showing that stomatal densities were increased for seven species and varieties by growth in progressively hotter and more arid climates, and Reed and Hirano (1931) found that stomatal density of a sun leaf of Eureka lemon (636 mm^{-2}) was nearly twice that of a shade leaf (322 mm^{-2}). Monselise (1951c), working with Shamouti orange, provided strong evidence that such differences were due to the failure of sun leaves to expand completely. Because of such effects it is important that stomatal density be measured only on mature fully expanded leaves. Very high densities (920 mm^{-2}) have been reported for young, small, lemon leaves (Reed and Hirano, 1931). Stomatal density also varies with position on the leaf; for samples always equidistant from the lamina edge and the midrib, but taken from the middle, tip, and base of the leaf, Reed and Hirano (1931) found values of 651, 637, and 580 mm^{-2}, respectively, for Eureka lemon, and 480, 453, and 399 mm^{-2}, respectively, for Valencia orange. On a more microscale, stomatal density is lower in the vicinity of large veins, oil glands, and trichomes (Hirano, 1931).

Venation in citrus is pinnately reticulate and shows the simple correlation venation pattern (Dede, 1962). The vascular system develops in such a way as to enclose rather than to lie above or below the numerous oil glands (Scott et al., 1948; Schneider, 1968). There is a question as to whether the leaf's vascular network provides adequate reticulation in view of the "mesophyll collapse" reported by Hilgeman (1963) in severely stressed orange leaves.

Mesophyll palisade has been described as comprising two layers of tightly packed cylindrical cells, with a third layer of shorter and less tightly packed cells sometimes present (Schneider, 1968). However, the tight packing of the palisade cells may be more apparent than real since Turrell (1947) found that more surface was exposed to the intercellular space by the palisade tissue than by the spongy mesophyll, the ratio being 2.4 in *Citrus limonia* and 3.1 in *Citrus grandis*. Intercellular spaces are largest in this region of the leaf. Turrell (1936) compared the ratio (R) of internal exposed surface area of leaves to the external surface, per volume of tissue for a number of species and genera, including *Citrus limonia* and *C. grandis*. The R values for these two species (17.2 and 22.2, respectively) were high and Turrell considered them typical of xeromorphic sun leaves. He correctly recognized the significance of this ratio in connection with gas exchange of leaves, but chose to emphasize the possibility that high R values would permit high transpiration rates. However, as the leaf diffusive resistance values already quoted indicate, transpiration rates are in fact remarkably low in citrus (see also Section III,B). A more reasonable interpretation of the high R values in citrus may be that they help to reduce internal resistance to fixation of CO_2 by providing a relatively large internal surface area for initial absorption of CO_2 by the mesophyll (Hoare and Barrs, 1974). This view is supported by the observation of Nobel et al. (1975) that higher photosynthetic rates in plants of *Plectranthus parviflora* grown under higher irradiances can be accounted for solely in terms of increasing R values.

Even in a relatively turgid citrus leaf, osmotic potential does not appear to rise above about -10 bars; the highest values found by Bielorai (1968) were -12.5 bars for sour orange and -10.4 bars for sweet lime leaves, falling to -24.0 bars and -22.8 bars, respectively, at a soil ψ of -17 bars. Osmotic potentials of -11.2 and -13.25 bars were measured for leaves of Lisbon and Genoa lemons, respectively, by Boiko and Boiko (1959), and of -18 and -23 bars, respectively, for young and old orange leaves by Levitt and Ben Zaken (1975). Such values are somewhat higher than those of leaves of many mesophytic species (Barrs, 1971a). Total ψ of Washington navel leaves varied from about -5 to -35 bars (Klepper and Ceccato, 1969); however, citrus can survive leaf ψ as low as -60 bars (Kaufmann, 1977a).

2. Fruit

Anatomy and morphology of citrus fruit were described by Schneider (1968); only those aspects relevant to water relations of the fruit will be considered here. Epicuticular wax is present, as on the leaf, but differences in composition and distribution occur (Baker et al., 1975; Schulman and Monselise, 1970). An important difference between leaves and fruits may be that wax production in all except the youngest leaves occurs at a rate sufficient to increase depth of surface wax for the first 4 months and to maintain an approximately constant level thereafter, but in fruit wax production is at times unable to keep up with fruit growth, resulting in peaks and troughs in wax thickness on fruit (Baker et al., 1975), or a steady decline in fruit wax thickness (Schulman and Monselise, 1970). On the microscopic level this is reflected in cracking and lifting of the wax film on fruits but not on leaves (Albrigo, 1972b). This is attributable to the relatively fast rate and short period of leaf expansion, which therefore involves stretching of less hardened and less structural wax than is the case with the fruit, which expands relatively rapidly over 6 months and continues to expand, albeit more slowly for an additional 6 to 8 months or more (Bain, 1958; Hilgeman et al., 1959). As discussed below, consequences for water loss may be considerable.

The epicuticular wax appears to be very important in reducing water loss from citrus fruits since total epicuticular wax was inversely correlated with post-harvest weight loss of Valencia oranges ($r^2 = 0.94$), and chemical removal of surface wax increased rates of fruit weight loss at least 300% (Albrigo, 1972a). Not only does the epicuticular fruit wax crack and lift but its distribution may be uneven, thereby increasing water loss from the fruit. Such losses can be minimized by preharvest, anti-transpirant plastic sprays that coat the fruit relatively continuously and hence improve orange peel quality by reducing dehydration and aging (Albrigo and Brown, 1970).

The extent and composition of the cuticle also differ between citrus leaves and fruits, and among species (Baker and Procopiou, 1975). Citrus fruits are unusual in having cuticles that are lighter and have less cutin content than corresponding leaf cuticles. This could be one reason why the epicuticular wax seems to be so important in citrus fruit. Cell division continues in the epidermis until the fruit is ripe and the cuticle is thin over recently divided cells (Schneider, 1968).

The epidermis contains functional stomata (Albrigo and Brown, 1970), together with accessory cells (Schneider, 1968) and, as in leaves, the stomatal pores are often plugged (Albrigo and Brown, 1970). However, stomata are comparatively infrequent in the fruit. Rokach (1953) found 74 mm^{-2} at the equator and 67 mm^{-2} at the ends of Shamouti oranges. Albrigo (1977) calculated the ratio of stomatal densities between navel orange leaves and fruit to range between 23 and 46, a range similar to that of the ratio of diffusive resistance between fruit and leaf of 13 to 56 that he found for Valencia trees. Actual fruit resistances varied between 67 and 157 sec cm^{-2}, and were higher on the northwest than on the southeast side of the tree. For late stage I (Bain, 1958) Valencia orange fruit, Elfving and Kaufmann (1972) found nighttime fruit and leaf resistances to be approximately equal at just over 100 sec cm^{-1}, but daytime fruit resistances were 30 to 40 sec cm^{-1}, compared with 10 to 20 sec cm^{-1} for those of leaves. The diurnal change in fruit resistance, although not as large as in leaves, was considerable, indicating that fruit stomata were highly active at this stage.

Kaufmann (1970a) found that osmotic potentials in all tissues of citrus fruits varied from -10 to -15 bars. Diurnal changes in osmotic potential of the juice vesicles appear to be minimized because these are anatomically isolated from the transpiration stream and are at the end of the carbohydrate pathway. Kaufmann found that osmotic potentials became less negative from the stem to the stylar end in the exocarp, mesocarp, endocarp, and vesicles, together with an additional osmotic gradient such that the osmotic potential of the vesicles was considerably lower than that of the pericarp in all parts of the fruit, the typical pericarp to vesicle gradient being 1 bar cm^{-1}. These gradients were to some extent eliminated when fruits were placed in a humid chamber overnight with their stems in water. Turgor pressure in the vesicles was linearly correlated with exocarp ψ. Extrapolation of this relationship indicated that vesicle turgor pressure would decline to zero at about -21 bars. A trend for a gradient of vesicle turgor pressure higher at the stem end than at the stylar end was also apparent. However, turgor pressure in the exocarp was not correlated with exocarp ψ and appeared to remain constant over the range of exocarp ψ's examined (about -4 to -11 bars). It is not clear whether this was an artifact of measurement associated with the presence of cell wall pectin contaminating exocarp samples, or whether turgor pressure was maintained by accumulation of osmotically active solutes in the exocarp as the fruit was stressed. The vesicles are anatomically somewhat isolated. They may be coated by cuticle, so water or solute transfer can only occur through the vesicle stalks, thus permitting the reported gradients of osmotic potential within the pulp. The ovules or seeds, but not the vesicles, are connected to a vascular network running in the central axis.

3. Stems

A detailed account of morphology, anatomy, and development of the citrus stem has been given by Schneider (1968). In the secondary thickened stem of *Citrus sinensis* the wood is diffuse porous, light yellow, and hard. Annual rings are terminated by rows of parenchyma cells that can vary in thickness and become discontinuous in development, making ring delineation difficult. Growth rings are distinguishable with some degree of certainty in subtropical climates, although it is possible for false rings to form (Hilgeman, 1977). Recent work suggests that the view that the vessels normally remain free of tyloses and wound gum may not be entirely correct since Nemec (1975) found plugging of vessels in older wood of *Citrus sinensis* even in healthy trees, although it was more frequent in diseased trees with sandhill or young tree decline.

Large differences in hydraulic conductivity have been reported for stems of rough lemon, grapefruit, and sour orange (De Villiers, 1939), the ratios of water transported under constant head for constant time being $4.4:2.66:1$, respectively. Analysis of De Villers' data shows that differences in hydraulic conductivity within species were significantly correlated with vessel diameter but not with xylem area or number or length of vessels. Furthermore, there were no significant differences among species in these anatomical characters, except between lemon and grapefruit in vessel length. This did not appear important since in grapefruit the correlation between hydraulic conductivity and vessel length was very low ($r^2 = 0.01$) and not significant. Thus, within species, only differences in vessel diameter appeared important in explaining differences in hydraulic conductivity, but between species none of the observed anatomical properties seemed important.

As De Villiers (1939) concluded, it appears that other as-yet-unknown factors, such as size of the perforations in the xylem vessels, may also be important in determining rates of water flow through citrus stems. In view of the large differences that she found in stem hydraulic conductivity (over 400%) among species, further research in this area appears worthwhile. It is interesting to note that Haas and Halma (1932) found transpiration rates for rooted cuttings of lemon to be greater than those for grapefruit, which in turn exceeded those of Valencia orange.

Interconnection of the vascular system in citrus stems appears to be adequate to allow water to flow relatively freely throughout the whole system. De Villiers (1939) found that dye injected into a stem at one point flowed upward and subsequently passed into all of the seven branches present 1 to 2 feet above the point of injection. In another experiment she found that dye applied to the cut end of a branch flowed downward in the xylem, and from the subsequent pattern of coloration of the stem xylem, she concluded that the xylem of one branch may be in connection with xylem strands on all sides of the tree. However, there was some evidence that dye transported in the youngest xylem was found mainly on one side of the tree and apparently moved around some of the branches without entering them.

4. Roots

Following seed germination, the primary root is the first organ to appear, growing rapidly downward to form a well-defined taproot (Schneider, 1968), although this may not always be true since Savage et al. (1945) found that in Morton citrange the taproot divided into several roots which tended to fan out and descend obliquely. However, it is likely that the taproot is lost at an early age in most cultivated citrus because it is usually severed when nursery trees are dug out prior to transplanting. The taproot is supplemented by large pioneer roots that branch and rebranch irregularly to form the root system framework. Castle (1980) recognizes a bimorphic root distribution in citrus since some of the smaller laterals formed from the pioneer roots are more or less vertically oriented and emerge from the crown of more horizontally arranged laterals to explore the deeper regions of the soil.

Fibrous roots, which are usually regarded as the principal pathway for uptake of minerals and water, occur in small bunches 20 to 30 cm or more long on the taproots of young seedlings and on pioneer roots of older trees. As roots branch repeatedly the diameters of the successive new fibrous roots decrease from about one to less than 0.5 mm. This results in development of a relatively shallow mat of fibrous roots that rapidly absorb applied nutrients and superficial soil water. The deeper lateral roots, which also carry some fibrous roots, are considered by Castle (1980) to prevent extreme stress from drought and take up nutrients that are not absorbed by the upper fibrous roots. Fibrous roots vary from shrivelled, brown, and corky to succulent, light yellow, and thread-like with actively growing white tips. Ford (1952) noted that the corky surface feeder roots resumed growth after rain; with numerous light-colored, white-tipped feeder roots evident within 24 hr. He considered that uptake of dissolved nutrients and water occurred only near their tips. However, this view may not be entirely correct since both water and nutrients can be absorbed by at least partially suberized citrus roots. Chapman and Parker (1942) found that, while periods of rapid nitrate uptake were usually associated with periods of active growth of the feeder roots, substantial amounts of nitrate were taken up by fibrous roots that had become brownish and partially suberized. Hayward *et al.* (1942) obtained direct potometric evidence for uptake of water by suberized roots of sour orange seedlings and proposed that there were enough lenticels on roots to account for the rates of uptake they observed. Although these rates were low, they suggested such water absorption could contribute significantly to total uptake because suberized roots are a large fraction of the whole root system, especially in winter months.

Many factors can affect the final form of the root system. Prominent among these is the rootstock itself. Savage *et al.* (1945) found considerable differences between 15 6-yr-old rootstocks budded to Parson Brown orange scions. Rough lemon and Cuban shaddock had the most extensive root systems, calamondin the deepest, and sweet lime and grapefruit the greatest abundance of fine fibrous roots. Ford (1954a) found that mature sweet orange had more feeder roots than four other species, but rooting depth was less than that of rough lemon or Cleopatra mandarin. In a later study Ford (1964) found that seedlings of rough lemon produced more feeder roots than those of Cleopatra mandarin or *Poncirus trifoliata,* with feeder roots particularly sparse in the latter species. Castle and Krezdorn (1975) found two extreme types of root systems, intensive and extensive, with intermediate forms in between, and that rooting depth was correlated with tree height. Interestingly, rough lemon exhibited extensive root development both laterally and vertically; Ford (1954a) had already suggested that the deeper rooting habit of rough lemon might account for the superior drought resistance of this rootstock, and Castle and Krezdorn (1977) later showed that leaf ψ was significantly higher for Orlando tangelo on rough lemon rootstock than on three other rootstocks which had shallower and sparser root systems. Similarly Albrigo (1977) found that leaf ψ

was usually higher for Valencia orange on rough lemon than on sour orange or on Carrizo citrange rootstocks; however the fruit had consistently lower soluble solids content and he suggested that this was the result of dilution arising from the improved water status of the tree on the rough lemon stock. The first measurements of differences in leaf ψ apparently arising from effects of the stock were reported by Bell et al. (1973). He found lower leaf ψ (higher stress) for Orlando tangelos on sour orange than on sweet lime rootstock.

Earlier work supports these observations, since rootstock effects on drought resistance, water status, and transpiration of scions have been reported. Spurling (1951) attributed the greater drought resistance of trees on citronella rootstock over sweet orange rootstock to the deeper and more ramified rooting habit of the former. Oppenheimer and Mendel (1934) and Mendel (1951) found that scions on rough lemon and on sweet lime rootstocks had higher transpiration rates than those on sweet orange, and that leaves from scions on rough lemon stocks had the lowest water saturation deficit. They also observed that, as available soil water declined, transpiration of trees on sour orange stocks decreased somewhat before that of trees on rough lemon or sweet lime stocks. Wallace et al. (1952) found that scion leaves on sour orange had a lower moisture content than those grown on rough lemon, sweet orange, or grapefruit rootstocks. Ongun and Wallace (1958c) found significant differences in transpiration rates among seven rootstocks with the same scion; again transpiration rate was highest for the rough lemon rootstock and lowest for sour orange, with grapefruit stock intermediate. These results are perhaps related to the work of De Villiers (1939), already discussed, showing a similar ranking of stem hydraulic resistance (rough lemon < grapefruit < sour orange) and to the observation of Haas (1936) that transpiration rates of rough lemon cuttings were significantly greater than those of grapefruit cuttings. How much these differences in rootstock water-supplying ability are due to differences in size and how much to differences in efficiency of water transport by different rootstocks remains to be resolved. The above data suggest that differences in rootstock efficiency may be important in water relations of citrus.

Root systems may be affected by scion as well as by rootstock. Ford (1952) and Calvert et al. (1977) found that concentrations and total amounts of feeder roots were considerably higher, regardless of depth, for grapefruit than for orange scions on rough lemon or sour orange stock, a result Ford expected because grapefruit trees were usually larger than orange trees of equal age. Other plant factors affecting citrus root systems are tree age and density of planting. Ford (1954a) found that for rough lemon stock the roots of 5-yr-old trees did not penetrate the soil below 9 feet but roots of 10- to 15-yr-old trees had entered the 9- to 13-feet zone, and roots of 20- to 30-yr-old trees were found in the 13- to 17-feet zone.

Boswell *et al.* (1975) found that root spread, circumference, and diameter were greater for widely spaced trees than for closely planted ones, which had overlapping root systems. Similar results were found by Kaufmann *et al.* (1972), who pointed out the adverse effects of root crowding in terms of competition for water and nutrients. Attempting to minimize root competition by supplementing the water supply of closely spaced trees with frequent irrigation caused root aeration problems (further detail in Section IV,B).

Soil temperature can have a profound effect on growth and water relations of citrus roots. As long ago as 1927, Girton showed that the minimum temperature for root development of citrus seedlings was about 12°C, the optimum 26°C, and the maximum 37°C. Subsequent work confirmed the general validity of these findings obtained in solution culture, and extended them to plants growing in soil (Haas, 1936; Ongun and Wallace, 1958a,b; Cary, 1971; Fawusi and Ormrod, 1975). However, the lower limit for root growth may have been set somewhat too high, because North and Wallace (1955) reported very slow root elongation at temperatures as low as 9°C in solution culture.

Responses of roots to soil temperature can be dramatic, leading to a doubling (Cary, 1971) or even almost a quadrupling (Ongun and Wallace, 1958b) of growth when root temperature is raised 6°C from 18 or 19°C, respectively. North and Wallace (1955) observed maximum root elongation at between 27 and 32°C under laboratory conditions. However, observations of significantly higher feeder root density on the southern than on the northern sides of trees in California suggest that soil temperature is no less important in the field (Cahoon *et al.*, 1961). Growth of stems and leaves is also favored by higher soil temperatures (Haas, 1936; Ongun and Wallace, 1958a,b; Fawusi and Ormrod, 1975; Liebig and Chapman, 1963; Khairi and Hall, 1976a), although the response of these organs is less and that of fruit is little or none (Cary, 1971).

Vinokur (1957) suggested that low temperature, apart from disturbing citrus water relations (discussed below), also affects citrus growth by reducing the rate of NO_3–N absorption by roots and by slowing translocation from leaves to roots, leading to reduced photosynthesis. However, Khairi and Hall (1976a) concluded that the positive effect of warm soil temperature on bud break and subsequent shoot growth in citrus was not associated with improved mineral nutrition or plant water balance. Instead, they suggested that soil temperature may have influenced bud break through change in hormonal balance in the shoot, possibly by temperature effects on synthesis and/or translocation of hormones from roots to shoots.

Hydraulic conductivity of citrus roots is to a large extent controlled by soil temperature. Bialoglowski (1936) found for rooted leafy lemon cuttings in nutrient solution that, when root temperature was the only variable, transpiration during the day was greatest at 25°C, declining slowly as temperature rose to 40°C and more abruptly at 45°C, and declining sharply as root temperature fell below 25°C. However, at night when transpiration was very low the rate was independent of root temperature over the range 0 to 40°C. Bialoglowski pointed out that Q_{10} for the effect of temperature on daytime transpiration was much higher than the effect of temperature on viscosity of water, and that this effect became increasingly apparent as 0°C was approached. He concluded that the effects of low root temperature on transpiration were too great to be explained in terms of increased viscosity of water alone and that activity of the entire root system was affected.

Similar results were obtained for citrus rooted in soil although the optimum temperature for transpiration was higher and varied among species. Thus, Haas (1936) found optimum root temperatures of 31°C for lemon and Valencia orange cuttings and 27°C for grapefruit cuttings. Vinokur (1957) noted that the daytime transpiration rate of lemon cuttings was considerably higher at a root temperature of 33 than at 13°C, and Ongun and Wallace (1958c) found that transpiration of Washington navel orange on several different rootstocks was reduced by root temperatures below 32°C. Cameron (1941), working with two young Valencia orange trees on sweet orange stock growing in a weighing lysimeter in a lath house, found that as root temperatures were reduced from 32 to 7°C there was an increasingly rapid decline in daytime transpiration and that there was a similar but very much smaller effect of root temperature on nighttime transpiration, which was only about 10% of daytime values. Cameron also observed wilting of leaves at a root temperature of about 7°C and showed that relative leaf saturation deficit for these leaves was more than twice as high as that for leaves when root temperature was about 29°C. The data appeared consistent with the view of Kramer (1940) that decreased water absorption at low root temperatures was due to decrease in permeability of root membranes combined with increase in viscosity of water. North and Wallace (1955) observed wilting of greenhouse-grown citrus, even though soil water availability was high, when soil temperatures were between 9 and 17°C. Low soil temperatures can have similar effects in the field. For example, Elfving et al. (1972) showed that midday leaf ψ's were sharply reduced by low soil temperature, although nighttime or sunrise ψ's were independent of soil temperature. Kaufmann (1975) compared relative resistance to water flow from soil to leaves for *Picea engelmannii* and citrus. This resistance was more sensitive to lower soil temperatures in citrus than in spruce, and especially in citrus, only a small fraction of the increased resistance was attributed to effects of increased viscosity of water at the lower soil temperatures. He concluded that root permeability to water flow is much more sensitive to low soil temperature in citrus than in spruce.

Ramos and Kaufmann (1979) demonstrated direct dependence of citrus root hydraulic resistance on root temperature. As shown in Fig. 3, root hydraulic resistance increased as root temperature was lowered from 35 to 5°C, and the rate of increase in resistance was greater than the rate of increase in viscosity of water and hence could not be attributed entirely to the latter. They suggested that their material did not acclimate to lower root temperatures because their experiments were completed quickly, and that this was why even 35°C appeared suboptimal for root hydraulic conductivity. Further work is needed to verify their suggestion that citrus roots can acclimate to low soil temperatures and to ascertain whether such capacity varies among rootstocks. Data of Ongun and Wallace (1958c) suggest that this may be the case. They found that, although transpiration of Washington navel scion on sour orange stock was significantly lower than that of the same scion on grapefruit stock at root temper-

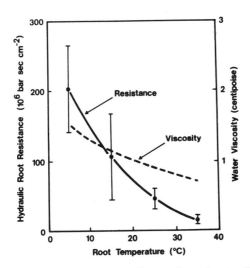

Fig. 3. Temperature has a much greater effect on flow of water through rough lemon roots than it does on viscosity of water, implying that temperature has direct effects on permeability of citrus root cell membranes to water. (After Ramos and Kaufmann, 1979.)

atures of 22 and 32°C, transpiration rates of the two scion/rootstock combinations did not differ significantly at 12°C.

Ramos and Kaufmann (1979) also investigated the effect of prior soil water stress on hydraulic resistance of lemon seedling roots. Water stress was imposed by watering plants every 2 or 3 days instead of every day to induce mild or severe water stress, as measured by nighttime leaf ψ's of -6, -11, and -20 bars. These regimens were imposed for 18 days; all plants were than finally watered and root resistances determined 5 hr later. Both mild and severe stress increased root hydraulic resistance significantly. However, there was no significant difference between effects of mild and severe stress, severe stress being associated with a slightly lower hydraulic resistance. Overall, water stress almost doubled root hydraulic resistance (221.0×10^6 bar sec cm^{-2} compared with 110.7×10^6 bar sec cm^{-2} for the control plants). The responsible mechanism is as yet unclear. Ramos and Kaufmann (1979) suggested that stress may have caused increased suberization of the cortical cell walls of the root, or that root membrane permeability was decreased. Further work is needed on these possibilities.

The anatomy of citrus roots has been described by Cossman (1940), Hayward and Long (1942), Schneider (1968), and Schneider and Baines (1968). One feature, only briefly considered by these authors, that may be of considerable physiological importance, is the presence of mucilage adhering to the surface of young feeder roots. Numerous microorganisms occur in this mucilage layer, which is about 2 μm thick. The origin of the layer as a secretion from the root or the microorganisms, or both, is as yet unresolved (Brams, 1969), but Oades (1978) suggests that mucilage may originate in both the root and microorganisms. Brams found that the mucilage layer contained 2–20% of the total specific root cations and suggested that it may act as a supply and a reservoir of nutrients for the plant. Oades suggested inter alia that root mucilage may be important in improving root soil contact (and hence uptake of water and ions) and protecting the root tip from desiccation.

The conditions under which root hairs form in citrus are poorly understood (Schneider, 1968). At one time it was considered that root hairs did not form at all (Cossman, 1940). However, it is now clear that primary root hairs develop normally from the epidermis and that secondary root hairs may develop from the underlying hypodermis as the epidermis breaks down (Hayward and Long, 1942). The secondary root hairs form in circular groups from hypodermal cells that enlarge by lengthening their anticlinal walls (Schneider and Baines, 1968). They are typically shorter and thicker than epidermal root hairs, and they probably function in absorption until their walls become suberized, as is also the case with persistent primary root hairs (Hayward and Long, 1942).

Factors that favor development of root hairs include good soil aeration, high soil temperature, and relatively low pH (Girton, 1927; Hayward and Long, 1942). Responses to soil water content are also important but appear to vary among species, since Cossman (1940) found that the longest root hairs were produced by *C. limetta* and *C. limonia* in the dry soil, but in wet soil the longest hairs were produced in *C. aurantium*.

Variations in shape and length of root hairs have been reported by Hayward and Long (1942); most are not long, about 50–150 μm, occasionally extending to 250 μm; others may be only short papillate projections, (Fig. 1). Hayward and Long (1942) pointed out that root hairs commonly vary in length from 0.15 to 10 mm; hence, citrus root hairs are at or below this lower limit. This has led to the view (Trumble, 1952) that citrus trees are comparatively inefficient in absorbing water partly because they lack abundant root hairs, which are the most efficient absorbers of soil water. In cereal crops root hairs typically grow along 10 to 30% of total root length (Greacen *et al.*, 1976). A comparable value is not available for citrus, but anatomical data suggest it is very much lower. The relative paucity of citrus roots is highlighted by the statement (Wallace and Nauriyal, 1958) that the total root system (about 10 to 20 miles long) of an orange tree, is less extensive than that of a single wheat plant. Certainly the leaf area of citrus is far greater than in wheat, emphasizing the relatively unfavorable root to shoot ratio of citrus. A deficiency is the lack of quantitative information on root density of citrus, a parameter crucial to understanding of crop water uptake (Tinker, 1976).

A further topic requiring clarification concerns the relative magnitudes and importances of various possible pathways for water uptake presented by citrus roots. In addition to the primary and secondary root hairs and hypodermal lenticels discussed above, attention should be given to the groups of hypodermal cells that may enlarge both radially and tangentially without initial periclinal divisions, rupturing the epidermis in the process. These, unlike the other hypodermal cells, do not become cutinized and may function in absorption (Hayward and Long, 1942).

As in the stem, blocking of root xylem vessels has been associated with "diseases" such as blight, young tree decline, and sandhill decline (Allen and Cohen, 1974; Young and Garnsey, 1977). However, attempts to isolate causal organisms have been unsuccessful (VanderMolen, 1978). Whether plug formation is an expression of premature senescence or a response to an invading pathogen is not known. Nemec *et al.* (1975) found that xylem vessels of citrus roots became blocked by lipids and gums.

Plugging occurred in both diseased and healthy trees, and root hydraulic conductivity was correlated with the degree of plugging in both. However, with the same amount of plugging, the effect was nearly twice as great in blighted as in healthy trees, suggesting that other factors may also be limiting water movement in diseased roots. One of these factors could be development of bands of vessels of reduced size along the cambium; these were found in 19% of diseased trees but were never found in healthy trees. Such bands reflect internal water stress at the time of their formation (Zahner, 1968) and perhaps represent a feedback response, aggravating the disease since water movement in citrus wood occurs principally adjacent to the cambium (Nemec et al., 1975). The gum appears to be a mucopolysaccharide and the lipid a resinous wood extractive. The origin of these plugs is uncertain. Nemec et al. suggested they originated in wood parenchyma cells and leaked into the xylem, but VanderMolen (1978) considered they originated from the middle lamella and primary walls of vessels.

Interconnections in the root system appear adequate to ensure that water applied to one part of the root system becomes available to the whole tree. This has been demonstrated both by experiments in which dyes and salts introduced into one root were transported to branches on all sides of the tree (De Villiers, 1939) and by experiments in which only part of the root system was irrigated. Hilgeman (1974) found that, provided irrigations were sufficiently frequent, applications to alternate sides of trees were adequate, resulting in no detrimental effects on tree growth and fruit yield. Furr and Taylor (1933) found that irrigation of only part of the root system, either by localized application of water or by cutting some roots before irrigation, did not cause differences in volume of fruit on opposite sides of trees. Instead, root pruning caused a reduction in hydration of the tree as a whole.

B. Acquisition, Transport, and Loss of Water by Trees

Water relations of citrus are influenced to a large extent by high resistances to water transport within the plant. These are associated, at least in part, with the low root:shoot ratio of citrus and its poorly developed root hairs (Section II,A,4); they are quantified in Section II,B,3. They may result in abnormally low leaf ψ's in citrus. Davies and Kozlowski (1975), working in a controlled environment with well-watered seedlings, observed minimum leaf ψ's of -16 bars in *Citrus mitis,* compared with -6 bars in *Fraxinus americana* and -4 bars in *Acer saccharum.* It seems likely (Kadoya, 1977) that these resistances play a central role in short-term cyclic oscillations in transpiration by citrus, described on p. 347.

Another feature of citrus is the sensitivity of its stomata to air humidity, which may well be associated with the high internal resistances noted above.

1. Short-Term Oscillations

Cyclic oscillations in transpiration with a period of a few minutes to a few hours occur in many species and genera including citrus, the rhythmic opening and closing of stomata being the immediate cause (Barrs, 1971b). Kriedemann (1968) found cyclic oscillations in transpiration rate when an attached orange leaf was exposed to a stream of dry air in an assimilation chamber. There was strong circumstantial evidence that these oscillations were caused by cyclic changes in stomatal aperture since net photosynthesis cycled in phase with the transpiration cycles (Fig. 4). Later Kriedemann (1971) established that the cyclic oscillations in photosynthesis of citrus were attributable to accompanying oscillations in both stomatal resistance (1.6 to 60 sec cm^{-1}) and residual resistance to CO_2 uptake and fixation by the leaf, and that the two were very highly linearly correlated ($r = +0.914$). How or why these two resistances should cycle

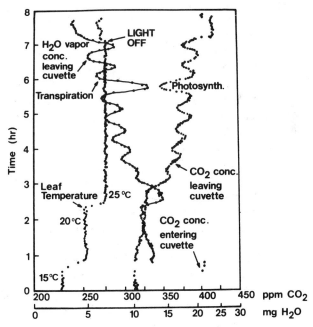

Fig. 4. A sudden change in leaf temperature has led to cyclic oscillations in gas exchange by this citrus leaf held under steady conditions in a cuvette. (E. Törökfalvy and P. E. Kriedemann, unpublished.)

together remains an interesting question. Closer examination showed that while this positive correlation held over the whole range of stomatal opening, the residual resistance only increased again during the stomatal closing cycle, when stomatal resistance had already increased to above 20 sec cm^{-1}. Other plant parameters observed to cycle were the CO_2 compensation value and leaf ψ, the latter averaging -9.32 ± 0.57 atm at the onset of stomatal closure and -6.04 ± 0.46 atm at the onset of stomatal opening.

As will be detailed below, other investigators reported cyclic changes in leaf thickness, fruit size, and stem diameter, presumably as a result of propagation of such cyclic oscillations in water status throughout the plant. Often, cyclic responses have been characterized entirely or largely on the basis of such changes in physical dimensions.

In Kriedemann's studies the period of the cycles was about 40 min., similar to that reported by Kadoya et al. (1975) and Kadoya (1977, 1978) for Citrus natsudaidai. However, Bravdo (1977) found that the duration of the cycle for Citrus sinensis varied widely and was a function of leaf age, e.g., 3.7, 30.3, and 160.7 min. for young, mature, and old leaves, respectively. A direct comparison of Bravdo's data with those of other investigators is difficult since he worked at the CO_2 compensation value, and this may have influenced the results. It is also not clear from Bravdo's study whether all three such cycles could have occurred on the same plant at the same time. Ehrler (1971) reported cyclic oscillation of stomatal aperture and leaf thickness in lemon leaves. Two leaves (age not stated) on the same plant cycled in phase, but again the conditions under which he carried out his experiment were unusual, since his plants were in darkness. He inferred from measurements of leaf temperature and leaf thickness that the stomata were opening for about 20 min. with peak opening at 10 min. after the onset of opening, the periods of opening being separated by 60–85 min. of closure.

The possibility that citrus leaves on the same plant may cycle asynchronously cannot be excluded because this occurs in other woody species (Teoh and Palmer, 1971). However, Kadoya et al. (1975) found that cycles of leaf thickness for three leaves (upper, middle, and lower) in the same Citrus natsudaidai plant were synchronous and in phase with cycles of stem thickness.

Cyclic oscillations in gas exchange in citrus can be initiated by sudden changes in leaf environment such as temperature (Fig. 4). Sudden changes in air humidity (Kriedemann, 1968; Possingham and Kriedemann, 1969; Hall et al., 1975; Camacho-B. et al., 1974a; Kadoya et al., 1975) or CO_2 concentration (Kriedemann, 1971; Kadoya, 1977) can also induce the onset or cessation of cycling. Changes in the root environment, notably soil temperature, can also influence oscillations

(Kadoya, 1977, 1978); responses seem slower but this may merely reflect the greater difficulty in imposing a step-function change in the soil than in the aerial environment.

Initiation of cyclic responses by changes in the environment suggests instability of the stomatal control system, especially since the cycles may persist for all of a photoperiod (Kadoya, 1978). However, occurrence of nocturnal cyclic oscillations in a steady environment indicates that an external trigger may not always be required (Ehrler, 1971).

Cyclic oscillation in the field has also been suspected, for instance, by Hilgeman et al. (1969) on the basis of irregularities in transpiration of orchard leaves and by Kadoya (1973), who reported cyclic fluctuations in fruit diameter, with a period of about 20 to 30 min., superimposed on a general increasing trend of fruit growth of young Satsuma mandarins. Marloth's (1950) dendrograph records showed small rapid changes in fruit diameter that were associated with hot dry winds, and Camacho-B. (1977) reported cyclic oscillations of leaf conductance after a hot dry breeze passed over a citrus orchard. Levy and Kaufmann (1976) observed cyclic oscillations in leaf conductance and photosynthesis of orchard trees from about 10.30 A.M. to 2.00 P.M. Their observations are reinforced by dendrograph records of cyclic oscillations in trunk diameter of a 30-yr-old tree with a period of about 1 hr during most of the daylight hours.

The phase relationships between the various plant parameters associated with cyclic oscillations have not been closely studied in citrus. Kadoya (1978) noted that cyclic oscillations in leaf thickness (a measure of leaf water content) lagged behind those in leaf temperature (a measure of transpiration rate) by about 10 min. Levy and Kaufmann (1976) found that the phase relationships were affected by root temperature, since cyclic changes in trunk diameter and xylem pressure potential (a measure of leaf ψ) lagged behind changes in leaf conductance when the root temperature was 5°C but not when it was 25°C. They suggested that this was due to the higher root resistance at the lower temperature (Section II,A,4), limiting the speed of response of tree water content to transpiration. These observations of persistent cycling at a root temperature of 5°C contrast with those of Kadoya (1978), who found that soil temperatures between about 10 and 16°C could inhibit cyclic responses, depending on the time of year. Possibly species differences are involved, since Kadoya worked with Citrus natsudaidai, but Levy and Kaufmann used navel orange on "Troyer" citrange rootstock and grapefruit budded to sour orange.

Cowan (1977) discussed models that sought to explain cyclic oscillations in plants generally, all based on the premise that oscillations were due to the properties of a loop in which rate of evaporation, through physiological processes, influences stomatal aperture that, in turn, affects rate of evaporation. However, both Cowan (1977) and Kadoya (1978) considered that these models did not provide completely satisfactory explanations of the phenomenon. Cowan considered that such models may require revision to incorporate adjustments of osmotic pressure in guard cells and possibly changes in the concentration of abscisic acid or related compounds in the leaf. It is likely that, as in other plants (Kriedemann *et al.,* 1972), citrus stomata respond to abscisic acid (ABA), since Davies and Kozlowski (1975) found a relatively rapid (about 60% reduction in transpiration accompanied by a tripling of leaf resistance when $10^{-4} M$ ABA was applied to *Citrus mitis* leaves.

Opinion is divided about the significance of stomatal oscillations. Hopmans' (1971) view is that they reflect failure of the stomatal control system. Cowan (1972) argued that, if cycling occurs commonly in nature, such failure is unlikely. Rather, he considers oscillations as exploratory excursions in approach to an optimal state. However, it is difficult to accept this view when such oscillations can be maintained throughout much of the "working day" of plants, as the data of Levy and Kaufmann (1976) suggest. Camacho-B. (1977) mentioned that stomatal oscillations may favorably affect water-use efficiency (WUE) (i.e., the ratio between CO_2 assimilation and transpiration). Although this suggestion can be reconciled with the sustained cycling that can occur in citrus, it lacks experimental support.

2. Diurnal and Seasonal Patterns

Diurnal patterns of citrus water use, like the short-term oscillations already discussed, are essentially cyclic and again result in associated cyclic changes in water status of the various tree organs. Once more hydraulic coupling between organs plays a part, and phase lags in changes in water status of different organs are again apparent. Such changes were characterized by early researchers in terms of changing organ water content and by later researchers, largely due to advent of the pressure chamber (Scholander *et al.,* 1964), in terms of changing organ ψ's.

Figure 5 shows an example with well-established Valencia orange trees on sour orange rootstock growing in a gravel soil in Arizona. The stomata opened relatively rapidly after break of day to give minimal leaf diffusive resistance at about 9.00 A.M., in some cases reaching values as low as 1 sec cm⁻¹. This indicated that full sunlight was not required for maximal stomatal opening, an observation confirmed by laboratory studies such as those of Hall *et al.* (1975) and Davies and Kozlowski (1974). Stomatal opening resulted in immediate onset of transpiration which peaked at about 1 P.M. This peak is perhaps reflected in the somewhat later peak in leaf relative saturation deficit, since Elfving *et al.* (1972)

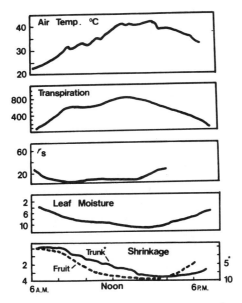

Fig. 5. Diurnal patterns in atmospheric conditions and Valencia orange tree moisture status under orchard conditions. Note time lag for expression of leaf moisture tension in trunk and fruit. Transpiration is expressed as mg H_2O g^{-1} hr^{-1}, stomatal resistance in sec cm^{-1}, leaf moisture as percent relative saturation deficit, trunk shrinkage is in microns, and fruit shrinkage is shown as percentage of turgid volume. (After Hilgeman et al., 1969.)

showed that, under otherwise favorable environmental conditions, there was a close correlation between leaf ψ and transpiration rate of citrus. Subsequently, leaf water deficit declined as stomata closed and transpiration rate fell as evening approached. The slowest region of the tree to respond to onset of water stress, as measured by a dendrograph, was the trunk, there being no response for about 2 hr, with the greatest shrinkage occurring while other organs were rehydrating. However, an interesting feature of the diurnal pattern of variation in trunk diameter is that despite the lag, response of trunk diameter is particularly sensitive to changes in transpiration rate. Hilgeman et al. (1969) showed, for example, that changes in trunk diameter reflect changes in cloud cover. Abrupt increases in trunk diameter also were recorded at the beginning of rainfall.

It is tempting to interpret the rather slow fruit shrinkage as a consequence of a ψ gradient between the leaves and the fruit and subsequent flow of water from the fruit to the leaves; such a view has long been held (Furr and Taylor, 1933; Rokach, 1953; Chaney and Kozlowski, 1971). However, Elfving et al. (1972) followed diurnal variations in diffusion resistance and ψ of both fruits and leaves, and concluded that, although differences between fruit and leaf ψ's occurred, the evidence did not permit an unequivocal statement that transpiring fruits are midday reservoirs of water for leaves. However, they conceded that diurnal diameter changes in nontranspiring fruit might be explained by flow of water from fruits to leaves. Elfving *et al.* (1972) criticized earlier experiments with fruit bearing branches detached from trees on the grounds that unnatural ψ gradients may have been induced so as to cause water movement from fruits to leaves. Similarly, Maotani *et al.* (1976) found that leaf ψ's on intact branches of Satsuma mandarin trees were similar whether fruits were present or not. However, in detached branches, the ψ of leaves on fruit-bearing branches exceeded that on branches without fruit. Diurnal fluctuations in fruit diameter were unaffected by the presence or absence of leaves. They concluded that the role of fruits as a water reservoir for leaves was negligible, at least in intact branches. However, the issue cannot yet be considered finally resolved because Hilgeman (1977) found that when transpiration from grapefruits was prevented by enclosing them in plastic bags, together with a small amount of water to ensure that the fruit did not transpire, they shrank only 31% less than adjacent unbagged control fruits. He calculated that translocation of water from the fruit to other parts of the tree accounted for 69% of diurnal fruit shrinkage. Furthermore, Trumble (1952) observed that December drop of young fruit associated with an exceptionally severe early summer period in Australia was markedly less (33%) in Valencia trees from which the mature fruits were not picked until after the end of November than in trees (53%) from which the mature fruits had been picked earlier. Trumble ascribed the lighter fruit drop in the late-picked trees to capacity of their mature fruits to supply water to young fruits. December drop was highest, 68% in Navel orange trees, which do not bear mature fruits at that time of year. Whatever the cause, there is no doubt that the amount of water in citrus fruit typically declines during the day. Because of the anatomical isolation of the fruit vesicles (Section II,A,2), much of this water is lost from the pericarp rather than from the pulp (Rokach, 1953).

The data in Fig. 5 should be interpreted with caution, since microscale effects, notably those due to movement of the sun around the tree, were important in influencing fruit diameter. Thus, fruit on the east side of the tree shrank rapidly in the morning, attaining maximum shrinkage by 11 to 12 A.M. Fruits on the south side showed maximum shrinkage between 2 and 4 P.M., and those on the northwest side did not show full shrinkage until near 5 P.M., when the sun shone directly on that part of the tree. While it is possible to plot average diurnal variation in organ water balance, this may be too simplistic, since all representatives of the same organ on any one tree may not be in the same state at any one time.

Similarly, Camacho-B. *et al.* (1974a) reported that leaf ψ was more negative by 2–5 bars on the sunny than on the shady side of the tree at midday, with accompanying lower leaf diffusion resistances (5–8 sec cm^{-1}) than in shaded leaves (8–12 sec cm^{-1}). These lower resistances, coupled with leaf temperatures that were higher by several degrees, caused transpiration gradients to be appreciably higher from leaves exposed to the sun than from shaded leaves.

Not only is position on the canopy surface important in this regard, but also position within the canopy. Thus, Machida and Maotani (1974) showed a difference of 2 or more bars in ψ between leaves in the outer surface of the canopy and those within the crown and that this difference was maintained over almost all of the daylight period. The marked attenuation of irradiance with depth in the citrus canopy that accounts for this observation was discussed in Section II,B.

Other environmental factors also modify the diurnal cycle of citrus water use and water relations. Elfving and Kaufmann (1972) found that the diurnal march of leaf ψ was sensitive to vapor pressure deficit (VPD), higher VPD's resulting in more negative leaf ψ's. Thus, for daily maximum VPD's of 3.1, 11.4, and 38.7 mm Hg, the corresponding leaf ψ's were -7.5, -10.3, and -16.0 bars, respectively. Effects of VPD may even override to some extent those of soil ψ. For example, Elfving and Kaufmann found that leaf ψ's at VPD's of 16.2 and 8.7 mm Hg were -11.1 and -10.1 bars, respectively, even though the corresponding soil ψ's were -0.2 and -0.3 bars. But, it should be borne in mind that VPD and radiation are strongly correlated. Hence at least part of this apparent relation could be due to response of leaf ψ to solar radiation (Smart and Barrs, 1973; Machida and Maotani, 1974). However, Elfving and Kaufmann's suggestion that at night VPD has a very small effect on citrus leaf ψ because leaf diffusion resistance is then high and transpiration low, is unexceptionable. Elfving and Kaufmann (1972) also indicated that VPD may have an appreciable effect on diurnal fluctuations in fruit diameter. When VPD was low such fluctuations were small or absent, but when VPD was high marked diurnal fluctuations in fruit diameter occurred. Lombard *et al.* (1965) also found "the evaporative power of the atmosphere" to affect fruit growth, but only when soil suction was high (100 cbar). Diurnal fluctuations in fruit ψ are typically less than those in leaf ψ. Elfving and Kaufmann (1972) reported that values of about 10 bars in fluctuations of leaf ψ were accompanied by changes in fruit ψ of only about 2 bars. A larger diurnal fluctuation in fruit ψ was found by Kaufmann (1970a). This may have occurred because the nighttime minimum in fruit ψ that Kaufmann found was appreciably higher. While Elfving and Kaufmann (1972) reported that minimum nighttime VPD was not correlated with nighttime fruit growth, Hales *et al.* (1968) used multiple regression analysis to show that nighttime recovery of fruit volume in Arizona was adversely affected by low nighttime humidity. A relative humidity of near 37% appeared to be the critical value below which fruit volume did not recover completely from daytime stress.

Dominant soil factors influencing the diurnal pattern of citrus water relations include soil ψ and soil temperature. Hilgeman *et al.* (1969) showed that, under the same atmospheric conditions, "dry" trees (soil ψ -0.76 bars) had lower transpiration rates than "wet" trees (soil ψ -0.29 bars) throughout the day. Despite this, both leaf water deficit and fruit shrinkage were consistently greater throughout the day for "dry" than for "wet" trees. Furthermore, fruit shrinkage continued longer on "dry" trees. However, soil water stress was not severe, since fruits of "dry" trees expanded overnight by 1.2% as against 2.1% for fruits of "wet" trees. Following irrigation of both groups of trees, transpiration rates, particularly in formerly "dry" trees, remained high throughout the day and leaf water deficits for the two groups became almost identical and varied similarly throughout the day. However, fruit shrinkage was now appreciably greater in the former "wet" trees than in previously "dry" trees and this was accompanied by greater overnight fruit expansion in the latter (4.6% compared with 2.6% for fruits of "wet" trees). Hilgeman *et al.* (1969) described these higher nocturnal rates of fruit enlargement and lower shrinkage during the day as typical responses of fruit following depletion of soil water. Such an effect is not unknown in other species and organs (Gates, 1955; Owen and Watson, 1956): perhaps some form of osmotic adjustment is involved. Whatever the explanation this potentially economically important phenomenon deserves further study.

Somewhat different results were obtained by Lombard *et al.* (1965). They found that soil ψ did not influence fruit and leaf growth until it had fallen from -0.20 bar to -1.0 bar. They considered that the large associated decline in soil capillary conductivity (from 15×10^{-4} to 0.5×10^{-4} cm hr^{-1}) exerted considerable influence on growth of these organs. Possibly the sparse production of root hairs in citrus would amplify such an effect, although Newman (1976) considered that the role of root hairs in absorption of water has been exaggerated. He pointed out that the soil provides by far the greater area for radial flow; even at the surface of roots of species with normal root hair development, only about 1% or less of the surface area is occupied by root hairs. Furthermore, Newman estimated that even when the soil is quite dry, most water moves radially in the soil rather than in the root hairs. Hales *et al.* (1968) emphasized the dominant influence of soil ψ on fruit growth. Of 10 meteorological variables studied, soil ψ showed by far the largest and most significant simple correlation with fruit growth. Maotani *et al.* (1977) found that initially as the soil dried the amplitude in diurnal fluctuation of fruit diameter increased, but when dawn leaf ψ had fallen to -12 bars, diurnal fluctuations in fruit diameter decreased again because fruits no longer recovered full turgor at night. Elfving and Kaufmann (1972) showed a decisive influence of declining soil ψ on the diurnal pattern of leaf ψ. As soil ψ declined, both afternoon minimum leaf ψ's and nighttime maximum leaf ψ's dropped increasingly faster. These investigators also showed that low soil temperature (7.5°C) greatly lowered the afternoon minimum in leaf ψ although the nighttime maximum was hardly affected.

While the diurnal pattern of water balance of citrus organs above ground has been widely studied, much less attention has been paid to the roots. Allen and Cohen (1974) followed diurnal variations in leaf and root ψ's of both healthy and diseased citrus trees. In all cases diurnal changes in leaf and root ψ's showed similar patterns with no perceptible phase differences. Blight did not affect root ψ, but leaf ψ fell much faster by day in blighted than in healthy trees, even though nighttime recovery appeared complete. Presumably as a result of this, there was a large temporary rise in stomatal resistance in leaves of blighted trees in early afternoon, reaching a peak at about 2 P.M. before declining and finally rising again as evening approached. Healthy trees did not show such a midafternoon peak in stomatal resistance. Responses were broadly similar in trees with young tree decline disease (Fig. 6) although daytime root ψ's were somewhat lower in diseased than in healthy trees. Overall, the observed ψ values and associated root–leaf ψ gradients imply that there was increased resistance to water flow in diseased trees and that the resistance was located principally in the trunk and main branches.

Few researchers have addressed the more onerous problem of characterizing seasonal changes in the water relations of citrus. Hilgeman *et al.* (1969), however, followed variations in apparent transpiration, leaf saturation deficit, fruit shrinkage, and trunk shrinkage through an entire season. Internal water stress was at a minimum in March and April, as evidenced by the low leaf water saturation deficits during that time, together with absence of fruit shrinkage and only slight trunk shrinkage. This was largely due to the low transpiration rates of the spring leaves together with much shedding of old leaves. However, in early May there quickly followed a period of high transpiration and large internal water deficits. Hilgeman *et al.* (1969) attributed this change to immature cell structure and high water content of new spring leaves. Another factor could well have been lower stomatal resistances in young leaves than in older ones (Kriedemann, 1971; Albrigo, 1977), once the stomata on the young leaves had become functional. High transpiration with large daily

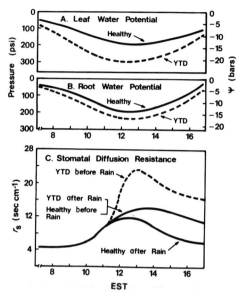

Fig. 6. Citrus roots, like leaves, show a diurnal variation in water potential; overall the patterns are very similar with no perceptible phase differences. Leaf water potential (A) is lowered more than root water potential (B) in trees affected by young tree decline (YTD), implying that YTD mainly restricts water flow in above ground parts of the tree. Consequences of YTD for stomatal resistance (C) can be considerable, especially before rain. (After Allen and Cohen, 1974.)

internal deficits in leaves, fruits, and trunks continued throughout June and July. Although cooler cloudy conditions lowered transpiration in August, internal stresses did not decrease accordingly. Apparently, even fairly low transpiration can be sufficient to generate considerable internal water stress (Hilgeman, 1977). Transpiration increased again on warmer days in September and early October, only to decrease once more with the onset of winter due to low temperatures and low vapor deficits. Initially, this led to reduction of internal water stress as measured by diurnal trunk shrinkage, but as winter advanced and soil temperatures fell, water stress increased again, as shown by both higher leaf saturation deficits and increased trunk shrinkage, even though apparent transpiration was now minimal. A January freeze even caused severe wilting of leaves.

Soil temperature influences water relations of citrus throughout the year. Hilgeman et al. (1969) found that the seasonal curve of estimated apparent leaf transpiration lagged behind the curve of estimated potential evapotranspiration, reflecting the influence of the seasonal rise and fall of soil temperature on leaf transpiration.

3. Hydraulic Resistances and Models

a. Hydraulic Resistance of the Whole Plant. Hydraulic resistance in the citrus plant is high compared with that of herbaceous species. A direct comparison was made by Hoare and Barrs (1974) for rooted cuttings of Washington navel, sweet pepper, and rock melon plants in nutrient solution in a controlled environment. From measurements of leaf ψ, solution osmotic potential, and transpiration rate they calculated the hydraulic resistance in citrus to be 22.4 bars gm^{-1} dm^2 hr, which was significantly greater than in sweet pepper or rock melon, 3.67 and 2.25 bars gm^{-1} dm^2 hr, respectively. Additional evidence comes from the work of Camacho-B. *et al.* (1974b), who found (Fig. 7) that under nonlimiting soil water conditions, leaf ψ of citrus and pear decreased as transpiration rate was increased by increasing the evaporative demand in a controlled environment. The herbaceous species sesame, sweet pepper, and sunflower were

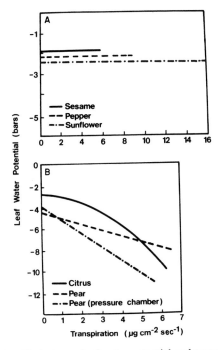

Fig. 7. The relationship between leaf water potential and transpiration differs between herbaceous (A) and woody (B) plants and reflects differences in their hydraulic conductivity. Soil water was nonlimiting, transpiration was varied by changing atmosphere dew point between 7 and 26°C, and air temperature varied between 20 and 33°C. (After Camacho-B. *et al.*, 1974b, and reproduced by permission of the American Society of Plant Physiologists from *Plant Physiol.*, **54**, 169–172.)

not similarly affected. These investigators concluded that the herbaceous species had efficient water transport systems, with a range of transpiration rates maintained by a small and constant ψ gradient. The woody species, on the other hand, had relatively inefficient water transport systems with leaf ψ dropping as transpiration rate increased.

However, some doubt surrounds the precise form of the relationship between leaf ψ and transpiration of citrus. The curve shown in Fig. 7 was obtained with greenhouse-grown seedlings transferred to a controlled environment for study, and implies that resistance to flow first decreased and then increased as transpiration rate increased; but Camacho-B. *et al.* (1974b) found the relationship to be linear for orchard trees, which implies a hyperbolic decline in resistance with increasing transpiration rate (Janes, 1970) over the whole range tested. Some resolution of this difficulty comes from these researchers' additional observation that if seedlings were first exposed to a number of water stress cycles, they responded much more as did orchard trees. Possibly this latter response reflects increased root hydraulic resistance attributable to greater suberization (Ramos and Kaufmann, 1979). Such an effect may be inevitable, but it could also be interpreted (Camacho-B. *et al.*, 1974b) as an adaptation to drier environments since the larger resistance to flow will cause lower leaf ψ and earlier stomatal closure. Nevertheless, even for orchard trees, the exact form of the relationship between leaf ψ and transpiration rate under nonlimiting soil conditions is uncertain since Elfving *et al.* (1972) found a concavely curvilinear relationship, rather than the linear relationship of Camacho-B. *et al.* (1974b). This implies faster decline in hydraulic resistance at higher transpiration rates than would occur if the relationship between leaf ψ and transpiration rate is linear. Camacho-B. *et al.* (1974b) stated that this discrepancy may have been due to differences in calibration curves used to convert pressure chamber readings to leaf ψ's.

In order to elucidate physical and biological principles relating to movement of water through citrus trees, the determination of an unequivocal relationship is desirable, particularly since the curvilinear relationship of Elfving *et al.* has been included in a conceptual model (discussed later). However, this may require some effort since similar discrepancies exist between the results of other studies for herbaceous species (Kaufmann, 1976). Mantell *et al.* (1979) supported the view that transport of water through citrus plants may be less efficient than through herbaceous species. When tritiated water was supplied to roots of sour orange seedlings growing in nutrient solution, the residence time for tritium in the leaves was about 30 hr, compared with published values of only 3–12 hr for a number of herbaceous species. However, as they pointed out, comparison with forest trees is more difficult since in these, published residence times varied widely, from 23 hr to 11 days.

Little unequivocal information is available concerning the effect of decreasing soil ψ on the total hydraulic resistance of citrus trees. For data obtained with plants in soil, e.g., Elfving *et al.* (1972), it is difficult to separate the effects of decreasing soil ψ on hydraulic conductivity of the soil from the effects on the hydraulic conductivity of the plant, although Newman (1969) considered that the former may not be important until ψ in the rhizosphere falls close to the wilting point. Hoare and Barrs (1974) used an alternative approach in which rooted Washington navel cuttings growing in nutrient solution were stressed by successive additions of polyethylene glycol. From the equilibrium rate of transpiration, leaf ψ, and ψ of the solution after each solution change, they calculated that the whole plant hydraulic resistance increased rapidly once the ψ of the solution around the roots had declined below -3 bars. This response contrasted with that of the herbaceous species rock melon and sweet pepper. Rock melon did not show any increase in hydraulic resistance as a result of the treatment (confirming technical validity), and sweet pepper showed only a relatively small and poorly maintained increase in hydraulic resistance once the ψ around the roots had fallen below -8 bars.

Castel (1978) reported a total hydraulic resistance of 116.7×10^9 J kg^{-1} m^{-1} sec^{-1} for rooted Valencia cuttings growing in nutrient solution, using techniques similar to those described by Landsberg *et al.* (1976) for partitioning between leaves, stems, and roots. The values he obtained were 18.3×10^9, 23.5×10^9, and 74.8×10^9 J kg^{-1} m^{-1} sec^{-1}, respectively, and their approximate ratios were 1:1.3:4.1. Castel reported that the total resistance was similar to that found by Hoare and Barrs (1974). Since both total hydraulic resistance (discussed above) and resistance to flow through leaves (Pospisilova, 1972; Boyer, 1974; Landsberg *et al.*, 1976) may vary with transpiration rate, Castel's values for both total and component hydraulic resistances cannot be applied to all situations. Nevertheless they are useful as a general guide.

b. Hydraulic Resistance of Roots. Slatyer (1967) suggested that resistance to flow through roots probably was the major plant hydraulic resistance because water in its passage through the plant has only to flow through cell membranes in the root endodermis. Castel's results given above are in general accord; he found citrus root resistance to be four times as high as leaf resistance, and three times as high as stem resistance. These ratios are similar to those for tomato, sunflower, and soybean (Jensen *et al.*, 1961; Boyer, 1971), although there may be some exceptions since Begg and

Turner (1970) found the greatest hydraulic resistance in tobacco to be in the leaf.

The magnitude of conductance of unstressed citrus roots was determined by Ramos and Kaufmann (1979), on the basis of endodermal surface area, as 1.06×10^{-7} cm bar^{-1} sec^{-1} (Section A,4). This value is similar to those found by Newman (1973) for roots of broad bean, dwarf bean, and sunflower, but about 5 and 14 times smaller than Newman found for maize and tomato, respectively. It is, therefore, not possible on the available evidence to generalize that root hydraulic resistance at least is lower in citrus than in herbaceous species, even though, as emphasized, this resistance is the principal component of citrus hydraulic resistance. However, the present situation is unsatisfactory, since by virtue of the techniques employed, e.g., Newman (1973), root resistance has mainly been determined at low flow rates and there is evidence (Janes, 1970) that hydraulic resistance is then considerably increased.

In view of the dominant role of citrus roots, it may be that the large increase in total hydraulic resistance of citrus in response to simulated soil water stress discussed above is due to increase in root resistance. This would be consistent with the results of Ramos and Kaufmann (1979), but it should be borne in mind that, even though stress was applied to the roots, the accompanying lowering of plant ψ could affect other component hydraulic resistances. For instance, as discussed earlier (Section II,A,3) citrus stems shrink during water stress and this could be associated (Rawlins, 1963; Johnson, 1977) with increased stem resistance. Such possible interactions emphasize difficulties in unravelling the component hydraulic resistances in the integrated and dynamic plant system.

One aspect of the work of Hoare and Barrs (1974) that bears directly on the results of Ramos and Kaufmann (1979) is that the former investigators, in their short-term experiments, did not find any significant after-effect of water stress on total hydraulic resistance. However, in their relatively longer-term experiments Ramos and Kaufmann (1979) did find such an effect on root hydraulic resistance. This apparent discrepancy may aid in interpreting Ramos and Kaufmann's results, since it is more consistent with their suggestion of increased root suberization than with their alternative suggestion of decreased permeability of root cell membranes as the cause of increased root resistance.

c. Hydraulic Resistance of Stems. As discussed in Section II,A,3, there is evidence (De Villiers, 1939) that stem hydraulic resistance varies among citrus species, and data of Castel (1978) suggest that although this is low it is not negligible. However, little is known about the comparative magnitudes of hydraulic resistances in stems of citrus and of other species.

The only information known to the authors that bears on this point is the (1918) observation of Farmer, cited by Heine (1971), that the relative conductivity of the stem of *Choisya ternata*, a relative of citrus, was the third lowest of 29 values obtained for a wide range of species. This contrasts dramatically with the values of Berger, cited by Heine (1971) for sunflower and grape stems, which were about 11 and 40 times higher, respectively.

Further indirect evidence for relatively high resistance to water flow in citrus stems comes from observations (Gale and Poljakoff-Mayber, 1964) of extremely low heat pulse velocities (about 3 to 8 cm hr^{-1}) in sour orange stems. These were similar to, and in fact slightly lower than, the corresponding range (about 3 to 13 cm hr^{-1}) these researchers found for *Pinus halepensis* seedlings. Heat pulse data collected for a range of tree species show that angiosperm forest trees usually have heat pulse velocities well in excess of those of conifers, reflecting differences in stem conductivities (Kramer and Kozlowski, 1960, 1979).

Other evidence for appreciable stem flow resistance in citrus comes from observations of Kaufmann (1975) of higher xylem pressure potentials (by 3 to 5 bars) in shaded than in sunlit citrus foliage, coupled with correspondingly lower transpirational flux densities in the shaded foliage. Kaufmann concluded that significant flow resistances existed after the pathway branched toward the sunlit and shaded portions of the crown. He also found that xylem pressure potentials of shaded foliage of *Picea engelmannii* were 1 to 4 bars less negative than those of sunlit foliage. The low stem conductivity of conifers has been noted above.

d. Hydraulic Resistance of Leaves. Movement of water through citrus leaves has received little attention. Mantell *et al.* (1979) found that when tritiated water was applied to roots of sour orange seedlings in nutrient solution the half-time required by leaf laminae to reach maximum label was about 8 hr, almost double that for the midveins. The possibility that there may be appreciable resistance to water flow through citrus leaves is raised by the observation of Torre and Zabala (1976) that water deficits increase progressively from the leaf base to the tip. Castel's (1978) data showed that leaf resistance was nearly half as great again as stem resistance.

e. Models. As Kaufmann (1977b) pointed out, both empirical and conceptual models have been applied to citrus water relationships. An example of the former is the work of Smart and Barrs (1973) discussed in Section III,B. Elfving *et al.* (1972) successfully applied the conceptual model of Gradmann (1928) and van den Honert (1948), which is essentially an

Ohm's law analog, to citrus water relationships. The basic premise of this model is that steady state flux of water through the soil–plant–atmosphere continuum is proportional to the ψ gradient and inversely proportional to the flow resistances. Phase change from liquid water to water vapor at the surface of the leaf cells complicates the model somewhat, but does not present insuperable difficulties (Rawlins, 1963). The model is essentially flux = $(\psi_{soil} - \psi_{leaf})$/resistance (soil-to-leaf), which can be rearranged to give $\psi_{leaf} = \psi_{soil} -$ (flux) [resistance (soil-to-leaf)]. This form, in which ψ_{leaf} is explicit, is the one used by Elfving *et al.* (1972). Kaufmann (1977b) pointed out that the right-hand side of the equation should include a gravitational head term. However, the absence of this term leads to little error in this instance. Flux was estimated from VPD/leaf diffusive resistance, (r_1), measured with a diffusive porometer. Under nonlimiting edaphic conditions (soil ψ above -0.3 bars and soil temperature 15°C or higher), leaf $\psi(Y)$, was related logarithmically to VPD/r_1, (X), by:

$$Y = -8.28 - 9.23 \log(X + 0.5), \quad (r = -0.92)$$

As discussed above, the curvilinearity in this relationship suggests that the associated hydraulic resistance from soil to leaf was not constant. This relationship held for data collected at various times of day and night and different times of the year. It also held for trees growing in contrasting climatic and edaphic environments, ranging from cool and humid coastal conditions with clay soil, to hot, dry desert conditions with sandy soil.

The establishment of such a relationship provides an elegant method of testing for limiting edaphic conditions, since the latter will cause obvious and quantifiable deviations from this relationship. Thus Elfving *et al.* (1972) showed that leaf ψ departed from the above relationship, becoming increasingly negative as soil temperature declined to 5°C, deviations first becoming apparent at a soil temperature of 13.5°C. Time of day influenced the effect of reduced soil temperature, since low soil temperatures caused large depressions of leaf ψ at midday, and smaller depressions at sunset, but were without effect on the value prevailing at dawn. This is readily understandable in terms of the magnitude of the accompanying flux values, which would have been so small at sunrise that the second term on the right-hand side of the equation would become negligible. These researchers similarly demonstrated that soil ψ's increasingly below -0.3 bars caused increasing depressions of leaf ψ from values predicted by the model. As Elfving *et al.* pointed out, such a model may be used not only to diagnose the presence of undesirable limiting factors in the soil–plant portion of the soil–plant–atmosphere continuum, but also to quantify differences between species or varieties in response to environmental stress.

The soundness of two aspects of the model was tested by Camacho-B. *et al.* (1974a). First, the model assumes that plant capacitance effects are negligible, i.e., there is no diurnal hysteresis in the relationship between leaf ψ and transpiration rate. Otherwise, there would be two leaf ψ values corresponding to the same transpiration rate, or *vice versa* and the model would be invalidated. This assumption was tested by taking measurements of both ψ_{leaf} and VPD/r_l throughout the day. When plotted, the data suggest a small hysteresis loop. However, width of the loop is less than the experimental error. Camacho-B. *et al.* (1974a) concluded that these data substantiated the model assumption that in citrus at least, the dynamic system of water flow in the soil–plant–atmosphere system can be described by a series of steady states.

Second, the validity of estimating transpiration from VPD/r_l was investigated by comparing estimates obtained by this method with direct gravimetric determinations of transpiration rate. Camacho-B. *et al.* (1974a) found that the estimated transpiration values were valid in the field because leaf-to-air temperature differences were rarely as high as 1°C, since sampling was restricted to leaves not exposed to direct solar radiation. However, in a controlled environment, larger leaf-to-air temperature differences were encountered and it was appropriate to replace VPD by vapor pressure gradient, requiring a knowledge of both VPD and leaf temperature.

In view of these validations of the model there appears to be considerable scope for it to be applied on a worldwide basis and to take in a far wider range of citrus species and cultivars than have so far been studied.

C. Physiology of Water Stress

1. Stomatal Responses

Stomata are the adjustable portals through which terrestrial higher plants gain CO_2 and, inevitably, due to the saturation vapor pressure in the intercellular spaces of the leaf (Farquhar and Raschke, 1978), simultaneously lose water vapor. They therefore constitute a critical interface between the inside of the leaf and the plant's aerial environment. This is underscored by the wide range of aerial environmental factors to which stomata respond (light, CO_2, humidity, temperature, wind) plus leaf water status. The latter reflects both the interplay of these aerial factors and the effects of edaphic factors. Complex feed-back (Raschke, 1975) and even feed-forward (Cowan, 1977) loops are involved in control of stomatal aperture. It is, therefore, not surprising that there often are interactions among factors influencing stomatal aperture. In citrus, as discussed below, this is particularly true for responses to temperature and humidity.

This section is principally concerned with effects of water stress on stomatal aperture. However, for the stomata to respond to water stress they must first be open. Citrus stomata open fully at less than full sunlight. However, the response curves obtained (Ehrler and Van Bavel, 1968; Davies and Kozlowski, 1974; Hall *et al.*, 1975) show considerable variation, probably reflecting the use of different species and experimental conditions. Davies and Kozlowski (1974) found that stomatal closure in *Citrus mitis* in response to a reduction in irradiance was faster than in four woody angiosperms and that opening in response to increased irradiance was among the slowest in *Citrus mitis*. Stomatal opening in response to light is indirect, occurring as a result of photosynthetic lowering of CO_2 concentration in the leaf intercellular spaces (Raschke, 1975).

The water stress that illuminated stomata experience arises from an imbalance between water supply and demand. The latter depends on the leaf-to-air vapor pressure gradient (VPG) and is a function of leaf temperature and atmospheric VPD. The former depends on the magnitude of liquid phase supply resistances (Rawlins, 1963), which in turn depend at least partially on soil water status and soil temperature (Section II,A,4). It will be convenient to consider first the effects of aerial factors on citrus leaf resistance.

So long as the leaf-to-air vapor pressure gradient is held constant, leaf temperatures in the range 20 to 41°C appear to have little or no effect on citrus stomatal resistance. Sheriff (1977) found no effect of leaf temperature in the range 30 to 40.5°C on stomatal resistance of *Citrus sinensis* seedlings. Hall *et al.* (1975), also using *C. sinensis* seedlings, found a slight reduction in stomatal resistance over the range 20 to 30°C, and Kaufmann and Levy (1976) found little effect of temperatures up to 37°C and a slight lowering of stomatal resistance of rough lemon seedlings at 41°C. However, since citrus stomatal resistance depends strongly on the leaf-to-air VPG the actual resistance at any particular leaf temperature will depend on the accompanying VPD of the air. Data of Hall *et al.* (1975) obtained in controlled environment studies (Fig. 8) illustrate this point well. The response to increased leaf temperature, a slight fall in leaf resistance, was very similar at low ($3.9 \pm 0.2 \mu g \, cm^{-3}$) humidity and moderate (10.3 ± 0.8 $\mu g \, cm^{-3}$) humidity differences between leaf and air, but the absolute levels were offset by about 1 sec cm^{-1}, the higher resistance occurring at the higher humidity difference. When leaf temperature was raised but absolute humidity external to the leaf was maintained constant (at $18.0 \pm 0.6 \, \mu g \, cm^{-3}$), leaf resistance increased linearly with leaf temperature

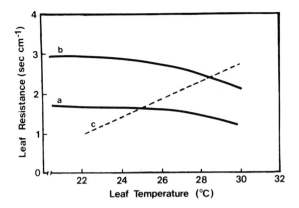

Fig. 8. Stomatal aperture in orange leaves does not decrease with rising temperature provided the vapor pressure gradient from the leaf to the air remains constant; however, apertures are smaller at a moderate than at a low vapor pressure gradient (line b versus line a). Stomatal apertures decrease proportionately if leaf temperature rises but absolute humidity external to the leaf is held constant, since this results in a progressive steepening of the leaf-to-air vapor pressure gradient (line c). (After Hall *et al.*, 1975.)

because the leaf-to-air humidity difference was increased with each increase in leaf temperature. These responses were not influenced by the direction of temperature change.

That the mode of action of increasing leaf temperature in increasing citrus leaf resistance was simply by increasing the leaf to air humidity difference is confirmed in Fig. 9, which shows that this response can be elicited either in this way or by decreasing external absolute humidity at constant leaf temperature (Hall *et al.*, 1975). A second interesting feature of Fig. 9 was the clear tendency for acclimation of leaves to continued daily exposure to a range of VPD's in such a way that the resistance becomes less sensitive to VPD. This acclimation process continued for about 6 days, when the relationship between leaf resistance and VPD became stable. Both these types of responses, i.e., for leaf resistance to increase with VPD and for this effect to become smaller over about 5 days, have been confirmed in controlled environment studies by Khairi and Hall (1976b) for seedlings of two more cultivars of *Citrus sinensis* and for seedlings of *Citrus paradisi*.

In a separate study (1976c), these researchers also showed the first response to occur in *Citropsis gabunensis*. The second (acclimation) response has so far only been reported for greenhouse-grown material subsequently exposed to daily VPD cycles in a controlled environment. It is not known whether orchard trees respond similarly, nor is the method by which this acclimation proceeds understood.

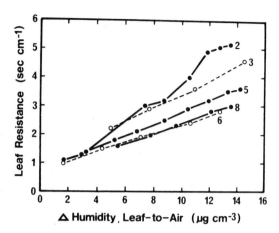

Fig. 9. Stomatal closing response in orange leaves to an increased leaf-to-air vapor pressure gradient is similar whether the gradient is steepened by raising leaf temperature (○) or by decreasing external absolute humidity (●). The numbers beside the lines mark progressive days on which responses to a range of vapor pressure gradients were observed; they show stomatal acclimation in the form of decreasing sensitivity to vapor pressure gradients, which takes 5 days to develop fully. (After Hall *et al.*, 1975.)

Sheriff (1977), working with excised shoots of seedlings of *Citrus sinensis* in a potometer, also found stomatal resistance to increase with increased VPD around the leaf. However, he found that transpiration declined as VPD increased, whereas Hall *et al.* (1975), on the second day of their experiment, found that transpiration rate, after an initial slight rise, remained constant as VPD increased. By the sixth day, due to acclimation, this response had become modified so that transpiration increased somewhat over the whole range of VPD increase tested. Perhaps this discrepancy arose because Sheriff used unacclimated plants that, by extrapolation of the results of Hall *et al.* (1975), could have shown more sensitivity than plants which had acclimated to increasing VPD for 2 days.

Citrus is not exceptional in showing decreased stomatal conductance as VPD increases. Kaufmann (1976) collated data showing that similar responses occur in sesame, sunflower, and *Prunus*. The response occurs also in conifers but is much smaller because initial conductances are very low. In a wide-ranging study of 26 genera and species, Sheriff (1977) recognized four different types of response of stomatal resistance to VPD. Nine other species and genera showed the citrus-type response. A second response was complete absence of an effect of VPD on leaf resistance, e.g., *Clianthus formosus*. The third response, decreased water uptake with decreased stomatal resistance, was typified by *Acacia aneura*. *Pelargonium* and *Eucalyptus* showed the fourth response with both stomatal resistance and water uptake increased with rise in VPD. Sheriff did not find any unifying principle of origin or leaf anatomy among these four responses. Thus the citrus-type response also occurred in perennials found exclusively in arid or semiarid regions and in cereals that normally grow in more mesic climates. Stomatal humidity responses did not appear to depend on leaf venation, stomatal pore depth, or size of substomatal cavities.

Because Sheriff was working with excised tissue, it seems that there must be sufficient hydraulic resistance in citrus leaves to cause water stress to develop within the stomata, resulting in reduced stomatal apertures in low humidities. A further requisite for effectiveness of such a mechanism is that the stomata themselves should transpire, allowing stomatal water content to fall before or at least faster than leaf water content. Lange *et al.* (1971) have proposed that such stomatal water deficits may result from peristomatal transpiration through ectodesmata. However, Edwards and Meidner (1979) suggested that guard cell transpiration proceeds from the surface of the guard cell beneath the pore, causing variations in matric potential of the guard cell wall in this region, which in turn influence stomatal ψ and, therefore, stomatal aperture. Possibly this could be an improvement since it would permit guard cell transpiration to decline as stomatal aperture is reduced, in turn reducing stomatal water deficit and so preventing further stomatal closure, whereas peristomatal transpiration would presumably continue unchecked as stomatal aperture declined.

For a plant such as citrus which has high "supply" resistances (Section II,B,3), stomatal sensitivity to VPD may well play a key role in maintenance of favorable leaf water relations and also explain the capacity of this crop to succeed in arid aerial environments. Both laboratory and field observations are consistent with this view. Thus Hall *et al.* (1975) measured citrus stomatal resistances (including a substantial boundary layer resistance) as low as 1 sec cm^{-1} in a controlled environment when VPD was low, but they found comparatively high values at other times. In well-watered rough lemon orchard trees there was approximate equality of xylem pressure potential and transpiration rate in both humid and dry air, as a result of greater stomatal closure in dry air (Kaufmann, 1977b). The picture that emerges is of citrus as a plant with a potentially high leaf diffusive capacity that is in practice seldom realized due to the strong feedback loop between humidity and the stomata. However, there are some difficulties with this view, because Khairi and Hall (1976c) found that, even under high humidity, stomatal resistance was higher in *Citropsis gabunensis,* a native of humid, tropical Africa than in *Citrus sinensis,* native to subhumid, subtropical Asia, and that stomatal resistances were similar in "*Eremolemon*" and lemon, despite the probable origin of the former as a hybrid between lemon and *Eremocitrus glauca,* native to hot semiarid regions of Australia.

A second stomatal adaptive response of greenhouse-grown plants is that prior exposure to soil water stress results in lower subsequent transpiration rates under well-watered conditions. This was shown by Kaufmann and Levy (1976) for rough lemon seedlings exposed to three cycles of water stress lasting 3 to 8 days, the plants being rewatered when soil ψ had fallen to -0.8 bar, compared with well-watered controls in which soil ψ was never allowed to fall below -0.05 bar. Throughout the subsequent range of VPDs investigated, stomatal conductance was lower by about 0.05 cm sec^{-1} in the prestressed than in the unstressed plants, although both sets of plants still showed the decline in leaf conductances with decreasing VPD already discussed. The consequences for transpiration were considerable at high VPD's since transpiration rate in the prestressed plants was roughly half that of the controls. This led to the initial wilting of unstressed seedlings when exposed to high VPD's, in contrast to the previously stressed plants, which remained turgid. Prior water stress thus led to more effective control of transpiration. However, at low VPD's differences in transpiration rate were far less marked. Prior exposure to high or low VPD had no modifying influence on this response. Again, it is not known if such a response occurs in orchard trees, and the mechanism of the response remains obscure. However, it may be that increased root resistance was a factor, since the stress treatment was similar to the one that Ramos and Kaufmann (1979) subsequently found to increase the root hydraulic resistance of rough lemon seedlings (Section II,A,4). Such increased root resistance could have lowered leaf ψ somewhat, thereby reducing stomatal conductance. Support for this suggestion comes from the observation (Camacho-B. *et al.,* 1974a) of a shift in the relationship between leaf ψ and transpiration in the required direction, after subjecting seedlings, in this case of *Citrus sinensis,* to a series of drying cycles (Section II,B,3).

Whether citrus leaf ψ can act in this way is not certain, since Camacho-B. (1977) stated that no clear influence of leaf ψ on citrus stomatal behavior has been found. Although there is relatively good evidence, reviewed above, for a stomatal feedback loop, based more or less exclusively on the stomatal water relations of citrus, it does seem unlikely that the common feedback loop based on ψ of the whole leaf (Rasche, 1975) should be completely inoperative in citrus. Certainly there is evidence that when citrus leaf ψ falls in response to a decline in ψ around the roots, stomatal resistance does increase, and a causal link seems reasonable. Hoare and Barrs (1974), working with rooted cuttings of Washington navel orange, found clear evidence for onset of stomatal closure when leaf ψ had fallen to about -13 bars, and the ψ around the roots had decreased to about -3 bars. During these experiments the plant aerial environment was held constant. Kaufmann and Levy (1976), who followed the relationship between xylem pressure potential and stomatal conductance with potted rough lemon seedlings through three soil drying cycles, showed that stomatal conductance declined as xylem pressure potential fell; the initial xylem pressure potential for onset of stomatal closure varied somewhat between cycles but was close to the -13 bars of Hoare and Barrs. However, the data of Kaufmann and Levy (1976) do show a considerable hysteresis loop on rewatering, associated with more rapid recovery in xylem pressure potential than in leaf conductance, which tends to blur somewhat the relationship between the two. Even so, results such as those of Allen and Cohen (1974) in which lowered leaf ψ due to increased plant supply resistances is associated with increased stomatal resistance and altered patterns of stomatal opening (Section II,B,2) are most simply interpreted in terms of a causal link between citrus leaf ψ and stomatal conductance. Furthermore, Stanhill (1972) found that canopy resistance of orchard trees was a sensitive function of soil ψ and he suggested that there would be an even closer relationship of canopy resistance with leaf ψ.

2. Photosynthetic Responses

Photosynthetic rates are governed by both stomatal and nonstomatal (mesophyll) resistances to inward diffusion and ultimate fixation of CO_2 in the leaf (Kriedemann, 1971). Because these two resistances may respond differently to environmental stress, the response of net photosynthesis in citrus may differ in some respects from the response, just discussed, of stomatal resistance alone.

Thus the temperature optimum for net photosynthesis in citrus appears to be relatively low, in contrast to the rather indeterminate temperature response of stomatal resistance. Khairi and Hall (1976b), working at constant VPD with two cultivars of *Citrus sinensis* and *C. paradisi* budded to Troyer citrange rootstock, found a temperature optimum for photosynthesis of only 22°C. They attributed this to the low temperature optimum of the mesophyll resistance. Working with rooted cuttings of orange, Kriedemann (1968) found a comparable temperature optimum of about 25°C, provided the air was humid, but in dry air the optimum was reduced to about 15°C. Khairi and Hall (1976b) found similar effects of increased VPD which they attributed to increased stomatal resistance. A marked effect of humidity was reported by Ono et al. (1978) for *Citrus unshiu*, under otherwise optimal conditions; increasing relative humidity from 28 to 86% almost doubled net photosynthesis. Species differences may also be important since Kriedemann (1968) found a somewhat higher temperature optimum of 30°C for rooted rough lemon cuttings in humid air, declining to 20°C in dry air, and Khairi and Hall (1976c) reported that photosynthetic rates in both *"Eremolemon"* and lemon were not reduced until temperatures above 30°C, accompanied with high VPD's, were reached. There was also evidence that these two latter species acclimated to these conditions. The apparent relatively low temperature optimum for photosynthesis in orange is perhaps surprising in view of the subtropical origin of citrus (Section I) and the optimum day temperature of 30°C for vegetative growth of Navel and Valencia orange trees (Reuther, 1973). Such a low temperature optimum for net photosynthesis in orange cannot be accepted as definite, since in a separate study of *Citrus sinensis* raised in a controlled environment under high irradiance, Khairi and Hall (1976c) found that net photosynthesis at 30°C and 8 mbars VPD was at least as high as at 22°C. This was also true for *Citrus gabunensis*. The photosynthetic rates obtained with *Citrus sinensis* in this work were exceptionally high and, according to Khairi and Hall, possibly reflect the growth conditions. The 8°C increase in temperature optimum for net photosynthesis of *Citrus sinensis* compared with their report (1976b) suggests that as in other species the temperature optimum depends on growth conditions. Hence, there appears to be a need for caution in applying the 22°C optimum to orchard trees.

Despite the evergreen habit of citrus, the possibility of a seasonal shift in optimum temperature for photosynthesis has not been widely investigated. However, Ono *et al.* (1978) found an optimum of about 28°C in midsummer for *Citrus unshiu* but much lower values in late spring and autumn. Such an effect seems likely for other species also since Pisek *et al.* (1967) reported cold limits for photosynthesis of −1.3 and −7.0°C in summer and winter, respectively, for *C. limon,* with corresponding tissue freezing temperatures of −3.2 to −4.2 and −5.6 to −7.2°C, respectively.

As well as acclimation of transpiration via modified stomatal behavior discussed above, citrus also shows acclimation of net photosynthesis. This has been detected by Khairi and Hall (1976b) using glasshouse-grown "Campbell Valencia" orange and "Frost Marsh" grapefruit. Acclimation consisted of significant day-by-day increases in both net photosynthesis and leaf conductance. As plants were subjected to the same temperature and humidity treatments on successive days, acclimation was more pronounced at higher temperatures and VPD's. Part of this effect was attributed to the higher light regimen in the controlled environment facility used to measure photosynthesis than in the greenhouse where the plants were grown. This aspect of acclimation involved decreased mesophyll resistance but no change in stomatal resistance, i.e., a strictly non-stomatal response. The stomatal component of acclimation could not be definitely associated with any environmental parameter or experimental procedure. Preliminary experiments suggested that stomatal acclimation to evaporative demand did not occur. This is perhaps surprising in view of the earlier report of Hall *et al.* (1975) showing acclimation of leaf resistance (a reduction) by repeated exposure to increased VPD.

Despite the fact that most citrus plantations are irrigated, comparatively little research has been done on effects of soil water status on citrus photosynthetic rate. Furthermore, such research began about 25 yr after similar studies were commenced on the apple crop. In 1965 Thompson *et al.* studied the effects of soil ψ, relative humidity, and temperature on gas exchange of recently expanded leaves of rough lemon in a controlled environment. Gas exchange declined precipitously when soil ψ fell below -0.5 bar. Further experiments established that the plants were much more sensitive to high temperatures when these were accompanied by low rather than by high humidities and low soil ψ. These experiments were conducted under a low (300 fc) light intensity making extrapolation to the field situation difficult. Kriedemann (1968) also showed that when photosynthesis in rough lemon was reduced by soil water stress, the effect was greater at low than at high humidity.

Diurnal patterns of response to soil water stress have been followed by Bielorai and Mendel (1969) working with greenhouse-grown, potted sour orange and sweet lime seedlings exposed to virtually full daylight, and four levels of soil water stress (-0.2, -1.2, -3.0, and -15 bars). Photosynthetic rates decreased in both species as soil ψ declined. Sweet lime, which had an appreciably higher photosynthetic rate than sour orange at -0.2 bar, showed a much larger reduction of photosynthesis between -0.2 bar and -1.2 bars than sour orange. This trend continued until finally at -15 bars, the rate in sweet orange was somewhat higher than that in sweet lime. Even under high soil ψ both species showed a tendency for a midday decline in net photosynthesis followed by a temporary increase in the afternoon. This basic pattern persisted at lower soil ψ's. Transpiration rates were also reduced as soil ψ fell, but were not as strongly affected as photosynthetic rates. Midday depression of transpiration was absent or less marked than in photosynthesis. Measurements of relative leaf resistances suggested that mesophyll resistance to inward diffusion of CO_2 increased faster than stomatal resistance, and indeed may have initiated the increase in stomatal resistance. This suggests the possibility that stomatal resistance in citrus may be only indirectly coupled with leaf ψ via mesophyll resistance. Results of Wong *et al.* (1979) obtained with a wide range of plant species strengthen this possibility.

Kriedemann (1971) reported a close correlation ($r = +0.914$) between these two resistances in orange and also (1968) obtained evidence for a nonstomatal effect of water stress on citrus photosynthesis when he found that water stress raised the CO_2 compensation value from 65 to between 220 and 350 ppm in lemon leaves.

Working with rooted Washington navel cuttings Hoare and Barrs (1974) found that about half the decline in net photosynthesis due to simulated soil water stress could be attributed to increased leaf mesophyll resistance which increased *pari passu* with stomatal resistance. However, at the most extreme ψ (about -8 bars) there was a tendency for greater sensitivity to water stress in mesophyll resistance than in stomatal resistance. Comparable results were obtained with two herbaceous species, rock melon and sweet pepper, since these also showed nonstomatal effects on net photosynthesis of water stress around the roots, and again about half the decline in photosynthesis could be attributed to increased mesophyll resistance. Orange appeared to be less sensitive to simulated soil water stress than rock melon since any lowering at all of ψ around the roots of rock melon reduced photosynthesis but this did not occur until ψ decreased to about -3 bars in citrus. In this respect sweet pepper responded similarly to orange but was possibly slightly more tolerant of lowered ψ around the roots. Recovery from water stress by orange and pepper the next day was not as complete as in melon, even though the latter showed the greatest sensitivity to stress of the three species. In orange recovery of photosynthesis and transpiration was approximately equal, but in melon and pepper there was greater recovery in photosynthesis than in transpiration. However, in rough lemon Thompson *et al.* (1965) found that net photosynthesis recovered faster than transpiration from water stress. Kriedemann (1968), also working with rough lemon, found substantial recovery of net photosynthesis within 40 min. after watering the plant. Ono *et al.* (1978) found that a soil pF of about two was optimal for net photosynthesis in *Citrus unshiu*. Kamota *et al.* (1974) working with the same species found photosynthesis did not decline until soil pF had fallen to 2.7–3.0 which was similar to the value for grain sorghum but higher than for cucumber, sweet corn, and rice (pF 2.2–2.3) and tomato, eggplant, sweet pepper, taro, and soybean (pF 2.3–2.6). Furthermore, net photosynthesis persisted in *C. unshiu* beyond pF 4.0 although it had fallen to zero in the other species by then. Overall, the available evidence suggests no undue sensitivity of net photosynthesis in citrus to soil water deficit.

3. Water-Use Efficiency (WUE)

The sensitivity of citrus stomata to VPD, discussed above, ensures that stomatal resistance is usually high compared with other crops (Mantell, 1977). In turn this leads to a tendency for WUE to be higher in citrus than in other C_3 crop species because the stomatal and mesophyll resistances in an unstressed leaf are more nearly equal, presenting a relatively large resistance to water loss compared with CO_2 uptake (Gaastra, 1959). Data of Hoare and Barrs (1974) illustrate this point; average stomatal resistance to CO_2 for melon, pepper, and Washington navel orange under the same controlled environment conditions were 1.46, 2.65, and 6.67 sec cm^{-1}, respectively. Corresponding mesophyll resistances were 4.52, 6.33, and 10.64, respectively; stomatal and mesophyll resistances for orange were significantly greater than for melon or pepper. Even though photosynthetic rates were the lowest in citrus, the ratio of stomatal to mesophyll resistance was the highest; thus, WUE expressed as the ratio of net photosynthesis to transpiration was significantly higher in orange (0.0155) than sweet pepper (0.0135) or rock melon (0.0134). Results of Holmgren et al. (1965) show this to be true of some other woody versus herbaceous species comparisons. However, Hoare and Barrs (1974) showed that this advantage of orange was not maintained during simulated soil water stress because the relatively low initial stomatal resistances in the other two species allowed faster increase in stomatal than in mesophyll resistance in these, leading to increased WUE, whereas in citrus the rates of change of the two resistances were more closely matched and WUE remained approximately constant. Bielorai and Mendel (1969) also did not find any increase in WUE of sour orange and sweet lime as soil water stress increased. In pepper and melon as well as in orange, Hoare and Barrs (1974) observed a final drastic fall in WUE at low simulated soil ψ because the mesophyll resistance then increased much faster than the stomatal resistance. This occurred at about the same simulated soil water stress (about -6 bars) in both melon and orange, but not until nearly -10 bars in pepper, suggesting greater drought resistance in the latter in terms of photosynthetic activity at least. Measurements taken the day after water stress had been relieved did not show significant differences in WUE among the three species. This was because mesophyll resistance had recovered further than stomatal resistance in melon and pepper, but not in orange.

These results taken together with those of Bielorai and Mendel (1969) do not support the view (Stanhill, 1972) that WUE will increase in citrus as soil ψ falls, principally because Stanhill assumed, apparently incorrectly, that declining soil ψ does not lead to increased mesophyll resistance in citrus. However, some support for Stanhill's prediction is provided by the work of Kamota et al. (1974) showing a gradual increase in WUE in Satsuma mandarin as soil water supplies declined. Also Kriedemann (1971) observed an increase in WUE during the course of the closing phase of cyclic oscillations in stomatal aperture in orange leaves, but comparison with the effects of soil water stress is not straightforward because ψ rose rather than fell as stomata closed during these cyclic movements. Further work appears desirable to settle this issue.

The effects of elevated temperature and low humidity on citrus WUE were evaluated by Khairi and Hall (1976b). They found that while actual WUE decreased under these conditions, intrinsic WUE, calculated for a constant VPD, increased with increasing VPD, indicating a possible adaptive mechanism for conditions of high evaporative demand. Differences in WUE between species reported by Khairi and Hall (1976c) lend support to this view. WUE was significantly higher in *Citrus sinensis* than in *Citrus gabunensis*, which is to be expected in view of the more humid native habitat of the latter, already noted. Additionally, WUE was higher in *"Eremolemon"* than in *"Lisbon"* lemon, inferred to be due to lower mesophyll resistance in *"Eremolemon"* than in lemon because photosynthetic rates were higher in the former even though stomatal conductances were similar. Again a higher WUE would be expected in *"Eremolemon"* than in lemon because of the probable origin of the former as a hybrid between lemon and *Eremocitrus glauca,* native to hot semiarid regions of Australia. Khairi and Hall (1976c) suggested that, in hot dry climates, high WUE may have a greater adaptive advantage than high rates of photosynthesis and growth for evergreen species. Despite this, acclimation, whereby repeated exposure of plants to a daily range of VPD's led to less sensitivity of stomata and of photosynthesis to higher VPD, did not lead to increased WUE.

On a diurnal basis, citrus WUE may drop near noon. In Satsuma mandarin, WUE was higher in the morning and afternoon than at midday (Kamota *et al.,* 1974). Inspection of diurnal curves of transpiration and photosynthesis measured by Bielorai and Mendel (1969) for sour orange and sweet lime suggests that this was also broadly true for these two species under both high and low soil moisture, since photosynthesis frequently showed a midday depression, but transpiration either did not decrease or showed a much smaller drop.

4. Vegetative Growth

a. Leaves. The most dramatic and obvious response of citrus to water stress is leaf abscission (Spurling, 1951; Mathews, 1972; Marsh, 1973; Kaufmann, 1977a); see also Section II,A,1. However, although abscission can be drastic, some form of adjustment apparently occurs since orchards abandoned for 5 yr may still carry leaves. Such leaf abscission is one process, the other being reduced top growth, by which citrus achieves a more favorable top:root ratio when undergoing water stress (Kaufmann, 1977a).

Other aspects of the response of citrus leaf growth to water stress have received less attention. Levy *et al.* (1978b) suggested that water stress reduces final leaf size in grapefruit. Observations of Monselise (1951c) of higher stomatal densities in sun than in shade leaves of Shamouti orange (Section II,A,1) are consistent with this view. However, there is little quantitative information about the relationship between citrus leaf water status and leaf growth. Maotani *et al.* (1977) found that growth of the fruit and trunk of Satsuma mandarin was more sensitive to leaf water stress than was leaf growth itself. Fruit growth ceased when dawn leaf ψ fell to about -8 bars but leaf growth persisted until a value of about -15 to -20 bars had been reached. Leaf initiation may even stop completely when periods between irrigations are long, followed by bursts of shoot growth when irrigation is resumed (Levy *et al.*, 1978b). Water stress during growth may also affect citrus leaf morphology, with leaves typically thick and leathery rather than succulent (Kaufmann, 1977a).

As a consequence of these responses canopy growth is likewise reduced by water stress. In a 14-year trial with grapefruit, for instance, Levy *et al.* (1978b) observed 12 and 29% reductions in canopy growth for plots irrigated every 30 or 40 days, respectively, compared with control plots irrigated every 14 to 18 days. Samish (1957) found that during 6 years of irrigation when the available soil water in the second foot fell to 29 or 13%, crown growth of grapefruit was reduced by 11 and 37%, respectively, compared with trees irrigated at 45% soil water. Similar responses were obtained for Valencia orange tree canopies (Hilgeman and Sharp, 1970; Hilgeman, 1977). Levy *et al.* (1978b) showed close correlation between final yield and canopy volume. However, Hilgeman (1956) and Hilgeman and Sharp (1970) found the optimal irrigation treatment to be a combination of high (March to July) and low (August to February) regimes. This resulted in the development of relatively small and compact but high-yielding trees. There was no advantage in maintaining a high soil water regimen throughout the year, since even though larger canopies resulted, there was no commensurate increase in yield. Mantell *et al.* (1976) imposed a similar high spring/low summer irrigation regimen with essentially the same results for Shamouti orange.

b. Stems. Like canopy growth, citrus stem growth is adversely affected by water shortage. Continued water stress can lead to twig dieback (Marsh, 1973) and, if sufficiently severe and prolonged, to death of branches and of the whole tree; fruit and leaves are normally affected first in this process (Mathews, 1972). Severe but temporary stress may induce false annual rings (Hilgeman, 1977). When stress is less severe, for instance, under inadequate irrigation regimes, trunk growth is generally proportionately reduced as less water is applied (Hilgeman, 1951; Hilgeman and Sharp, 1970). Reduced trunk growth may be the first measurable response of citrus to reduced soil water availability (Hilgeman, 1977), although in the short term this may be only a reversible shrinkage. Reductions of 9% in trunk circumference after about 20 years on inadequate irrigation were reported by Cahoon *et al.* (1961) for oranges and 1 to 25% in trunk cross sectional area for grapefruit after 6 (Samish, 1957), or 14 yr (Levy *et al.*, 1978b). However, it should be kept in mind that, as noted above for canopies, irrigation regimes that optimize fruit production may be suboptimal for trunk growth.

Trunk growth sometimes is not reduced under low levels of irrigation. This occurs in years when fruit yields are reduced even in plots receiving adequate irrigation (Hilgeman, 1974). Such response points to competition for photosynthates between trunk and fruit, which results in annual variations in trunk growth usually being inversely related to fruit yield, although in certain years growth may be modified by climate. An inverse relationship between trunk growth and fruit yield may also be apparent on a longer term basis. However, it is possible that, as trees age, a larger portion of the available photosynthate is diverted from vegetative growth to fruiting (Hilgeman and Sharp, 1970). There is also evidence that prior water stress may have a delayed retarding effect on trunk growth (Halma, 1935; Samish, 1957).

Using a dendrograph, Hilgeman (1963) showed that daily growth of Valencia orange stems in Arizona was reduced and afternoon shrinkage increased as soil ψ fell to -0.65 bars at 18 in. depth between irrigations in May and early June. However, during the rest of the year, climatic and/or physiological conditions appeared to exert greater effects than soil water content within the range studied. The lowest seasonal growth occurred in late June at temperatures of 42–44°C, with high VPD. Maximum growth occurred in September, when evaporation was markedly lower than at any earlier period during the growing season. High daily shrinkage occurred in October with lower temperatures but with evaporation similar to that in September. Neither trunk growth nor afternoon shrinkage bore a simple relationship with soil water status or climatic conditions; both depend on a complex of factors varying throughout the season.

Hilgeman (1963) showed that a very light shower (0.02 in.) at noon or even cloud stopped trunk shrinkage temporarily. Abrupt reduction in transpiration caused by rain appeared to induce greater hydration of the tree than did flood irrigation. The latter could result from an intrinsically high root hydraulic resistance (Section II,B,3) exacerbated by temporary anaerobiosis (Kaufmann, 1977b).

c. Roots. Although the anatomy of citrus roots has been well documented, (see Section II,A,4), little is known in quantitative terms about the effects of water stress on their growth, doubtless because of the technical difficulties involved in obtaining such information. Efforts to study citrus roots directly *in situ* seem to have been limited to the work of Waynick and Walker (1930), who constructed root observation pits in citrus groves, and Monselise (1947), who relocated attached roots into bottles filled with soil and subsequently buried in the ground to facilitate recovery and measurement of the roots. Waynick and Walker stated that dry soil severely limited root growth, a view endorsed by Marsh (1973) but in neither case supported by quantitative data. Monselise (1947) concluded that a soil ψ of about -7.5 to -8.0 bars seriously hampered root growth, and that root growth ceased when soil ψ and root osmotic potential were equal. His conclusion is supported by Cossman's (1940) observation that the roots of sweet lime stock which Monselise used, had osmotic potentials varying between -6.9 and -9.6 atm in well-watered and dry soils, respectively. Further work in this area is needed.

Soil water stress may lead to increased depth of rooting (Marsh, 1973). Such an effect was inferred by Samish (1957) from observations of increased depth of water extraction by lightly irrigated grapefruit trees, although he considered response to result from improved aeration of the heavy clay soil he was working with rather than from direct effects of soil water deficit. Such responses may be limited by effects on root growth of lower soil temperatures at greater depths (Reuther, 1973). Even when rooting depth is not increased by water stress the proportion of feeder roots of orange may be increased (Hilgeman and Sharp, 1970; Cahoon *et al.*, 1961, 1964b). This may be accompanied by an apparent increase (Cahoon *et al.*, 1961) or decrease (Hilgeman and Sharp, 1970) in total weight of feeder roots, but in both cases the net effect of water stress is reduction in the top-to-root ratio, due to large decreases in top growth which occur at the same time.

Finally, if some of the roots of citrus remain in soil containing easily available water, then other roots in dry soil are not damaged (Marsh, 1973). Presumably this is achieved by transfer of water from the wet root to the dry root, as has been reported in ash trees (Slavikova, 1967).

5. Reproductive Growth

The first step in reproductive growth is initiation of the flower bud. According to Magness (1953), fruit bud initiation is generally increased in fruit trees if water shortage occurs during the period in which initiation can be effected. This appears to apply to citrus, since a period of water stress favored subsequent flowering in the Philippines, and such periods of stress are deliberately imposed to induce flowering of lemons at commercially desirable times in Sicily and Israel (Monselise and Halevy, 1964). Several investigators emphasized that blooming can occur at any time of the year if a dry period is followed by heavy irrigation (Erickson, 1968; Cassin *et al.*, 1969; van der Weert *et al.*, 1973).

Flower bud initiation is apparently mediated by changes in hormone balance. Magness (1953) suggested that formation of flower-inducing hormones resulted from accumulation of sugars in water-stressed trees. Consistent with this view are the findings that carbohydrate contents were increased in leaves of orange trees receiving a high/low irrigation regimen, and that fruit production was maintained at a high level in these trees despite the reduced winter period of irrigation (Hilgeman and Sharp, 1970). A more recent alternative view (Nir *et al.*, 1972) is that the period of water stress serves to reduce the gibberellic acid level in the plant and that this is an essential precondition for flower initiation. Supporting evidence for this view comes from their observations that CCC, generally regarded as an antigibberellin, can replace water stress and induce flower formation in *"Eureka"* lemon trees, but that exogenous gibberellin can inhibit flower formation. The mechanism by which water stress reduces gibberellic acid levels may simply be via inhibition of root growth. These researchers point out that root pruning, which decreases water absorption, induced flower bud formation in citrus (Iwasaki *et al.*, 1959). Finally Nir *et al.* suggested that there is a requirement for gibberellic acid levels to rise once more to induce growth of generative branches, in general, and pedicels of flowers, in particular. This is achieved by resumption of root growth upon subsequent rainfall or irrigation.

Citrus is unusual since, unlike most deciduous fruit crops, it is self-thinning (Jones and Cree, 1965). Many flower buds and flowers drop before fruit is set as a result of weakening of tissue in a preformed abscission zone at the point of attachment of the base of the ovary to the disc or of the pedicel to the twig (Erickson, 1968). Water stress at this time is most undesirable since it leads (Magness, 1953; Shmueli *et al.*, 1973; van der Weert *et al.*, 1973; Shalhevet *et al.*, 1976) to increased flower and young fruit drop ("June" drop in the northern hemisphere, "December" drop in the southern hemisphere), which if sufficiently severe can lead to complete crop failure.

The crucial role of a favorable tree water balance in assuring good fruit set was shown by Costa (1979). Either misting or applying an anti-transpirant spray shortly after bloom of navel orange increased both percentage of fruit retained and leaf ψ. Timing of the application required care since late application at full expansion of spring flush leaves caused fruit drop. The critical plant ψ for satisfactory fruit set may vary among species; Costa found greater fruit set in Satsuma mandarin than in navel orange, even though leaf ψ was higher in the latter.

Increase in citrus fruit size is principally achieved during the second phase of growth, about 70% of total fruit volume growth occurring at that time (Hilgeman *et al.*, 1959). Hence, while water stress should be avoided at blooming and fruit set, it is still undesirable at this second stage, since rates of fruit expansion and final fruit size may otherwise be reduced (Marloth, 1950; Hilgeman *et al.*, 1959; Shalhevet *et al.*, 1976). Thus, Hilgeman (1974) found that fruit on trees irrigated on both sides grew at an average rate of 0.70 cm^3 day^{-1} compared with 0.61 cm^3 day^{-1} for trees irrigated on alternate sides only. Beutel (1964) observed that when soil ψ fell to -0.45 to -0.65 bars and -0.65 to -0.80 bars, respectively, fruit growth rates of navel oranges and lemon were about 60% and 20% as much as when soil ψ was in the range of -0.10 to -0.40 bars. Furthermore, fruits grew fastest immediately after an irrigation and growth slowed just before irrigation. Irrigation itself stimulated fruit growth, since growth in the next 3 days was about 30% greater than in the subsequent 3-day period, even though tensiometers read about -0.10 bars during both periods. Under ideal conditions oranges increased in circumference about 0.031 in. per day, lemons about 0.022 in. per day. Waynick (1928) found that fruit volume was reduced nearly 30% below the average under a low-irrigation regimen and pointed out that this could represent 100% difference in financial return.

On the basis of the foregoing it is not surprising that increased irrigation generally leads to increased fruit size. However, a possible complicating factor is that of crop size, since there is a tendency, presumably because of increased competition between fruits for photosynthates, for heavier crops to be accompanied by smaller fruits (Marloth, 1950), although this may not be the sole cause (Waynick, 1929). Whereas high irrigation rates, unless excessive (Marsh, 1973; Shmueli *et al.*, 1973), lead to increased yields (Samish, 1957; Bielorai and Levy, 1971; Bielorai *et al.*, 1973; Levy *et al.*, 1978b), this is not normally realized in practice where fruit size is simultaneously reduced. Nevertheless, it is possible for a low irrigation rate to produce larger fruit than a moderate irrigation rate, as was found by Hilgeman (1951) for Valencia oranges. This occurred because the drier regimen led initially to greater fruit drop, leaving fewer fruits that were, therefore, able to grow longer. Even so, growth of these fruits was not greater than that of fruit from heavier-yielding, well-watered trees.

A heavy crop also leads to competition with other organs of the plant for photosynthate and may lead to both reduced vegetative development (Lenz, 1967) and a drastic (five-fold) reduction in root starch content, which may in turn lead to reduced root activity and increased water stress in the tree crown. Jones *et al.* (1975) suggested that such an effect may be associated with reduced yield in the following year, and hence is one predisposing factor in biennial bearing (Jones and Cree, 1965; Moss and Muirhead, 1971a).

While yields are generally reduced by soil water stress, few attempts have been made to quantify this relationship. Bielorai and Levy (1971) found that grapefruit yields were significantly reduced when the number of drought days (days when soil ψ was below one bar) exceeded 60. When the number of such days did not exceed 60 and soil ψ did not fall below -4 to -5 bars, yields were within 10% of the maximum. Maotani et $al.$ (1977) noted that Satsuma fruits stopped growing when dawn leaf ψ dropped to -8 bars, contrasting with leaf growth that persisted until -15 to -20 bars. Minimum leaf ψ and maximum leaf diffusive resistance measured between 1 and 2 P.M. were also related to fruit growth, which stopped when these reached -17 bars and -16 sec cm^{-1}, respectively.

Although soil water stress is the principal factor affecting fruit growth rates, others such as soil temperature and atmospheric humidity also play a role; diurnal aspects of these have been discussed (Section II,B,2). One aspect not previously covered is the interaction between soil water status and maximum air temperature (Fig. 10). When soil ψ was in the range of -0.10 to -0.40 bars optimum temperature for growth of fruit of navel orange and lemon was about 97°F, but when soil ψ fell to -0.50 and -0.80 bars the optimum temperature declined to about 92°F in lemon and 87°F in orange. Possibly this 5°F difference reflects a lower drought tolerance in orange than in lemon. The response described by Beutel (1964) probably includes effects of increased VPD as well as response to temperature since his observations were taken in the field.

Citrus fruits typically grow most at night (Spurling, 1951; Reuther, 1973) when water deficits are low (Section II,B,2). However, Kaufmann (1970b) pointed out that the fruit itself may tend to counter such expansion, since he found pericarp strength to be greatest at low temperatures and high fruit ψ's, which occur at night. Commercial significance of this apparently has not been investigated.

Citrus production must be considered in terms of quality as well as quantity. Where citrus is produced under irrigation, these objectives may to some extent be disparate since higher quality fruit is usually produced under an irrigation regimen that is somewhat suboptimal for yield. Indeed, certain disorders are directly attributable to excessive turgor in the fruit. Thus Cahoon et $al.$ (1964a) found that oleocellosis occurred in turgid desert lemons which had an immature rind and so could not withstand normal picking and handling. This condition is the result of an interaction between soil suction and VPD levels (Fig. 11). Control is effected by picking only when Fig. 11 predicts that the combined levels of these two factors will not result in an excessive spotting index. As a further check

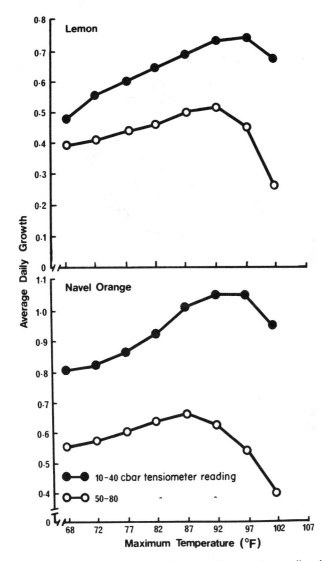

Fig. 10. Temperature optimum for citrus fruit growth depends on soil moisture status, decreasing as soil dries. This effect is more marked in navel orange than in lemon. Average daily growth is in $\frac{1}{32}$-in. circumference. (After Beutel, 1964.)

the pressure necessary to release oil from the rind can be measured and referred to a calibration curve relating this pressure to the spotting index. This rind oil rupture pressure is particularly convenient because in one measurement it combines the effects of both VPD and soil ψ on fruit

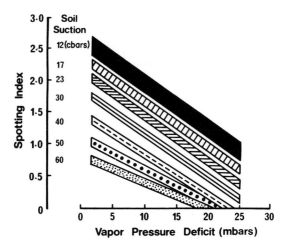

Fig. 11. Desert lemons can show oleocellosis (breakdown of epidermal cells, measured by increased spotting index) at high fruit turgor, due to the combined influence of high humidity and abundant soil moisture. Potential for damage diminishes as turgor is reduced by increasing vapor pressure deficit or decreased soil moisture availability. Index of 3 corresponds to 6 or more necrotic spots per fruit. (Reproduced by permission from Cahoon *et al.*, 1964a.)

turgor. Similarly, Da Cunha *et al.* (1978) showed that stylar-end breakdown in "Tahiti" lime was quantitatively related to fruit ψ, and also to rind oil release pressure. Incidence of this breakdown was about 40% in the most turgid fruit ($\psi \gtrsim 4$ bars, rind oil release pressure <2.0 kg) and declined linearly with declining fruit ψ and increasing rind oil release pressure until a minimum susceptibility of 1–2% was reached at $\lesssim -11$ bars and $\gtrsim 4.5$ kg. Deliberately reduced fruit turgor was seen as one approach to reducing stylar-end breakdown.

High irrigation regimes may also favor a number of skin blemishes, including wind scar and scab (Koo and Reese, 1977). However, these effects were not considered to be important in commercial production of citrus. Furthermore, these researchers found that irrigation reduced the incidence of plugging and rumple in lemons. Similarly, Raciti and Scuderi (1971) found that irrigation reduced creasing in oranges.

Other aspects of citrus fruit quality are more definitely affected by irrigation regimes. By judiciously irrigating alternate sides of Valencia orange trees (980 mm yr^{-1}), compared with wetting all sides of the trees (1720 mm yr^{-1}), Hilgeman (1974) avoided significant yield reduction but improved fruit quality in terms of both percentage of total soluble solids and percentage of total acid. Later, Hilgeman (1977) showed that a similar effect could be obtained by conventional irrigation (i.e., irrigation on all sides of trees) in which late summer water stress was induced by combining a high March-to-July regime with a low August-to-February regimen. The advantages of this regime were again significant increases in percentage of total soluble solids and percentage of total acids. Disadvantages were a decrease in fruit size and increase in peel thickness, which was principally due to stress in October. However, stress in this month also accounted for 69% of the increase in total soluble solids. Furthermore, the high/low regime, by providing fewer irrigations, increased WUE because of reduced evaporative losses during and immediately following each irrigation. Similar results were found for Shamouti orange (Mantell *et al.*, 1976). Reduced summer irrigations resulted in increased numbers of smaller fruits per tree, with thicker peel and juice having higher levels of sugar, citric acid, and vitamin C, together with a higher total soluble solids-to-acid ratio.

Despite these findings, improving fruit quality by reducing irrigation requires caution since too-low irrigation regimens depress both fruit yield and quality. Samish (1957) found that his lowest irrigation regimen (when available soil water in the second foot fell to 13%) adversely affected fruit quality in grapefruit by a reduction in size, increase in rind thickness, and changed taste due to a lower sugar–acid ratio which caused an appearance of immaturity, even though the actual amounts of both sugar and acid in the juice were increased. Shalhevet *et al.* (1976) reported that low irrigation regimes reduced fruit size, peel thickness, and percentage of juice in grapefruit and Shamouti orange. However, no apparent differences were found in fruit quality on trees receiving the optimal regime or wetter treatments. Although orange fruit quality can be improved by reducing summer irrigation as discussed above, this approach requires caution with other citrus fruits. Levy *et al.* (1978a) found that summer water stress in grapefruit trees caused high titratable acidity levels in the fruit in the following winter, making them less palatable. Such aftereffects of water stress were apparent for up to 6 months.

Climate and location of fruit on the tree affect fruit quality, again probably via tree or fruit water status. Fruits grown in moister climates in the United States had thinner rinds and more juice than those in drier climates (Cooper *et al.*, 1963). However, regional differences in soluble solids and acid content apparent earlier, had largely disappeared by the time fruit was harvested.

Sites and Reitz (1949) found that percentage of soluble solids in mature fruit was highest for more exposed fruit than fruit near the top of the tree; fruit with the lowest percentage of soluble solids were "inside" the tree; other fruits were intermediate. They concluded that intensity and duration of irradiance received by the fruit were the principal factors affecting percentage of soluble solids. Albrigo (1977) attributed these effects to ψ differences between fruits due to canopy microclimate differences and height of the fruit in the tree.

III. ORCHARDS

A. MACROCLIMATE AND GLOBAL DISTRIBUTION

Major commercial citrus areas are located on the edges of the tropics, i.e., between 20° and 40° N and S latitudes (Reuther, 1977). Temperatures tolerated range from 10 to 35°C, with an optimum between 20 and 30°C. Such conditions typify Mediterranean type climates with wet winters and dry summers and are compatible with the cultural requirements of citrus orchards (Bielorai et al., 1973). Nevertheless, trees obviously possess some drought- and heat-resisting characteristics, as documented around 330 B.C. by Europeans who visited Iran (see Chapot, 1975). In that region trees endure maximum temperatures above 40°C for at least 120 days each year, plus several days over 50°C. Sunburn can occur on fruits and branches, although grapefruit trees receive some protection via their pendant canopy (Reuther, 1973).

Climate, especially thermal conditions plus water supply, has a strong influence on tree vigor and fruit quality. Species and cultivar distinctions around the world with respect to environmental preferences were summarized by Nakagawa (1969) in terms of crop response to "heat units" (sum of daily mean temperatures above 10°C during the growing season). Satsuma orange required 2400–3000 such units and failed to mature if the value was below 1900–2200 units. By contrast, grapefruit performed satisfactorily above 4200–4500, whereas Valencia orange showed a wide tolerance and occupied regions conferring between 2100 and 5100 such units. Although Washington navel oranges are also widely distributed, their apparent preference is for Mediterranean type climates with heat unit accumulations between 2500 and 3000.

Temperature and evaporation conditions have additional and more subtle effects on fruit growth. Figure 12 provides comparative data for grapefruit and Washington navel orange growing under high temperature and evaporative demand at Riverside, California (United States), compared with uniformly mild conditions in the small Andean Valley at Medellín in Colombia. At Riverside, the seasonal range in temperature is 13–36°C with precipitation 276 mm; by contrast, Medellin provides a mean annual temperature of 21.4°C with diurnal fluctuation of only 11–14°C. Climatic effects on fruit growth (Fig. 12) are self-evident. Notwithstanding

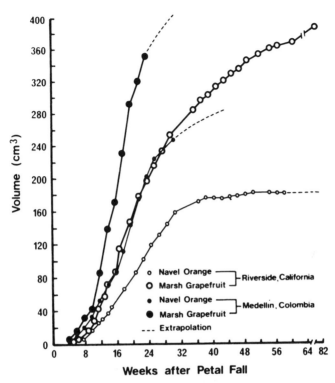

Fig. 12. Milder climatic conditions, and especially lower evaporative demand in Colombia, promote faster growth of both Marsh Grapefruit and Navel orange compared with southern California. (Reproduced from Reuther and Rios-Castano, 1969.)

irrigation, evaporative demand at Riverside is obviously much greater with clear consequences for fruit growth. Since fruit enlargement is turgor-dependent, it is most likely mediated via tree/fruit water relations.

B. MICROCLIMATE AND WATER USE

Physical principles underlying WUE in citrus orchards center on transfer of radiant energy, momentum, latent heat, and sensible heat. Additional factors such as tree density, canopy architecture, foliar characteristics, moisture supply, and soil type then impinge on these physical parameters in determining the orchard's heat budget.

1. Aerial Factors

Reception and dissipation of radiant energy are of central significance, with net radiation accounting for almost three-quarters of the energy utilized for evapotranspiration; reflection from unshaded soil accounts for the remainder (Ben-Asher and Sammis, 1978).

Irrigated citrus trees are typified by dense canopies with low albedo, but high "optical density"—i.e., the canopy is a highly effective trap for radiant energy. In terms of reflectivity of photosynthetically active radiation, Stanhill *et al.* (1966) cited an average albedo for 30-yr-old irrigated citrus with 80% ground cover of 0.17—a value lower than the 0.26 suggested by Monteith (1959) as a mean for agricultural crops showing complete ground cover, and much lower than the 0.37 recorded by Stanhill *et al.* (1966) for desert vegetation.

Heavy attenuation of shortwave radiation within the canopy has been detailed by Kalma and Stanhill (1972), who provide mathematical models of microclimatic relationships. Mean annual values of net radiation flux density took the form $Q_* = 0.80 \ K\downarrow - 130$ cal cm^{-2}day^{-1} ($K\downarrow$ refers to incident global radiation). As established by Greene and Gerber's (1967) earlier measurements in orange trees using a paired net radiometer, the outermost 0.3 to 1.0 m of the canopy shaded inner regions to such an extent that less than 1% of $K\downarrow$ penetrated beyond 1.3 m. Over 90% of the available light was intercepted by the outermost meter of foliage, and most of it was absorbed within the first 0.3 m (see also Figs. 2 and 5 in Kalma and Stanhill, 1972). Marsh seedless grapefruit show even stronger attenuation of sunlight and, as reported by Monselise (1951a), interior regions received less than 1% outside radiation. Paradoxically, such absorption appears to exceed the tree's photosynthetic requirements, especially under exposed conditions in semiarid environments. Monselise (1951b), for example, demonstrated morphological adaptation in sweet lime seedlings to 30% reduction in solar radiation in Israel that increased relative growth rate from 1.523 gr/100 gr/day under exposed conditions, to 2.055 gr/100 gr/day under moderate shade. Although full sunlight provides additional energy which is then dissipated in faster evapotranspiration without commensurate effects on photosynthesis, reproductive development is nevertheless enhanced (Shmueli *et al.*, 1973). Improved yield, even at the expense of vegetative growth and water use, is then an acceptable "trade-off."

Despite relatively low albedo and strong absorptivity, evapotranspiration from highly productive irrigated citrus under semiarid conditions is constrained by "tree" factors. For example, Stanhill (1972) outlined the case for a 35-yr-old irrigated Shamouti orange orchard with 4.5 m high trees on a 4 × 4 m spacing that covered 90% of the soil surface (Leaf Area Index = 7). Using Penman's heat balance and aerodynamic equation in the form

$$E = \frac{\Delta \ Q_* + \gamma \ E_a}{\Delta + \gamma/R_a/(R_a + R_s)}$$

annual *potential* evapotranspiration ($R_s = 0$) was 344.4 cm, where *actual* evapotranspiration ($R_s = 1.62$) was 84.1 cm yr^{-1}. For comparative interest, class-A pan evaporation, directly measured, was 156.9 cm yr^{-1}.*

Clearly, the biological resistances embodied in R_s can exert significant control over water loss and are largely attributable to canopy factors (80%), with bare soil resistance under tree contributing a further 20% (Stanhill, 1972). R_s therefore reflects changes in the tree's stomatal resistance or hydraulic conductivity, which are in turn linked to soil/root factors and evaporative conditions. Accordingly, Stanhill (1972) demonstrated an increase in canopy resistance as soil moisture diminished with R_s approaching an asymptotic value of *ca.* 2 sec cm^{-1} as ψ_{soil} in the upper 80 cm of the profile fell below -6 bars. In that instance, R_s depended on both soil moisture status and radiant load. Prolonged exposure to full sun also induced partial stomatal closure, and hence an increase in R_s (see earlier discussion in Section II,C,1). Further observations on biological resistance by Haseba and Takechi (1966) demonstrated an asymmetric change in their physiological factor (ϵ) during the forenoon compared with the afternoon on clear days, but not on cloudy days. Leaf resistance was the component responsible and values ranged between 5 and 18 sec cm^{-1}.

Since canopy resistance (R_s) is considerable and Kalma and Fuchs (1976) implied that internal tree resistance is an order of magnitude greater than atmospheric resistance, R_s will suppress actual evapotranspiration well below potential values. Irrigation requirements therefore hinge on this factor regardless of whether watering schedules are based on heat balance equations or are derived empirically from pan evaporation. Taking the first approach, Van Bavel *et al.* (1967) provided an expression for potential evaporation (E_o) in the form of a modified Penman equation. Potential evaporation (E_o) was computed from daily values of net radiation, sensible air temperature, vapor pressure deficit, and daily wind run.

Values for E_o derived in this way under arid conditions of Arizona are shown as seasonal patterns in Fig. 13. Due to substantial leaf canopy resistance (R_s in Fig. 13), actual rates of evapotranspiration from the orange orchard (E in Fig. 13) are reduced to about 50% of potential values.

Meteorologically obtained values for R_s during the summer range between 3.5 and 4.5 sec cm^{-1}, whereas direct estimates lie between 2.5 and 10.0 sec cm^{-1}. In this same context, Van Bavel *et al.* (1967) emphasized that seasonal variation in R_s (Fig. 13) is not fortuitous but ap-

---

*E, evaporation rate, is expressed in units of water depth evaporated per unit of ground surface per unit time; Q_* is the radiation balance of the evaporating surface expressed in similar units. E_a represents the drying power of the atmosphere, Δ is the slope of the saturation vapor pressure curve at mean air temperature, γ is the psychrometric constant. Gaseous diffusive resistance to water vapor is represented by R_a (external) and R_s (tree canopy).

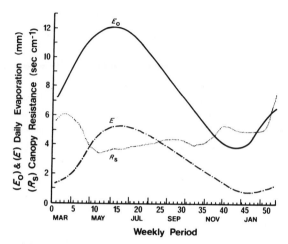

Fig. 13. Potential evaporation (E_0) derived from a modified Penman equation is substantially greater than actual rates of evapotranspiration (E) due to substantial resistance offered by trees (R_s) to water movement. (Reproduced by permission from Van Bavel *et al.*, 1967.)

pears related to canopy leaf age. The rise in late autumn is associated with abscission of old leaves, which continues through winter and reaches a peak around April and tapers off through late spring and summer. Leaf canopy effects on R_s may well be exacerbated during winter and early spring by greater root resistance to water uptake—a temperature-dependent process as discussed earlier (Section II,A,4).

2. Edaphic Factors

While evapotranspiration is coupled directly to aerial microclimate, and especially to net radiation, soil moisture held within the root zone obviously confers some buffering capacity. Consequently, water loss from an orchard must be matched by irrigation which varies in level and frequency according to both aerial and edaphic factors. Apart from its role in buffering the tree's moisture status against wide fluctuations in evaporative demand, the soil acts as a sink for deep percolation as well as an additional source of evaporative loss and radiant energy.

Kalma and Stanhill (1969) estimated evaporative losses from bare soil to account for 26% of annual evapotranspiration from a mature orange plantation in the central coastal plain of Israel. By comparison, annual loss to deep drainage from the upper 2.8 m of soil amounted to 20% of total water income. Above ground losses had been estimated by the vertical energy balance method based on the equation

$$E = (Q + S)/[L (1 + \beta)]$$

where E is evaporation from bare soil (cm^{-2} min^{-1}), Q is the radiation balance (g cal cm^{-2} min.$^{-1}$), S is the soil heat flux (g cal cm^{-2} min.$^{-1}$), L is the latent heat of vaporization at mean soil temperature (gr cal cm^{-3}) and β is the dimensionless Bowen ratio derived from the relationship $\beta = (\Delta t/\Delta e)\gamma$, where Δt is temperature gradient in °C, Δe is vapor pressure gradient in mbars, and γ is the psychrometer constant $= 0.61$ mbar °C^{-1}.

Calculated in this way, evaporation losses were linearly related to moisture content of the soil's surface layer. Kalma and Stanhill (1969) therefore advocated irrigation practice that would keep the upper soil layer dry, provided trees were not adversely affected. Ben-Asher and Sammis (1978) subsequently addressed this question because of their concern that in trickle irrigated orchards, transpirational loss from relatively isolated trees in bare soil may be enhanced by small-scale advective energy from unshaded areas between rows. If so, the desired advantage in water use would be offset by greater radiant load on individual trees, i.e., a micro-oasis effect.

Accordingly, Ben-Asher and Sammis (1978) analyzed orchard energy balance over wetted versus dry areas within a 300 ha lemon grove planted to a 7 × 7 m grid on sandy soil near Yuma, Arizona. Irrigation was supplied daily from trickle sources (ca. 320 liter/tree^{-1}) while global, reflected, and net radiation were measured simultaneously above sunlit soil and the wetted complex (tree plus moist ground). Soil heat flux, Bowen ratio, and albedo were determined, and the energy contribution from dry to wet components of the orchard was measured directly with a net radiometer placed in the tree and held normal to the beam of energy reflected from the surrounding dry soil (i.e., the calculated view angle).

Net radiation accounted for 70% of the energy dissipated by evapotranspiration whereas energy reflected from unshaded soil contributed the remaining 30%. Despite this substantial latter component, evapotranspiration from the orchard and its ratio to class-A pan evaporation were smaller than previously recorded, i.e., 1.10 in spring compared with 0.53 in summer. Clearly, the transfer of energy from dry to wet areas via reflection, reradiation, and advective processes, despite the attendant increase in evapotranspiration, does not carry a specific disadvantage as far as trickle irrigation is concerned.

Although soil type does not appear to limit transfer of relationships between evapotranspiration and microclimate from one region to another (Shalhevet and Bielorai, 1978), water lost from the profile through deep percolation must be debited against input when constructing a water balance equation. In keeping with a shallow-rooted tree of tropical origin, at least two-thirds of an orchard's transpirational supply is derived from the top 25–30 cm of the profile (West and Perkman, 1953), so that percolation

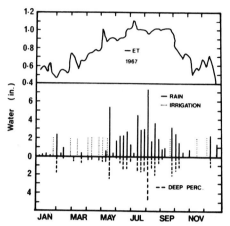

Fig. 14. Deep percolation losses are maintained at a relatively small component of a citrus orchard's water budget, even on sandy soils, by judicious irrigation. (Reproduced from Rogers and Bartholic, 1976.)

out of the root zone may become important especially on porous soils. Data generated by Rogers and Batholic (1976) for an orange grove at Arcadia, Florida, on a deep sandy soil indicate relative magnitudes of evapotranspiration, rainfall, irrigation, and deep percolation (see Fig. 14). Soil water-holding capacity within the wet zone was estimated as equivalent to 10–12 cm. Figure 14 shows annual trends in components of the water budget equation which can also be presented in the form:

$$S = R - RD + IRR - ET - DP$$

where S is water stored in the soil, R is daily rainfall, RD represents runoff, IRR irrigation, ET evapotranspiration, and DP deep percolation of water below the root zone. Data arranged over weekly periods provide annual trends as shown in Fig. 14. During summer, rainfall exceeded ET, deep percolation was substantial, but irrigation was unwarranted. Nevertheless, occasional irrigations were needed in late fall and winter to meet evapotranspirational demand. The number of applications varied between 3 and 18 depending on weather and soil. Data obtained during 1967, for example, indicated an above-average rainfall of 1520 mm and yet 18 irrigations of 25 mm were required when a 50-mm soil water holding capacity was inserted into the water budget equation. Over the subsequent year, rainfall was less, but better distributed, and only 14 irrigations were required. Under average conditions with an 80-mm soil water holding capacity, number of irrigations varied from 6 to 11 and net water required varied from 250 to 460 mm. Deep percolation reached 650 to 800 mm, which led to a net recharge for the aquifer that year.

IV. IRRIGATION

In arid or semiarid regions of the world, irrigation is the most crucial as well as the most expensive component in citrus production. As early as 1647 Ferrarius published the rather perceptive observation that citrus trees need abundant moisture, but cannot endure stagnant soil water. Moreover, varying requirements for irrigation as a function of seasonal conditions and soil type were recognized.

Tree growth and productivity are affected differently by moisture supply. While biomass accumulation (hence, vegetative growth) is a continuing process, fruit production is a seasonal culmination of reproductive development followed by successful fruit set and enlargement. Moderate water stress favors reproductive development, whereas fruit set and enlargement are turgor dependent. Productivity is, therefore, a function of both timing and amount of irrigation as outlined below.

A. FRUIT PRODUCTION

Under adequate management, irrigation boosts orchard production by increase in fruit number and size. As reported by Suzuki and Kaneko (1970), Satsuma orange trees growing in pots of loam soil in Japan showed a 50% increase in yield, for the same fruit number per tree, when ψ_{soil} was kept between -0.4 and -1.14 bar compared with -9.74 to -19.52 bars on sparsely watered controls. These trees appeared to reach both full growth and fruiting potential when ψ_{soil} was kept above -2.5 bars with fruit growth serving as an early index of soil moisture tension. Huberty and Richards (1954) had previously reported a similar finding for navel orange; a 30% reduction in yield preceded even marginal effects on tree vigor. Cahoon et al. (1961) offer comparable findings for Washington navel on sweet orange rootstock under furrow irrigation on a 3- versus 6-week schedule. During 1934–1957 more frequent irrigation resulted in 20% more crop, notwithstanding fewer deep roots compared with trees on the 6-week schedule. Stolzy et al. (1963) documented a similar response for Washington navels established just prior to 1934 in a Ramona sandy loam soil near Riverside on a 3- versus 6-week roster, which was subsequently rescheduled in 1957 to coincide with minimum ψ_{soil} (indicated by tensiometers) of 20 and 100 cbars, respectively. Data gathered during 1934–1963 showed consistently higher yield under more frequent irrigation up to 1957, i.e., 250–300 cf. 200–250 lb. tree $^{-1}$ with general improvement in orchard performance following introduction of tensiometers. After that event, trees previously on a 6-week schedule seemed better adapted to moisture tension than trees irrigated with a 3-week frequency. Using $\psi_{soil} = 100$ cbars as the lower threshold before irrigating, yields were 250–270 cf. 210–230 lb. tree^{-1} for 6- and 3-week frequency, respectively.

Notwithstanding internal limitations on fixation and distribution of photosynthetic products in citrus trees (Possingham and Kriedemann, 1969), biomass gain commonly shows close positive correlation with transpirational losses (Stanhill, 1960). Moreover, as common physical processes are involved, both transpiration and evapotranspiration are related to class-A pan evaporation (Stanhill, 1961). Accordingly, the effect of crop moisture status on yield should also be reflected in its comparative rate of evapotranspiration (i.e., relative to class-A pan). Taking E_t/E_o as the ratio of seasonal evapotranspiration to seasonal class-A pan evaporation, Shalhevet and Bielorai (1978) demonstrated a positive correlation between relative yield (Y) and water use for both irrigated row crops and orchards. Defining Y (treatment yield) as a percentile of the highest yield obtained in each year of an experiment, the following relationships were derived from a total of 20 yield experiments over the period 1958–1974 on cotton, sorghum, and grapefruit (see Fig. 15). Overhead sprinklers were used in most experiments, although some cotton was grown under furrow irrigation.

$$\text{Cotton } Y = -6.0 + 199.2 \ (E_t/E_o) \qquad r = 0.944$$
$$\text{Sorghum } Y = -29.3 + 227.2 \ (E_t/E_o) \qquad r = 0.937$$
$$\text{Grapefruit } Y = \ \ \ 37.2 + 102.2 \ (E_t/E_o) \qquad r = 0.730$$

In both irrigated row crops (Fig. 15) the correlation was close, and about 90% of the variation in yield was attributed to differences in potential evapotranspiration and depth of irrigation water. For grapefruit, only 50% of yield variation could be so attributed. Compared with annual row crops, the perennial nature of citrus makes the relationship more tenuous because irrigation timing can be crucial for reproductive development, fruit set, and fruit enlargement. Moreover, cropping behavior in one season also influences root extension and top growth that has a carryover effect on yield in successive seasons.

Shalhevet and Bielorai (1978) also identified a fundamental difference between the two field crops and citrus in terms of yield response to soil moisture status. When row crop dry matter production was plotted as a function of E_t/E_o (Fig. 15) the line passed close to the origin (especially for cotton), but grapefruit yield showed sufficient positive intercept on the abscissa to imply a substantial evaporation component, plus the need for a

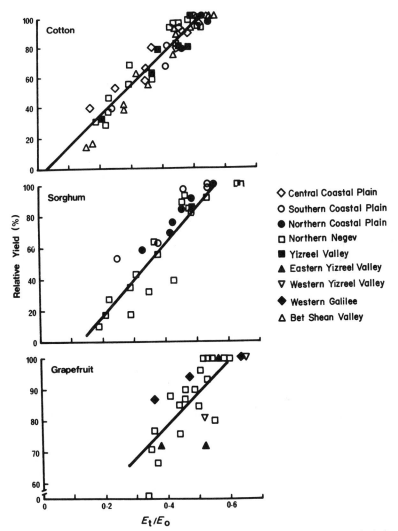

Fig. 15. Relative yield of cotton, sorghum, and grapefruit as a function of "irrigation adequacy" (as inferred from the ratio of seasonal evapotranspiration E_t to class-A pan evaporation E_o) emphasize lower water-use efficiency in citrus. (Adapted from Shalhevet and Bielorai, 1978, and reproduced by permission of The Williams & Wilkins Co., from *Soil Sci.* **125**, 240–247.)

threshold quantity of water to produce some biomass as a prerequisite for marketable yield. In addition, the slopes for cotton and sorghum in Fig. 15 are similar, i.e.. the increase in their relative yield is double that of relative transpiration; for grapefruit the slope is only 1:1. Low photosynthetic capacity of citrus foliage, as discussed previously in Section II, would be a major limiting factor in this context.

B. Tree Responses to Stress and Irrigation Scheduling

Orchardists seek to achieve a balance between tree growth and optimum productivity by judicious management of irrigation, and citrus trees provide one clear illustration of the axiom that total photosynthetic activity, and hence generation of biomass, does not necessarily correlate with economic yield. Although irrigated trees undoubtedly grow faster (see, for example, Sites *et al.*, 1951; Hilgeman *et al.*, 1959), continuous abundant soil moisture encourages dominance of vegetative growth and leads to production of fruits that are high in juice, low in soluble solids, as well as acid and of poorer keeping quality (Sites *et al.*, 1951). Irrigation appears to have a diluting effect. Hilgeman *et al.* (1959) demonstrated a curvilinear relationship between the amount of rainfall and extent of total fruit growth of Valencia oranges in Arizona. Ultimate size of fruit was determined by soil moisture supply while rate of fruit enlargement showed a rapid, although transient response to irrigation. During the phase of rapid fruit enlargement, 2.55 mm of water produced a 1.55 ml gain in fruit volume whereas 25.5 mm of irrigation led to only 4 ml gain in enlargement; hence, the curvilinear relationship referred to above.

Fruits are, however, closely coupled "hydraulically" with the tree's vascular network, and together with trunk growth, enlarging fruits can provide early indication of impending stress. Hilgeman (1963) obtained a dendrograph record of trunk growth in Valencia orange trees under rigorous conditions in Arizona, where orchards sometimes experienced extremely desiccating conditions, i.e., 47°C and relative humidity of 8%. Natural precipitation had a greater effect on trunk expansion than free water in the root zone following summertime irrigation. Apparently, the milder meteorological conditions associated with rainfall predisposed trees toward a sustained response. Given the situation in which trunk growth and shrinkage reflected interplay between evaporative demand, tree conditions, and soil moisture, Hilgeman (1963) stressed the need for additional criteria in scheduling irrigation, although diurnal patterns in trunk diameter suggested that trees began "losing ground" once root zone soil moisture tension exceeded 65 cbars.

Root growth does, however, persist and trees watered infrequently develop more extensive deep roots. Cahoon *et al.* (1961), for example, identified irrigation frequency as a factor in root development and excluded effects of rootstock, soil type, and crop load. Using Washington navel on sweet orange rootstock under furrow irrigation, between 1934–

1957, a 3-week interval led to fewer deep roots, especially in the 0.6–1.0 m zone, compared to the 6-week interval, although the orchard generated 20% more crop. Notwithstanding this compensatory response, soil moisture is probably the main edaphic factor limiting root growth during the irrigation season and, under summer conditions, Monselise (1947) found that root growth fluctuation in Shamouti oranges on sweet lime rootstocks in light sandy soil followed the rhythm of irrigation. Fluctuations in root growth were very similar to those in soil moisture. Once ψ_{soil} fell below about -7.5 to -8.0 bars, root growth ceased.

Adverse effects of restricted root growth at low ψ_{soil} on tree moisture status may well be exacerbated by changes in their hydraulic conductivity. Ramos and Kaufmann (1979) demonstrated that either mild or severe long term stress (watering frequency 2 and 3 days, respectively) induced increase in hydraulic root resistance of *Citrus jambhiri* (Lush) after several drying cycles. Increased resistance was not due simply to decreased root growth but to changes in permeability of root cell membranes or to increased suberin deposition in walls of cortical cells (see Sections II,A,4 and II,B,3, for further details).

Despite these adverse reactions to intensified moisture stress in roots, trunk, and fruits, studies such as those cited by Hilgeman and Sharp (1970) under arid conditions in Arizona on Valencia orange, have not established clear advantages in maintaining high availability of soil moisture throughout the entire year. Indeed, once the crop for a particular season has been set, moderate moisture stress is conducive to higher soluble solids in harvested fruit, thinner peels, and overall improvement in yield, which probably is associated with decreased vegetative growth.

Since irrigation can be such a decisive management tool, scheduling becomes important. Knowing that adequate moisture during the post-bloom phase is crucial for successful fruit set (see, e.g., Jones and Cree, 1965; Moss and Muirhead, 1971a,b); Hilgeman and Sharp (1970) achieved similar yields in Valencia orange despite wide differences in total water supplied. Timing was the key!

Using a flooding system between ridges with a large basin in each row on a calcareous sandy loam, Hilgeman and Sharp (1970) provided 15 irrigations on a 14- to 20-day cycle that delivered 172 cm of water per year and compared tree performance with that from three other schedules, including a system of only nine irrigations but with high summer frequency. In that case, although only 118 cm of water were applied, yield was neither significantly nor consistently affected. Productivity from these two treatments over the period 1948–1968 varied between 50 and 200 kg tree^{-1} and bears some analogy to orchard performance at Riverside where Jones and Cree (1965) observed large variations among seasons over a

38-yr period with Washington navels. Yields which ranged from ca. 26 to 150 kg tree^{-1} appeared negatively correlated with the previous year's crop and with daily maximum temperature following bloom. Comparable findings in Australia led Moss and Muirhead (1971a,b) to resolve the interacting effects of climatic and tree factors relating to yield in both Washington navel and late Valencia. In their experience, temperature and evaporation rates during November were the most consistent predictors of yield. After allowing for effects due to tree age and yield the previous season, 44.5% of the remaining variability in late Valencia was attributed to temperature, with 32.1% due to evaporative conditions. Corresponding figures for Washington navel were 52.0 and 39.4%, respectively (G. I. Moss, personal communication).

Further evidence for tree moisture status and evaporative conditions as major factors in "June drop" in navel oranges in the San Joaquin Valley of California was provided by Brewer *et al.* (1973). Under adverse summer conditions when air temperature exceeded 35°C, overhead irrigation reduced air temperature 5 to 8°C and raised relative humidity 8 to 12% with an attendant increase in ψ_{leaf} of between 5 and 7 bars. Premature fruit drop was reduced and yield increased. Similarly Koo and Sites (1955) showed highly significant yield response to post-bloom irrigation of three varieties of orange in Florida. Increases from 9.5 to 33.5% were recorded.

Given sufficient information to time irrigation according to critical phases in a tree's seasonal development, the next task is to gauge orchard water requirement *vis-a-vis* evaporative demand.

In terms of biological indicators trees should be uniquely placed because they are continuously "solving" their own water balance equation, and, in principle, an appropriate physiological parameter is all that should be required. Kaufmann (1969) suggested fruit turgor as such an index and relying on the inherent turgidity of fruit on a well-watered tree, he devised an ingenious method of measuring the width of a gap formed in a fruit's equatorial plane following partial incision. Gap width decreased as soil moisture became less available, and the space reopened as fruit turgor was restored following irrigation. A close curvilinear relationship between ψ_{leaf} and gap width (mm) was demonstrated for both Valencia and Washington navel on sweet orange rootstock. However, change in slice gap width as a function of changes in ψ_{leaf} from -4 to -16 bars was only 2 mm (i.e., varied from about 4 to 6 mm) and was not considered a reliable guide for scheduling. Data of J. Blackwell (personal communication) indicated an equally limited range of variation in slice gap width in limes growing under trickle irrigation on the central Pacific island of Niue (lat. 20°S).

Leaf ψ, also considered a possible biological index, has the advantage that both photosynthesis and transpiration bear some relationship to stomatal aperture, which in turn is coupled to ψ_{leaf} (see Section II,C for further details). However ψ_{leaf} is highly dynamic and subject to wide variation according to both incident radiation and soil moisture status. As demonstrated by Smart and Barrs (1973) in their comparative study of diurnal patterns in ψ_{leaf} on peaches, prunes, citrus, and grapes under irrigation, insolation had a dominating effect on leaf moisture status and, depending on species, between 76 and 96% of the variation in ψ_{leaf} could be ascribed to this factor. Utilization of such data in irrigation scheduling would require comprehensive study of ψ gradients in the soil–plant–atmosphere continuum as a function of microclimate (especially insolation), stage of growth, crop load, and soil type. Ultimately, forecast weather conditions might then enable prediction of minimum values for ψ_{leaf} and, given some knowledge of threshold values for leaf moisture tension above which gas exchange was adversely affected, irrigation requirements could be calculated (Smart and Barrs, 1973).

Ideally, water should be supplied in advance of requirements as described above, but present models of the soil–plant–atmosphere continuum either are not sufficiently comprehensive or are hindered in their application by lack of physical homogeneity in the orchard (especially large area groves in uneven terrain) (Kalma, 1970). Consequently, practical approaches tend to be retrospective; i.e., an orchard's evapotranspirational demand, derived from either micrometeorological models or evaporimeters, is subsequently matched by an appropriate amount of applied water.

Given the inherent difficulty in applying energy budget techniques to heterogeneous plant communities, empirical determinations of evapotranspiration are most realistic. One of the classic examples comes from a 1000-acre citrus watershed near Riverside, California (see Davis et al., 1969; Kalma and Fuchs, 1976). In this enormous "lysimeter," water sources and percolation losses were measured, so that evapotranspiration could be obtained by difference. Mean annual evaporation from vegetation and soil in this watershed over 9 years was 720 mm, and varied from 22 mm during January to 102 mm in August. Comparative data for the period 1967–1971 showed that regional evapotranspiration was only 38% of total evaporation from a screened class-A pan.

Applying a similar water balance approach but on a more manageable scale in Israel, Kalma (1970) inferred water loss from a detailed survey of soil moisture changes. An orange orchard planted to a 4 × 4 m grid evapotranspired at an annual mean rate equivalent to 54% of total evaporation from a screened class-A pan. Daily rates varied from 1.0 mm in midwinter to 4.4 mm in midsummer. In Arizona, Van Bavel et al. (1967) estimated cumulative evapotranspiration from mature orange orchards (crop + soil) as 66% of annual evaporation from a class-A pan. A water budget was derived from measurements of rainfall, irrigation, and ψ_{soil}. Daily rates of evaporation varied from 0.9 mm in midwinter to 5.3 mm in midsummer.

Taking the ratio between measured evapotranspiration (E_t) and evaporation from a free water surface (E_o), defined by Penman (1948) as an indication of crop factor (f), then the relationship $E_t = fE_o$ serves as a guide for irrigation scheduling. Potential evaporation (E_o) can be predicted from meteorological data (using, e.g., Penman's formula) or derived empirically using direct measurements of evaporation from a class-A pan (E_p) using a relationship such as $E_o = 0.8 E_p$ (Fleming, 1964a). Crop factor (f) values for the two closely monitored orchards cited above would therefore be 0.54/0.8 = 0.68 and 0.66/0.8 = 0.83 for Israel and Arizona, respectively. In the interests of simplicity, some investigators use the more direct relationship between E_t and E_p referred to as pan coefficient (K_p) where $E_t = K_p E_p$. In that instance, K_p values for the two orchards would be 0.54 and 0.66, respectively.

Variations in tree size, orchard layout, soil management, and weed control introduce further complications but these influences can be accommodated. Moreover, seasonal changes in transpiration response also impinge on the value ascribed to f (as may be inferred from Fig. 13). Accordingly, "crop factor" will be primarily a reflection of E_t/E_o but "weighted" by horticultural considerations.

Based on cultural conditions in Southern Australia (25°S–40°S, Mediterranean-type climate on light textured soils) Fleming (1964b) suggested f values of 0.7 during the growing season, 0.6 in autumn, and 0.5 in winter. Mean seasonal coefficient for citrus was 0.65 and compared with 0.75 for cotton, 0.80 for deciduous orchards, 0.90 for summer forage crops, and 1.10 for rice.

Clearly, citrus orchards transpire less water per unit land surface than other agricultural crops, a distinction already alluded to by Jensen and Haise (1963). These lower rates of evapotranspiration have been attributed to the combined influence of high gaseous diffusive resistances in leaves and low hydraulic conductivity in the tree's vascular network as discussed previously in Section II, and a distinction recognized by Stephen Hales as early as 1727!

C. Irrigation Systems and Soil Management

Orchard topography, soil permeability, and regional climate, but especially capital outlay, influence the initial choice of irrigation system. However, when water quality deteriorates or supply diminishes and recovery costs escalate, particular attention is given to WUE and hence to development of more sophisticated forms of water distribution and associated soil management practices.

In its simplest form, orchard irrigation relies on gravitational flow with a surface application via either general flooding, basin flooding, or furrows. General flooding calls for smooth or very gently sloping topography with coarse soil where borders limit lateral spread. A relatively high rate of water flow is maintained and, although capital investment is low, high input of skilled labor is required to operate the system. At best, general flooding provides decidedly nonuniform distribution of water.

If orchard soils are of low porosity, basin flooding or ring watering may be used. In this sytem one or up to five trees are surrounded by a small levee bank and water is introduced from a tank wagon and ponded. Since the wetted area is more readily circumscribed, and only the area immediately adjacent to the root system is moistened, some improvement in both WUE and root zone distribution is achieved.

Furrow irrigation as practiced on soils of medium porosity also results in partial surface wetting because successive ridges and grooves may be spaced up to 2 m apart with length depending on soil type and slope. Capital investment is again low, but some managerial skill is required to operate the system as well as to integrate furrow preparation with other facets of orchard management.

More sophisticated sprinkler irrigation systems that operate under pressure provide a new range of options. Land with irregular topography can be irrigated with minimum disturbance to topsoil while excess runoff, and hence erosion, can be virtually eliminated. Highly permeable soils such as sands and deep loams referred to in Section III can be irrigated with sufficient precision to avoid major losses through deep percolation. High costs associated with land preparation, leveling, and ditch maintenance are also avoided. Sprinkler systems can consist of portable main lines and laterals with water emerging from perforated pipes, or via rotating sprinklers that are spaced to achieve an overlapping circular wetting pattern.

Where frequent irrigation is required and labor is short, portable networks based on aluminum pipes have given way to fixed arrays with water distribution via plastic pipe. Stationary systems rely on either relatively few overhead sprinklers working at high pressure ($20–30 \text{ kg cm}^{-2}$) to generate a large wetted area (radius up to 15 m), or a higher density of low level undercanopy sprinklers each of which covers about 50 m^2. Foliar uptake of dissolved solutes, and especially Cl^- ions has encouraged the move from overhead sprinklers to undertree types (Calvert and Reitz, 1966). Moreover, low-discharge sprinklers (less than 200 liters per hr^{-1} at

15 kg cm^{-2}) that have been designed for fixed undertree installations, also allow use of smaller diameter polyvinyl chloride and polyethylene pipe with attendant savings in capital outlay and water use.

Either form of sprinkler irrigation (and especially overhead sprays) involves wetting a large proportion of the land surface, and as discussed in Section III evaporation from wetted areas between trees lowers irrigation efficiency without conferring major microclimatic benefits. Undertree microjets (small static sprinklers), therefore, represent some additional improvement in terms of irrigation efficiency, while dripper systems that discharge water onto the ground through one or more emitters adjacent to each tree represent a further advance (Myers, 1977).

Data presented in Table II emphasize the advantage of drip irrigation in terms of establishment cost, engineering demands, and irrigation efficiency. Where orchards are established under drip irrigation and root systems occupy a limited volume of soil which is confined to the wetted zone, WUE is further enhanced (Leyden, 1975a). Moreover, drippers enable the use of poorer quality water than could be tolerated under overhead sprinklers because direct foliar uptake, and hence leaf injury, is avoided (Calvert and Reitz, 1966). However, solute accumulation is unavoidable when a limited volume of soil is kept at relatively high moisture content and greatest accumulation will occur at the wet surface due to continuous evaporation. Some accumulation will also occur within the profile at the wet–dry interface (see Yaron *et al.*, 1973). In a comparative trial of strip flooding, basin flooding, and drip irrigation at Weslaco, Texas, Leyden (1975b) reported considerable accumulation of salt in the 0–2.5 cm layer. Values for electrical conductivity were 2.5, 2.6, and 26.5 mmho for strip, basin, and drip irrigation, respectively. Equivalent values in the 2.5–5.0 cm layer were 1.2, 1.2, and 3.8. Clearly, surface solute accumulation was confined to the proximity of the emitter and fell from 26.5 mmho near the outlet to 5.0 mmho within 60 cm. Fortunately, these salts were

TABLE II

COMPARATIVE COST AND EFFICIENCY OF PRESSURIZED IRRIGATION SYSTEMS[a]

Item	Drip	Undertree spray	Overhead sprinkler
Water distribution cost ($/ha)	865	1235	2965
Operating pressure (kg cm^{-2})	8.4	10.5	38.7
Irrigation efficiency (%)	95	85	70

[a]Data from Myers (1977).

also highly mobile and were leached out of the profile by the average annual rainfall of 640 mm. Electrical conductivity of saturation extracts was then less than 2 mmho.

Given the positive linear relationship between daytime evaporation from bare soil and the moisture content of upper layers as described by Kalma and Stanhill (1969), subsurface irrigation which would leave the upper soil layer dry offers even greater potential for economy in water use. Working in a 35-yr-old Shamouti orange orchard on loamy red-brown soil near Rehovot, Israel, Kalma and Stanhill (1969) estimated soil evaporation losses as 10% of evaporation from a screened class-A pan with subsurface irrigation, compared with 13% with undertree sprinkler application. This difference implied a 32 mm reduction in evaporation from bare soil during the irrigation season and represented a 7% saving in irrigation water requirement.

Notwithstanding such theoretical advantages, subsurface trickle irrigation has not yet gained general acceptance. Buried pipes are difficult to maintain; solutes tend to accumulate in upper layers of the soil; and blocked emitters or orifices often are not detected until trees dependent on that particular source are showing adverse signs of moisture stress. Recognizing these shortcomings, Brown et al. (1974) nevertheless maintained a successful operation for at least 8 years on alluvial soils in southern Louisiana. Supplementary irrigation was required for only two periods, mid-May to mid-June and mid-October to mid-November, to complement otherwise satisfactory rainfall (average 1651 mm yr^{-1}). Water was applied by pressurizing the system (buried to about 46 cm) long enough (4–6 hr) to produce moist patches at the soil surface along the length of each row. The system was designed to supply 2.5 cm of water in about 4 hr to individual rows of trees without wetting adjacent rows which were used as nonirrigated controls. Yield of both citrus varieties under test was increased; 7% in Washington navel and 10% in Owari Satsuma. Average weight per fruit showed a significant increase although percent juice in fruits was slightly reduced. On balance, the authors favored above ground installation of trickle irrigation over the subsurface method.

Sound soil management is critical for the long-term viability of citrus orchards and constitutes a key facet in overall irrigation practice. Under flood or furrow systems, the orchardist can maintain clear intertree space by repeated cultivation, as is required for furrow formation, or by application of herbicides, where basins or broad based furrows are used.

Soil properties, notably soil texture, influence rooting depth (Shalhevet et al., 1976). On a deep well-drained fine sand, Ford (1954a) reported that 15% of the feeder roots of rough lemon were below 9 feet and that they could penetrate to more than 17 feet. Soil water status can be a decisive influence since citrus roots will not grow into or exist long in a water-saturated soil. Hence, when the water table is within a few feet of the surface, the roots are confined to a shallow zone and tree growth is unsatisfactory (Ford, 1954b). Some evidence exists for a difference in tolerance to temporary flooding between rootstocks since Ford (1964) found that rough lemon was more tolerant than sour orange. He also noted that flooding damage is greater when it is accompanied by low soil pH.

An explanation for differences in tolerance of poor soil aeration between citrus rootstocks was given by Luxmoore et al. (1971), who suggested that increased root porosity enabled greater transport of oxygen from shoots to roots. Root porosity (volume of gas space per unit volume) was significantly lower in rough lemon (7.2%) seedlings than in sweet orange (10.9%) or "Troyer" citrange (9.4%) seedlings. They also showed that roots from "Troyer" citrange trees grown in nutrient solution exhibited higher porosity than roots from trees grown in sand. This suggested that not only may root porosity differ between rootstocks but that adaptive changes in root porosity can be induced by the root environment. Luxmore et al. suggested that environmental factors such as soil temperature and even light intensity may also affect root porosity. Further work in this area appears desirable.

Even under nonflooded conditions, better aeration associated with less frequent irrigation permitted citrus roots (grapefruit on sour orange stock) to grow twice as deep as roots of trees receiving more frequent irrigation, and to be more effective in removing deeper soil water (Samish, 1957). Under copious irrigation the top:root ratio was greatly increased due to increased top growth and reduced root growth. However, citrus is not normally grown for increased root production and over a 7-yr period, the accompanying fruit yields under medium and scant irrigation declined to 53 and 32%, respectively, compared with yield under the copious irrigation. Similar results were obtained for Washington navel orange on sweet orange stock (Cahoon et al., 1961).

Methods of irrigation may affect feeder-root production. Biely et al. (1958) found that sprinkler irrigation increased production of orange feeder roots by 50%, but that under furrow irrigation root production decreased 36%. Spurling (1951) attributed the inadequacy of furrow irrigation to effects of unevenness of water application especially on light soils. Rapid penetration of water with little lateral movement from the furrows resulted in over-watering of the row centers under the furrows and a dry strip of soil under the trees. The former restricted root development by causing excessive leaching of soil nutrients and was most important at the top end of the rows. The latter was most important toward the end of the irrigation run, where 8- to 10-foot widths of soil sometimes were not wetted during irrigation.

By analogy with patterns of water and salt distribution which develop under tricklers, furrow irrigation can also lead to great variability in salt concentration and distribution. Directly beneath furrows salt concentration is low, but local concentrations tend to form under ridges, a distinction in solute levels that can extend down the profile for over a meter. As Harding and Ryan (1961) encountered in their long-term fertilizer trial, salt distribution was inversely related to water distribution. While percolation of irrigation water was primarily downward, capillary movement of water from furrows toward raised edges and subsequent evaporation, led to high accumulation of salts on furrow crests.

As with trickle systems, if rainfall is insufficient to leach surface salts through the profile and past the root zone during periods of slow growth and low evaporative demand, trees may be adversely affected. For example, Heller et al. (1973) describe the case of a 10-yr-old orchard of Shamouti orange trees on sour orange rootstocks where soil Cl^- ion concentration of only 7 mEq liter^{-1} (saturation extract) was enough to depress yields 18% below productivity on soil Cl^- around 3.5 mEq liter^{-1}. Similarly, Bielorai et al. (1978) showed that grapefruit yield was influenced by soil Cl^- ion concentration in the Negev. Productivity showed a linear decline (15% decrease from 5 to 15 mEq liter^{-1}). Under more extreme conditions described by Yaron et al. (1973) where wheat was sown into cotton fields previously trickle irrigated but not adequately leached, wheat seedlings failed to emerge along salinized strips formed at the junction of previously wetted zones.

Bare cultivation as required for furrow irrigation can lead to both localized solute accumulation (described above) and impaired leaching due to reduced permeability. The preferred alternative described by Jones et al. (1961) is non-tillage combined with chemical weed control. A change from clean cultivation of Washington navel orchards around Riverside to non-tillage almost doubled the net rate of water intake from 2.43 acre inches per acre per irrigation in 1953–1959 to 4.19 (same units) over the period 1955/1959 following discontinuation of the earlier practice. Presumably, improved soil structure was a major contributing factor although winter cover cropping did not appear to be beneficial because infiltration rates stayed between 2.55 and 2.97 (same units) under that soil management system over the same period.

Fertilizer type and placement also affect root production and distribution. Cahoon et al. (1959) showed that for Washington navel on sweet orange rootstock, more roots were produced for each of four 1-foot increments of soil depth in response to urea than to four other nitrogen sources (manure, calcium nitrate, ammonium sulfate, sodium nitrate). When applied N was confined to half the ground area by injection of anhydrous ammonia, fewer roots developed on the unfertilized area and many in the fertilized band. Smith (1965) speculated that this effect could be interpreted in terms of a feedback mechanism in which the underlying assumption was that carbohydrates were returned in greater abundance from the leaves to those roots providing most minerals to the leaves originally, thus favoring growth of such roots. Not all fertilizers stimulate citrus root growth. Smith (1965) found that after 5 years of treatment, feeder root concentrations in the top five feet of soil declined 31% in response to nitrate of soda, showed little change in response to ammonium nitrate, but increased 38% in response to anhydrous ammonia. The poor response to nitrate of soda was attributed to toxic effects of sodium.

Application of nitrogen fertilizer can show additional interactive effects with soil management. Jones et al. (1961), noted that both yield and net water intake rate were least when $NaNO_3$ and $(NH_4)_2SO_4$ were used as N sources. Application of gypsum offset this detrimental effect which was also partly corrected by changing from clean cultivation to non-tillage. Taking the two extremes of clean cultivation in combination with $NaNO_3$ fertilizer versus nontillage in combination with $Ca(NO_3)_2 \cdot 2 H_2O$ as a N source, yields (pounds of fruit per tree per year) were 71 and 232, respectively. Similarly, the apparent paradox of improved soil stability and infiltration rate under repeated $(NH_4)_2SO_4$ application on fine-textured citrus soils at Griffith, NSW, Australia, was also rationalized in terms of Ca^{2+} ion status. Cary and Evans (1972) suggested in situ liberation of $CaSO_4$ (gypsum is a well known soil stabilizing agent) following progressive acidification, as the process responsible.

Clearly, advanced irrigation practice must sustain orchard productivity by matching water and nutritional requirements in a way that causes minimum soil disturbance. Pressurized systems based on trickle or microjet outlets seem to approach this ideal as well as providing some scope for management of tree growth and reproductive development via manipulation of the wetted zone. In that instance, a confined volume of adequately wetted soil limits root growth which in turn acts as a constraint on tree growth. Reproductive development, and hence heavy cropping on small but well nourished trees, is thereby fostered.

REFERENCES

Abdalla, K. M., Stino, G. R., El-Azzouni, M. M., and Mohsen, A. M. (1978). Anatomical features of the stomatal apparatus in some citrus species and cultivars. Egypt. J. Hortic. 5, 1–7.

Aharoni, N. (1978). Relationship between leaf water status and endogenous ethylene in detached leaves. Plant Physiol. 61, 658–662.

Albrigo, L. G. (1972a). Distribution of stomata and epicuticular wax on oranges as related to stem and rind breakdown and water loss. J. Am. Soc. Hortic. Sci. 97, 220–223.

Albrigo, L. G. (1972b). Ultrastructure of cuticular surfaces and stomata of developing leaves and fruit of the 'Valencia' orange. *J. Am. Soc. Hortic. Sci.* **97,** 761–765.

Albrigo, L. G. (1977). Rootstocks affect 'Valencia' orange fruit quality and water balance. *Proc. Int. Soc. Citric., 1977* Vol. 1, pp. 62–65.

Albrigo, L. G., and Brown, G. E. (1970). Orange peel topography as affected by a preharvest plastic spray. *HortScience* **5,** 470–472.

Allen, L. H., Jr., and Cohen, M. (1974). Water stress and stomatal diffusion resistance in citrus affected with blight and young tree decline. *Proc. Fla. State Hortic. Soc.* **87,** 96–101.

Anonymous (1979). World citrus outlook for 1985. *Citrograph* **64,** 232–237.

Bain, J. M. (1958). Morphological, anatomical and physiological changes in the developing fruit of the 'Valencia' orange *Citrus sinensis* (L.) Osbeck. *Aust. J. Bot.* **6,** 1–25.

Baker, E. A., and Procopiou, J. (1975). The cuticle of *Citrus* species. Composition of the intracuticular lipids of leaves and fruits. *J. Sci. Food Agric.* **26,** 1347–1352.

Baker, E. A., Procopiou, J., and Hunt, G. M. (1975). The cuticles of *Citrus* species. Composition of leaf and fruit waxes. *J. Sci. Food Agric.* **26,** 1093–1101.

Barmoew, C. R., and Biggs, R. H. (1972). Ethylene diffusion through citrus leaf and fruit tissue. *J. Am. Soc. Hortic. Sci.* **97,** 24–27.

Barrs, H. D. (1971a). Osmotic pressure: Spermatophytes. *In* "Respiration and Circulation" (P. L. Altman and D. S. Dittmer, eds.), pp. 713–721. Fed. Am. Soc. Exp. Biol., Bethesda, Maryland.

Barrs, H. D. (1971b). Cyclic variations in stomatal aperture, transpiration and leaf water potential under constant environmental conditions. *Annu. Rev. Plant Physiol.* **22,** 223–236.

Begg, J. E., and Turner, N. C. (1970). Water potential gradients in field tobacco. *Plant Physiol.* **46,** 343–346.

Bell, W. D., Bartholic, J. F., and Cohen, M. (1973). Measurement of water stress in citrus. *Proc. Fla. State Hortic. Soc.* **86,** 71–75.

Ben-Asher, J., and Sammis, J. W. (1978). Radiation and energy balance of a trickle-irrigated lemon grove. *Agron. J.* **70,** 568–572.

Beutel, J. A. (1964). Soil moisture, weather and fruit growth. *Calif. Citrogr.* **49,** 372.

Bialoglowski, J. (1936). Effect of extent and temperature of roots on transpiration of rooted lemon cuttings. *Proc. Am. Soc. Hortic. Sci.* **34,** 96–102.

Bielorai, H. (1968). Estimation of the osmotic potential of citrus seedling leaf sap at different soil-moisture levels by means of refractometry and conductimetry. *Ann. Bot. (London)* [N. S.] **32,** 473–477.

Bielorai, H. (1980). The prospects of water management for optimum citrus growth and production. *Proc. Int. Soc. Citric., 1978* pp. 227–233.

Bielorai, H., and Levy, J. (1971). Irrigation regimes in a semi-arid area and their effects on grapefruit yield, water use and soil salinity. *J. Agric. Res. Isr.* **21,** 3–12.

Bielorai, H., and Mendel, K. (1969). The simultaneous measurement of apparent photosynthesis and transpiration of citrus seedlings at different soil moisture levels. *J. Am. Soc. Hortic. Sci.* **94,** 201–204.

Bielorai, H., Levin, I., and Assaf, R. (1973). Irrigation of fruit trees. *Ecol. Stud.* **5,** 397–409.

Bielorai, H., Shalhevet, J., and Levy, Y. (1978). Grapefruit response to variable salinity in irrigation water and soil. *Irrig. Sci.* **1,** 61–70.

Biely, M. I., Wallace, A., and Kimball, M. H. (1958). Some factors influencing feeder-root weights and distribution in citrus. *In* "Tree Physiology Studies at U.C.L.A." (A. Wallace, ed.), Spec. Rep. No. 1, pp. 16–29. Dep. Hortic. Sci., University of California, Los Angeles.

Boiko, L. A., and Boiko, L. A. (1959). A method for determining the osmotic pressure of cell sap using collodion sacks. *Sov. Plant Physiol. (Engl. Transl.)* **6**, 639–641.

Boswell, S. B., McCarty, C. D., and Lewis, L. N. (1975). Tree density affects large-root distribution of 'Washington' navel orange trees. *HortScience* **10**, 593–595.

Boyer, J. S. (1971). Resistance to water transport in soybean, bean and sunflower. *Crop Sci.* **11**, 403–407.

Boyer, J. S. (1974). Water transport in plants: Mechanism of apparent changes in resistance during absorption. *Planta* **117**, 187–207.

Brams, E. (1969). The mucilaginous layer of Citrus roots—its delineation in the rhizosphere and removal from roots. *Plant Soil* **30**, 105–108.

Bravdo, B. (1977). Oscillatory transpiration and CO_2 exchange of citrus leaves at the CO_2 compensation concentration. *Physiol. Plant.* **41**, 36–41.

Brewer, F. R., Aljibury, F., Optiz, K., and Hench, K. (1973). Reduction of "June Drop" on navel oranges with thermostatically controlled overtree sprinklers. *HortScience* **8**, 272.

Brown, R. T., Constantin, R. J., and Brand, H. J. (1974). Effect of irrigation on production and quality of citrus. *La. Agric.* **18**, 14–15.

Burrows, F. J., and Milthorpe, F. L. (1976). Stomatal conductance in the control of gas exchange. *In* "Water Deficits and Plant Growth" (T. T. Kozlowski, ed.), Vol. 4, pp. 103–152. Academic Press, New York.

Cahoon, G. A., Morton, E. S., Jones, W. A., and Garber, M. J. (1959). Effects of various types of nitrogen fertiliser on root density and distribution as related to water infiltration and fruit yields of Washington navel oranges in a long-term fertiliser experiment. *Proc. Am. Soc. Hortic. Sci.* **74**, 289–299.

Cahoon, G. A., Huberty, M. R., and Garber, M. J. (1961). Irrigation frequency effects on citrus root distribution and density. *Proc. Am. Soc. Hortic. Sci.* **77**, 167–172.

Cahoon, G. A., Grover, B. L., and Eaks, I. L. (1964a). Cause and control of oleocellosis on lemons. *Proc. Am. Soc. Hortic. Sci.* **84**, 188–198.

Cahoon, G.A., Stolzy, L. H., Garber, M. J., and Morton, E. S. (1964b). Influence of nitrogen and water on the root density of mature Washington navel orange trees. *Proc. Am. Soc. Hortic. Sci.* **85**, 224–231.

Calvert, D. V., and Reitz, H. J. (1966). Salinity of water for sprinkler irrigation of citrus. *Proc. Fla. State Hortic. Soc.* **79**, 73–78.

Calvert, D. V., Ford, H. W., Stewart, E. H., and Martin, F. G. (1977). Growth response of twelve citrus rootstock-scion combinations on a spodosol modified by deep tillage and profile drainage. *Proc. Int. Soc. Citric., 1977* Vol. 1, pp. 279–284.

Camacho-B., S. E. (1977). Some aspects of the stomatal behaviour of citrus. *Proc. Int. Soc. Citric., 1977* Vol. 1, pp. 66–69.

Camacho-B., S.E., Kaufmann, M. R., and Hall, A. E. (1974a). Leaf water potential response to transpiration by Citrus. *Physiol. Plant.* **31**, 101–105.

Camacho-B., S.E., Hall, A. E., and Kaufmann, M. R. (1974b). Efficiency and regulation of water transport in some woody and herbaceous species. *Plant Physiol.* **54**, 169–172.

Cameron, S. H. (1941). The influence of soil temperature on the rate of transpiration of young orange trees. *Proc. Am. Soc. Hortic. Sci.* **38**, 75–79.

Cary, P. R. (1971). Growth of 'Washington Navel' and 'Late Valencia' orange cuttings as affected by root temperature, nutrient supply and crop load. *Hortic. Res.* **11**, 143–155.

Cary, P. R., and Evans, G. N. (1972). Long-term effects of soil management treatments on soil physical conditions in a factorial citrus experiment. *J. Hortic. Sci.* **47**, 81–91.

Cassin, J., Bourdeaut, J., Fougue, A., Furon, V., Gaillard, J. P., Le Bourdelles, J., Montagut, G., and Moreuil, C. (1969). The influence of climate upon the blooming of citrus in tropical areas. *Proc. Int. Citrus Symp., 1st, 1968* Vol. 1, pp. 315–323.

Castel, J. C. (1978). Some aspects of gas exchange and water relations in orange cuttings. M.Sc. Thesis, University of New South Wales.

Castle, W. S. (1980). Citrus root systems: Their structure, functions, growth and relationship to tree performance. *Proc. Int. Soc. Citric., 1978* pp. 62–69.

Castle, W. S., and Krezdorn, A. H. (1975). Effect of citrus rootstocks on root distribution and leaf mineral content of 'Orlando' tangelo trees. *J. Am. Soc. Hortic. Sci.* **100,** 1–4.

Castle, W. S., and Krezdorn, A. H. (1977). Soil water use and apparent root efficiencies of Citrus trees on four rootstocks. *J. Am. Soc. Hortic. Sci.* **102,** 403–406.

Chaney, W. R., and Kozlowski, T. T. (1971). Water transport in relation to expansion and contraction of leaves and fruits of Calamondin orange. *J. Hortic. Sci.* **46,** 71–81.

Chapman, H. D., and Parker, E. R. (1942). Weekly absorption of nitrate by young, bearing orange trees growing out of doors in solution culture. *Plant Physiol.* **47,** 366–376.

Chapot, H. (1975). The citrus plant. *In* "Citrus" (E. Häfliger, ed.), Ciba-Geigy Agrochem. Tech. Monogr. No. 4, pp. 6–13. Ciba-Geigy Ltd., Basel, Switzerland.

Ciha, A. J., and Brun, W. A. (1975). Stomatal size and frequency in soybeans. *Crop. Sci.* **15,** 309–313.

Cooper, W. C., Peynado, A., Furr, J. R., Hilgeman, R. H., Cahoon, G. A., and Boswell, S. B. (1963). Tree growth and fruit quality of Valencia oranges in relation to climate. *Proc. Am. Soc. Hortic. Sci.* **82,** 180–192.

Cossman, K. F. (1940). Citrus roots: Their anatomy, osmotic pressure and periodicity of growth. *Palest. J. Bot., Rehovot Ser.* **3,** 65–104.

Costa, J. T. A. (1979). Spring environmental stress and fruit set of navel oranges, *Citrus sinensis* (L.) Osbeck. *Diss. Abstr. Int. B* **39,** 3074B.

Cowan, I. R. (1972). An electrical analogue of evaporation from, and flow of water in plants. *Planta* **106,** 221–226.

Cowan, I. R. (1977). Stomatal behaviour and environment. *Adv. Bot. Res.* **4,** 117–228.

Cowan, I. R., and Milthorpe, F. L. (1971). Permeability to water: Plant structures. *In* "Respiration and Circulation" (P. L. Altman and D. S. Dittmer, eds.), pp. 705–709. Fed. Am. Soc. Exp. Biol., Bethesda, Maryland.

Da Cunha, G. A. P., Davenport, T. L., Soule, J., and Campbell, C. W. (1978). Fruit turgor influences susceptibility of 'Tahiti' lime to stylar-end breakdown. *J. Am. Soc. Hortic. Sci.* **103,** 622–625.

Davies, W. J., and Kozlowski, T. T. (1974). Stomatal responses of five woody angiosperms to light intensity and humidity. *Can. J. Bot.* **52,** 1525–1534.

Davies, W. J., and Kozlowski, T. T. (1975). Effects of applied abscisic acid and plant water stress on transpiration of woody angiosperms. *For. Sci.* **21,** 191–195.

Davis, S., Bingham, F. T., Shade, E. R., and Grass, L. B. (1969). Water relations and salt balance of a 1000-acre citrus watershed. *Proc. Int. Citrus Symp., 1st, 1968* Vol 3, pp. 1771–1777.

Dede, R. (1962). Foliar venation patterns in the Rutaceae. *Am. J. Bot.* **49,** 490–497.

De Villiers, E. (1939). The xylem anatomy, relative water conductivity and transport of dyes in citrus stems. *S. Afr. J. Sci.* **36,** 291–313.

Edwards, M., and Meidner, H. (1979). Direct measurement of turgor pressure potentials. IV. Naturally occurring pressures in guard cells and their relation to solute and matric potentials in the epidermis. *J. Exp. Bot.* **30,** 829–837.

Ehrler, W. L. (1971). Periodic nocturnal stomatal opening of *Citrus* in a steady environment. *Physiol. Plant.* **25,** 488–492.

Ehrler, W. L., and Van Bavel, C. H. M. (1968). Leaf diffusion resistance, illuminance and transpiration. *Plant Physiol.* **43,** 208–214.

Elfving, D. C., and Kaufmann, M. R. (1972). Diurnal and seasonal effects of environment on plant water relations and fruit diameter of Citrus. *J. Am. Soc. Hortic. Sci.* **97**, 566–570.

Elfving, D. C., Kaufmann, M. R., and Hall, A. E. (1972). Interpreting leaf water potential measurements with a model of the soil-plant-atmosphere continuum. *Physiol. Plant.* **27**, 161–168.

Erickson, L. C. (1968). The general physiology of citrus. In "The Citrus Industry" (W. Reuther, L. D. Batchelor, and H. J. Webber, eds.), Vol. II, pp. 54–63. Div. Agric. Sci., University of California, Berkeley.

Farquhar, G. D., and Raschke, K. (1978). On the resistance to transpiration of the sites of evaporation within the leaf. *Plant Physiol.* **61**, 1000–1005.

Fawusi, M. O. A., and Ormrod, D. P. (1975). Temperature effects on the growth and zinc nutrition of sweet orange. *Hortic. Res.* **15**, 1–8.

Ferrarius, J. B. (1647). "Hesperides; sive de malorum aureorum cultura et usu libri quattuor." Rome.

Fleming, P. M. (1964a). Evaporimeter relationships at Griffith, N. S. W. *Civ. Eng. Trans. Inst. Eng. Aust.* **CE6**, 15–24.

Fleming, P. M. (1964b). A water budgeting method to predict plant response and irrigation requirements for widely varying evaporation conditions. *Trans. Int. Congr. Agric. Eng., 6th, 1964* Vol. II, pp. 66–77.

Ford, H. W. (1952). The distribution of feeder roots of orange and grapefruit trees on rough lemon rootstock. *Citrus Mag.* **14**, 22–23.

Ford, H. W. (1954a). The influence of rootstock and tree age on root distribution of citrus. *Proc. Am. Soc. Hortic. Sci.* **63**, 137–142.

Ford, H. W. (1954b). Root distribution in relation to the water table. *Proc. Fla. State Hortic. Soc.* **67**, 30–33.

Ford, H. W. (1964). The effect of rootstock, soil type, and soil pH on citrus root growth in soils subject to flooding. *Proc. Fla. State Hortic. Soc.* **77**, 41–45.

Furr, J. R., and Taylor, C. A. (1933). The cross-transfer of water in mature lemon trees. *Proc. Am. Soc. Hortic. Sci.* **30**, 45–51.

Gaastra, P. (1959). Photosynthesis of crop plants as influenced by light, carbon dioxide, temperature and stomatal diffusion resistance. *Meded. Landbouwhogesch. Wageningen* **59**, 1–68.

Gale, J., and Poljakoff-Mayber, A. (1964). Effect of soil moisture stress on the correlation between heat pulse velocity and transpiration. *Plant Cell Physiol.* **5**, 447–455.

Gates, C. T. (1955). The response of the young tomato plant to a brief period of water shortage. II. The individual leaves. *Aust. J. Biol. Sci.* **8**, 215–230.

Gates, D. M., and Tantraporn, W. (1952). The reflectivity of deciduous trees and herbaceous plants in the infrared to 25 microns. *Science* **115**, 613–616.

Girton, R. E. (1927). The growth of citrus seedlings as influenced by environmental factors. *Univ. Calif., Berkeley, Publ. Agric. Sci.* **5**, 83–117.

Gradmann, H. (1928). Untersuchungen über die Wasserverhältnisse des Bodens als Grundlage des Pflanzenwachstums. *Jahrb. Wiss. Bot.* **69**, 1–100.

Greacen, E. L., Ponsana, P., and Barley, K. P. (1976). Resistance to water flow in the roots of cereals. *Ecol. Stud.* **19**, 86–100.

Greene, B. A., and Gerber, J. F. (1967). Radiant energy distribution in citrus trees. *Proc. Am. Soc. Hortic. Sci.* **90**, 77–85.

Haas, A. R. C. (1936). Growth and water losses in citrus as affected by soil temperature. *Calif. Citrogr.* **21**, 467, 479.

Haas, A. R. C., and Halma, F. F. (1932). Relative transpiration rates in citrus leaves. *Bot. Gaz. (Chicago)* **93**, 466–473.

Hales, S. (1727). "Vegetable Staticks." Oldbourne, London.

Hales, T. A., Mobayen, R. G., and Rodney, D. R. (1968). Effects of climatic factors on daily 'Valencia' fruit volume increase. *Proc. Am. Soc. Hortic. Sci.* **92,** 185–190.

Hall, A. E., Camacho-B., S. E., and Kaufmann, M. R. (1975). Regulation of water loss by citrus leaves. *Physiol. Plant.* **33,** 62–65.

Halma, F. F. (1935). Trunk growth and the water relation in leaves of citrus. *Proc. Am. Soc. Hortic. Sci.* **32,** 273–276.

Harding, R. B., and Ryan, T. M. (1961). Soil salinity and water distribution under nontillage and tillage in a differentially fertilised irrigated citrus orchard. *Proc. Am. Soc. Hortic. Sci.* **77,** 155–166.

Haseba, T., and Takechi, O. (1966). Studies of transpiration in relation to the environment. I. Diurnal change in transpiration from citrus leaves. *J. Agric. Meteorol. (Tokyo)* **22,** 1–6.

Hayward, H. E., and Long, E. M. (1942). The anatomy of the seedlings and roots of the Valencia orange. *U. S., Dep. Agric., Tech. Bull.* **786,** 1–31.

Hayward, H. E., Blair, W. M., and Skaling, P. E. (1942). Device for measuring entry of water into roots. *Bot. Gaz. (Chicago)* **104,** 152–160.

Heine, R. W. (1971). Hydraulic conductivity in trees. *J. Exp. Bot.* **22,** 503–511.

Heller, J., Shalhevet, J., and Goell, A. (1973). Response of a citrus orchard to soil moisture and soil salinity. *Ecol. Stud.* **4,** 409–419.

Hilgeman, R. H. (1951). Irrigation of Valencia oranges. *Calif. Citrogr.* **36,** 370, 372.

Hilgeman, R. H. (1956). New irrigation plan for orange trees. *Calif. Citrogr.* **41,** 455.

Hilgeman, R. H. (1963). Trunk growth of Valencia orange in relation to soil moisture and climate. *Proc. Am. Soc. Hortic. Sci.* **82,** 193–198.

Hilgeman, R. H. (1974). Irrigation of 'Valencia' orange trees by wetting alternate sides. *Congr. Mund. Citric., 1st, 1973* Vol. 1, pp. 265–269.

Hilgeman, R. H. (1977). Response of citrus trees to water stress in Arizona. *Proc. Int. Soc. Citric., 1977* Vol. 1, pp. 70–74.

Hilgeman, R. H., and Sharp, F. O. (1970). Response of 'Valencia' orange trees to four soil water schedules during 20 years. *J. Am. Soc. Hortic. Sci.* **95,** 739–745.

Hilgeman, R. H., Tucker, H., and Hales, T. A. (1959). The effect of temperature, precipitation, blossom date and yield upon the enlargement of Valencia oranges. *Proc. Am. Soc. Hortic. Sci.* **74,** 266–279.

Hilgeman, R. H., Ehrler, W. L., Everling, C. E., and Sharp, F. O. (1969). Apparent transpiration and internal water stress in Valencia oranges as affected by soil water, season and climate. *Proc. Int. Citrus Symp. 1st, 1968* Vol. 3, pp. 1713–1723.

Hirano, E. (1931). Relative abundance of stomata in citrus and some related genera. *Bot. Gaz. (Chicago)* **92,** 296–310.

Hoare, E. R., and Barrs, H. D. (1974). Water relations and photosynthesis amongst horticultural species as affected by simulated soil water stress. *Proc. Int. Hortic. Congr. 19th, 1974* Vol. III, pp. 321–334.

Holloway, P. J., and Baker, E. A. (1970). The cuticles of some angiosperm leaves and fruits. *Ann. Appl. Biol.* **66,** 145–153.

Holmgren, P., Jarvis, P. G., and Jarvis, M. S. (1965). Resistances to carbon dioxide and water vapour transfer in leaves of different plant species. *Physiol. Plant.* **18,** 557–573.

Hopmans, P. A. M. (1971). Rhythms in stomatal opening of bean leaves. *Meded. Landbouwhogesch. Wageningen* **71-3,** 1–86.

Huberty, M. R., and Richards, S. J. (1954). Irrigation tests with oranges. *Calif. Agric.* **8,** 8–15.

Iwasaki, T., Owada, A., and Iiya, Y. (1959). Studies on the differentiation and development of the flower bud in citrus. I–VI. *Bull. Tokai-Kinki Natl. Agric. Exp. Stn., Hortic.* **5,** 1–76.

Janes, B. E. (1970). Effect of carbon dioxide, osmotic potential of nutrient solution and light-intensity on transpiration and resistance to flow of water in pepper plants. *Plant Physiol.* **45**, 95–103.

Jeffree, C. E., Johnson, R. P. C., and Jarvis, P. G. (1971). Epicuticular wax in the stomatal antechamber of Sitka spruce and its effects on the diffusion of water vapour and carbon dioxide. *Planta* **98**, 1–10.

Jensen, M. E., and Haise, H. R. (1963). Estimating evapotranspiration from solar radiation. *J. Irrig. Drainage Div. Am. Soc. Civ. Eng.* **89**, 15–41.

Jensen, R. D., Taylor, S. A., and Wiebe, H. H. (1961). Negative transport and resistance to water flow through plants. *Plant Physiol.* **36**, 633–638.

Johnson, R. P. C. (1977). Can cell walls bending round xylem vessels control water flow? *Planta* **136**, 187–194.

Jones, W. W., and Cree, C. B. (1965). Environmental factors related to fruiting of Washington navel oranges over a 38-year period. *Proc. Am. Soc. Hortic. Sci.* **86**, 267–271.

Jones, W. W., and Steinacker, M. L. (1951). Seasonal changes in concentrations of sugar and starch in leaves and twigs of citrus trees. *Proc. Am. Soc. Hortic. Sci.* **58**, 1–4.

Jones, W. W., Cree, C. B., and Embleton, T. W. (1961). Some effects of nitrogen sources and cultural practices on water intake by soil in a Washington navel orange orchard and on fruit production, size and quality. *Proc. Am. Soc. Hortic. Sci.* **77**, 146–154.

Jones, W. W., Embleton, T. W., and Coggins, C. W., Jr. (1975). Starch content of roots of 'Kinnow' mandarin trees bearing fruit in alternate years. *HortScience* **10**, 514.

Kadoya, K. (1973). Studies on the translocation of photosynthates in satsuma mandarin. IV. Diurnal fluctuations of fruit growth and some environmental conditions. *J. Jpn. Soc. Hortic. Sci.* **42**, 215–220.

Kadoya, K. (1977). Studies of the hydrophysiological rhythms of citrus trees. II. Seasonal effects of carbon dioxide on the cyclic fluctuations of leaf thickness. *J. Jpn. Soc. Hortic. Sci.* **46**, 153–157.

Kadoya, K. (1978). Studies on the hydrophysiological rhythms of citrus trees. III. The effect of soil temperature on cyclic fluctuations in leaf thickness. *J. Jpn. Soc. Hortic. Sci.* **47**, 167–171.

Kadoya, K., Kameda, K., Chikaizumi, S., and Matsumoto, K. (1975). Studies on the hydrophysiological rhythms of citrus trees. I. Cyclic fluctuations of leaf thickness and stem diameter of natsudaidai seedlings. *J. Jpn. Soc. Hortic. Sci.* **44**, 260–264.

Kalma, J. D. (1970). Some aspects of the water balance of an irrigated orange plantation. Ph.D. Thesis, Volcani Inst. Agric. Res., Bet Dagan, Israel.

Kalma, J. D., and Fuchs, M. (1976). Citrus orchards. *In* "Vegetation and the Atmosphere" (J. L. Monteith, ed.), Vol. 2, pp. 309–328. Studies. Academic Press, New York.

Kalma, J. D., and Stanhill, G. (1969). Transpiration, evaporation and deep drainage losses from an orange plantation. *Isr. J. Agric. Res.* **19**, 11–24.

Kalma, J. D., and Stanhill, G. (1972). The climate of an orange orchard: Physical characteristics and microclimate relationships. *Agric. Meteorol.* **10**, 185–201.

Kamota, F., Ban, Y., and Shimura, K. (1974). Studies on photosynthesis and transpiration of vegetable crops. I. Differential response to soil moisture in photosynthesis and transpiration among crops. *Bull. Veg. Ornamental Crops Res. Stn., Ser. A* **1**, 109–139.

Kaufmann, M. R. (1968). Evaluation of the pressure chamber method for measurement of water stress in citrus. *Proc. Am. Soc. Hortic. Sci.* **93**, 186–190.

Kaufmann, M. R. (1969). Relation of slice gap width in oranges and plant water stress. *J. Am. Soc. Hortic. Sci.* **94**, 161–163.

Kaufmann, M. R. (1970a). Water potential components in growing citrus fruits. *Plant Physiol.* **46**, 145–149.

Kaufmann, M. R. (1970b). Extensibility of pericarp tissue in growing citrus fruits. *Plant Physiol.* **46**, 778–781.

Kaufmann, M. R. (1975). Leaf water stress in Engelmann spruce: Influence of the root and shoot environments. *Plant Physiol.* **56**, 841–844.

Kaufmann, M. R. (1976). Water transport through plants: Current perspectives. *In* "Transport and Transfer Processes in Plants" (I. F. Wardlaw and J. B. Passioura, eds.), pp. 313–327. Academic Press, New York.

Kaufmann, M. R. (1977a). Citrus under minimum water conditions. *Citrograph* **62**, 277–278.

Kaufmann, M. R. (1977b). Citrus—a case study of environmental effects on plant water relations. *Proc. Int. Soc. Citric., 1977* Vol. 1, pp. 57–62.

Kaufmann, M. R., and Levy, Y. (1976). Stomatal response of *Citrus jambhiri* to water stress and humidity. *Physiol. Plant.* **38**, 105–108.

Kaufmann, M. R., Boxwell, S. B., and Lewis, L. N. (1972). Effect of tree spacing on root distribution of 9-year old 'Washington' navel oranges. *J. Am. Soc. Hortic. Sci.* **97**, 204–206.

Khairi, M. M. A., and Hall, A. E. (1976a). Effects of air and soil temperatures on vegetative growth of citrus. *J. Am. Soc. Hortic. Sci.* **101**, 337–341.

Khairi, M. M. A., and Hall, A. E. (1976b). Temperature and humidity effects on net photosynthesis and transpiration of citrus. *Physiol. Plant.* **36**, 29–34.

Khairi, M. M. A., and Hall, A. E. (1976c). Comparative studies of net photosynthesis and transpiration of some citrus species and relatives. *Physiol. Plant.* **36**, 35–39.

Kishore, V., and Chand, J. N. (1972). Citrus canker in Haryana. *Haryana Agric. Univ. J. Res.* **2**, 124–127.

Klepper, B., and Ceccato, R. D. (1969). Determinations of leaf and fruit water potential with a pressure chamber. *Hortic. Res.* **9**, 1–7.

Koo, R. C. J., and Reese, R. L. (1977). Influence of nitrogen, potassium and irrigation on citrus fruit quality. *Proc. Int. Soc. Citric., 1977* Vol. 1, pp. 34–38.

Koo, R. C. J., and Sites, J. W. (1955). Results of research and response of citrus to supplemented irrigation. *Soil Sci. Soc. Fla., Proc.* **15**, 180–190.

Kramer, P. J. (1940). Root resistance as a cause of decreased water absorption by plants at low temperatures. *Plant Physiol.* **15**, 63–79.

Kramer, P. J., and Kozlowski, T. T. (1960). "Physiology of Trees." McGraw-Hill, New York.

Kramer, P. J., and Kozlowski, T. T. (1979). "Physiology of Woody Plants." Academic Press, New York.

Kriedemann, P. E. (1968). Some photosynthetic characteristics of citrus leaves. *Aust. J. Biol. Sci.* **21**, 895–905.

Kriedemann, P. E. (1971). Photosynthesis and transpiration as a function of gaseous diffusive resistances in orange leaves. *Physiol. Plant.* **24**, 218–225.

Kriedemann, P. E., Loveys, B. R., Fuller, G. L., and Leopold, A. C. (1972). Abscisic acid and stomatal regulation. *Plant Physiol.* **49**, 842–847.

Landsberg, J. J., Blanchard, T. W., and Warrit, B. (1976). Studies on the movement of water through apple trees. *J. Exp. Bot.* **27**, 579–596.

Lange, O. L., Lösch, R., Schulze, E.-D., and Kappen, L. (1971). Responses of stomata to changes in humidity. *Planta* **100**, 76–86.

Lenz, F. (1967). Relationships between the vegetative and reproductive growth of Washington navel orange cuttings (*Citrus sinensis* L. Osbeck). *J. Hortic. Sci.* **42**, 31–39.

Levitt, J. (1972). "Responses of Plants to Environmental Stresses." Academic Press, New York.

Levitt, J., and Ben Zaken, R. (1975). Effects of small water stresses on cell turgor and intercellular space. *Physiol. Plant.* **34**, 273–279.

Levy, Y., and Kaufmann, M. R. (1976). Cycling of leaf conductance in citrus exposed to natural and controlled environments. *Can. J. Bot.* **54**, 2215–2218.

Levy, Y., Bar-Akiva, A., and Vaadia, Y. (1978a). Influence of irrigation and environmental factors on grapefruit acidity. *J. Am. Soc. Hortic. Sci.* **103**, 73–76.

Levy, Y., Bielorai, H., and Shalhevet, J. (1978b). Long-term effects of different irrigation regimes on grapefruit tree development and yield. *J. Am. Soc. Hortic. Sci.* **103**, 680–683.

Leyden, R. F. (1975a). Drip irrigation on bearing citrus trees. *J. Rio Grande Val. Hortic. Soc.* **29**, 31–36.

Leyden, R. F. (1975b). Comparison of three irrigation systems for young citrus trees. *J. Rio Grande Val. Hortic. Soc.* **29**, 25–29.

Liebig, G. F., and Chapman, H. D. (1963). The effect of variable root temperatures on the behaviour of young Navel orange trees in a greenhouse. *Proc. Am. Soc. Hortic. Sci.* **82**, 204–209.

Lombard, P. B., Stolzy, L. H., Garber, M. J., and Szuszkiewicz, T. E. (1965). Effects of climatic factors on fruit volume increase and leaf water deficit of citrus in relation to soil suction. *Soil Sci. Soc. Am., Proc.* **29**, 205–208.

Luxmoore, R. J., Stolzy, L. H., Joseph, H., and DeWolfe, T. A. (1971). Gas space porosity of citrus roots. *HortScience* **6**, 447–448.

Machida, Y., and Maotani, T. (1974). Studies on leaf moisture stress in fruit trees. I. Evaluation of the pressure chamber method for estimating the leaf water potential of satsuma trees. *J. Jpn. Soc. Hortic. Sci.* **43**, 7–14.

Magness, J. R. (1953). Soil moisture in relation to fruit tree functioning. *Rep. Int. Hortic. Congr., 13th, 1952* Vol. 1, pp. 230–239.

Mantell, A. (1977). Water use efficiency of citrus: Room for improvement? *Proc. Int. Soc. Citric., 1977* Vol. 1, pp. 74–79.

Mantell, A., Meirovitch, A., and Goell, A. (1976). The response of Shamouti orange on sweet lime rootstock to different irrigation regimes in the spring and summer (Progress report, 1973–1974). *Spec. Publ., Agric. Res. Org. Bet Dagan* No. 58, pp. 1–16.

Mantell, A., Monselise, S. P., and Goldschmidt, E. E. (1979). Movement of tritiated water through young citrus plants. *J. Exp. Bot.* **30**, 155–164.

Maotani, T., Yamatsu, K., and Machida, Y. (1976). Studies on leaf water stress in fruit trees. IV. The role of satsuma fruits as water reservoirs for the leaves. *Bull. Fruit Tree Res. Stn., Ser. E* No. 1, pp. 51–57.

Maotani, T., Machida, Y., and Yamatsu, K. (1977). Studies on leaf water stress in fruit trees. VI. Effects of leaf water potential on growth of satsuma trees. *J. Jpn. Soc. Hortic. Sci.* **45**, 329–334.

Marloth, R. (1950). Citrus growth studies. II. Fruit growth and fruit internal quality changes. *J. Hortic. Sci.* **25**, 235–248.

Marsh, A. W. (1973). Irrigation. *In* "The Citrus Industry" (W. Reuther ed.), Vol. III, pp. 230–279. Div. Agric. Sci., University of California, Berkeley.

Mathews, I. (1972). Drought damage to citrus trees—resultant problems and their treatment. *Citrus Grow. Sub-Trop. Fruit J.* No. 464, pp. 11–13.

Mendel, K. (1951). Orange leaf transpiration under orchard conditions. Part III. Prolonged soil drought and the influence of stocks. *Palest. J. Bot., Rehovot Ser.* **8**, 45–53.

Meyer, B. S., and Anderson, D. B. (1952). "Plant Physiology." Van Nostrand-Reinhold, Princeton, New Jersey.

Miskin, K. E., and Rasmusson, D. C. (1970). Frequency and distribution of stomata in barley. *Crop Sci.* **10**, 575–578.

Monselise, S. P. (1947). The growth of citrus roots and shoots under different cultural conditions. *Palest. J. Bot., Rehovot Ser.* **6**, 43–54.

Monselise, S. P. (1951a). Light distribution in citrus trees. *Bull. Res. Counc. Isr., Sect. D* **1**, 36–53.

Monselise, S. P. (1951b). Growth analysis of citrus seedlings. I. Growth of sweet lime seedlings in dependence upon illumination. *Pales. J. Bot., Rehovot Ser.* **8**, 54–75.

Monselise, S. P. (1951c). Some differences between sun and shade leaves of citrus trees. *Palest. J. Bot., Rehovot Ser.* **8**, 99–101.

Monselise, S. P., and Halevy, A. H. (1964). Chemical inhibition and promotion of citrus flower bud induction. *Proc. Am. Soc. Hortic. Sci.* **84**, 141–146.

Monteith, J. L. (1959). The reflection of short-wave radiation by vegetation. *Q. J. R. Meteorol. Soc.* **85**, 386–392.

Moss, G. I., and Muirhead, W. A. (1971a). Climatic and tree factors relating to the yield of orange trees. I. Investigations on 'Washington Navel' and 'Late Valencia' cultivars. *Hortic. Res.* **11**, 3–17.

Moss, G. I., and Muirhead, W. A. (1971b). Climatic and tree factors relating to the yield of orange trees. II. Interactions with cultural and nitrogen fertiliser treatments. *Hortic. Res.* **11**, 75–84.

Myers, J. M. (1977). Functional performance of irrigation systems for orchard crops in Florida. *Fla. State Hortic. Soc., Proc.* **90**, 258–260.

Nakagawa, Y. (1969). Studies on the favourable climatic environments for fruit culture. VI. Analysis of climatological conditions in the major citrus production areas in the world. *Bull. Hortic. Res. Stn., Ser. A* **8**, 73–94.

Nemec, S. (1975). Aging characteristics of wood of healthy and sandhill-declined *Citrus sinensis*. *Can. J. Bot.* **53**, 2712–2719.

Nemec, S., Constant, R., and Patterson, M. (1975). Distribution of obstructions to water movement in citrus with and without blight. *Fla. State Hortic. Soc., Proc.* **88**, 70–75.

Newman, E. I. (1969). Resistance to water flow in soil and plant. II. A review of experimental evidence on the rhizosphere resistance. *J. Appl. Ecol.* **6**, 261–272.

Newman, E. I. (1973). Permeability to water of the roots of five herbaceous species. *New Phytol.* **72**, 547–555.

Newman, E. I. (1976). Water movement through root systems. *Philos. Trans. R. Soc. London, Ser. B* **273**, 463–478.

Nir, I., Goren, R., and Leshem, B. (1972). Effects of water stress, gibberellic acid and 2-chloroethyltrimethylammoniumchloride (CCC) on flower differentiation in 'Eureka' lemon trees. *J. Am. Soc. Hortic. Sci.* **97**, 774–778.

Nobel, P. S., Zaragoza, L. J., and Smith, W. K. (1975). Relation between mesophyll surface area, photosynthetic rate and illumination level during development for leaves of *Plectranthus parviflorus* Henckel. *Plant Physiol.* **55**, 1067–1070.

North, C. P., and Wallace, A. (1955). Soil temperature and citrus. *Calif. Agric.* **9**, 13.

Oades, J. M. (1978). Mucilages at the root surface. *J. Soil Sci.* **29**, 1–16.

Ongun, A. R., and Wallace, A. (1958a). Inorganic composition of small Washington navel orange trees on different rootstocks grown in a glasshouse with different root temperatures. *In* "Tree Physiology Studies at U.C.L.A." (A. Wallace, ed.), Spec. Rep. No. 1, pp. 75–83. Dep. Hortic. Sci., University of California, Los Angeles.

Ongun, A. R., and Wallace, A. (1958b). Influence of root temperature on the growth of small Washington navel orange trees on three different rootstocks in a glasshouse. *In* "Tree Physiology Studies at U.C.L.A." (A. Wallace ed.), Spec. Rep. No. 1, pp. 84–86. Dep. Hortic. Sci., University of California, Los Angeles.

Ongun, A. R., and Wallace, A. (1958c). Transpiration rates of small Washington navel orange trees grown in a glasshouse with different rootstocks and at different root temperatures. *In* "Tree Physiology Studies at U.C.L.A." (A. Wallace ed.), Spec. Rep. No. 1, pp. 87–103. Dep. Hortic. Sci., University of California, Los Angeles.

Ono, S., Kudo, K., and Daito, H. (1978). Studies on the photosynthesis and productivity of satsumas (*Citrus unshiu*). I. Effect of environmental factors on photosynthetic activity. *Bull. Shikoku Agric. Exp. Stn.* **31**, 147–157.

Oppenheimer, H. R., and Mendel, K. (1934). Some experiments on water relations of citrus trees. *Hadar* **7**, 2, 3, 6.

Owen, P. C., and Watson, D. J. (1956). Effect on crop growth of rain after prolonged drought. *Nature (London)* **177**, 847.

Penman, H. L. (1948). Natural evaporation from open water, bare soil, and grass. *Proc. R. Soc. London Ser. A* **193**, 120–146.

Pisek, A., Larcher, W., and Unterholzner, R. (1967). Kardinale Temperaturbereiche der Photosynthese und Grenztemperaturen des Lebens der Blätter verschiedener Spermatophyten. I. Temperaturminimum der Nettoassimilation, Gefrier-und Frostschadensbereiche der Blätter. *Flora (Jena), Abt. B* **157**, 239–264.

Pospisilova, J. (1972). Variable resistance to water transport in leaf tissue of kale. *Biol. Plant.* **14**, 293–296.

Possingham, J. V., and Kriedemann, P. E. (1969). Environmental effects on the formation and distribution of photosynthetic assimilates in citrus. *Proc. Int. Citrus Symp., 1st, 1968* Vol. 1, pp. 325–332.

Possingham, J. V., Kriedemann, P. E., and Loveys, B. R. (1980). Photosynthesis and Growth in Citrus. Proceedings of an International Symposium on Citriculture, Bangalore, December 1977, Hort. Soc., India.

Raciti, G., and Scuderi, A. (1971). The effect of hydric factors on the occurrence of "creasing". *Ann. Ist. Sper. Agrum., 1970/1971* Vols. 3/4, pp. 263–286.

Ramos, C., and Kaufmann, M. R. (1979). Hydraulic resistance of rough lemon roots. *Physiol. Plant.* **45**, 311–314.

Raschke, K. (1975). Stomatal action. *Annu. Rev. Plant Physiol.* **26**, 309–340.

Rawlins, S. L. (1963). Resistance to water flow in the transpiration stream. *Conn., Agric. Exp. Stn., New Haven, Bull.* **664**, 69–85.

Reed, H. S., and Hirano, E. (1931). The density of stomata in citrus leaves. *J. Agric. Res.* **43**, 209–222.

Reuther, W. (1973). Climate and citrus behavior. *In* "The Citrus Industry" (W. Reuther, ed.), Vol. III, pp. 281–337. Div. Agric. Sci., University of California, Berkeley.

Reuther, W. (1977). Citrus. *In* "Ecophysiology of Tropical Crops" (P. deT. Alvim and T. T. Kozlowski, eds.), pp. 409–439. Academic Press, New York.

Reuther, W., and Rios-Castano, D. (1969). Comparison of growth, maturation and composition of citrus fruits in subtropical California and tropical Colombia. *Proc. Int. Citrus Symp., 1st, 1968* Vol. 1, pp. 277–300.

Rogers, J. S., and Bartholic, J. F. (1976). Estimated evapotranspiration and irrigation requirements for citrus. *Soil Crop Sci. Soc. Fla., Proc.* **35**, 111–117.

Rokach, A. (1953). Water transfer from fruits to trees in the Shamouti orange tree and related topics. *Palest. J. Bot., Rehovot Ser.* **8**, 146–151.

Samish, R. M. (1957). Irrigation requirements of a young citrus orchard. *Ktavim (Engl. Ed.)* **7**, 123–139.

Sapra, V. T., Hughes, J. L., and Sharma, G. C. (1975). Frequency, size and distribution of stomata in Triticale leaves. *Crop Sci.* **15**, 356–358.

Savage, E. M., Cooper, W. C., and Piper, R. B. (1945). Root systems of various citrus rootstocks. *Proc. Fla. State Hortic. Soc.* **58**, 44–48.

Schneider, H. (1968). The anatomy of citrus. *In* "The Citrus Industry" (W. Reuther, L. D.

Batchelor, and H. J. Webber, eds.), Vol. II, pp. 1–85. Div. Agric. Sci., University of California, Berkeley.

Schneider, H., and Baines, R. C. (1968). *Tylenchulus semipenetrans:* parasitism and injury to orange tree roots. *Phytopathology* **54**, 1202–1206.

Scholander, P. F., Hammel, H. T., Hemmingsen, E. A., and Bradstreet, E. D. (1964). Hydrostatic pressure and osmotic potential in leaves of mangroves and some other plants. *Proc. Natl. Acad. Sci. U.S.A.* **52**, 119–125.

Schönherr, J., and Schmidt, H. W. (1979). Water permeability of plant cuticles: Dependence of permeability coefficients of cuticular transpiration on vapour pressure saturation deficit. *Planta* **144**, 391–400.

Schulman, Y., and Monselise, S. P. (1970). Some studies on the cuticular wax of citrus fruits. *J. Hortic. Sci.* **45**, 471–478.

Scora, R. W. (1975). On the history and origin of citrus. *Bull. Torrey Bot. Club.* **102**, 369–375.

Scott, F. M., Schroeder, M. R., and Turrell, F. M. (1948). Development, cell shape, suberization of internal surface, and abscission in the leaf of the Valencia orange, *Citrus sinensis. Bot. Gaz. (Chicago)* **109**, 381–411.

Shalhevet, J., and Bielorai, H. (1978). Crop water requirement in relation to climate and soil. *Soil Sci.* **125**, 240–247.

Shalhevet, J., Mantell, A., Bielorai, H., and Shimshi, D. (1976). Irrigation of field and orchard crops under semi-arid conditions. *Int. Irrig. Inf. Cent. Publ.* No. 1.

Sheriff, D. W. (1977). The effect of humidity on water uptake by, and viscous flow resistance of, excised leaves of a number of species: Physiological and anatomical observations. *J. Exp. Bot.* **28**, 1399–1407.

Shmueli, E., Bielorai, H., Heller, J., and Mantell, A. (1973). Citrus water requirement experiments conducted in Israel during the 1960's. *Ecol. Stud.* **4**, 339–350.

Sites, J. W., and Reitz, H. J. (1949). The variation in individual Valencia oranges from different locations of the tree as a guide to sampling methods and spot-picking for quality. I. Soluble solids in the juice. *Proc. Am. Soc. Hortic. Sci.* **54**, 1–10.

Sites, J. W., Reitz, H. J., and Deszyck, E. J. (1951). Some results of irrigation research with Florida citrus. *Proc. Fla. State Hortic. Soc.* **64**, 71–79.

Slatyer, R. O. (1967). "Plant-Water Relationships." Academic Press, New York.

Slavikova, J. (1967). Compensation of root suction force within a single root system. *Biol. Plant.* **9**, 20–27.

Smart, R. E., and Barrs, H. D. (1973). The effect of environment and irrigation interval on leaf water potential of four horticultural species. *Agric. Meteorol.* **12**, 337–346.

Smith, P. F. (1965). Effect of nitrogen source and placement on the root development of Valencia orange trees. *Proc. Fla. State Hortic. Soc.* **78**, 55–59.

Spurling, M. B. (1951). Water requirements of citrus. *Bull. Dep. Agric., South Aust.* No. 419, pp. 1–19.

Stanhill, G. (1960). The relationship between climate and the transpiration and growth of pastures. *Proc. Int. Grassl. Congr., 8th, 1960* pp. 293–296.

Stanhill, G. (1961). A comparison of methods of calculating potential evapotranspiration from climatic data. *Isr. J. Agric. Res.* **11**, 159–171.

Stanhill, G. (1972). Recent developments in water relations studies: Some examples from Israel citriculture. *Proc. Int. Hortic. Congr. 18th, 1970* Vol. 4, pp. 367–379.

Stanhill, G., Hofstede, G. J., and Kalma, J. D. (1966). Radiation balance of natural and agricultural vegetation. *Q. J. Roy. Meteorol. Soc.* **92**, 128–140.

Stolzy, L. H., Taylor, O. C., Garber, M. J., and Lombard, P. B. (1963). Previous irrigation treatments as factors in subsequent irrigation level studies in orange production. *Proc. Am. Soc. Hortic. Sci.* **82**, 199–203.

Suzuki, T., and Kaneko, M. (1970). The effect of suction pressures in the soil solution during summer on growth and fruiting of young satsuma orange trees. *J. Jpn. Soc. Hortic. Sci.* **39**, 99–106.

Tan, G.-Y., and Dunn, G. M. (1975). Stomatal length, frequency and distribution in *Bromus inermis* Leyss. *Crop Sci.* **15**, 283–286.

Teoh, C. T., and Palmer, J. H. (1971). Nonsynchronized oscillations in stomatal resistance among sclerophylls of *Eucalyptus umbra*. *Plant Physiol.* **47**, 409–411.

Thompson, I. R., Stolzy, L. H., and Taylor, O. C. (1965). Effect of soil suction, relative humidity and temperature on apparent photosynthesis and transpiration of rough lemon (*Citrus jambhiri*). *Proc. Am. Soc. Hortic. Sci.* **87**, 168–175.

Tinker, P. B. (1976). Transport of water to plant roots in soil. *Philos. Trans. R. Soc. London, Ser. B* **273**, 445–461.

Torre, G. A., and Zabala, L. M. (1976). Comportamiento de la intensidad transpiratoria y el déficit hídrico en plantas jóvenes de cîtricos. *Cienc. Agropec., Ser. I* No. 23, pp. 1–9.

Trumble, H. P. C. (1952). Improved irrigation programmes for citrus. *J. Dep. Agric. South Aust.* **56**, 133–139.

Tukey, H. B., Jr. (1970). The leaching of substances from plants. *Annu. Rev. Plant Physiol.* **21**, 305–324.

Turrell, F. M. (1936). The area of the internal exposed surface of dicotyledon leaves. *Am. J. Bot.* **23**, 255–263.

Turrell, F. M. (1947). Citrus leaf stomata: Structure, composition and pore size in relation to penetration of liquids. *Bot. Gaz. (Chicago)* **108**, 476–483.

Turrell, F. M., Austin, S. W., and Perry, R. L. (1962). Nocturnal thermal exchange of citrus leaves. *Am. J. Bot.* **49**, 97–109.

Van Bavel, C. H. M., Newman, J. E., and Hilgeman, R. H. (1967). Climate and estimated water use by an orange orchard. *Agric. Meteorol.* **4**, 27–37.

van den Honert, T. H. (1948). Water transport in plants as a catenary process. *Discuss. Faraday Soc.* **3**, 146–153.

VanderMolen, G. E. (1978). Electron microscopy of vascular obstructions in citrus roots affected with young tree decline. *Physiol. Plant Pathol.* **13**, 271–274.

van der Weert, R., Lenselink, K. J., and van Sloten, D. H. (1973). Citrus yield in relation to soil moisture. *Landbouwproefstn. Suriname, Bull.* **90**, 1–35.

Vinokur, R. L. (1957). Influence of temperature of the root environment on root activity, transpiration and photosynthesis of leaves of lemon. *Sov. Plant Physiol. (Engl. Transl.)* **4**, 268–273.

Wallace, A., and Nauriyal, J. P. (1958). Some growth habits of Valencia orange trees. *In* "Tree Physiology Studies at U.C.L.A." (A. Wallace, ed.), Spec. Rep. No. 1, pp. 46–56. Dep. Hortic. Sci., University of California, Los Angeles.

Wallace, A., Naude, C. J., Mueller, R. T., and Zidan, Z. I. (1952). The rootstock-scion influence on the inorganic composition of citrus. *Proc. Am. Soc. Hortic. Sci.* **59**, 133–142.

Waynick, D. D. (1928). Factors concerned in the growth of Valencia orange. *Calif. Citrogr.* **13**, 200.

Waynick, D. D. (1929). Effect of late thinning on growth of Valencia oranges. *Calif. Citrogr.* **14**, 509.

Waynick, D. D., and Walker, S. J. (1930). Rooting habits of citrus trees. *Calif. Citrogr.* **15**, 201, 238–239.

Webber, H. J. (1943). Plant characteristics and climatology. *In* "The Citrus Industry" (H. J. Webber, ed.), Vol. II, pp. 41–70. Univ. of California Press, Berkeley.

West, E. S., and Perkman, O. (1953). Effect of soil moisture on transpiration. *Aust. J. Agric. Res.* **4**, 326–333.

Wong, S. C., Cowan, I. R., and Farquhar, G. D. (1979). Stomatal conductance correlates with photosynthetic capacity. *Nature (London)* **282**, 424–426.

Yaron, B., Shalhevet, J., and Shimshi, D. (1973). Patterns of salt distribution under trickle irrigation. *Ecol. Stud.* **4**, 389–394.

Young, R., and Meredith, F. (1971). Effect of exposure to subfreezing temperatures on ethylene evolution and leaf abscission in citrus. *Plant Physiol.* **48**, 724–727.

Young, R. H., and Garnsey, S. M. (1977). Water uptake patterns in blighted citrus trees. *J. Am. Soc. Hortic. Sci.* **102**, 751–756.

Zahner, R. (1968). Water deficits and growth of trees. *In* "Water Deficits and Plant Growth" (T. T. Kozlowski, ed.), Vol. 2, pp. 191–254. Academic Press, New York.

CHAPTER 6

APPLE ORCHARDS

J. J. Landsberg

LONG ASHTON RESEARCH STATION, LONG ASHTON, BRISTOL, ENGLAND

and

H. G. Jones

EAST MALLING RESEARCH STATION, EAST MALLING, MAIDSTONE, KENT, ENGLAND

I. INTRODUCTION

The objectives of research on water deficits and plant growth must be to understand how water deficits develop and how they affect the opera-

Water Deficits and Plant Growth, Vol. VI
Copyright © 1981 by Academic Press, Inc.
All rights of reproduction in any form reserved.
ISBN 0-12-424156-5

tion of the physiological processes that contribute to plant growth. However, although this knowledge may be of considerable academic interest, it is of no practical value unless we also understand how these physiological processes affect the performance of the plant.

Apple trees are complex perennial plants and water deficits affect them in a number of ways, depending on the stage in the annual growth cycle at which the deficits occur, and on the magnitude of the deficits. Since apple trees are cultivated for their fruit, and our concern is with the effects of deficits on fruit production, it is important to understand the annual production cycle of the trees and the way the various physiological processes which constitute this cycle contribute to fruit production.

The growth pattern of a plant is determined by the way in which assimilates are allocated to various organs and tissues. It has been shown (Heim *et al.*, 1979) that partitioning of assimilates in apple trees is strongly dependent upon the number of fruits present on a tree. Studies on water relations of apple trees must, therefore, have as their objective the identification of the effects of water deficits on production of fruits, on production of assimilates, and on patterns of assimilate partitioning.

The aim of this chapter is to present and discuss current knowledge of how water deficits develop in apple trees, how accurately these deficits can be calculated, and how they affect growth and productivity of apple trees. The several processes involved will be linked by means of simple mathematical models of the flow of water through the soil–plant–atmosphere system and of the effects of water deficits on physiological processes. We also consider the effects of water deficits on fruit quality and the questions of drought resistance and adaptation.

II. GROWTH AND REPRODUCTIVE CYCLE

The growth and reproductive cycle of a bearing apple tree is depicted diagrammatically in Fig. 1. We outline here the main events that take place during the cycle, with some comments on the way in which they may be affected by water stress. Detailed discussion of the effects of water deficits on particular physiological processes, and on vegetative and reproductive growth, is postponed to later sections.

The process of apple bud morphogenesis has been analyzed by Landsberg and Thorpe (1975), who described it by a model. Bud development begins shortly after full bloom and proceeds through the summer. The buds, which consist of shortened axes bearing "leaf formations" at each node, will not produce floral primordia unless they have produced about 20 nodes by the end of the growing season (Abbott, 1977). Those that do

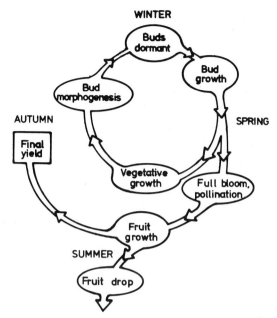

Fig. 1. Schematic outline of the growth and reproduction cycle of an apple tree.

(fruit buds) are usually formed on spurs—short lateral shoots which may or may not carry fruits. The number of buds that produce floral primordia will be reduced if the tree is carrying a large number of fruits.

When morphogenesis is complete apple buds become dormant. A period of chilling is usually required before growth can begin but severe water stress may substitute for it (H. G. Jones, unpublished). Landsberg (1974) has suggested that dormancy is induced by accumulation of inhibiting substances, the breaking of dormancy by cold being brought about by their breakdown, leading to a switch in the balance of growth substances in favor of growth promoting substances. However, there is an alternative suggestion from Russian work (Kolomiec, cited by Salter and Goode, 1967) to the effect that growth of flower buds follows reduction in cell sap concentration. This would explain the effect of water stress and the occasional occurrence of out-of-season blossom. In the arid steppe region of Russia, Garjugin (1964) found that winter irrigation improved the differentiation of fruit buds, as well as growth and fruiting of apples, plums, cherries and apricots.

Dormancy is a physiological condition and is quantitative; in suitable temperatures bud growth will commence after a relatively short period of cold but will not proceed at the maximum rate, at any given temperature, if the cold period has not been long enough (Landsberg, 1974). With post-dormancy rise in temperatures, growth proceeds through bud swelling and opening with leaves appearing about 30 days before the flowers open. The leaves subtending flower clusters are usually smaller than those produced by vegetative buds. Transpiration begins shortly after the leaves unfold. Water stress at this time will have adverse effects on leaf growth and possibly on ovule fertility, although there appears to be no information on this point. Modlibowska (1961) found that severe preblossom water stress caused cell sap concentration to increase and that buds in which the water content was reduced to less than 74% were less susceptible to frost damage.

At full bloom each bud gives rise to a cluster of flowers that, in most cultivars, require cross-pollination by insects. The effectiveness of pollination depends on many factors, one of which is ovule fertility. Hot, dry conditions at this time lead to massive loss of flowers. Early stage fruitlet losses are normally heavy, particularly if initial set is heavy, these losses being exacerbated by water stress (see Section VI,C). Fruit drop continues for about 1 month after pollination; there may be subsequent losses but the number of fruits which a tree will carry to harvest is usually determined within, say, 6 weeks of pollination.

Shoot growth commences at about the time of full bloom so that leaf area often increases steadily until shoot growth stops in the latter part of the season—about January in the southern hemisphere and August in the northern hemisphere. On occasions, however, the older leaves may start to fall before this date, giving rise to a long period of nearly constant leaf area (Palmer and Jackson, 1977). The pattern of leaf area development is illustrated by Heim et al. (1979). The leaf area attained by apple trees obviously depends on the size of the tree, and on the pruning regime. Canopy structure and leaf area of orchards are described in Section IV.

From the time of ovule fertilization, apple fruitlets show an initial, relatively slow weight increase, followed by a period of exponential growth, then a declining rate of weight increase until harvest. Cortical cells divide rapidly during the exponential phase. In England, Denne (1960) found differences between the cultivars Millers' Seedling, in which cell division continued for about 12 weeks after full bloom, and Cox's Orange Pippin, in which it continued for only 6–7 weeks. In a similar study carried out in New Zealand (Denne, 1963) cell division in Cox continued for only about 30 days; the difference (from England) was explained by the higher temperatures in New Zealand. Growth after cell division is caused by assimilate uptake and cell expansion and is strongly affected by water stress (see Section VI).

III. CLIMATIC REQUIREMENTS

Apples are grown in a wide range of environments, from Denmark throughout Europe to Northern Italy, in Israel, South Africa, Australia, New Zealand, and Tasmania, in various South American countries, and in North America from Ontario to California. Their optimum climatic requirements might be summarized as a cool-to-cold winter followed by a rapid rise in temperature in spring, with little likelihood of frost as the flowers grow and no early season water stress. In areas where frosts are likely to occur shortly before or during flowering, full bloom can be delayed (to reduce the risk of frost damage) by use of sprinkler irrigation during late winter in order to lower bud temperatures by evaporative cooling (Anderson, 1974). High radiation during the growing season will ensure high rates of photosynthesis, hence large amounts of assimilates. Post-harvest relief of water stress will increase nitrogen mobilization in the soil and encourage the final stages of bud morphogenesis and production of carbohydrate reserves, to be mobilized for bud growth in the spring.

Climatic data from a number of apple-producing regions are presented in Fig. 2, where the extent to which they conform to the "ideal" situation outlined above can be assessed.

The transpiration curves in Fig. 2B represent calculated water loss from the trees of a "standard" modern orchard and, since they do not include losses from vegetation between the tree rows or from bare soil, they represent the minimum water requirements in each region. Where deficits (in relation to rainfall) are indicated, there will almost certainly be periods of quite severe water stress.

IV. EVAPORATION FROM ORCHARDS

A. Canopy Structure

As a prelude to discussing evaporation from orchards and how to calculate it, it is necessary to consider briefly the types and arrangements of trees found in commercial orchards. These vary from the old-fashioned large bush trees, which may have been planted at spacings of 10 × 15 m (i.e., 60–70 trees per hectare), to the modern intensive orchard systems with 2500 or more trees per hectare. Arrangement also varies: hedgerow orchards, fairly typical of modern production systems, consist of trees planted in rows in which the foliage is virtually continuous; row width is likely to be about 4 to 5 m with the trees spaced 2 to 3 m in the row, giving populations in the range 600–1000 trees per hectare. Dwarf bush or

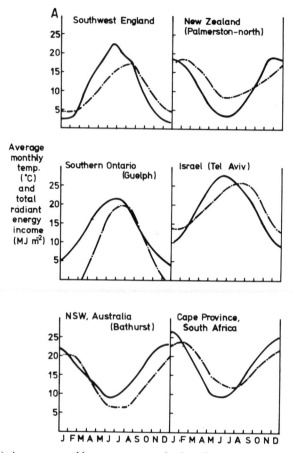

Fig. 2. (A) Average monthly temperatures (broken lines) and total monthly radiant energy income (solid lines) for six apple growing regions. (B) Total monthly transpiration (curves) calculated using the equations and relationships given in the text, and average monthly rainfall (histograms) for the growing seasons of the six regions in (A). The transpiration values do not represent total water use by orchards.

pyramid trees may be planted at similar spacings and intensive systems may use single or double rows, spaced 3 to 4 m apart with the trees about 1 m apart in the row. There are also various training systems using wires, e.g., vertical and oblique cordons, T-frame systems, etc. The tendency to use smaller trees is dictated by both management and productivity considerations. Even higher density plantings have been proposed (Hudson, 1973) but are not yet commercially viable.

Grass or grass–clover swards in interrows are common, but the modern tendency is towards overall herbicide treatments. Clearly the presence

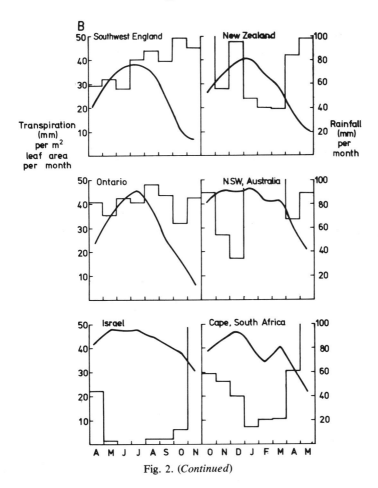

Fig. 2. (*Continued*)

of grass will make an enormous difference to rates of evaporation from orchards. Data on water loss from orchards are scarce (see, however, Black and Mitchell, 1970; Vink, 1971; Goode, 1975); therefore, most of the subsequent discussion is concerned with rates of water loss from the trees alone. Most of the information available pertains to evaporation from the leaves of apple trees, though evaporation from fruits is discussed in Section VII,A. Evaporation from stems will not be considered, although there are reports that it can be a significant source of water loss (e.g., Antipov, 1977). Since stems have relatively few functional stomata, they exercise little control over water loss.

Leaf area and its distribution are important factors determining evaporation from orchards, largely through radiation interception (see p. 427).

The total amount and distribution of foliage are strongly dependent on orchard age, rootstock, and the pruning system adopted (e.g., Jackson, 1970), with the annual maximum leaf area* per tree ranging from less than 1 m² in the year after planting to several hundred square meters at maturity (Jackson, 1975). Because of the slow growth of apple trees and the need to maintain access for spraying machinery, in many orchards about one-third of the ground area is not shaded by apple leaves. Maximum leaf area density is often found some distance from the trunk. For very large "standard" trees, Heinicke (1964) reported a maximum vertically summed leaf area index (LAI) of 5.02 between 8 and 12 ft (2.5 and 3.5 m) from the trunk while the LAI 4 ft (1.2 m) from the center was only 3.02. For mature hedgerow orchards in the United Kingdom, LAI values range from less than one for Cox on the rootstock MM106 (Landsberg *et al.,* 1975) to 2.45 for Golden Delicious on rootstock M2 (Jackson, 1975). Corresponding values for bush trees in the United Kingdom have not been found to exceed 2.6, though values of over 4.5 have been reported for orchards in Washington State in the United States (Heinicke, 1963, 1964). As an approximation, however, many orchards in full leaf have an LAI of about unity. Leaf area per tree (A, m²) may be estimated from trunk girth (G) by the relationship $A = cG^b$. Obviously the values of the coefficients depend on tree vigor, pruning practice, and time of year (Byass, 1968; Holland, 1968; Landsberg, 1980). Landsberg found that $c = 0.0023$ and $b = 2.8$ provided good estimates of maximum leaf area for a wide range of trees.

B. ORCHARD MICROCLIMATE

Evaporation rates depend on energy balance and exchange processes. A convenient formula for their calculation is that derived by Monteith (1965):

$$\lambda E = \frac{sQ_1 + \rho c_p D/r_a}{s + \gamma(2 + r_1/r_a)} \tag{1}$$

where ρ is the density of air, c_p is the specific heat of air at constant pressure, γ is the psychrometric constant, s is the slope at air temperature of the saturation vapor pressure/temperature curve, and D is the vapor pressure deficit of the air. The boundary layer resistance (r_a) reflects the efficiency of exchange processes between the surface and the bulk air; r_1 here denotes resistance to transfer of water vapor from leaf to air. Equa-

---

*All leaf areas refer to one side of the leaf.

tion (1) does not allow for changes in the long-wave radiation balance of leaves as stomata close and leaf temperatures rise. To allow for this effect, Jones (1976) proposed the replacement of Q_1 by the net isothermal radiation (Q_1^*), which is the net radiation that would be received by the canopy at air temperature. This leads to

$$\lambda E = \frac{sQ_1^* + \rho c_p D/r_{hr}}{s + \gamma(2 + r_1/r_{ha})} \tag{2}$$

where r_{hr} is the parallel resistance to heat loss by long-wave radiation (r_r) and sensible heat transfer (r_{ha}). The resistance r_r is given by

$$r_r = \rho c_p/(4 \, \sigma \, \epsilon \, T^3) \tag{3}$$

where σ is the Stefan-Boltzmann constant, ϵ is emissivity, and T is the temperature.

Equations (1) and (2) can be used, with r_1 estimated from equation (9) (Section IV,C) and the micrometeorological relationships outlined later, to calculate transpiration rates per unit leaf area from apple trees.

1. Radiation Interception

There have been a number of studies of the radiation balance of apple orchards. Proctor *et al.* (1972) measured the components of the radiation balance of a 10-yr-old McIntosh tree in Ontario and found the average reflection coefficient to be 0.17. Net radiation (Q_n) was linearly related to solar radiation (Q_s):

$$Q_n = a + bQ_s \tag{4}$$

with $a = -98$ W m^{-2}, and $b = 0.80$. This is comparable to results of Landsberg *et al.* (1973) who, working with a hedgerow orchard in England, found the reflection coefficient to vary between 0.13 and 0.19. From their results $a = -14.7$ W m^{-2} and $b = 0.67$.

Suckling *et al.* (1975) measured the transmission of global and photosynthetically active radiation within a dwarf apple orchard and found that absorption of solar radiation increased from about 0.15 in early June to a maximum of 0.35 in mid-August. Palmer (1977) expanded a model by Jackson and Palmer (1972), with which they calculated the interception of light by hedgerow orchards, to calculate the absorption of light within canopies. Palmer used a complex empirical expression for his extinction coefficient. Charles-Edwards and Thorpe (1976) produced a more elegant and general model for radiation interception by hedgerow orchards, in which the orchard canopy was described by a series of ellipsoids, defined by semi-axes in the x, y, and z directions. Irradiance at a point P from the direction of a point Q was given by

$$I_P = I_Q e^{-\frac{1}{2}Fkl}$$ (5)

where I_P and I_Q are irradiances in the direction QP at the points P and Q, respectively, F is the leaf area density (leaf area per unit volume of canopy), k is a canopy extinction coefficient, and l is the path length of radiation within the canopy. Taking $k = 1$ they obtained good agreement with measurements. This model has been expanded by Thorpe *et al.* (1978) to calculate the irradiance, in two wave-band intervals (400–700 nm and 700–3000 nm) received by leaves in each of 10 volume elements, summed to give the power intercepted by a whole tree or length of row. The model was also developed to calculate the net radiation exchange per unit leaf area (Q_1). This involves information on boundary layer and stomatal (leaf) conductances (see Sections IV,B,2 and IV,C).

Thorpe *et al.* (1978) were also able to simulate the diurnal course of transpiration rate by a single young Golden Delicious tree, and Thorpe (1978) tested the radiation interception model in a hedgerow orchard and on other young trees. He derived useful empirical relationships between Q_1 and solar radiation or intercepted short-wave radiation. The relationship between Q_1 and Q_s was

$$Q_1 \simeq 0.34\, Q_s - 5$$ (6)

which is in agreement with Butler (1976), who stated: "In general the ratio Q_1/Q_n can be taken to be about 0.5." In view of the values given by Suckling *et al.* (1975) for absorption of global radiation by a dwarf orchard (0.15 to 0.35), we can take $Q_1 \simeq 0.3\, Q_s$ as a working approximation for most types of modern apple orchards.

2. Exchange Processes

In aerodynamic terms apple orchards are exceedingly rough, nonhomogeneous surfaces. Often they are also relatively small, so that a boundary layer representative of the orchard as a whole is not likely to become established to any depth. It is therefore seldom likely to be useful to try to apply standard aerodynamic theory (e.g., as outlined by Thom, 1975) to the transfer of mass and momentum between an orchard and the overlying air. Randall (1969) made wind speed measurements in and above an orchard from which he calculated the parameters of the log-linear profile, but many more such measurements would be required, over a number of orchard types of large area, before any useful generalizations about transfer processes from orchards could be made.

More progress has been made by treating the exchange of heat and mass in terms of exchange at the single tree/leaf level, analogous to the treatment of radiation balance. Landsberg and Powell (1973) used small apple trees in a wind tunnel and an artificial tree to examine the boundary layer resistance to mutual transference. The following equation is derived from their results:

$$r_a = 58\sigma^{0.56} \left(\frac{d}{u}\right)^{0.5} \tag{7}$$

where d (m) is a characteristic leaf dimension, σ is a measure of foliage density given by the ratio of leaf area to the frontal (silhouette) area of the tree, and u (m sec^{-1}) is wind speed. With the value 58 given for the constant, equation (7) gives r_a in sec m^{-1}. The expression may be used for heat and water vapor transfer, although for accurate work corrections should be made for the differences in diffusivity. Landsberg and Powell's work covered a range of σ values from 1.0 to about 3. Landsberg et al. (1975) used equation (7) successfully to solve the leaf energy balance equations for trees for which $\sigma \simeq 4$, and Thorpe and Butler (1977) tested it in the field and found that it gave good results with $\sigma = 4.6$. Very roughly σ can be taken as equal to the leaf area per tree.

To use equation (7), which is applied to all the leaves on a tree, we require the wind speed at the level of the foliage. Landsberg et al. (1973) measured wind speeds in a hedgerow orchard and found that they could be described by the one-dimensional analytical profile given by Thom (1971) and by Landsberg and James (1971). The equation for the profile is

$$u(z) = u(h) \left(1 + \Phi \left(1 - \frac{z}{h}\right)\right)^{-2} \tag{8}$$

where $u(z)$ and $u(h)$ are wind speeds at any height (z) and at the canopy top ($z = h$), and ϕ is a composite parameter incorporating the drag and momentum exchange characteristics of the canopy (see Landsberg and James, 1971). Landsberg and Jarvis (1973) showed that, as we would expect, ϕ increases as foliage density increased. In-crop profiles tend to linearity as ϕ decreases; when $\phi = 0.1$ the profile is essentially a straight line. In hedgerow-type orchards ϕ will vary with wind direction; Landsberg et al. (1973) obtained a value of $\phi = 1.2$ when the wind was almost normal to the rows (the trees were about 2.5 m high in rows 4.8 m apart). Other measurements in the same orchard (J. J. Landsberg, unpublished) showed ϕ could fall to 0.38 when the wind was blowing almost along the rows. Measurements in an old bush orchard, with trees about 4 m high planted on a 6.1-m square grid, gave values of ϕ of about 0.5, with little variation.

Wind speeds at the canopy top $[u(h)]$ are lower than those measured in standard meteorological enclosures because of the absorption of momentum by the canopy. An analytical solution for $u(h)$ would be possible if sufficient information on wind profiles above orchards was available (see earlier comment), but in the absence of such information it is possible to estimate $u(h)$ from standard wind speed (or wind run) measurements (u_s). Landsberg (1977), using measurements made over many months, found straight line relationships between $u(h)$ and u_s for hedgerow and bush orchards, with slopes of 0.61 and 0.96, respectively. However, these should be treated with care since they depend on the size of the orchard, surrounding terrain, and other factors.

3. The Role of Stomata

The diffusion of CO_2 into a leaf from the outside air and its absorption by wet cell surfaces within the leaf are essential first steps in photosynthesis. However, the inevitable consequence of these processes is the loss of water vapor along the same pathway—through the stomata—in the reverse direction. It is generally accepted that control of water loss is achieved by changes in stomatal aperture, and the mechanisms of stomatal functioning have been the subject of a number of recent reviews (e.g., that of Burrows and Milthorpe (1976 and others cited by them). In dealing with stomatal control of water vapor loss we will use both resistance (r_l) and conductance $(g_l = 1/r_l)$ terminology as convenient.

C. Responses of Stomata to Environmental and Plant Factors

Stomatal aperture, and hence stomatal resistance to gas exchange, may be affected by light, temperature, air humidity, CO_2 concentration, leaf water status, the presence or absence of fruits, and by mineral nutrition. The mechanisms of these responses are discussed in many papers (see the review by Burrows and Milthorpe, 1976), and we will again restrict our attention to the empirical information available on apple trees.

As background we note that apple leaves are hypostomatous, having between 2×10^4 and 6×10^4 stomata per cm^2 on the abaxial surfaces (Slack, 1974; Beakbane and Majumder, 1975; West and Gaff, 1976, and other authors cited in these papers). There is considerable variation from leaf to leaf and from point to point within a leaf, and it appears that the number of functional stomata increases from the time of leaf emergence, reaching a maximum when the leaves are 4 to 6 weeks old (Slack, 1974; Watson and Landsberg, 1979). Slack (1974) found the mean pore length was about 20 μm and the mean distance between aperture centers about 55 μm. She found small but significant differences between scion varieties, with stomatal density higher in Cox's Orange Pippin than Golden Delicious. Beakbane and Majumder (1975) found significantly higher stomatal densities in leaves from more invigorating rootstocks.

Most of the information on responses of apple stomata to environmental factors comes from two studies—those of West and Gaff (1976) and Warrit et al. (1980). To describe the effects on stomatal conductance of photon flux density (Q_p) and the leaf-to-air vapor pressure gradient (D_l), Thorpe et al. (1980) developed an empirical model from the data of Warrit et al. The equation of the model is

$$g_1 = \frac{(1 - \alpha D_l)}{(1 + \beta/Q_p)} g_r \tag{9}$$

where g_r is a reference conductance and the values of α and β are established empirically. Thorpe et al. gave values for g_r between 3.8 and 8.7 mm sec^{-1}, with α between 0.02 and 0.033 and β between 59 and 79 μE m^{-2} sec^{-1}.

Landsberg et al. (1975), working with porometers in an orchard, found that maximum conductances were reached at a photon flux density of about 400 μE m^{-2} sec^{-1} (175 W m^{-2}). Warrit et al. (1980) fitted a rectangular hyperbola [incorporated in equation (9)] to leaf chamber data defining the changes in g_1 of Golden Delicious leaves with Q_p; this indicated maximum stomatal opening when Q_p was greater than about 200 E m^{-2} sec^{-1}. From their leaf chamber work West and Gaff (1976) found minimum resistances (maximum conductances) of cv. Granny Smith leaves at a light intensity of 15 W m^{-2}, which seems extraordinarily low.

Warrit et al., who examined the effects of temperature with D_l held constant, found no consistent temperature effect on g_1. West and Gaff's observations that leaf resistance increased (g_1 decreased) with increasing leaf temperature, may be confounded by the effects of leaf-to-air vapor pressure difference (D_l), which must increase as T_l increases if D is held constant.

Plotting transpiration rate (E) against leaf-to-air vapor concentration gradient (χ_l), West and Gaff found that E increased with χ_l up to $\chi_l = 12$–14 gm m^{-3}; it then decreased. Farquhar (1978) has demonstrated that a simple linear model of the effect of humidity on stomatal conductance [as in equation (9)] is sufficient to explain reductions in transpiration rate produced by increasing evaporative demand in terms of a feed-forward response. This occurs when a change in the environmental factors causing a change in transpiration rate causes a change in the conductance *independent* of the change in transpiration rate. The only hypothesis that is consistent with feed-forward is that direct cuticular loss of water from guard cells and subsidiary cells (peristomatal transpiration) influences stomatal aperture (see references cited by Farquhar; also Jarvis, 1980).

Landsberg and Butler (1980) examined the effects of stomatal response in apples to humidity and leaf temperature, using equation (9) in the leaf energy balance equations and solving numerically for T_1. They found that, especially at higher values of Q_1, transpiration rate reached a clear maximum well within the range in which stomatal response to D_1 was linear. Both feed-back and feed-forward mechanisms are operative: transpiration rates increase as D (and hence D_1) increases, but when stomata begin to close and transpiration to decrease in response to increasing D_1 (feed-forward), the closure leads to an increase in T_1 and hence a further increase in D_1, and more rapid closure. There may also be a feed-back operating via leaf water potential.

It is possible to calculate the transpiration rates shown in Fig. 3 (from West and Gaff) using equation (9). The calculations were carried out as follows: since $E = \chi_1/(r_1 + r_a)$, we assumed a standard value for r_a (30 sec m^{-1}) and solved for r_1 using a value of E (45 mg m^{-2} sec^{-1}) read from West and Gaff's Fig. 6. Converting the corresponding value of χ_1 to D_1 and inserting this in equation (9) with the light flux density (equivalent to about 1000 μE m^{-2} sec^{-1}) given, we obtained a solution for g_r (= 7.6 mm sec^{-1}). Now using a range of χ_1 values spanning that shown in West and Gaff's Fig. 6 for each of which r_1 was calculated from equation (9), we calculated

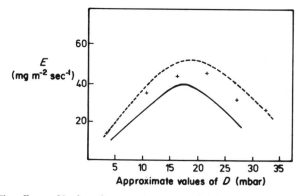

Fig. 3. The effects of leaf-to-air vapor pressure gradients (D) on leaf transpiration rates (E). The curves encompass the measurements made by West and Gaff (1976); crosses are values calculated using equation (9). (Based on Fig. 6 in West and Gaff, 1976.)

the values of E shown as crosses on Fig. 3. (We obtained r_l values between 244 and 719 sec m^{-1}.) The calculations are not precise since the conversion of χ_l to D_l was based on the assumption that $T = 23°C$, the chamber temperature usually used by West and Gaff, who varied leaf temperatures and air humidity (D) to achieve different values of χ_l but did not give the values used. However, the procedure demonstrates that the parabolic relationship between E and χ_l shown by West and Gaff does not necessarily indicates a lack of stomatal response to humidity until transpiration begins to decline; it provides further support for the feed-forward hypothesis and also seems to validate the model of Thorpe $et\ al.$ (1980), since the parameters obtained from West and Gaff's (1976) data fall within the range given by Thorpe $et\ al.$ The calculated stomatal resistances yielded transpiration rates within the range given by West and Gaff (see Fig. 3).

In their study West and Gaff tested the effect of χ_l on transpiration rates in CO_2-free air and found that, although an asymptote was reached, E did not decline appreciably with increasing χ_l. Warrit $et\ al.$ showed a linear relationship between internal CO_2 concentration (C_i) and ambient concentration (C_a). They attributed much of the light response of stomata to changes in C_i.

With regard to leaf water potential (ψ_l), West and Gaff found no effect on stomatal resistance until ψ_l fell to about -1.8 MPa; there was then a rapid rise in r_l as ψ_l decreased to about -2.4 MPa. Warrit $et\ al.$ found threshold values at which g_l decreased abruptly, varying between -2.0 and -2.6 MPa, and Lakso (1979) observed a strong seasonal effect with the threshold decreasing from -1.0 to -4.0 MPa between May and September. Since Landsberg $et\ al.$ (1975) found little effect of ψ_l on g_l when $\psi_l > -1.1$ MPa, they were able to split their field data into two response groups. When $\psi_l \leq -1.4$ MPa the maximum values of g_l reached were less than half those reached when $\psi_l > 1.1$ MPa. There is no evidence, in any of the research reviewed, of interaction between D_l and ψ_l in their effects on apple stomata.

An interesting study of the effect of fruiting on transpiration rates and stomatal opening in apple leaves was made by Hansen (1971), who measured the rate of water loss by fruiting and defruited trees grown in pots. Over three measurement periods Hansen found that trees with fruits transpired, on average, 0.76 liters of water per 24 hr in summer, while defruited trees lost 0.38 liters. Autumn measurements showed the corresponding losses to be 0.20 and 0.05 liters. On a relative size scale stomatal apertures in the trees with fruits were 0.45 and those without fruit 0.17. Tunsuwan and Bünemann (1973) showed that the presence of fruit increased the stomatal aperture on fruiting spur leaves, but had no clearcut effect on stomatal aperture of extension leaves. Measurements in a Bramley orchard in Kent (Jones and Higgs, 1980) demonstrated that conductances of extension leaves on fruiting branches can be twice as high as on nonfruiting branches.

Summarizing this section, apple stomata respond to light in much the same way as those of most other mesophytic plants: stomatal conductance is reduced by low air humidity, this reduction being (apparently) independent of changes in ψ_l; there is a fairly abrupt reduction in conductance at some critical value of ψ_l, which may vary with conditions under which the tree has been grown. Large increases in CO_2 concentration cause stomatal opening. Quantitative descriptions of the responses to light and humidity can be used, in conjunction with micrometeorological relationships, to calculate transpiration rates of apple trees. The stomatal conductance of apple leaves is increased by the presence of fruits. Further discussion of the adaptive responses to stress shown by stomata is found in Section VIII.

V. WATER FLOW THROUGH THE SOIL-PLANT SYSTEM; PLANT WATER POTENTIAL

As discussed in Section VI, plant water status determines many physiological responses. Leaf water potential, which is frequently used as a measure of water status, is dependent not only on the water status of other parts of the system (particularly the soil), but also on the evaporative demand and stomatal aperture and on flow resistances in the transport pathway (see Jarvis, 1975). Unfortunately, much of the literature on orchard crops is rather difficult to interpret in terms of plant water status since many authors have only measured some aspect of soil water status, neglecting the role of the atmospheric environment and the plant's control of water loss.

We shall make use of some simple models of water flow in the soil–tree–atmosphere system as a basis for our discusssion of variation in water potential (ψ) at various points within apple trees both diurnally and during the season.

Using an analogue of Ohm's Law, as used by van den Honert (1948) and many others, the water flow pathway through the soil–tree–atmosphere system can be represented by flow through a series of hydraulic resistances, as shown in Fig. 4. The water potentials at different points in the system can then be described by the following system of equations:

$$E = \frac{\psi_s - \psi_l}{R_s + R_r + R_{st} + R_l} = \frac{\psi_s - \psi_r}{R_s}$$

$$= \frac{\psi_r - \psi_{st}}{R_r} = \frac{\psi_{st} - \psi_x}{R_{st}} = \frac{\psi_x - \psi_l}{R_l} \tag{10}$$

where E is the flux of water through the system, and ψ_s, ψ_r, ψ_{st}, ψ_x, and ψ_1, respectively, refer to the water potentials in the bulk soil, at the surface of the roots, at the base of the stem, at the base of the petiole, and in the bulk of the mesophyll cells of the leaf. The resistances R_s (soil), R_r (root), R_{st} (stem), and R_1 (leaf) refer to the hydraulic resistances between these sites, as shown in Fig. 4. The plant resistance R_p is the sum of the last three. In this model, since E is not likely to be directly affected by normal changes in the liquid phase resistances, the flux is driven by a current generator dependent only on environment and stomatal conductance. Although the model may give an adequate description of the system when in a steady state, it involves great simplifications. For instance, no account is taken of branching in the pathway (see, e.g., Denmead and Millar, 1976), and distributed systems have been lumped into single components. An example of this latter approximation is the assumption of a single root-surface potential, even though the roots may not all be at the same potential.

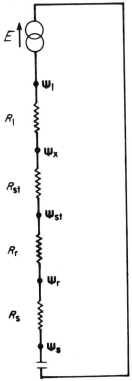

Fig. 4. Resistance network used as a basis for the analysis of water flow through the soil–tree system. E is the flux water, R denotes resistance, and ψ denotes potential. Abbreviations as in text.

Where water availability in different parts of the root zone changes—for instance, when the surface layers dry out—this could lead to apparent changes in the pathway through the roots and hence the root resistance (Landsberg and Fowkes, 1978). Another problem is that several of the resistances are often apparently flow-dependent (Brouwer, 1954; Stoker and Weatherley, 1971; Fiscus and Kramer, 1975; Weatherley, 1979).

A more important limitation of this simple resistance model, however, is that E varies rapidly with changes in environmental conditions so that it is no longer possible to use a steady-state solution for the flow system and dynamic components such as capacitances must be included (see, e.g., Powell and Thorpe, 1977; Jones, 1978a). The capacitance (C) of tissue is defined by

$$C = dW/d\psi \tag{11}$$

where W is tissue water content. Incorporation of the capacitances of the various tissues of the transpiring tree, which may cause changes in ψ to lag behind changes in transpiration rate, and of the radial resistance ($R_1 - R_4$) to water exchange between the capacitors (e.g., bark) and the main transpiration stream gives the model in Fig. 5.

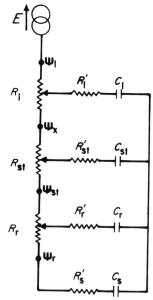

Fig. 5. Resistance (R)–capacitance (C) network representing water movement through the soil–tree system. Note the additional resistances (cf. Fig. 4) between the storage regions [leaves (l), stems (st), roots (r), and soil (s)] and the main flow pathway.

Flow models employing a single constant resistance and capacitance (e.g., Jones, 1978a; Wallace, 1978) can be written in the form

$$E = (\psi_s - \psi_l)/R - Cd\psi_l/dt \tag{12}$$

Wallace solved for ψ_l by assuming that E changes linearly over short (hourly) intervals, which gives a linear first-order differential equation in $\Delta\psi$ ($=\psi_s - \psi_l$), the solution of which is

$$\Delta\psi(t) = RE(t_1) + mRt - mR^2C + (k/C)\exp(-t/RC) \tag{13}$$

where k is a constant of integration determined by the initial boundary conditions and m is the rate of change of E. An alternative approach to fitting this type of model, suggested by Jones (1978a), involves the use of cubic splines to fit the observed trends in E. This has the advantage that these trends can be fitted more precisely, using all the available measurements, whenever they were made. Although the parameters of simple resistance/capacitance models such as those of Jones (1978a) and Wallace (1978) can be successfully estimated by statistical fitting, the model in Fig. 5 has too many parameters for them to be determined by statistical procedures (Monteith, 1980). Powell and Thorpe (1977) used a simple version of that model with eight parameters estimated independently (from anatomical measurements) and were able to make quantitative estimates of water potentials in apple. However, the results were probably no better than would have been obtained with much simpler models. Powell and Thorpe (1977) also derived an analytical solution for their model, but it depended on a simple driving function—they assumed that diurnal evaporation could be approximated by a sine function.

A more reliable approach to the calculation of ψ is to estimate resistances and capacitances from observed relationships between flow rate and ψ and tissue water content and ψ. It is difficult to choose a consistent baseline for expressing resistance and capacitances, since various authors have used ground area, leaf area, tree number, stem cross section, or fresh weight as their bases. Furthermore, capacitance values are very dependent on the material used. The first two bases are probably the most useful since they are most often used for evaporation studies. Therefore, an attempt has been made to convert all resistances quoted to these bases, although since capacitances are so dependent on tissue volume they are expressed as percentage change in water content per MPa. Estimates of capacitances in apple trees are presented in Table I; the range of the values is caused by the fact that capacitance changes with tissue water content.

TABLE I

Capacitance Estimates for Apple Tissue

	% Fresh weight/MPa	Reference
Leaves		
Cox	2.5	Landsberg *et al.* (1976)
Golden Delicious	3.4–5.8	Jones and Higgs (1979b)*
Empire	3.0	Davies and Lakso (1979)
Bark	7.6	Powell and Thorpe (1977)
Fruit[a]	5.5–20.0	H. G. Jones and K. H. Higgs
		(unpublished data)

[a]Immature Golden Delicious.

The degree of hysteresis in the relationship between ψ_l and transpiration rate provides information on the value of the capacitances and associated storage resistances. In some studies, diurnal hysteresis in this relationship was small (e.g., Landsberg *et al.*, 1975), suggesting that the time constants (equal to RC) for exchange between flowing xylem sap and the water storage tissue are either very short or very long in relation to diurnal changes. However, Powell (1976) observed significant hysteresis (see Figs. 6 and 8), whereas Powell and Thorpe (1977) estimated time constants as long as 230 min. for trunk bark. It is relevant to note here that Jones (1978a) found the fit of his dynamic model to wheat to be greatly improved if the constant resistance was replaced by a variable resistance that increased with decreasing flow rate.

A. Soil Resistance and Root Distribution

The soil resistance to water uptake has been partitioned into "rhizosphere" and "pararhizal" components corresponding, respectively, to the

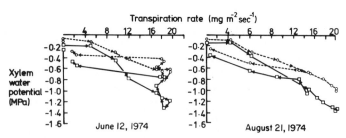

Fig. 6. Changes in xylem water potential in apples (Cox's Orange Pippin), early and later in the season, in relation to transpiration rates. Data represented by solid lines and squares were obtained from well-irrigated trees; dotted lines and circles from trees in relatively dry soil. (Unpublished data of D. B. B. Powell.)

resistance to water movement in the immediate vicinity of the root and resistance to water movement from a zone of moist soil to the root zone (Etherington, 1967; Newman, 1969). The pararhizal component is primarily determined by the depth of the root zone in relation to water supply and by the soil type and water content. The rhizosphere component, although affected by soil type and water content, is largely determined by root density (Gardner, 1960; Cowan, 1965; Newman, 1969).

The size of the pararhizal and rhizosphere resistances can be estimated from a knowledge of the distribution of roots in the soil profile. Many clear diagrams and photographs of excavated apple root systems are available for trees grown on different rootstocks in a range of soils (Rogers, 1932; Rogers and Vyvyan, 1934; Coker, 1958; Levin et al., 1973) and at a range of tree spacings (Atkinson et al., 1976). Commonly the apple root system consists of a horizontal scaffold of major roots radiating from the trunk in the top soil, with vertical "sinkers" to deeper horizons, where they may proliferate. The spread is largely dependent on planting density, though it is greater in sandy soils and with more vigorous rootstocks (Rogers and Vyvyan, 1934). The spread of roots is usually greater than that of the branches (Atkinson, 1977).

The depth of the root system is particularly important to the pararhizal resistance. As in many other crops the majority of apple tree roots are found within the top 25–50 cm of the soil (Rogers and Vyvyan, 1934; Coker, 1958; Levin et al., 1973). The exact proportion of roots at lower levels depends on soil depth, soil type, aeration, nutrient availability, and plant spacing. With a range of close plantings (0.17–11.1 trees m^{-2}, Atkinson et al., 1976), the percentage of total root weight below 50 cm was greater in the denser plantings, reaching 25% at the close spacing. Workers in the United Kingdom have reported that at least some roots from each apple tree usually penetrate to the full depth of the soil profile (Rogers and Vyvyan, 1928, 1934; Coker, 1958; Atkinson et al., 1976), even with bedrock as deep as 3 m. On occasions, roots have been observed penetrating fissures in the bedrock itself (Atkinson et al., 1976). Rogers and Vyvyan (1934) and Coker (1958) did not observe large differences in root depth between various vigorous and dwarfing rootstocks. Development of deep root systems by apple trees, as in other species, depends on adequate drainage (Coker, 1958; Webster, 1978). Localized concentrations of roots at particular depths are often associated with zones of high nutrient availability (Atkinson, 1974).

Root density in the top 10 cm or so of the soil tends to be lower under clean cultivation than with a grass cover (Coker, 1959), while overall herbicide treatment increases yet further the apple root population near the soil surface (Atkinson and White, 1976). Although Goode and Hyrycz

TABLE II

REPRESENTATIVE ESTIMATE FOR L_A (m m^{-2}) FOR APPLES AND OTHER SPECIES

	L_A	Reference
Apples		
Field (26-yr-old)	2.8×10^2	Atkinson (1973)
Field (5-yr-old)	$0.8–4.3 \times 10^2$	Atkinson *et al.* (1976)
Root laboratory	$3.6–2.4 \times 10^2$	Atkinson and Wilson (1979)
Field (10–11-yr-old trees)	$1.6–9.5 \times 10^{2a}$	Rogers and Vyvyan (1934)
Herbs: Gramineae	$2–40 \times 10^4$	Newman (1969)
Herbs: Non-*Gramineae*	$5–30 \times 10^3$	Newman (1969)
Woody plants	$\sim2–11 \times 10^3$	Newman (1969)

[a] Assuming average weight of 1 m of root = 3.5 gm.

(1964) reported only small effects of irrigation on the total amount of fibrous roots in an apple orchard, the density in the top 15 cm was twice as high for the irrigated as for the control trees. Similarly, Levin *et al.* (1973) reported that ample irrigation in Israel increased the proportion and the absolute number of roots in the top 30 cm of the soil, with very few roots (8%) being found below about 60 cm. In a dry treatment 7% of the roots were found below 120 cm. Gorin (1963) has reported that irrigation can dramatically increase root density.

Because of this variability in root distribution it is not possible to generalize about the pararhizal resistance. The rhizosphere or root–soil interface resistance is, however, more amenable to estimation since it is largely a function of root density (Newman, 1969) and for the rest of the discussion will be equated with R_s (equation 10). There are no direct estimates of the rhizosphere resistance (R_s) for field-grown apples but it can be estimated for apples in soil at different water potentials using Newman's results and the values of root length per unit of ground area $[L_A \ (\text{m}^{-1})]$ for apple trees given in Table II, where mean values for other types of plant are included for comparison. The data for apples indicate that L_A tends to be rather lower than for many herbaceous species (particularly Gramineae) (cf. Newman, 1969). Unfortunately, the results from much of the earlier work on excavation of apple root systems were expressed in terms of fresh weight, so the conversions to L_A are only approximate and it is likely that L_A is usually underestimated since, at least with excavation, it is notoriously difficult to extract all the fine roots. A further difficulty is that excavation is usually done during the winter when there is negligible new root production and many of the summer population of fine roots have atrophied.

Rearrangement of the equations given by Newman (1969), which were based on Gardner's (1960) model, give

$$R_s \simeq \frac{1}{2\pi K L_A} \ln\left[\frac{2}{d}\frac{\sqrt{z}}{\pi L_A}\right] \tag{14}$$

where K is the soil hydraulic conductivity at ψ_s, d is the root diameter, and z is the depth of the root zone. The values presented in Table III were calculated using equation (14) and a curve describing the dependence of K on ψ_s for Pachappa sandy loam (Gardner, 1960). The table shows rhizosphere resistances for a range of typical L_A and ψ_s values. Since K at any ψ_s may vary by up to an order of magnitude for different soils, R_s will vary similarly.

The estimates of R_s in Table III are at best very approximate, even for the soil type in question, because of soil and root heterogeneity. A further cause of uncertainty is that, at least in stress conditions, root–soil contact is frequently reduced when plants are under stress and the roots shrink (Huck et al., 1970; Tinker, 1976; Weatherley, 1979). Faiz (1973) has estimated that root shrinkage may double the value of R_s.

B. Root and Plant Resistance

Since no in vivo estimates of R_r are available for apples, the best available estimates of root resistances are those of Landsberg et al. (1976) and Davies and Lakso (1979), derived from the difference between the observed resistance of whole trees with their roots in either moist soil or water and the resistance calculated when water uptake was directly through the base of the stem (cut under water). Values are tabulated in Table IV and

TABLE III

VALUES FOR R_s (MPa s m^{-1}) FOR A RANGE OF VALUES OF L_A AND ψ_s, CALCULATED USING EQUATION (14) AND APPROPRIATE HYDRAULIC CONDUCTIVITIES FOR PACHAPPA SANDY LOAM[a]

MPa	L_A (m^{-1})					
	1×10^2	5×10^2	1×10^3	5×10^3	1×10^4	5×10^4
-0.01	8.9×10^2	1.4×10^2	6.4×10^1	8.5	3.8	4.1×10^1
-0.05	3.4×10^5	5.5×10^4	2.5×10^4	3.3×10^3	1.5×10^3	1.6×10^2
-0.1	1.8×10^6	2.9×10^5	1.3×10^5	1.7×10^4	7.8×10^3	8.3×10^2
-0.3	1.5×10^7	2.4×10^6	1.1×10^6	1.4×10^5	6.3×10^4	6.7×10^3
-0.5	3.0×10^7	4.8×10^6	2.2×10^6	2.9×10^5	1.3×10^5	1.4×10^4
-1.0	7.8×10^7	1.2×10^7	5.5×10^6	7.4×10^5	3.3×10^5	3.6×10^4
-1.6	1.6×10^8	2.5×10^7	1.1×10^7	1.5×10^6	6.7×10^5	7.1×10^4
-2.5[b]	3.0×10^8	4.8×10^7	2.2×10^7	2.9×10^6	1.3×10^6	1.4×10^5

[a] Gardner (1960).
[b] Extrapolated value.

TABLE IV
FLOW RESISTANCES IN APPLE TREES

Material and source[a]	Leaf area per tree (m²)	Resistances		
		Plant basis MPa sec $m^{-3}/10^7$	Leaf area basis MPa sec $m^{-1}/10^7$	Ground area basis MPa sec $m^{-1}/10^7$
		$R_{soil} + R_{plant}$		
A	6^b	2–3.3	12–20	8.4–14
B	6^b	2–13	8–12	5.6–8.4
C	6.5	0.8	5.2	3.8
D	5.4	0.6	3.2	2.0
E	5.4	0.7	3.8	2.3
F	6^b	0.75	4.5	6.5
G	0.4–0.6	7–33	32–13.2	—
H	0.7	2.4	1.7	—
J	1.4	1.2	2.1	—
		$R_{soil} + R_{root}$		
A	6^b	1.4–2.8	8.4–17	6–12
B	6^b	1.1–1.7	6.6–10	4.6–7.0
F	6^b	0.7	4.3	3.0
		R_{root}		
C	6.5	0.5	3.3	2.4
G	0.4–0.6	3.9–17.5	2.0–8.8	—
	0.1	5.1	0.51	—
	0.07	1.8	0.13	—
		R_{stem}		
C	6.5	0.4	2.6	3.5
G	0.4–0.6	2.6–16.6	1.3–8.3	—
		R_{leaf}		
A	6^b	0.2–1.1	1.2–6.6	0.7–4.5
B	6^b	0.3–0.9	1.8–5.4	1.1–3.9
C	6.5	0.3	2.0	1.1
G	0.4–0.6	3.6–8.0	1.8–4.0	—

[a] A, B—8-yr-old Cox on MM106; A, droughted, B, irrigated (Powell, 1974). C—9-yr-old Cox on MM106; irrigated (Landsberg et al., 1976). D, E—9-yr-old Cox on MM106; D, irrigated, E, droughted (Landsberg et al., 1975). F—9-yr-old Cox on MM106; irrigated (Powell and Thorpe, 1977). G—2-yr-old Golden Delicious on MM106; in pots (Landsberg et al., 1976). H, J—3/H) and 4(J) yr-old Golden Delicious on MM106, in 50-liter containers (D. B. Powell and A. Berger, unpublished data). K, L—2-yr-old Empire on MM106, in pots, glasshouse; L, outdoors (Davies and Lakso, 1979).

[b] Estimated.

averaged between 40 and 68% of the total plant resistance. The estimated resistances were rather lower for orchard trees than for young potted trees. Powell and Thorpe (1977) estimated root + soil resistance for an orchard tree as 96% of the total flow resistance, though the soil component may have been significant. These root resistances are likely to be significantly greater than R_s for soil water potentials above about -1.0 MPa (see Tables II and III).

The main component of the root resistance to water flow through the plant (Fig. 4) is the radial transfer through the cortex (Scott-Russell, 1977). It is likely that the axial component of the root resistance in apples is small; this is indicated by the vessel diameters (Reidl, 1937) and the Poisseuille–Hagen equation for flow in tubes, and from studies of the hydraulic conductivity of apple root systems with an applied pressure differential (Baxter and West, 1977). This may contrast with the situation in some cereals, where the axial resistance can be significant (Passioura, 1972).

Water transfer across the root cortex probably occurs largely in the apoplast (Weatherley, 1963; Russell, 1977), although as suberization of the endodermis proceeds, the water has to enter the symplast to pass the Casparian band. This may explain why water uptake is high in the white (unsuberized) zone near the root tip and declines markedly away from the tip as browning progresses and suberin increases (Graham et al., 1974; Russell, 1977). However, in secondarily thickened older roots after cortical degeneration, at least in the anatomically similar roots of cherry, water absorption can be nearly as rapid as by white roots (Atkinson and Wilson, 1979). This high permeability may be attributable to cracks between the corky surface cells (see photograph in Atkinson, 1972, p. 45). Since apple roots turn brown within a few weeks in summer (Head, 1968), it follows that a large part of the total water flux must enter the plant via the older roots, so the soil resistance calculations above, which were based on total L_A, are likely to be reasonable.

A large body of evidence indicates that the root resistance is dependent on flow rate, being highest at the lowest flows (Brouwer, 1954; Stoker and Weatherley, 1971; Jones, 1978a). A similar effect was observed by Landsberg et al. (1976), although Landsberg et al. (1975) found no significant change in R_p (and presumably R_t) over a wide range of fluxes for orchard trees (see also Fig. 6), and Baxter and West (1977) reported linear relationships between flow through isolated root systems and the pressure applied. Various mechanisms have been proposed to account for this phenomenon (e.g., Fiscus, 1975; Powell, 1978; Landsberg and Fowkes, 1978), but it seems unlikely that a single mechanism is involved. Several methods for further partitioning R_p have been tried. One approach is to measure the water potentials at the intermediate points in Fig. 5 and use equation (10). A good estimate of ψ_x can be obtained by enclosing the leaf in a plastic bag for at least 30 min. before measurement of ψ_l in a pressure chamber (Powell, 1974). In this time ψ_l equilibrates with ψ_x at the point of attachment. Estimates of R_l derived from these measurements are given in Table IV; it is of interest that Landsberg *et al.* (1976) observed similar sensitivity of R_l to E as did Boyer (1974).

Estimates of ψ_{st} at the base of the tree are more difficult but it is probably close to ψ_x, since water potential gradients along branches and down the stem are usually less than 0.1 MPa m^{-1} (Jones and Higgs, 1979a) and are independent of branch diameter (Powell and Thorpe, 1977). Therefore, R_{st} is likely to be small (Table IV). A complication is that apple cultivars are usually grafted onto a rootstock and a significant flow resistance may occur within the graft union. Anatomical measurements indicate, however, that good vessel continuity exists through the union in apples with no significant attendant flow resistance (D. S. Skene, personal communication).

Another approach to the estimation of the components of R_p involves calculation of hydraulic resistances from anatomical measurements (Furr, 1928; Powell and Thorpe, 1977). This, however, is only appropriate for R_{st} and the axial component of the root resistance, being more difficult to apply to the other resistances. Ewart (1905) and Malhotra (1931) provide data on lengths of xylem vessels in apples, which are similar to those in other deciduous species, averaging 24 cm in length (range 15–34 cm). These long vessels provide further support for the assumption that the longitudinal flow resistance in the stem is small.

Experimental estimates of the longitudinal specific conductivity of apple wood have been made using 15–20 cm lengths of excised wood (Farmer, 1919; Furr, 1928). The specific conductivity with a 30-cm Hg pressure differential averaged 42 cm^3 cm^{-2} (15 min.)$^{-1}$ (30 cm Hg)$^{-1}$ (range 18–73) [= 1.18 × 10^{-4} m^3 m^{-2} sec^{-1} MPa^{-1} (range 0.51–2.05)]. The conductivity was greatest in the trunk, decreasing to the branch ends. This agrees with Furr's (1928) observation that the cross-sectional area of the trunk was smaller than the total of the branches. Estimates of hydraulic conductivities of apple wood (expressed on a xylem area basis) made by Baxter and West (1977) range from 0.13 × 10^{-12} m^2 for fine roots through 1.5 − 4.5 × 10^{-2} for trunk and twigs to 18.2 × 10^{-12} m^2 for main roots.

C. DIURNAL AND SEASONAL TRENDS IN WATER POTENTIALS

As in other plants diurnal trends in ψ_l in apple are strongly dependent on environment (Fig. 6). In general the diurnal minimum ψ_l occurs in the early afternoon at the time of maximum evaporation rate (Landsberg *et al.*, 1975; Powell, 1974; Goode and Higgs, 1973). As expected ψ_{bark}, estimated from trunk diameters, lags behind ψ_l (Powell and Thorpe, 1977) and does not fall as low as ψ_l. In well-irrigated trees, ψ_l values commonly fall to between -0.6 and -2.0 MPa in the United Kingdom (Goode and Higgs, 1973; Powell, 1974; Landsberg *et al.*, 1975; Goode *et al.*, 1979; Jones and Higgs, 1979b), whereas the leaves of droughted trees can reach -2.5 to -3.0 MPa (Jones and Higgs, 1979b; H. G. Jones, unpublished data). Stomatal closure and leaf fall generally tend to prevent more severe stresses developing in field-grown trees, in contrast to the situation with pot-grown material where stresses develop more rapidly. Leaf water potentials of -4.5 MPa have been recorded in trees grown in containers and subjected to severe water shortages (Chapman, 1973).

In general the diurnal minimum water potential decreases during the season even if trees are irrigated (Goode and Higgs, 1973; Powell, 1974), although the effect is clearer in unirrigated trees as the soil dries. The diurnal minimum in Goode and Higgs' study declined from -1.0 to -2.0 MPa in irrigated trees and from -1.5 to -3.0 MPa in unirrigated trees between June and September. The reason for this change in irrigated trees is not clear (cf. similar observations for wheat—Jones, 1977), but changes in evaporative demand relative to root activity may be a factor.

The diurnal maximum water potential is largely dependent on soil water status, and occurs at or near dawn (Goode and Higgs, 1973; Powell, 1974; Figs. 6 and 8).

Fruit water potential lags behind ψ_l, with considerable damping (Goode *et al.*, 1979) owing to the high capacitance and larger resistance to water flow through the peduncle. These authors reported that the damping of diurnal fluctuations in ψ_{fruit} to a minimum ψ_{fruit} of -1.6 MPa when ψ_l was -2.4 MPa, while the maximum ψ_{fruit} was -0.2 MPa where ψ_l had returned to zero. There was a linear relationship between ψ_{fruit} and ψ_l. Similar results have been reported by Chapman (1971) with different cultivars. Fruit expansion also varies diurnally and has been related to leaf water potential (Powell and Thorpe, 1977), being most rapid in the early morning and often becoming negative when ψ_x is low during the day (see also Fig. 8). Changes in fruit volume lag behind those in trunk girth by several hours.

VI. EFFECTS OF DEFICITS ON GROWTH AND REPRODUCTION

The level of ψ in a plant that constitutes a deficit, in the sense of affecting the functioning of the plant, can seldom be precisely identified. Various physiological processes differ in their sensitivity, and the responses also vary with age or condition of the plant.

The time scale of water deficits varies from minutes to months or even years, and the effects of the deficits on plants will range from completely reversible to irreversible changes in physiological functioning or in growth patterns. Equating "deficit" with any significant reduction in water potential, we will be mainly concerned here with the diurnal course of water potentials and with time scales of weeks. As a general principle the time scale considered (in terms of deficits) should be related to the response time of the process under consideration, e.g., if we are concerned with growth processes where the response can only be clearly observed after a period of days or weeks, then we must consider the water status of the plant over that period.

On the longer time scale, since ψ varies continuously in space and time, the problem arises as to what value of water potential should be used as a basis for analysis. Plant and soil water potentials are likely to be in (approximate) equilibrium at dawn, although this is by no means certain; in dry conditions, when soil hydraulic conductivity is very low and plant water potentials (ψ_{plant}) have reached very low levels during the preceding day, overnight recovery by the plant may not restore ψ_{plant} to the same level as ψ_{soil}. Nevertheless the dawn value of ψ_{plant}—if available—provides a reasonable basis for analysis, since it is less sensitive to environmental conditions than the diurnal minimum ψ.

Although ψ is the most common and convenient measure of water status in trees, it may not always be the most relevant in determining physiological responses to drought. For instance the turgor pressure in particular cells controls processes such as cell extension and stomatal movement (Hsiao, 1973), and there is also evidence that photosynthesis is affected more by changes in cell volume (or turgor) than by changes in water potential (Jones, 1973). For this reason it is useful to be able to convert water potentials to measures such as relative water content. Appropriate data for apples have been given by Powell and Blanchard (1976) and Jones and Higgs (1979b). Similarly, conversions between ψ_l and leaf osmotic potential (ψ_π), such as those given by Goode and Higgs (1973) and Goode et al. (1979) allow calculation of turgor pressure.

In considering the effects of water deficits on growth we are essentially considering their effects on production and allocation of dry matter. This is complicated in apples by the influence of the fruits themselves on these processes. Dry matter production per unit leaf area tends to be highest in those trees carrying most fruit (Maggs, 1963; Avery, 1970; Heim et al., 1979). This may be related to the increase in stomatal conductance induced by the presence of fruits (see Section IV,C).

Heim et al. (1979) have clearly demonstrated the influence of fruits on assimilate partitioning; Fig. 7 is redrawn from their paper and shows that the fraction (p_j) of the total dry weight gain (ΔW) over any time interval

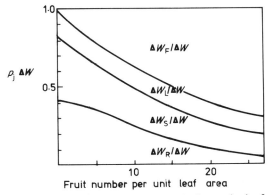

Fig. 7. The influence of fruits on assimilate partitioning. p_j is the fraction of the dry weight gain (ΔW) that goes to plant component j (e.g., $p_F \Delta W = \Delta W_F$). $\Delta W_F/\Delta W$, $\Delta W_L/\Delta W$, $\Delta W_S/\Delta W$, and $\Delta W_R/\Delta W$ denote the proportions of dry weight going to fruits, leaves, stems, and roots, respectively. (Redrawn from Heim *et al.*, 1979.)

(Δt) that goes to plant component j depends on fruit number per unit leaf area (X). As X increases the fraction of ΔW that goes to fruits (p_F) increases and the fraction that goes to the other components, particularly roots and stems, necessarily decreases. The increase in p_F is asymptotic to a limiting value, which appears to be about 0.75. If $p_j(X)$ is the value of p_j at any value of X, the following equation holds:

$$\frac{\Delta W}{\Delta t} = [p_L(X) + p_{St}(X)_R + p_F(X)] \frac{\Delta W}{\Delta t} \qquad (15)$$

It is clear that water deficits may affect growth by altering ΔW through effects on leaf area, stomatal conductance, or possibly—although we have no evidence of this—directly on photosynthesis or respiration. If this reduction were not accompanied by nonproportional effects on p_{St}, p_F, and p_R then the growth pattern of the tree would remain unaltered. However, there is some evidence (Maggs, 1961) that watering regimes affect the distribution of dry matter. The most obvious effect (not included in Maggs' data) is on p_F; if fruit numbers are reduced by water stress, X will be altered and hence the whole pattern of growth. Some of the data considered below provide information on the interactions between total dry matter production and dry matter partitioning.

A. PHYSIOLOGICAL PROCESSES

As in other species, water deficits in apples lead to large increases in concentration of the plant growth regulator abscisic acid (ABA) (Davies and Lakso, 1978). Bingham (1972) reported a 25-fold increase in ABA concentration (from 12 to 300 ng gm^{-1}) as ψ_l decreased from -0.4 to -2.6 MPa in pot-grown apple seedlings, although Warrit (1977) observed only a five-fold increase in ABA concentration over this range of ψ_l. It is widely accepted that ABA plays a major part in stress-induced stomatal closure in plants (e.g., Mansfield, 1976; Section IV,C), and Goode et al. (1978a) attempted to exploit this to minimize water stress and improve yields by spraying ABA onto apple trees as an antitranspirant. The sprays caused stomatal closure, reduced transpiration, and increased water potential, but the effects were transitory and did not influence yield.

In addition to its role in stomatal closure, ABA inhibits growth and bud break in apple (Yadava and Dayton, 1972; Powell, 1975; Singha and Powell, 1978); there has also been speculation (Jones, 1978b) that it can act as a general stress hormone, causing a syndrome of changes in plant growth and development that might all be regarded as adaptive in that they fit the plant better for reduced water availability (see Section VIII). Davies and Lakso (1978) provide some weak evidence that increases in ABA levels are primarily determined by leaf turgor and not ψ_l. High temperature stress, which induced plant water stress, has also been regarded as decreasing cytokinin production by apple roots (Gur, 1976). (Cytokinins are often antagonistic to ABA in their effects.)

Water deficits reduce the rate of carbon dioxide assimilation by apple leaves (Heinicke and Childers, 1935; Allmendinger et al., 1943; Schneider and Childers, 1941; Cervenka, 1967; Kumashiro and Tateishi, 1967), although Schneider and Childers and Cervenka found that moderate water deficits actually enhanced photosynthesis rates. Most of the reduction in photosynthesis can probably be attributed to stomatal closure, since in those cases where appropriate measurements were made stomatal conductance or transpiration rate declined at least as fast as net photosynthesis. Lakso (1979) observed a good linear relationship between stomatal conductance and net photosynthesis in both field and potted trees. There is sometimes a lag of several days in recovery of photosynthesis after relief of stress (Allmendinger et al., 1943; Schneider and Childers, 1941), but this may be due to a delay in the recovery of stomatal functioning; H. G. Jones (unpublished data) observed that stomatal conductances of potted trees took up to 4 days to return to control values at at the end of a 2-week drought period during which leaf water potentials declined to -3.1 MPa. There was complete recovery of ψ_l within 24 hr, so the lag in recovery of photosynthesis does not necessarily provide evidence that the intercellular processes of photosynthesis were inhibited by the stress (cf. Lakso, 1979). There is evidence that, at least in the later part of the season, stress has to be quite severe before photosynthesis is significantly affected. Lakso (1979) reported substantial photosynthesis with ψ_1 as low as -6.0 MPa, whereas Kriedemann and Canterford (1971) found that normal photosynthesis rates of pear leaves were maintained down to -3.0 to -3.5 MPa.

Leaf respiration rates are also affected by plant water deficits less severe than permanent wilting, though the details of the response are variable. Schneider and Childers (1941) reported a progressive increase in respiration rate as the soil dried out, while Carlier (1961) found a reduction in respiration rate (which was greatest in the youngest leaves) as the soil dried. Further studies are required to determine the conditions that cause these different responses.

Drought is known to increase cell sap concentration and decrease osmotic potentials of apple leaf sap (see Modlibowska, 1961). These effects may be related to the observed increases in reducing sugars and decreases in polysaccharides in apple tissue (Kushnirenko and Kornesku, 1972).

B. VEGETATIVE GROWTH

1. Leaves

Simons (1956) found that leaf size was reduced by water stress and that the effects were different in different cultivars. The length and width of palisade cells were significantly reduced by stress (see also Kumashiro and Tateishi, 1967).

A great deal of information on the effects of water deficits on leaf growth can be obtained from data presented by Chapman (1973), who watered young trees of two cultivars (Delicious and Granny Smith) grown in pots in a glasshouse, daily, weekly, fortnightly, and thrice-weekly. The data illustrate well the ways in which changes in the amount of assimilate going to leaves, induced by water stress, may manifest themselves. At the end of the growing period (30 weeks after applying treatments at about the time of leaf emergence) leaf area on the Delicious trees watered weekly was 60% lower than on those watered daily, while in Granny Smith the reduction was 40%. Reductions in leaf area in the trees watered fortnightly were 82% and 61% in Delicious and Granny Smith, respectively. The percentage reductions in dry matter production with weekly watering were exactly the same as the reductions in leaf area, but fortnightly watering resulted in dry matter reductions of over 90%, relative to daily watering, compared to leaf area reductions of about 80% and 60%, respectively, for the two cultivars.

The dry weight per unit leaf area (specific leaf weight) of the leaves of well-watered Golden Delicious trees was markedly lower than that of trees subjected to severe stress. There was no consistent change in specific leaf weight in Granny Smith.

2. Extension Shoots

Although shoots and stems (trunks) are treated as an entity in equation (15) they are very often considered separately in studies on the growth of apple trees. Maggs (1961) found that "droughting" (a low water regime) reduced the proportion of assimilate going to "new stem" by about 5% but did not reduce the proportion of the total dry matter increment which went to "old stem." The watering regimes imposed by Chapman (1973) did not lead to differences in shoot number per tree but caused considerable difference in shoot length. With both his cultivars, watering regime had little effect on shoot growth for about 10 weeks, i.e., the initial rate of shoot growth was about the same in all cases; the differences in final shoot length were the result of a decline in the rate of growth of shoots on stressed trees after this time, while those on well-watered

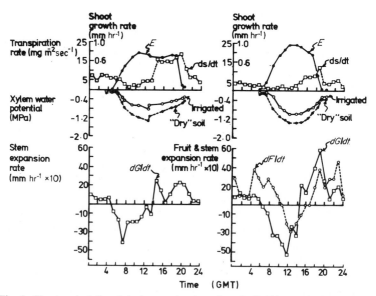

Fig. 8. Short-period (hourly) changes in shoot length (ds/dt), trunk girth (dG/dt), and fruit diameter (dF/dt) in relation to transpiration rate (E) and xylem water potential over 2 days. The growth rate measurements were made on the irrigated trees. (Unpublished data of D. B. B. Powell.)

trees continued to grow for much longer. The same pattern has been observed by G. Heim (unpublished data) and by Kongsrud (1969), who applied six watering treatments to potted apple trees (cv. Lobo). These induced a shoot growth ratio (in terms of total length) of about 2 : 1 between well-watered trees and those receiving least water. The proportion of growth made in the early part of the season increased with decreasing (average) soil moisture potentials. Kongsrud's treatments caused about 2.5-fold differences in dry matter increments between the wettest and driest regimes. In an irrigation experiment (Goode and Hyrycz, 1964) well-watered trees produced more shoots than unwatered trees but there was no difference in total shoot length. (There was no evidence that any of the trees in this experiment suffered severe water stress.)

The data considered so far have all reflected cumulative effects of water deficits on leaf and shoot growth. Figure 8 illustrates short period responses of shoot, girth, and fruit growth to changes in xylem water potential (ψ_x). Shoot growth rates (ds/dt, mm hr^{-1}), measured as described by Powell (1976), are plotted against time on the upper part of the diagrams and may be compared with the time course of transpiration rate (E) and xylem water potential. It is clear that there is no simple relationship between ds/dt and ψ_x.

3. Trunk Growth

In general, as would be expected, water deficits tend to reduce trunk girth and adequate water increases it (Goode and Hyrycz, 1964; Assaf *et al.* 1974, 1975; Hewett, 1976). Taerum (1964) used this as a basis for an irrigation control experiment. Measurements by Goode and Hyrycz (1964) over a number of years showed that the effects of reduced water supplies on trunk girth are cumulative. They also present data showing that in years when the trees carried large numbers of fruits trunk growth in all irrigation treatments was depressed; when fruit loads were light trunk growth was generally much greater, that of trees receiving less water being smaller than well-watered trees but still much greater than trunk growth of well-watered trees carrying many fruits.

Diurnal variations in trunk girth (G) are shown in Fig. 8, where the daily shrinkage documented by Powell and Thorpe (1977) is apparent. In this case there is clearly a relationship between dG/dt and $d\psi_x/dt$.

4. Root Growth

The variability of root distribution, and the effects of irrigation on it, have been discussed in Section V,A. In general water stress increases the proportion of new dry matter going to root rather than to shoot tissues, though the absolute amount of root growth may be decreased by stress.

For example Maggs (1961) found that fewer roots were produced with "low water" than with "high water" treatments, though the percentage of the dry weight increment that went to the roots was higher in the "low water" treatment. In Kongsrud's (1969) irrigation experiments early drought gave lower root/top ratios than did late drought, whereas Cripps (1971) found that water stress caused much more root proliferation in his containers than occurred in well-watered treatments and consistently increased root/top ratios.

C. REPRODUCTIVE GROWTH

We include under this heading the effects of water deficits on bud morphogenesis and flower and fruit numbers.

The question of whether water deficits affect bud morphogenesis was discussed in detail by Salter and Goode (1967), who summarized the evidence available to them: (1) in wet summers extension growth continues late, apparently at the expense of flower buds (mechanisms not defined); (2) when lightly pruned trees are irrigated to promote vigor this does not appear to reduce flower bud formation; (3) flower bud formation is prevented by severe drought and reduced by moderate drought.

In a study by Powell (1976) on Cox's Orange Pippin the number of flower clusters produced in 1974, expressed as a ratio of the number produced in 1973, ranged from 0.8 in trees droughted in spring 1973 to about 0.2 in trees well-watered in 1973. It is likely, however, that this was not a direct enhancement of flower cluster production by drought, but an effect of fruit load in 1973 on flower cluster production. The fruit load on the droughted trees in 1973 was greatly reduced. Luckwill (1970) observed that summer drought tends to increase flower production in maiden trees, this being associated with early cessation of terminal growth. Russian work cited by Salter and Goode (1967) indicated that high cell sap concentrations lead to the formation of floral primordia; when concentrations remain low buds remain vegetative.

When the water status of trees was improved by misting (Goode et al., 1979) over a period of four summers there was a very considerable increase in production of flower clusters. Kongsrud (1969) showed that the number of flower clusters per tree decreased as the soil water potential at which the trees were watered decreased. On balance, therefore, the evidence suggests that bud morphogenesis is more likely to proceed to the point where floral primordia are produced if trees are not subject to significant water deficits (which confirms Salter and Goode's point 3).

As pointed out in Section II fruit number depends, in the first place, on the number of flowers present and the effectiveness of pollination.

However, once successfully pollinated the likelihood that a fruitlet will remain on the tree is influenced by the number of fruitlets present—high fruitlet numbers (relative to tree size) lead to high losses (Llewellyn, 1968)—and by growing conditions, particularly tree water status. This is clearly illustrated by data from Powell (1974), who used soil covers to induce stress in 9-yr-old Cox trees in England. Fruit set, as a percentage of flower number, ranged from 8.5% in trees that received no water between the end of March and June, through 15.7% in trees watered by rain to 24% in well-irrigated trees. Powell also showed that the reduction in drop was caused by an increase in the number of fruitlets per cluster retained, with a considerable reduction in the numbers of clusters that retained no fruitlets. Similar, if less convincing, results were obtained by Kongsrud (1969). Assaf et al. (1974, 1975) provide data on fruit drop in irrigation experiments in Israel, showing it to be consistently inversely related to the amount of water applied.

VII. IRRIGATION AND ITS EFFECTS

A. APPLE QUALITY

Climate and tree water relations during the growing season have been implicated in many aspects of harvest quality and the susceptibility of apples to storage disorders. In spite of this there is surprisingly little evidence for direct effects of water stress on these parameters (van Zyl, 1970). In general, extremes of rainfall and of temperature are detrimental to many aspects of fruit quality, including flavor, shape, appearance, and storage behavior (Anonymous, 1963; Sharples, 1973).

The most obvious and well-documented effect of tree water status on an aspect of fruit quality is on fruit size, which, other things being equal, tends to be reduced in dry years and increased by irrigation (Goode and Ingram, 1971; Assaf et al., 1974, 1975; Goode et al., 1975, 1978b; Hewett, 1976; Powell, 1976). Increases in fruit size come about through the increase in ΔW (equation 15); if the values of $p_j(X)$ remain unaltered relative to those in trees suffering water deficits, this would result in increased growth of all components. In the United Kingdom late irrigation has been found particularly effective in increasing mean fruit size (Goode et al., 1978b), though in other climates (e.g., Israel—Assaf et al., 1976) the effects are sometimes small. Attempts to increase fruit size using other means of reducing plant stress, such as ABA sprays to close stomata (Goode et al., 1978b) or "Vaporgard" antitranspirant sprays (Weller and Ferree, 1978), have not been successful, even though water deficits were reduced by these treatments. However, Miller (1979) reported that antitranspirants could increase fruit size in trees in dry soil. Stress reduction by misting also tended to increase fruit size (Good *et al.*, 1979).

The effects of water status on fruit size are complicated by the interaction of size with fruit load. For instance, stress early in the season can indirectly increase final fruit size by reducing fruit set and fruit load (e.g., Powell, 1976), while stress in the previous season can affect fruit load by reducing (Salter and Goode, 1967) or increasing (Lees, 1926; Swarbrick and Luckwill, 1953; Powell, 1976) flower bud production, though flower bud production also depends on fruit load. That water stress commonly increases the proportion of smaller fruits is illustrated by data of Powell (1976) (see Fig. 9).

Many other aspects of quality are related to fruit size, and therefore may be indirectly influenced by tree water relations. For example larger fruit tended to have lower acidity (e.g., Assaf *et al.*, 1976) and they may be more susceptible to storage rots (Edney, 1973), bitter pit (though see below), breakdown, and other storage disorders (Sharples, 1973), and less susceptible to shrivel (Wilkinson and Fidler, 1973).

Reduced water supply during the growing season leads to an increase in skin cracking and russet, both of which reduce storage life by increasing the rate of water loss from stored fruit or allowing infection by rotting fungi. The fine cracking reported by Goode *et al.* (1975) and russet, can be reduced by irrigation or misting (Goode *et al.*, 1978b, 1979). Apples tend to shrivel more readily following a dull, wet growing season, but Fidler and North (1969) reported a three-fold reduction in density of openings (lenticels and stomata) in Cox's Orange Pippin apples with irrigation. Another problem in humid areas is "rain cracking," in which absorption of water through the skin in wet seasons can lead to swelling of the cortical layers and subsequent skin damage (Pierson *et al.*, 1971).

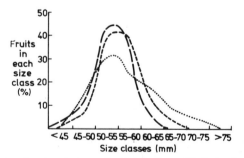

Fig. 9. The effects of water stress on the distribution of fruit size in well-irrigated trees (dotted curve), trees watered by rainfall only (short dashed curve), and trees subjected to water stress by covering the soil (long dashed curve). (Drawn from data by Powell, 1976.)

The so-called corking disorders of apples have frequently been associated with drought during the growing season (see Faust and Shear, 1968). One group of these disorders includes "drought spot," so named by Mix (1916), but now known to be primarily caused by boron deficiency (Faust and Shear, 1968). Bitter pit has also been related to suboptimal water supply, but the effects are most likely to be indirect, acting through an influence on fruit calcium content. Water deficits reduce net calcium uptake into the fruit, particularly later in the season, when high transpiration rates or large water deficits may even cause withdrawal of calcium from the fruit (Wilkinson, 1968; Mix and Marschner, 1976; Hanger, 1979; Tromp, 1979). In a dry year a high proportion of the total calcium is taken up by the fruit in the first month to 6 weeks after petal fall, in contrast to the situation in more humid seasons or climates when uptake may continue until harvest. The variable effects of irrigation on bitter pit development (see Faust and Shear, 1968) are probably explicable in terms of calcium content of the fruit.

The importance of fruit water relations in the development of calcium related disorders is largely related to the transport of calcium into the fruit. Since most translocation of calcium occurs via the transpiration stream in the xylem (see, e.g., Hanger, 1979), the rate of evaporation from apple fruits is an important factor regulating calcium accumulation. There are, however, few published data on transpiration water loss from apple fruits, and those measurements that have been made involved very unnatural (and unknown) conditions inside the measurement cuvettes (Heinicke, 1929), so are of little value. Recent, more precise measurements of diffusion conductances (g_f) on the surface of apple fruits at intervals through the season (Jones and Higgs, 1980) indicate that the conductance decreases from around 1.7×10^{-3} m sec^{-1} before flowering to less than 1×10^{-4} m sec^{-1} at harvest (Fig. 10). This decreased conductance is offset by increasing surface area (Fig. 10). Most water loss that does occur appears to be from the eye end of the fruit, since conductances from this area were frequently above 0.5 mm sec^{-1}.

Another disorder particularly prevalent in dry summers is superficial scald. Although the evidence for a drought effect on scald is entirely based on statistical correlations, useful predictions of scald occurrence can be made on the basis of soil moisture deficits (Fidler, 1956; Wilkinson and Fidler, 1973; Sharples, 1973). Without controlled experiments, however, it is notoriously difficult to separate the effects of water deficits from high temperatures and high radiation, since these variables all tend to be correlated. Other disorders have also been statistically related to rainfall. For instance *Gloeosporium* and other rots increase with increasing summer rainfall, probably because of the increased probability of orchard disper-

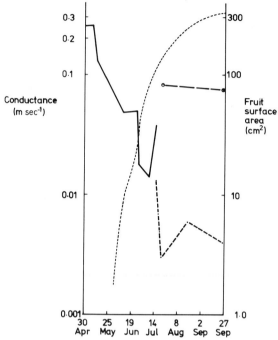

Fig. 10. Changes in the conductance of fruit surfaces with fruit growth. From April to July measurements were made on whole fruits (solid line); later measurements (July to September, dashed line) were made on part of the fruits. The two ringed points connected by dashes represent measurements made on the calyx end of the fruits. Fruit surface area is shown by the dotted line. Note that the conductance scale is logarithmic.

sal in wet seasons (Pierson *et al.*, 1971; Sharples, 1973). Similarly, misting has been reported to increase the incidence of soft rots (Goode *et al.*, 1979), though irrigation reduced incidence.

B. GENERAL EFFECTS

The effects of irrigation on apple tree growth and productivity can be understood in terms of the mechanisms and responses to relief of water deficits discussed previously. Unfortunately, in the reports of many irrigation experiments, information on environmental conditions was insufficient to allow assessment of the degree or the duration of stress to which the trees were subjected. Although it would be impractical, in commercial practice, to base irrigation schedules on plant water potential, it would be valuable if the principles and equations outlined in Sections IV and V were utilized, together with information on soil-water-holding capacity, in the design of experiments. Treatments might then be based on the expected degree and duration of water stress, the models would be checked by measurement of plant water potential made at appropriate times and analyzed in terms of weather data, and the results, in terms of tree performance, interpreted on the basis of the degree and duration of stress. Irrigation recommendations would then be made on the basis of water deficits likely to be experienced in any particular region. An alternative approach has been adopted with some success by Levin et al. (1980). This involves irrigating when the mean diameter of the fruits in an orchard ceases to increase linearly with time. Other possibilities that have been used for apples include use of either decreased leaf infiltration or decreased sap osmotic potential as means for scheduling irrigation.

Although the responses to be expected from irrigation are clear from Section IV, we provide here a brief general review of the results obtained in some irrigation experiments carried out in the field and with trees in containers. The conclusions that emerge confirm those reached in Section VI.

In the United States, irrigation during a drought, to maintain soil moisture at 80% of that available to a depth of 30 cm in the soil, almost doubled the total yield of early flowering cultivars and increased by 46% the number of fruits with diameters greater than 60 mm. The total yield of a late cultivar increased by only 12%. When the drought ended, the fruit on nonirrigated plots increased rapidly in size (O'Grady, 1959). Vegetative growth increased in well-watered plants, which had lower contents of soluble and insoluble nitrogen; these were higher in low moisture plants. Neubauer (1966) demonstrated increases in crown volume (shoot growth, leaf area), trunk diameter, total yield, and weight of fruits with increasing water applications, and also found that clover-grass swards reduced the effects of irrigation in relation to clean cultivated treatments. This is to be expected because of water loss from the orchard by transpiration from the swards. The adverse effects of grass swards were also demonstrated by Peerbooms (1968), who showed that high nitrogen applications reduced the quality of fruits. There is a need for more precise and quantitative work on the rates of water loss from orchards with vegetation in the interrows so that the equations presented in Section IV can be extended to allow calculation of total water loss from such orchards. A first approximation could be made by estimating the rate of water loss from the swards using the net radiation for the orchard as a whole (see Landsberg et al., 1973), estimates of wind speed and humidity at the level of the sward, and the combination evaporation equation [equation (1)]. However, this should be checked by micrometeorological measurements and direct measurements (possibly using small lysimeters) of the rate of water loss by the swards.

Pot-grown trees (Kumashiro and Tateishi, 1967) showed the responses we would expect, in terms of tree and fruit growth, when maintained at varying soil moisture levels. Fiedler (1967), also using container-grown trees, showed that increasing water supplies resulted in growth increases but also in nonproportional increases in water consumption. Reduced water supply, particularly in conjunction with increased nutrition, significantly reduced fruit set. High water supply (Fiedler and Weier, 1971a) enhanced fruit production by reducing fruit drop after flowering and in June. As low water supply reduced fruit number, individual fruit weights and diameters increased [cf. equation (15)]; low water supply to bearing trees increased flowering the following year (cf. Section VI,C). Other results of Fiedler and Weier (1971b) conform to expected patterns: low water supplies reduced nutrient uptake and vegetative growth; vegetative growth on nonbearing trees was better than in bearing ones. Results that also conform to those reviewed previously are presented by Dvořák (1968), Goode (1975), and other authors.

VIII. DROUGHT RESISTANCE AND ADAPTATION

There is a wealth of literature on the comparative "drought resistance" of different rootstock and scion varieties of apple, particularly from Russia (e.g., Filippov, 1962; Rom, 1965a; Misic and Gavrilovic, 1969; Černyševa, 1970), although no consistent definition of drought tolerance has been adopted. Common definitions, where explicitly stated, are expressed either in terms of plant survival (e.g., Demcěnko, 1968), the ability of a plant to retain its leaves after a drought period (e.g., Blasse, 1960), or in terms of yield maintenance. Many trials have provided information on varietal differences in yield response to irrigation. In some cases irrigation has increased yields without affecting yield ranking of cultivars (Friedrich, 1967), while in other cases significant genotype x environment (water supply) interactions have been detected (Rom, 1965b). Stability of yield or growth over a range of water supply conditions is also frequently regarded as a component of drought tolerance (though in many cases high absolute yields would be preferable). Unfortunately, it is a common observation that stability of yield is associated with a low yield potential (Jones, 1977).

It is not usually reported whether variations in drought tolerance are attributable to differences in stress avoidance [e.g., rooting depth or stomatal sensitivity to stress (Jones, 1978b)] or whether differences exist in stress tolerance (in terms of differing sensitivity of physiology to given levels of water stress). Mamaeva (1970) has proposed that both mechanisms may be important in drought tolerance differences between apple cultivars. There are several reports that dwarfing rootstocks, particularly M9 (Misic and Gavrilovic, 1969; Moiseev *et al.*, 1970; Razlivalova, 1974), are especially valuable in conferring drought resistance on apple scion varieties. Friedrich (1967) and Carlson (1967) also demonstrated that yield stability was better over a range of water regimes with M9 than with other rootstocks such as M1 and M4. Drought resistance has been frequently associated with a high bound water content or high "water retaining capacity" (e.g., Černyševa, 1970; Mamaeva, 1970; Razlivalova, 1974).

There is much evidence that apple trees can adapt to periods of water stress by reduction in total leaf area. Total leaf area can be decreased through increased leaf drop (e.g., Filippov, 1962; Jones and Higgs, 1979a), reduced rate of production of new leaves (e.g., Chapman, 1973), or reduced leaf size (Chapman, 1973). These responses, together with increased cuticle thickness (Kumashiro and Tateishi, 1967), increased stomatal resistance (Section IV,C), and increased root/shoot ratios (Section VI,B,4) all tend to conserve water. They may all be mediated by stress-induced increases in the endogenous growth regulator ABA. Other reports that apple trees adapt to periods of drought include evidence for increases in root drought hardiness (Wildung *et al.*, 1973), and the evidence that cell osmotic concentrations can increase with stress (Goode and Higgs, 1973; Davies and Lakso, 1978; Jones and Higgs, 1979b). These authors reported decreases of up to 0.4 MPa in osmotic potential in their unirrigated compared with irrigated trees. This "osmotic adjustment" or "adaptation" (see Turner, 1979) can help maintain turgor at suitable levels to maintain normal growth, even when leaf water potentials decline.

IX. CONCLUDING REMARKS

It should, ideally, be possible to calculate the water status of plants at any time from weather data. In the case of apple trees, considerable progress has been made although, to paraphrase Churchill: "This is not the beginning of the end, it may only be the end of the beginning . . ." There remains a need for many more measurements of stomatal resistance, of the resistances in the flow pathways through the plant, and of capacitance. There is also a need for hydrological data such as estimates of the amount of water intercepted by trees and lost by direct evaporation, and measurements of water use by orchards as a whole. These will be required to check the water balance calculations which must be made to estimate soil water content and hence water potential in the root zone.

However, as said at the beginning of this chapter, even precise knowledge about plant water status and the factors affecting it is of little practical value unless it can be coupled with knowledge about how water deficits affect the physiological processes that contribute to plant growth and development and hence the performance of the whole plant. Despite the extensive research of recent years, progress in these areas has been slow. It is not always clear which water status parameter should be used as a basis for the analysis of particular physiological processes and there have been few, if any, attempts to investigate how the integral of plant water potential and time affects the state of the tree at any time. This is important, since the state at any time is the end result of the physiological processes operating over a period of time. Assuming that the rate of those processes (dP/dt) is affected by plant water status (ψ_{plant}), we may write the general equation:

$$\frac{dP}{dt} = f(\psi_{plant}), \text{ hence } P(t) = \int_{t=0}^{t} f(\psi_{plant}) \, dt$$

The problem is to determine the functional form of $f(\psi_{plant})$ for particular processes.

Progress in the future may be more rapid if the questions asked of experiments are stated in these terms and set within the framework of models of the growth and development cycle of apple trees and the effects of environmental factors on their water relations, and on all the other processes involved. These may consist of a number of submodels, written at a number of levels, but the ultimate objective must always be the development of well-tested mechanistic models, which are useful both for the analysis of crop responses and as an aid to management decisions and yield prediction.

REFERENCES

Abbott, D. L. (1977). Fruit bud formation in Cox's Orange Pippin. *Long Ashton Res. Stn., Univ. Bristol, Rep.* pp. 167–176.

Allmendinger, D. F., Kenworthy, A. L., and Overholser, E. F. (1943). The carbon dioxide intake of apple leaves as affected by reducing the available soil water to different levels. *Proc. Am. Soc. Hortic. Sci.* **42**, 133–140.

Anderson, J. L. (1974). Delaying bloom and summer cooling, overhead misting. *Proc. Wash. State Hortic. Assoc.* **70**, 87–91.

Anonymous (1963). "The Effect of Weather and Climate upon the Keeping Quality of Fruit," Tech. Note 53. WMO.

Antipov, N. I. (1977). Water exchange of stems, leaf petioles and leaf blades. *Fiziol. Rast.* **24**, 785–789.

Assaf, R., Bravdo, B., and Levin, I. (1974). Effects of irrigation according to water deficit in two different soil layers, on the yield and growth of apple trees. *J. Hortic. Sci.* **49**, 53–64.

Assaf, R., Levin, I., and Bravdo, B. (1975). Effect of irrigation regimes on trunk and fruit growth rates, quality and yield of apple trees. *J. Hortic. Sci.* **50**, 481–493.

Assaf, R., Levin, I., and Bravdo, B. (1976). "Apple Response to Water Regimes and Hedgerow Training," Grower Assoc. Bull. No. 69. Israeli Growers' Assoc.

Atkinson, D. (1973). Root studies. *Annu. Rep., East Malling Res. Stn., Kent* pp. 46–60.

Atkinson, D. (1974). Some observations on the distribution of root activity in apple trees. *Plant Soil* **40**, 333–342.

Atkinson, D. (1977). Some observations on the root growth of young apple trees and their uptake of nutrients when grown in herbicided strips in grassed orchards. *Plant Soil* **49**, 459–471.

Atkinson, D., and White, G. C. (1976). Soil management with herbicides—the response of soils and plants. *Proc. Br. Crop Prot. Conf.* pp. 873–884.

Atkinson, D., and Wilson, S. A. (1979). The root-soil interface and its significance for fruit tree roots of different ages. *In* "The Soil-Root Interface" (J. L. Harley and R. Scott-Russell, eds.), pp. 259–271. Academic Press, New York.

Atkinson, D., Naylor, D., and Coldrick, G. A. (1976). The effect of tree spacing on the apple root system. *Hortic. Res.* **16**, 89–105.

Avery, D. J. (1970). Effects of fruiting on the growth of apple trees on four rootstock varieties. *New Phytol.* **69**, 19–30.

Baxter, P., and West, D. (1977). The flow of water into fruit trees. I. Resistances to water flow through roots and stems. *Ann. Appl. Biol.* **87**, 95–101.

Beakbane, A. B., and Majumder, P. K. (1975). A relationship between stomatal density and growth potential in apple rootstocks. *J. Hortic. Sci.* **50**, 285–289.

Bingham, G. E. (1972). Stomatal response in field corn (*Zea mays,* L.) and apple (*Malus sylvestris*). Ph.D. Thesis, Cornell University, Ithaca, New York.

Black, J. D. F., and Mitchell, P. D. (1970). Soil water use from an apple orchard under various soil management systems. *Aust. J. Exp. Agric. Anim. Husb.* **10**, 209–213.

Blasse, W. (1960). Der Einfluss unterschiedlicher Bodenpflegeverfahren im Obstbau auf den Verlauf der Bodenfeuchtigkeit 1958 und 1959. *Dtsch. Gartenbau* **7**, 102–105.

Boyer, J. S. (1974). Water transport in plants: Mechanisms of apparent changes in resistance during absorption. *Planta* **117**, 187–207.

Brouwer, R. (1954). The regulating influence of transpiration and suction tension on the water uptake by roots of intact *Vicia faba* plants. *Acta Hortic.* **3**, 264–312.

Burrows, F. J., and Milthorpe, F. L. (1976). Stomatal conductance in the control of gas exchange. *In* "Water Deficits and Plant Growth" (T. T. Kozlowski, ed.), Vol. 4, pp. 103–152. Academic Press, New York.

Butler, D. R. (1976). Estimation of the transpiration rate in an apple orchard from net radiation and vapour pressure deficit measurements. *Agric. Meteorol.* **16**, 277–289.

Byass, J. B. (1968). Equipment and methods for orchard spray application research. II. The geometry of apple trees. *J. Agric. Eng. Res.* **13**, 358–369.

Carlier, G. (1961). Etude des variations de l'intensité respiratoire des feuilles de pommier sur pied, provoquées par le desséchement et la réhumectation du sol. *C. R. Hebd. Seances Acad. Sci.* **253**, 898–900.

Carlson, R. F. (1967). Growth response of several rootstocks to soil water. *HortScience* **2**, 108–110.

Černyševa, A. I. (1970). Some physiological indices of apple varieties in relation to drought resistance. *Sb. Nauchn. Rab., Vses Nauchno-Issled. Inst. Sadovod.* **14**, 219–224.

Červenka, K. (1967). A study of the influence of light intensity and water deficit on assimilation of CO_2 by apple trees. *Ved. Pr. Ovocn. Vyzk. Ustavu Ovocn. Holovousich* **3**, 79–88.

Chapman, K. R. (1971). Plant water status of apple trees and its measurement in the field. 6. Diurnal variations in the water potential of apple leaves and fruits of three varieties at two stress levels. *Queensl. J. Agric. Anim. Sci.* **27**, 199–203.

Chapman, K. R. (1973). Effect of four water regimes in performance of glasshouse-grown nursery apple trees. *Queensl. J. Agric. Anim. Sci.* **30**, 125–135.

Charles-Edwards, D. A., and Thorpe, M. R. (1976). Interception of diffuse and direct-beam radiation by a hedgerow apple orchard. *Ann. Bot. (London)* [N.S.] **40**, 603–613.

Coker, E. G. (1958). The root development of black currants under straw mulch and clean cultivation. *J. Hortic. Sci.* **33**, 21–28.

Coker, E. G. (1959). Root development of apple trees in grass and clean cultivation. *J. Hortic. Sci.* **34**, 111–121.

Cowan, I. R. (1965). Transport of water in the soil-plant-atmosphere system. *J. Appl. Ecol.* **2**, 221–239.

Cripps, J. E. L. (1971). The influence of soil moisture in apple root growth and root : shoot ratios. *J. Hortic. Sci.* **46**, 121–130.

Davies, F. S., and Lakso, A. N. (1978). Water relations in apple seedlings: Changes in water potential components, abscisic acid levels and stomatal conductances under irrigated and non-irrigated conditions. *J. Am. Soc. Hortic. Sci.* **103**, 310–313.

Davies, F. S., and Lakso, A. N. (1979). Water stress responses of apple trees. II. Resistance and capacitance effects as affected by greenhouse and field conditions. *J. Am. Soc. Hortic. Sci.* **104**, 395–397.

Demčenko, V. G. (1968). Transplants with interstocks are cheaper. *Sadovodstvo (Kiev)* **4**, 26.

Denmead, O. T., and Millar, B. D. (1976). Water transport in wheat plants in the field. *Agron. J.* **68**, 297–303.

Denne, M. P. (1960). The growth of apple fruitlets and the effect of early thinning on fruit development. *Ann. Bot. (London)* [N.S.] **24**, 397–406.

Denne, M. P. (1963). A comparison between fruits of Cox's Orange Pippin from Kent, England and Auckland, New Zealand. *N.Z. J. Bot.* **1**, 295–300.

Dvořák, J. (1968). The effect of different soil moisture contents on the development of the apple variety Golden Pearmain. *Rostl. Vyroba* **14**, 983–990.

Edney, K. L. (1973). Fungal disorders. *In* "The Biology of Apple and Pear Storage" (J. C. Fidler, B. G. Wilkinson, K. L. Edney, and R. A. Sharples, eds.), Res. Rev. 3, pp. 133–172. Commonw. Agric. Bur., Fainham Royal, Bucks, England.

Etherington, J. R. (1967). Soil water and the growth of grasses. II. Effects of soil water potential on growth and photosynthesis of *Alopecunis pratensis*. *J. Ecol.* **55**, 373–380.

Ewart, A. J. (1905). The ascent of water in trees. *Philos. Trans. R. Soc. London* **198**, 41–85.

Faiz, S. M. A. (1973). Soil-root water relations. Ph.D. Thesis, University of Aberdeen.

Farmer, J. B. (1919). On the quantitative differences in the water conductivity of the wood in trees and shrubs. II. The deciduous plants. *Proc. R. Soc. London, Ser. B* **90**, 232–250.

Farquhar, G. D. (1978). Feedforward responses of stomata to humidity. *Aust. J. Plant Physiol.* **5**, 787–800.

Faust, M., and Shear, C. B. (1968). Corking disorders of apples: A physiological and biochemical review. *Bot. Rev.* **34**, 441–469.

Fidler, J. C. (1956). Scald and weather. *Food Sci. Abstr.* **28**, 545–554.

Fidler, J. C., and North, C. J. (1969). Production of volatile organic compounds by apples. *J. Sci. Food Agric.* **20**, 521–526.

Fiedler, W. (1967). Untersuchungen über die Wirkung variierter Wasser-und Nährstoffversorgung auf Apfelbäume in Gefässen. *Tagungsber., Dtsch. Akad. Landwirtschaftswiss. Berlin* **93**, 63–71.

Fiedler, W., and Weier, B. (1971a). Untersuchungen über die Wirkung variieter Wasser-und Nährstoffversorgung auf Apfelbäume in Gefässen IV. Einfluss unterschiedlicher Düngung und Bewässerung auf den Triebzuwachs. *Arch. Gartenbau* **19**, 349–356.

Fiedler, W., and Weier, B. (1971b). Untersuchungen über die Wirkung variierter Wasser-und Nährstoffversorgung auf Apfelbäume in Gefässen IV. Einfluss unterschiedlicher Düngung und Bewässerung aud Blütenausatz, Fruchtbildung under Ertrag. *Arch. Gartenbau* **19**, 357–369.

Filippov, L. A. (1962). The response of apple varieties to drought under Moldavian conditions. *Bot. Zh.* **47**, 821–829.

Fiscus, E. L. (1975). The interaction between osmotic and pressure induced water flow in plant roots. *Plant Physiol.* **55**, 917–922.

Fiscus, E. L., and Kramer, P. J. (1975). General model for osmotic and pressure-induced flow in plant roots. *Proc. Natl. Acad. Sci. U.S.A.* **72**, 3114–3118.

Friedrich, G. (1967). Zusatzbewässerung und Ertragsbildung im Obstbau. *Tagungsber. Dtsch. Akad. Landwirtschaftswiss. Berlin* **93**, 93–103.

Furr, J. R. (1928). The relation between vessel diameter and flow of water in the xylem of apple. *Proc. Am. Soc. Hortic. Sci.* **25**, 311–320.

Gardner, W. R. (1960). Dynamic aspects of water availability to plants. *Soil Sci.* **89**, 63–73.

Garjugin, G. A. (1964). Winter irrigations guarantee yields. *Sadovodstvo (Kiev)* **1**, 25.

Goode, J. E. (1975). Water storage, water stress and crop responses to irrigation. *In* "Climate and the Orchard" (H. C. Pereira, ed.), pp. 51–62. Commonw. Agric. Bur., Farnham Royal, United Kingdom.

Goode, J. E., and Higgs, K. H. (1973). Water, osmotic and pressure potential relationships in apple leaves. *J. Hortic. Sci.* **48**, 203–215.

Goode, J. E., and Hyrycz, K. J. (1964). The response of Laxton's Superb to different soil moisture conditions. *J. Hortic. Sci.* **39**, 254–276.

Goode, J. E., and Ingram, J. (1971). The effect of irrigation on the growth, cropping and nutrition of Cox's Orange Pippin apple trees. *J. Hortic. Sci.* **46**, 195–208.

Goode, J. E., Fuller, M. M., and Hyrycz, K. J. (1975). Skin-cracking of Cox's Orange Pippin apples in relation to water stress. *J. Hortic. Sci.* **50**, 265–269.

Goode, J. E., Higgs, K. H., and Hyrycz, K. J. (1978a). Abscisic acid applied to orchard trees of Golden Delicious apple to control water stress. *J. Hortic. Sci.* **53**, 99–103.

Goode, J. E., Higgs, K. H., and Hyrycz, K. J. (1978b). Nitrogen and water effects on the nutrition, growth, crop yield and fruit quality of orchard-grown Cox's Orange Pippin apple trees. *J. Hortic. Sci.* **53**, 295–306.

Goode, J. E., Higgs, K. H., and Hyrycz, K. J. (1979). Effects of water stress control in apple trees by misting. *J. Hortic. Sci.* **54**, 1–11.

Gorin, T. I. (1963). Transpiration of fruit trees in summer. *Vestn. S-kh. Nauki (Alma-Ata)* **8**, 114–117.

Graham, J., Clarkson, D. T., and Sanderson, J. (1974). Water uptake by the roots of marrow and barley plants. *ARC Letcombe Lab. Rep*, pp. 9–12.

Gur, A. (1976). Responses of apple trees to various kinds of root stress. *Isr. J. Bot.* **25**, 100.

Hanger, B. C. (1979). The movement of calcium in plants. *Commun. Soil Sci. Plant Anal.* **10**, 171–193.

Hansen, P. (1971). The effect of fruiting upon transpiration rate and stomatal opening in apple leaves. *Physiol. Plant.* **25**, 181–183.

Head, G. C. (1968). Seasonal changes in the diameter of secondarily thickened roots of fruit trees in relation to growth of other parts of the tree. *J. Hortic. Sci.* **43**, 275–282.

Heim, G., Landsberg, J. J., Watson, R. L., and Brain, P. (1979). Ecophysiology of apple trees: Dry matter production and partitioning by young Golden Delicious trees in France and England. *J. Appl. Ecol.* **16**, 179–194.

Heinicke, A. J. (1929). A method for studying the relative rates of transpiration of Apple leaves and Fruits. *Proc. Am. Soc. Hortic. Sci.* **26**, 312–314.

Heinicke, A. J. , and Childers, N. F. (1935). The influence of water deficiency in photosynthesis and transpiration of apple leaves. *Proc. Am. Soc. Hortic. Sci.* **33**, 155–159.

Heinicke, D. R. (1963). The micro-climate of fruit trees. II. Foliage and light distribution patterns in apple trees. *Proc. Am. Soc. Hortic. Sci.* **83**, 1–11.

Heinicke, D. R. (1964). The microclimate of fruit trees. III. The effect of tree size on light penetration and leaf area in Red Delicious apple trees. *Proc. Am. Soc. Hortic. Sci.* **85**, 33–41.

Hewett, E. W. (1976). Irrigation of apple trees in Nelson, New Zealand. *J. Agric. Res.* **19**, 505–511.

Holland, D. A. (1968). The estimation of total leaf area on a tree. *Annu. Rep. East Malling Res. Stn., Kent* pp. 101–107.

Hsiao, T. C. (1973). Plant responses to water stress. *Annu. Rev. Plant Physiol.* **24**, 519–570.

Huck, M. G., Klepper, B., and Taylor, H. M. (1970). Diurnal variations in root diameter. *Plant Physiol.* **45**, 529–530.

Hudson, J. P. (1973). Maximum intensity orchards—growing fruits without trees. *Proc. Int. Hortic. Congr., 18th, 1970* Vol. 1, p. 56.

Jackson, J. E. (1970). Aspects of light climate within apple orchards. *J. Appl. Ecol.* **7**, 207–216.

Jackson, J. E. (1975). Effects of light intensity on growth, cropping and fruit quality. *Commonw. Bur. Hortic. Plant. Crops (G.B.), Res. Rev.* No. 5, pp. 17–25.

Jackson, J. E., and Palmer, J. N. (1972). Interception of light by model hedgerow orchards in relation to latitude, time of year and hedgerow configuration and orientation. *J. Appl. Ecol.* **9**, 341–357.

Jarvis, P. G. (1975). Water transfer in plants. *In* "Heat and Mass Transfer in the Biosphere" (D. A. de Vries and N. H. Afghan, eds.), Part I, pp. 270–294. Scripta Book Co., Washington, D. C.

Jarvis, P. G. (1980). Stomatal response to water stress in conifers. *In* "Adaptation of Plants to Water and High Temperature Stress" (N. C. Turner and P. J. Kramer, eds.), pp. 105–122. Wiley (Interscience), New York.

Jones, H. G. (1973). Photosynthesis by thin leaf slices in solution. II. Osmotic stress and its effects on photosynthesis. *Aust. J. Biol. Sci.* **26**, 25–33.

Jones, H. G. (1976). Crop characteristics and the ratio between assimilation and transpiration. *J. Appl. Ecol.* **13**, 605–622.

Jones, H. G. (1977). Aspects of the water relations of spring wheat (*Triticum aestivum* L.) in response to induced drought. *J. Agric. Sci.* **88**, 267–282.

Jones, H. G. (1978a). Modelling diurnal trends of leaf water potential in transpiring wheat. *J. Appl. Ecol.* **15**, 613–626.

Jones, H. G. (1978b). How plants respond to stress. *Nature (London)* **271**, 610.

Jones, H. G., and Higgs, K. H. (1979a). Nutrition: Water relations. *Annu. Rep. East Malling Res. Stn., Kent* p. 242.

Jones, H. G., and Higgs, K. H. (1979b). Water potential—water content, relationships in apple leaves. *J. Exp. Bot.* **30**, 965–970.

Jones, H. G., and Higgs, K. H. (1980). Water relations. *Annu. Rep. East Malling Res. Stn., Kent* (in press).

Kongsrud, K. L. (1969). Effects of soil moisture tension on growth and yield in blackcurrants and apples. *Acta Agric. Scand.* **19**, 245–257.

Kriedemann, P. E., and Canterford, R. L. (1971). The photosynthetic activity of pear leaves (*Pyrus communis* L.). *Aust. J. Biol. Sci.* **24**, 197–205.

Kumashiro, K., and Tateishi, S. (1967). The effect of soil moisture on the tree growth, yield and fruit quality of Jonathan apples. *J. Jpn. Soc. Hortic. Sci.* **36**, 9–20.

Kushnirenko, M. D., and Kornesku, A. S. (1972). Water and nitrogen metabolism in apple leaves in relation to nutrition and soil moisture. *Vodn. Rezhim Rast. Razlichn. Vla-goobespechennosti* pp. 73–94.

Lakso, A. N. (1979). Seasonal changes in stomatal response to leaf water potential in apple. *J. Am. Soc. Hortic. Sci.* **104**, 58–60.

Landsberg, J. J. (1974). Apple fruit bud development and growth, analysis and an empirical model. *Ann. Bot. (London)* [N.S.] **38**, 1013–1024.

Landsberg, J. J. (1977). Studies on the effect of weather on the growth and production cycle of apple trees. *J. R. Agric. Soc. Engl.* **138**, 116–133.

Landsberg, J. J. (1980). Limits to apple yields imposed by weather. *In* "Opportunities for Increasing Plant Yields" (R. Hurd and P. V. Biscoe, eds.) (in press).

Landsberg, J. J., and Butler, D. R. (1980). Stomatal response to humidity: Implications for transpiration. *Plant, Cell, & Environ.* **3**, 29–33.

Landsberg, J. J., and Fowkes, N. D. (1978). Water movement through plant roots. *Ann. Bot. (London)* [N.S.] **42**, 493–508.

Landsberg, J. J., and James, G. B. (1971). Wind profiles in plant canopies: Studies on an analytical model. *J. Appl. Ecol.* **8**, 729–741.

Landsberg, J. J., and Jarvis, P. G. (1973). A numerical investigation of the momentum balance of a spruce forest. *J. Appl. Ecol.* **10**, 645–655.

Landsberg, J. J., and Powell, D. B. B. (1973). Surface exchange characteristics of leaves subject to mutual interference. *Agric. Meteorol.* **12**, 169–184.

Landsberg, J. J., and Thorpe, M. R. (1975). The mechanisms of apple bud morphogenesis: Analysis and a model. *Ann. Bot. (London)* [N.S.] **39**, 689–699.

Landsberg, J. J., Powell, D. B. B., and Butler, D. R. (1973). Microclimate in an apple orchard. *J. Appl. Ecol.* **10**, 881–896.

Landsberg, J. J., Beadle, C. L., Biscoe, P. V., Butler, D. R., Davidson, B., Incoll, L. D., James, G. B., Jarvis, P. G., Martin, P. J., Neilson, R. E., Powell, D. B. B., Slack, M., Thorpe, M. R., Turner, N. C., Warrit, B., and Watts, W. R. (1975). Diurnal energy, water and CO_2 exchanges in an apple (*Malus pumila*) orchard. *J. Appl. Ecol.* **12**, 659–684.

Landsberg, J. J., Blanchard, T. W., and Warrit, B. (1976). Studies on the movement of water through apple trees. *J. Exp. Bot.* **27**, 579–596.

Lees, A. H. (1926). Influence of summer rainfall and previous crop on fruiting of apples. *J. Pomol.* **5**, 178–194.

Levin, I., Bravdo, B., and Assaf, R. (1973). Relation between root distribution and soil water extraction in different irrigation regimes. *Ecol. Stud.* **4**, 351–359.

Levin, I., Assaf, R., and Bravdo, B. (1980). Irrigation, water status and nutrient uptake in an apple orchard. *In* "Mineral Nutrition of Fruit Trees" (D. Atkinson, J. E. Jackson, R. O. Sharples, and W. M. Waller, eds.). Butterworth, London.

Llewellyn, F. W. M. (1968). The log (−log) transformation in the analysis of fruit retention records. *Biometrics* **24**, 627–638.

Luckwill, L. C. (1970). The control of growth and fruitfulness of apple trees. *In* "Physiology of Tree Crops" (L. C. Luckwill and C. V. Cutting, eds.), pp. 237–254. Academic Press, New York.

Maggs, D. H. (1961). Changes in the amount and distribution of increment induced by contrasting watering, nitrogen and environmental regimes. *Ann. Bot. (London)* [N.S.] **25**, 353–361.

Maggs, D. H. (1963). The reduction in growth of apple trees brought about by fruiting. *J. Hortic. Sci.* **38**, 119–128.

Malhotra, R. C. (1931). A contribution to the physiology and anatomy of tracheae, with special reference to fruit trees. I. Influence of tracheae and leaves on the water conductivity. *Ann. Bot. (London)* **45**, 593–620.

Mamaeva, Z. G. (1970). The water retaining capacity of the leaves of apple varieties during wilting. *Sb. Tr. Aspir. Molodykh Nauchn. Sotr., Vses. Nauchno-Issled. Inst. Rastenievod.* **15**, 474–477.

Mansfield, T. A. (1976). Stomatal behaviour: Chemical control of stomatal movements. *Philos. Trans. R. Soc. London, Ser. B* **273**, 541–550.

Miller, S. S. (1979). Effect of pre-harvest anti-transpirant sprays on the size and quality of 'Delicious' apples at harvest. *J. Am. Soc. Hortic. Sci.* **104**, 204–207.

Misic, P. D., and Gavrilovic, M. D. (1969). A study of the relationship between Malling rootstocks and some apple varieties. *Arh. Poljopr. Nauke* **20**, 17–30.

Mix, A. J. (1916). Cork, drought spot and related diseases of the apple. *N. Y., Agric. Exp. Stn., Ithaca, Bull.* **426**, 473–522.

Mix, G. P., and Marschner, H. (1976). Calciumgehalte in Früchten von Paprika, Bohne, Quitte und Hagebutte im Verlauf des Fruchwachstums. *Z. Pflanzenernaehr. Bodenkd.* **5**, 537–549.

Modlibowska, I. (1961). Effect of soil moisture on frost resistance of apple blossom, including some observations on 'ghost' and 'parachute' blossoms. *J. Hortic. Sci.* **36**, 186–196.

Moiseev, N. N., Kemkina, A. G., and Karpova, E. I. (1970). The effect of dwarfing rootstocks on the drought resistance of apple cultivars. *Tr. Kaz. Gos. S-kh. Inst.* **13**, 76–80.

Monteith, J. L. (1965). Evaporation and environment. *Symp. Soc. Exp. Biol.* **19**, 205–234.

Monteith, J. L. (1980). The coupling of plants to the atmosphere. *In* "Plants and Their Atmospheric Environment" (J. Grace, E. D. Ford, and P. G. Jarvis, eds.). Blackwell, Oxford (in press).

Neubauer, S. (1966). Der Einfluss de Bewässerung auf die vegetative und generative Entwicklung van Apfelniederstämmen. *Obstbau, Berlin* **6**, 21–22.

Newman, E. I. (1969). Resistance to water flow in soil and plant. II. A review of experimental evidence on the rhizosphere resistance. *J. Appl. Ecol.* **6**, 261–272.

O'Grady, L. J. (1959). Orchard irrigation. *Pomol. Fruit Grow. Soc., Annu. Rep.* **78**.

Palmer, J. W. (1977). Diurnal light interception and a computer model of light interception by hedgerow apple orchards. *J. Appl. Ecol.* **14**, 601–614.

Palmer, J. W., and Jackson, J. E. (1977). Seasonal light interception and canopy development in hedgerow and bed system apple orchards. *J. Appl. Ecol.* **14**, 539–549.

Passioura, J. B. (1972). The effect of root geometry on the yield of wheat growing on stored water. *Aust. J. Agric. Res.* **23**, 745–752.

Peerbooms, H. (1968). Bodembehandelingsproef te Horst. *Fruiteelt* **58**, 604–606.

Pierson, C. F., Ceponis, M. J., and McCulloch, L. P. (1971). Market diseases apples, pears and quinces. *U.S., Dep. Agric., Agric. Handb.* **376**, 1–112.

Powell, D. B. B. (1974). Some effects of water stress in late spring on apple trees. *J. Hortic. Sci.* **49**, 257–272.

Powell, D. B. B. (1976). Some effects of water stress on the growth and development of apple trees. *J. Hortic. Sci.* **51**, 75–90.

Powell, D. B. B. (1978). Regulation of plant water potential by membranes of the endodermis in young roots. *Plant, Cell, & Environ.* **1**, 69–76.

Powell, D. B. B., and Blanchard, T. W. (1976). The quantitative analysis of water potential/symplastic water volume curves. *J. Exp. Bot.* **27**, 597–607.

Powell, D. B. B., and Thorpe, M. R. (1977). Dynamic aspects of plant-water relations. *In* "Environmental Effects on Crop Physiology" (J. J. Landsberg and C. V. Cutting, eds.), pp. 259–279. Academic Press, New York.

Powell, L. E. (1975). Some abscisic acid relationships in apple. *Riv. Ortoflorofruttic. Ital.* **59**, 424–432.

Proctor, J. T. A., Kyle, W. J., and Davies, J. A. (1972). The radiation balance of an apple tree. *Can. J. Bot.* **50**, 1731–1740.

Randall, J. M. (1969). Wind profiles in an orchard plantation. *Agric. Meteorol.* **6**, 439–452.

Razlivalova, M. V. (1974). The effect of clonal rootstocks on apple tree drought resistance. *Mater. Nauchno-Tekh. Konf. Posvyashch. 25-Letiyu Lesokhoz. Fak. Kaz. S-kh. Inst.* pp. 77–79.

Reidl, H. (1937). Bau und Leistung des Wurzelholzes. Jahrls. für wissenschaftliche Botanik **85**, 1–75.

Rogers, W. S. (1932). Root studies. II. The root development of an apple tree in a wet clay soil. *J. Pomol.* **10**, 219–227.

Rogers, W. S., and Vyvyan, M. C. (1928). Root studies. I. Root systems of some ten-year-old apple trees and their relation to tree performance. *Annu. Rep. East Malling Res. Stn., Kent* Suppl. II, pp. 31–43.

Rogers, W. S., and Vyvyan, M. C. (1934). Root studies. V. Rootstock and soil effects on apple root systems. *J. Pomol. Hortic. Sci.* **12**, 110–150.

Rom, R. C. (1965a). Malling Merton III apple rootstock exhibits tolerance to heat and drought. *Fruit Var. Hortic. Dig.* **19**, 56.

Rom, R. C. (1965b). Apple irrigation studies. *Arkh. Farm Res.* **14**, 8.

Russell, R. S. (1977). "Plant Root Systems." McGraw-Hill, New York.

Salter, P. J. and Goode, J. E. (1967). Crop responses to water at different stages of growth. *Commonw. Bur. Hortic. Plant. Crops (A. B.), Res. Rev.* No. 2.

Schneider, G. W., and Childers, N. F. (1941). Influence of soil moisture on photosynthesis, respiration and transpiration of apple leaves. *Plant Physiol.* **16**, 565–583.

Sharples, R. O. (1973). Orchard and climatic factors. *In* "The Biology of Apples and Pear Storage" (J. C. Fidler, B. G. Wilkinson, K. L. Edney, and R. O. Sharples, eds.), Res. Rev. 3, pp. 175–225. Commonw. Agric. Bur., Farnham Royal, Bucks, England.

Simons, R. K. (1956). Comparative anatomy of leaves and shoots of Golden Delicious and Jonared apple trees grown with high and low moisture supply. *Proc. Am. Soc. Hortic. Sci.* **68**, 20–26.

Singha, S., and Powell, L. E. (1978). Response of apple buds cultured *in vitro* to Abscisic acid. *J. Am. Soc. Hortic. Sci.* **103**, 620–622.

Slack, E. M. (1974). Studies of stomatal distribution on the leaves of four apple varieties. *J. Hortic. Sci.* **49**, 95–103.

Stoker, R., and Weatherley, P. E. (1971). The influence of root system on the relationship between the rate of transpiration and depression of leaf water potential. *New Phytol.* **70**, 547–554.

Suckling, P. W., Davies, J. A., and Proctor, J. T. A. (1975). The transmission of global and photosynthetically active radiation within a dwarf apple orchard. *Can. J. Bot.* **53**, 1428–1441.

Swarbrick, T., and Luckwill, L. C. (1953). The factors governing fruit bud formation. *In* "Science and Fruit" (T. Wallace and R. W. Marsh, eds.), pp. 99–109. University of Bristol, Bristol, England.

Taerum, R. (1964). Effects of moisture stress and climatic conditions on somatal behaviour and growth in Rome Beauty apple trees. *Proc. Am. Soc. Hortic. Sci.* **85**, 20–32.

Thom, A. S. (1971). Momentum absorption by vegetation. *Q. J. R. Meteorol. Soc.* **93**, 494–500.

Thom, A. S. (1975). Momentum, mass and heat exchange of plant canopies. *In* "Vegetation and the Atmosphere" (J. L. Monteith, ed.), Vol. 1, pp. 57–109. Academic Press, New York.

Thorpe, M. R. (1978). Net radiation and transpiration of apple trees in rows. *Agric. Meteorol.* **19**, 41–57.

Thorpe, M. R., and Butler, D. R. (1977). Heat transfer coefficients for leaves on orchard apple trees. *Boundary-Layer Meteorol.* **12**, 61–73.

Thorpe, M. R., Saugier, B., Auger, S., Berger, A., and Methy, M. (1978). Photosynthesis and transpiration of an isolated tree: Model and validation. *Plant, Cell, & Environ.* **1**, 279–284.

Thorpe, M. R., Warrit, B., and Landsberg, J. J. (1980). Responses of apple leaf stomata: A model for single leaves and a whole tree. *Plant, Cell, & Environ.* **3**, 23–27.

Tinker, P. B. (1976). Transport or water to plant roots in soil. *Philos. Trans. R. Soc. London, Ser. B* **273**, 445–461.

Tromp, J. (1979). The intake curve for calcium into apple fruits under various environmental conditions. *Commun. Soil Sci. Plant Anal.* **10**, 325–335.

Tunsuwan, T., and Bünemann, G. (1973). Spaltöffnungshalten bei Apfelbaümen mit und ohne Früchte. *Gartenbauwissenschaft* **38**, 109–115.

Turner, N. C. (1979). Drought resistance and adaptation to water deficits in crop plants. *In* "Stress Physiology in Crop Plants" (H. Mussell and R. C. Sharples, eds.), pp. 343–372. Wiley, New York.

Van den Honert, T. H. (1948). Water transport in plants as a catenary process. *Discuss. Faraday Soc.* **3**, 146–153.

van Zyl, H. J. (1970). Water and fruit quality. *Dec. Fruit Grower* **20**, 143–146.

Vink, N. (1971). Measured and estimated water use by a citrus and an apple orchard. *Econ. Rurale Libanaise* **36**, 62–82.

Wallace, J. S. (1978). Water transport and leaf-water relations in winter wheat crops. Ph.D. Thesis, University of Nottingham, Nottingham, England.

Warrit, B. (1977). Studies on stomatal behaviour in apple leaves. Ph.D. Thesis, University of Bristol, Bristol, England.

Warrit, B., Landsberg, J. J., and Thorpe, M. R. (1980). Responses of apple leaf stomata to environmental factors. *Plant, Cell, & Environ.* **3**, 13–22.

Watson, R. L., and Landsberg, J. J. (1979). The photosynthetic characteristics of apple leaves (cv. Golden Delicious) during their early growth. *In* "Photosynthesis and Plant Development" (C. R. Marcell, H. Clijsters, and M. van Poucke, eds.), pp. 39–48. Junk Publ., The Hague.

Weatherley, P. E. (1963). The pathway of water movements across the root cortex and leaf mesophyll of transpiring plants. *In* "The Water Relations of Plants" (A. J. Butler and F. H. Whiteland, eds.), pp. 85–100. Blackwell, London.

Weatherley, P. E. (1979). The hydraulic resistance of the soil-root interface—a cause of water stress in plants. *In* "The Soil-Root Interface" (J. L. Harley and R. Scott-Russell, eds.), pp. 275–286. Academic Press, New York.

Webster, D. H. (1978). Soil conditions associated with absence or sparse development of apple roots. *Can. J. Plant Sci.* **58**, 961–96.

Weller, S. C., and Ferree, D. C. (1978). Effect of a pinolen-base antitranspirant on fruit growth, net photosynthesis, transpiration and shoot growth of 'Golden Delicious' apple trees. *J. Am. Soc. Hortic. Sci.* **103**, 17–19.

West, D. W., and Gaff, D. F. (1976). The effect of leaf water potential, leaf temperature and light intensity on leaf diffusion resistance and the transpiration of leaves of *Malus sylvestris. Physiol. Plant.* **38**, 98–104.

Wildung, D. K., Weiser, C. J., and Pellett, H. M. (1973). Temperature and moisture effects on hardening of apple roots. *HortScience* **8**, 53–55.

Wilkinson, B. G. (1968). Mineral composition of apples. IX. Uptake of calcium by the fruit. *J. Sci. Food Agric.* **19**, 646–647.

Wilkinson, B. G., and Fidler, J. C. (1973). Physiological disorders. *In* "The Biology of Apple and Pear Storage" (J. C. Fidler, B. G. Wilkinson, K. L. Edney, and R. O. Sharples, eds.), Res. Rev. 3, pp. 63–131. Commonw. Agric. Bur., Farnham Royal, Bucks, England.

Yadava, V. L., and Dayton, D. F. (1972). Effect of exogenously supplied abscisic acid on a vigorous clonal apple rootstock. *HortScience* 7, 261–262.

CHAPTER 7

TEA PLANTATIONS

G. R. Squire and B. A. Callander

DEPARTMENT OF PHYSIOLOGY AND ENVIRONMENTAL STUDIES, UNIVERSITY OF
NOTTINGHAM SCHOOL OF AGRICULTURE, SUTTON BONINGTON, LOUGHBOROUGH,
LEICESTERSHIRE, ENGLAND

I. INTRODUCTION

Tea has been cultivated for at least 3000 years. The origin of its various botanical forms and the likely course of its spread before the

Water Deficits and Plant Growth, Vol. VI
Copyright © 1981 by Academic Press, Inc.
All rights of reproduction in any form reserved.
ISBN 0-12-424156-5

plantation era were discussed by Kingdon-Ward (1950). Its growth habit varies between two forms: the "China" type, a large shrub with small, upright leaves, and the "Assam" type, a medium-sized tree with large horizontal leaves (Barua, 1970; Hadfield, 1974b). In plantations, both forms of tea are cultivated into flat-topped bushes about 1 m high, but the difference in size and orientation of leaves still exists. About one-half of the world's crop is harvested in India, Sri Lanka, and China (Food and Agriculture Organization, 1979), but tea is a very important crop in many other countries. Up to the 1960's, experience and intuition formed what was known of the climatic requirements of tea. In a detailed review of earlier work, Carr (1972) concluded: (1) that the growth of shoots stops when daily mean temperature falls below about 13°C, and is optimum in the range 18–30°C, (2) that plantations need at least 1150 mm of rainfall a year, preferably well distributed; and (3) that shoots do not grow rapidly in dry air.

Tea bushes are raised in nurseries either from seed or by vegetative propagation, and transferred to a permanent site in the field when 2 to 5 yr old. Complete ground cover by the canopy can be achieved after a further 2 or 3 yr if bushes are irrigated during the dry season. Bushes are trained to form a plucking table—a flat, usually horizontal surface, below which is a dense layer of mature foliage about 30–50 cm deep. Young shoots grow from the axils of leaves on the plucking table and are harvested when 10–15 cm long. The table rises gradually because not all new leaves are harvested and, after usually 2 or 3 yr, the whole canopy is pruned to remove all the younger branches and many of the leaves. Cycles of harvesting and pruning may continue for 50 yr and, in some areas, plantations 100 yr old are harvested regularly. With local variations and adaptations, this basic cultural sequence is practiced in all tea areas of the world (Eden, 1965; Harler, 1964).

Much of the research in the major areas of tea production has been directed to improving the quality of tea, its method of manufacture, the control of its pests and diseases, and its basic agronomy. The hydrological cycle in tea plantations was studied in detail during the 1960's and 1970's, mainly in Africa, but there have been few studies on the physiological effects of shortage of water, even though annual drought, which affects plantations in many parts of the world, interrupts the flow of crop and sometimes causes defoliation or death of bushes. Work on water relations of tea plants, again mainly in Africa, began in the 1960's, gaining momentum with the development of research tools such as the pressure chamber. Most of the work which lies within the scope of this chapter has been done in Kenya, Tanzania, Malawi, northeast India, Japan, and the U.S.S.R. (Georgia).

II. HYDROLOGY

High annual rainfall is regarded as a major climatic requirement of tea, at least 1150 mm according to Carr (1972) and exceeding 1400 mm according to Harler (1964) and Eden (1965). How this precipitation is used by tea bushes depends on the relative size of terms in the equation of the hydrological balance: Precipitation = Evaporation + Outlofw + Increase of stored water. Outflow encompasses storm runoff, drainage (baseflow), and deep drainage.

A. PRECIPITATION

The only input to the water budget—precipitation—is almost entirely rainfall. Hail occurs in Kenya and Assam and, toward the extreme latitudes of the tea-growing areas such as Georgia, U.S.S.R., snow occurs in winter, but the significance of such precipitation lies more in the damage caused to tea than in its contribution to the hydrological cycle. Occult precipitation by interception of mist was reckoned by Eden (1965) to represent a significant input of water during the dry season, but the amount so precipitated has not been quantified.

Annual rainfall in tea areas ranges from less than 700 mm in Chipinga, Zimbabwe, where irrigation is required for economic yields (Harler, 1964) to more than 5000 mm in Ceylon (Eden, 1965). Harler (1966) suggested that rainfall exceeding 4500 mm is associated with decrease in yield, but this may be an effect of temporary waterlogging. Mann (1935) emphasized the importance of avoiding soil waterlogging for all but the shortest periods. Harler (1964) stated that the climax vegetation of all tea areas is forest; in rainfall records at Ranchi, Mauritius, and Darjeeling he found no evidence that the change in land use had affected either the amount or distribution of precipitation. Similarly, catchment studies in Kenya revealed no significant change in the annual rainfall to a 380-ha catchment, in relation to a forested control, after replacement of the indigenous forest by tea (Blackie, 1981).

The distribution of rainfall over the year depends on latitude. The general pattern is for equatorial regions to have two rainy seasons and two dry seasons (usually of different severity). At higher latitudes the seasons merge into one dry and one wet season, then finally into the summer and winter of temperate climates, where variation in rainfall is much less marked than variation in temperature. The monsoon of southeast Asia affects, to various degrees, the climates of almost all tea-growing lands: it influences the climates of Africa and of the Black Sea area of Europe, profoundly affects those of Japan, Formosa, Ceylon, and Southern India, and dominates those of China, Indo-China, and most of India. However, local topography superimposes its own effects on local climate and further generalizations are neither justifiable nor useful. Harler (1964) gives a brief description of annual rainfall and temperature distributions for specific stations in the major tea growing areas.

B. EVAPORATION

After precipitation, evaporation is the largest component of the water budget. The rate of evaporation from any crop is determined by the interaction of factors associated with the crop and factors associated with the microclimate; the amount of water used by a crop is, therefore, not under the unique control of either the crop or the microclimate. Statements about the evaporative water use, E, of a crop are therefore most informative when expressed in relation to an index that quantifies the evaporative demand of the microclimate alone. Such an index is the potential evaporation, E_0, which is the predicted evaporation from a free water surface exposed to the same microclimate as the crop. The departure of the ratio E/E_0 (the crop factor) from unity indicates the extent to which plant factors are controlling evaporation. The Penman (1948) formula for potential evaporation, with the modification of McCulloch (1965) that takes account of the dependence of the psychrometer constant on altitude, has been found to give the most realistic estimate of open water evaporation for tropical East Africa (Dagg, 1965). In this chapter, E_0 will refer to this form of the Penman potential evaporation.

Table I presents the reported ratios of actual in relation to potential evaporation. All data represent work carried out in Africa: there is little information from other tea areas. In Japan, Yanase (1973) measured transpiration from tea bushes in large cuvettes, but the results are difficult to reconcile with the associated meteorological data: in particular, transpiration often exceeded water loss from an (unspecified) evaporation gauge, and sometimes exceeded the evaporation equivalent of the incoming solar radiation even when advection was negligible.

A crop factor of 0.8–0.9 is typical of tea in the wet season in Africa and is similar to values of other well-watered crops with complete ground cover (Monteith, 1973). The variation of E/E_0 under different extremes of drought shows no consistent pattern, partly because the variable used by most workers, soil moisture deficit, is an inadequate index of drought. Physiological causes of seasonal changes in E/E_0 are discussed in Section III,C, but one purely physical cause is the absence, in times of drought, of rainfall intercepted and reevaporated by the canopy.

TABLE I

Ratios of Actual to Potential Evaporation for Mature Tea With Complete Ground Cover

Source	Location	Method	Averaging time	Index of drought	Range of index	E/E_0^e
Laycock (1964)	Mlanje, Malawi	Soil water balance[a]	4–5 days	Soil moisture deficit	Field capacity to wilting point	0.85
Dagg (1970)	Kericho, Kenya	Soil water balance[b]	1 month	Season	Wet	0.9
					Hot/dry	0.56
Wang'ati and Blackie (1971)	Kericho, Kenya	Soil water balance[b]	1 month	Season	Wet	0.75–0.85
Willatt (1971)	Mlanje, Malawi	Soil water balance[a]	2 weeks	Soil moisture deficit	0–200 mm	0.85–0.90
					200–350 mm	0.85–0.13
Fordham (1973)	Mlanje, Malawi	Micrometeorological energy budget	2 days	Season	End of dry season to beginning of wet season	0.52[f]
Willatt (1973)	Mlanje, Malawi	Soil water balance[a]	2 weeks	Soil moisture deficit	0–80 mm	0.7–0.9
					380 mm	0.3
Carr (1974)	Ngwazi, S. Tanzania	Soil water balance[c]	3 dry seasons	Season	Dry	0.34
Blackie (1981)	Kericho, Kenya	Catchment water balance	11 years	(Annual mean)		0.83
Callander and Woodhead (1981a)	Kericho, Kenya	Micrometeorological energy balance	26 days / 2 days	Soil moisture deficit	0–200 mm	0.76
					370 mm	0.52
Cooper (1981)	Kericho, Kenya	Soil water balance[d]	3 months / 1 month	Season	Wet	0.97
					Dry (soil moisture deficit up to 330 mm)	0.56
Eeles (1981)	Kericho, Kenya	Soil water balance[d]	1 month	Soil moisture deficit	0–100 mm	0.77
					100–200	0.79

[a–d] Soil water balance constructed using (a) gravimetric sampling, (b) lysimeter, (c) gypsum resistance units, and (d) neutron moisture meter.

[e] E/E_0 is the ratio of actual to potential evaporation.

[f] Value calculated from information supplied in reference.

Evaporation of Rainfall Intercepted by the Canopy

All values for water use in wet seasons shown in Table I refer to the mean rate of water loss from the canopy, without differentiating between water transpired through the plants, and intercepted water evaporated from leaf surfaces. In drought, evaporation from the canopy consists only of transpired water.

The contribution of intercepted water, I, to total evaporation is given by

$$I = sN_w$$

where s is the mean canopy storage, and N_w is the frequency of wetting. Only two estimates of s are available in the literature, both referring to "Assam" tea around Kericho, Kenya. Cooper (1981) measured s as the residual in a determination of the "precipitation budget" (storage = rainfall − throughfall − stemflow). In the presence of considerable scatter, particularly at large amounts of precipitation, he found a mean value for s of 2.5 mm. From energy balance measurements at a site adjacent to Cooper's, Callander (1978) also reported a *maximum* canopy storage (rain or dew) of about 2 mm but an average of only 0.8 mm. Nevertheless, this last value implies an annual total interception of nearly 200 mm for the tea of Kericho which, on average, experiences 240 rain days annually (Blackie, 1981). In the absence of similar measurements in other tea areas, a tentative estimate of canopy storage of around 1 mm might be applied to plantations of "Assam"-type tea; whether this value can be applied to "China"-type tea with its narrower, more erect leaves, is uncertain. Assuming 200 occasions of wetting annually and s equal to 1 mm, the ratio I/E_0 would be 0.12 for Mulanje, Malawi, and 0.14 for Kericho, Kenya, for Fort Portal, Uganda, and for Marikitanda, Tanzania (E_0 taken from the 1977 annual reports of the Tea Research Stations at these locations). These ratios are approximate and would be higher in the wetter months, but they quantify the magnitude of the change in crop factor that would arise from the absence of interception. That interception is at least a significant component of evaporation has been demonstrated by Blackie (1981), who obtained a significantly better fit of predicted to observed water use when canopy storage was included in the mathematical model. Nevertheless, better estimates of canopy storage, and of the rate of evaporation of intercepted water, are required before its role can be more accurately and generally quantified.

C. OUTFLOW

The high rainfall of tea areas means that many plantations occupy regions that give rise to rivers. Because these rivers are essential to the surrounding lower and drier areas, it is important to know what effect plantations have on streamflow, both in annual distribution and total amount. Long-term catchment studies designed to provide such information have been carried out only in Kenya (Pereira, 1962; Blackie, 1972a; Blackie and Edwards, 1981).

Baseflow arises from water, whose potential always remains above field capacity, percolating from the surface of the soil down to the water table before appearing as outcrops of the water table in streams and rivers. This flow, maintained over the whole year by reserves of soil moisture above field capacity, even during drought, averaged 40% of 2021 mm mean annual precipitation for a mature estate (Blackie, 1981). However, baseflow is a residual in the soil water balance, so there is no a priori reason for it to bear any fixed relation to any other component of the hydrological budget.

Surface runoff from plantations during and after heavy rain is undesirable because it causes erosion and because it reduces the amount of rainfall available to plants for transpiration. Blackie (1972a) found that mature tea lost only 1% of annual precipitation as surface runoff, a value similar to that of the forest is replaced. For a newly planted estate, mean runoff was held down to only 2% of annual precipitation by extensive measures designed to combat erosion and to promote percolation of rainwater into the soil (Blackie, 1972a; Pereira, 1973). During individual storms of over 50 mm it reached 6%. Detailed studies of the effects of various intercrops and mulches in controlling surface runoff (Othieno, 1975; Othieno and Laycock, 1977) show both the problem of runoff and erosion from bare soil when the tea is young, and the effective control imposed by bushes when they have developed about 60% of ground cover (Fig. 1). Blackie (1972a) suggests that for soils at Kericho, Kenya, tea gardens provide more efficient flood control than the indigenous forest.

D. STORAGE

The amount of "available" water (usually defined as that held at potentials between -0.1 and -15 bars) that can be stored within a soil depends on the depth and type of soil. The soil should be freely draining (Mann, 1935) and, to avoid waterlogging, fairly deep; otherwise, there are no unifying characteristics over the range of soil types in which tea is grown (Eden, 1965). The following water contents refer to that stored at field capacity. The deep soils of Kericho, Kenya, have an available water content of 0.22 by volume (Othieno, 1973), which represents a total available water content of about 1300 mm in a 6-m profile. In Southern Tanzania, Carr (1974) reported available water of 415 mm in the top 4.5 m, with an available water content of 0.14 in the first 0.5 m, 0.08

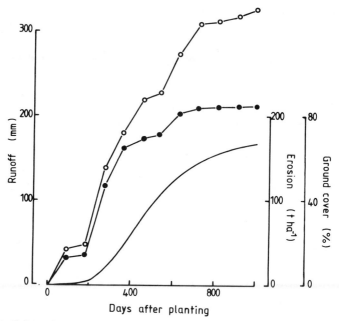

Fig. 1. Soil erosion (●) and runoff of rainwater (○) in relation to ground cover (——) during the establishment of a tea plantation in Kenya. (Adapted from original data of Othieno, 1975, and Othieno and Laycock, 1977.)

between 0.5 and 2 m, and 0.09 between 2 and 4.5 m. For the sandy clays of Mulanje, Malawi, Laycock and Wood (1963a) reported that the available water content was 0.15, while Willatt (1973) indicated a value of 0.13 for the first 1.5 m, and a total content of 550 mm in the top 4 m. In contrast Van der Laan (1971) found three soil types in the tea areas of Bangladesh to have a total water content of only 125–175 mm.

E. Hydrological Balance

From the discussion of the various components of the water budget, an annual hydrological balance can be composed (Table II). Although it is an idealized table, drawing from experiments performed at as wide a range of sites as possible, it is heavily weighted towards African conditions simply by the relative number of investigations carried out there. The comments beside each component of the water balance indicate the means of calculating its value.

The minimum water requirement for tea plantations is that which provides for transpiration alone, i.e., 0.70–0.75 annual E_0. Where there

TABLE II

IDEALIZED ANNUAL WATER BALANCE FOR MATURE TEA[a]

Component	Amount (mm)	Amount relative to precipitation	Comments
Precipitation	1700	—	Typical for East African tea areas
Evaporation			
Transpiration	1050	0.62	Transpiration = $0.75\ E_0$
Interception	200	0.12	200 raindays, canopy storage 1 mm
Streamflow			
Baseflow	430	0.25	Balance of water budget
Runoff	20	0.01	Assumed 1% of precipitation
Change in soil moisture	±30	±0.02	

[a] Annual E_0 is assumed to be 1400 mm, a typical value for many tea areas.

was rainfall there would also be the intercepted component of evaporation, which brings total evaporation in Table II to 1250 mm. The minimum water requirement for plantations is determined, therefore, by the evaporative demand of the atmosphere. Annual totals of E_0 for many areas in which tea is currently grown lie in the range 1300 to 1600 mm, so rainfall should on average be at least 85% of this value—a figure which compares closely with minimum values of 1150 mm and 1400 mm suggested previously by, respectively, Carr (1972) and Harler (1964), and Eden (1965).

III. FACTORS AFFECTING EVAPORATION

Evaporation is important both in the hydrological balance of catchments and in the heat and water balance of leaves. In a hydrological study in Kenya (Section II), mature tea plantations evaporated about two-thirds of annual rainfall. Transpiration through stomata accounted for 80–85% of evaporation, the rest being evaporation of rainwater intercepted by leaves. When rainfall was frequent, plantations evaporated at a rate determined largely by the atmospheric environment; the ratio of actual to potential evaporation (E/E_0) was constant at about 0.8–0.9 in the face of changing weather. In many areas, this ratio is equivalent to daily rates of evaporation of 3–5 mm (Busin, 1970; Callander and Woodhead, 1981a; Yanase, 1973), although higher values of 7–8 mm have been reported (Willatt, 1973; Yanase, 1973). During parts of the year when there was little rainfall, E/E_0 fell to about 0.5, and in another study in Malawi, to 0.13 (Table I). Such a large fall in E/E_0 reflects an important change in the partitioning of available energy at the canopy surface between sensible and latent heat. Dry seasons in Africa are likely, therefore, to have serious effects on the heat and water balance of leaves, especially when lack of rainfall coincides with hot air and bright sunlight.

Rapid evaporation occurs when stomata open during the day to permit diffusion of carbon dioxide from the atmosphere to the intercellular space of leaves. Because of the hydraulic resistance to the flow of water through plants, rapid evaporation usually causes a fall in the water potential (ψ) of leaves which may be sufficient to reduce the rate of photosynthesis and expansion of cells. However, rapid evaporation also has beneficial effects: in particular it cools the evaporating surface. If stomata close when radiation is high and transpiration fast, water may be conserved but leaves will become hotter. For plants growing in hot, dry environments, the balance between "good" and "bad" consequences of evaporation may determine life or death. For tea plantations, the balance may determine the production of crop and the profitability of cultural practices such as irrigation and shade.

There have been few studies on evaporation from tea leaves. Some studies measured the water lost from soil, others the state of water in the atmosphere, and still others examined intermediate processes such as stomatal behavior. But no single study has attempted to include the soil, the plant, and the atmosphere. This is largely because most work on tea has been done by individuals based at isolated stations in climates that cause rapid deterioration of instruments needed to examine the finer details of water relations. However, several botanists and physicists have studied different aspects of the water relations of tea plantations, and Sections III,A,B, and C are an attempt to piece their work together. First, as a reference for further discussion, the climates of several tea areas are compared with emphasis on different characteristics of drought. Second, the importance of evaporation to surface temperature is illustrated from micrometeorological studies in Africa. Finally, factors such as dry soil and dry air, responsible for reducing E/E_0, are examined.

A. DROUGHT IN TEA AREAS

Temperature is the major climatic factor determining the rate of growth of tea (Section IV,B), and the limits to commercial cultivation are set by temperature regimes (Harler, 1966). Within the zone of suitable temperature, the sites of plantations are determined by several factors such as type of soil, availability of labor, and amount and distribution of rainfall. Because of the annual cycle of tropical weather, many plantations are subjected to a dry season lasting between a few weeks and 6 months

(Section II,A). These dry seasons interrupt the flow of crop and sometimes cause scorch, leaf-fall, and death of young bushes. Historically, the severity of a dry season has been measured in terms of soil moisture deficit (Section II,B), but the atmospheric conditions are equally important. Several examples of areas with different types of drought are given in Table III. At Mufindi in Tanzania, the dry season lasts 6 months, but the air is relatively cool and saturation deficits (SD) are small. In contrast, the dry season at Mulanje in Malawi is shorter and the air is hotter and drier. At Anamallais in the Cardamom Hills of Southern India, the dry season is intermediate between Mulanje and Mufindi. All three of these sites contrast with Georgia, U.S.S.R., where it is too cold for growth in winter, and water shortage affects the crop only during relatively short rainless spells in summer. Although the monsoon climates, for example, at Assam in northeast India, have a definite dry season, dry atmospheres are not limited to the dry season, and large SD's may occur during the rains because the air is very hot. Therefore, in terms of SD, the midday atmosphere during the rains in Assam may be similar to or drier than that during the dry season in Tanzania.

B. EVAPORATION AND SURFACE TEMPERATURE

1. Radiation

Most tea plantations experience the large radiant fluxes of equatorial regions, these fluxes being enhanced at the high altitudes of some estates. Solar irradiance can exceed 1000 W m^{-2} (Blackie, 1972b; Ripley, 1967; Fordham, 1973; Yanase, 1973; Squire, 1977), and local peak values may approach the solar constant (1360 W m^{-2}) when reflection from clouds occurs. Estate tea in Kenya reflects about 20% of the total incoming shortwave radiation (Blackie, 1972b; Ripley, 1967). The net amount of longwave radiation emitted to the sky varies with surface temperature and with air temperature and humidity, but the measurements of Ripley (1967) and Blackie (1972b) suggest that it is typically 20% of S. The flat and dense canopy of the plucking table limits the amount of radiation penetrating the canopy: in India, Hadfield (1974b) found that for a range of tea varieties radiation was reduced by 99% within 30 cm of the plucking table, and in Malawi, Green (1971) found that only 5% of incoming radiation reached the ground. Consequently, below-canopy components of the energy balance—soil heat, and changes in stored heat and chemical energy— should be small. This was confirmed at Kericho, Kenya, by Callander and Woodhead (1981), who found that, overall, the sum of the below-canopy energy fluxes was 4% of net radiation. Thus, almost all the net irradiance,

TABLE III

CHARACTERISTICS OF DROUGHT IN SELECTED TEA AREAS[a]

Site	Latitude, longitude, altitude	Period of year	Rainfall	Characteristic weather			References
				Mean max. saturation deficit (mbars)	Mean max. temperature (°C)	E_0 (mm day⁻¹)	
Mufindi, Tanzania	8°33' S; 35°10' E; 1890 m	Dry season, 6 months, May to October	Less than 10 mm month⁻¹	10–20	18–24	3–5	Carr (1974)
Kericho, Kenya	0°22' S; 35°21' E; 2178 m	Dry season, 3–4 months, December to mid-March	Variable; sometimes low (30 mm month⁻¹)	25–35	23–27	4–6	Carr (1976); Othieno (1978a,b)
Mulanje, Malawi	16°05' S; 35°37' E; 650 m	Dry season, 3–4 months, mid-August to mid-November	15–30 mm month⁻¹	30–40	26–33	5–7	Ellis (1971); Laycock (1964)
Tocklai, Assam India	26°17' N; 94°12' E; 80 m	(1) Wet season; (2) Dry season, 3–4 months, December to February	Plentiful 10–30 mm month⁻¹	(15)[b] (15)	32 23–26	5	Dutta (1971); Hadfield (1968); Harler (1966)
Anamallais, southern India	ca. 10° N; ca. 77° E; 1060 mm	Dry season, 3 months, January to March	Less than 10 mm month⁻¹	(15–30)	25–30	—	Harler (1966); UPASI (1973)
Krasnodar region, U.S.S.R.	ca. 45° N; ca. 39° E; 0–100 m	Summer, June to August	Rainless periods of 2–6 weeks	(20–25)	(30)	—	Filippov (1968); Filippov (1969); Filippov and Busin (1969)

[a] Characteristics of the weather are averages intended to illustrate differences between sites. In particular seasons, drought may be more severe. [b] Figures in brackets have been calculated from data contained in the references.

equivalent to about 60% of S, is available at the tea canopy for partitioning between fluxes of latent and sensible heat.

2. Fluxes of Heat and Water Vapor

That the available energy at the canopy can exceed 600 W m^{-2} underlines the importance of evaporation in keeping leaf temperature below injurious levels. The major source of water vapor in a mature canopy is the substomatal cavities of transpiring leaves. By altering stomatal apertures, the plant can alter the resistance to, and hence the rate of, loss of water vapor through the stomata. For unit horizontal area of canopy, the resistance of many leaves transpiring together can be combined in the canopy resistance (Monteith, 1973). When stomatal surfaces are the largest source of water vapor, then the pattern of canopy resistance will follow closely the pattern of the stomatal resistance of individual leaves. Callander and Woodhead (1981b) have calculated the canopy resistance of tea from micrometeorological measurements over a wide range of atmospheric conditions. It had a typical value of about 80 sec m^{-1}, decreased with irradiance, and increased with SD (Fig. 2). They also found a sys-

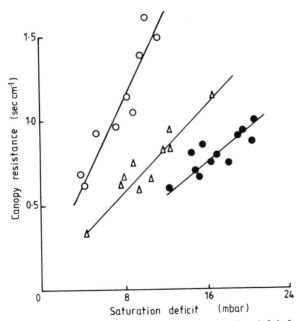

Fig. 2. The relation between canopy resistance and saturation deficit for tea in Kenya during the wet season. The data have been separated into three intervals of net radiation: ●, 600–650 W m^{-2}; △, 350–400 W m^{-2}; ○, 100–150 W m^{-2}. (Adapted from Callander and Woodhead, 1981b.)

tematic increase in canopy resistance in the dry season. Most of this increase could be explained by larger SD, but the remainder of the increase (about 10%) was attributed to the reduction in transpiring leaf area that commonly occurs during the dry season.

Between the canopy surface and the free airstream, the diffusion of water vapor is restricted by the aerodynamic resistance to transfer. This is the only resistance affecting the loss of heat from the canopy, and its magnitude is determined mainly by the aerodynamic roughness, quantified by the roughness length, z_0, and the windspeed. Callander and Woodhead (1981a) report a measured value for z_0 of 0.20 (± 0.07) m. They also measured the zero plane displacement as 0.7 (± 0.1) m (about 0.6 of the height of the crop), and found the aerodynamic resistance to transfer between the crop and 2 m above the canopy to be typically 20 sec m^{-1}.

The Bowen ratio, β, the ratio of sensible to latent heat flux, results from the interplay of the driving "forces" on, and various resistances to, the exchange of heat and water vapor. Therefore, the Bowen ratio cannot be considered as representative of a particular crop. Nevertheless, typical values for a crop are useful indices. In Malawi Fordham (1973) found β to lie in the range 0.2 to 0.5 in the wet and dry seasons, respectively; in Kenya, Callander and Woodhead (1981a) found β to be typically 0.3 in the wet season, increasing to unity or greater in one particular dry season. Thus, in the wet season, almost 50% of the incoming solar energy was transformed into latent heat and only 10–15% into sensible heat at the crop canopy; in the dry season, fluxes of latent and sensible heat were similar in magnitude, each amounting to about 30% of irradiance. The diurnal pattern of the energy budget measured over "Assam"-type tea in Kenya during wet and dry seasons is shown in Fig. 3.

The effect of change in β on the leaf-to-air temperature difference was not measured in these studies in Kenya. The difference in temperature between the canopy and the air is proportional to the convective heat flux density and the aerodynamic resistance to heat transfer (Monteith, 1973, Chapter 7). From the values of heat flux and aerodynamic resistance given above, it can be calculated that, for a Bowen ratio of 0.3, mean leaf temperature would be about 0.3 C° higher than air temperature for every additional 100 W m^{-2} of solar irradiance. For a Bowen ratio of unity, the temperature difference would rise to 0.6 C° per 100 W m^{-2}; and in the extreme case of complete stomatal closure, the temperature difference might rise to 1.2 C° per 100 W m^{-2}. These rough calculations show the mean difference in leaf-to-air temperature that might arise: individual leaves would be expected to show variation about the mean.

Accurate measurements of leaf temperature have seldom been made on tea bushes. Both Hanapi and Reesinck (1959), using thermocouples,

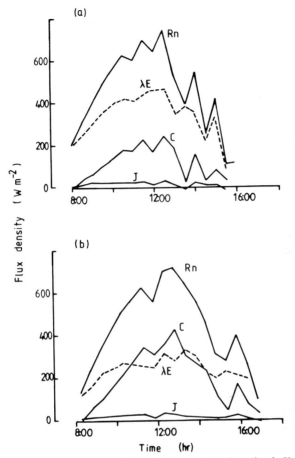

Fig. 3. Diurnal pattern of energy balance of a mature tea plantation in Kenya (a) during the wet season, and (b) the dry season. Rn, net radiation; C, sensible heat; λE, latent heat; J, below-canopy fluxes. (From Callander, 1978.)

and Fordham (1973), using thermistors, reported leaf-to-air differences of 2–4 C° in bright sunlight, whereas Green (1971) using thermistors, and Hadfield (1968), who did not give his method, regularly observed much higher differences of + 10C° or more. Although these high differences are theoretically possible, they are likely to occur only when stomata are completely closed in bright sunlight. More detailed information obtained from fine-wire thermocouples or infrared thermometers is required before any firm conclusions can be drawn about the typical range of leaf-to-air differentials on tea bushes.

If large differentials of 10C° or more do exist, it will be difficult to interpret their effect on tea leaves. For leaves in an air temperature of 35°C, such a difference would increase the leaf-to-air SD by about 40 mbars above the atmospheric SD, which itself would be of the order of 15 mbars during the monsoon in Assam and 30–40 mbars during the dry season in Malawi. Any observed response by tea leaves during a period of high temperature could be caused either by above-optimal temperature itself, or by a very large saturation deficit. For tea growing near the upper limit of its temperature range, evaporative cooling will be particularly important. Because of inconsistencies in the observed relation between stomatal aperture and shoot water potential (ψ_{shoot}) (Section III,C,1) it is not known how in general tea regulates the balance between temperature and conservation of water. However, in one study on irrigated tea during the dry season in Malawi (Squire, 1977, 1978), evaporative cooling was maintained at the expense of ψ, which dropped low enough to reduce the extension of shoots. The importance of evaporative cooling was demonstrated by applying a thin layer of liquid silicone rubber to the stomatal undersurface of leaves. When irradiance was above 1000 W m^{-2} and air temperature was 32–33°C, the part of the leaf covered by rubber became scorched within 1 to 2 hrs.

C. Causes of Reductions in the Ratio of Actual to Potential Transpiration (E/E_0)

When E/E_0 is reduced, more energy is partitioned into sensible heat. In consequence, leaves may overheat and scorch, and shoot growth may slow down or stop. Most studies listed in Table I gave values of E/E_0 of 0.8–0.9 during wet seasons in Africa. These values include evaporation of water that was intercepted by the canopy and prevented from entering the soil. It was concluded from two estimates in Kenya that the intercepted component of evaporation might be equivalent to about 0.15 E_0. Little or no rain falls during drought, so absence of the intercepted component would reduce E/E_0 from an average of 0.85–0.90 to 0.70–0.75. Any further reduction in E/E_0 must be caused by a reduction in canopy conductance. Since canopy conductance is determined by both leaf area index (LAI) and stomatal conductance (Monteith, 1965; Szeicz and Long, 1969), the effects of stomatal closure on E/E_0 can be assessed accurately only if LAI is known. Unfortunately, stomatal conductance and leaf area of tea have never been measured in detail at the same time. Most measurements have concentrated on stomata.

1. Diurnal Patterns of Stomatal Behavior

Observations of stomatal behavior at different sites and sometimes at one site are often conflicting, and there is no consistent description of stomatal behavior for tea bushes. The diurnal course of canopy resistance was measured from micrometeorological fluxes in Kenya (Callander and Woodhead, 1981b): resistance remained more or less constant until about 14:00 hr, after which it fell steadily toward dusk. A roughly symetrical diurnal course of leaf resistance and stomatal aperture, with a maximum around midday, was found during the wet season in Malawi by Squire (1978), using diffusion porometry and silicone rubber impressions, and by Fordham (1973), using a portable viscous flow porometer. The minimum resistance of the stomatous undersurface found by Squire was typically 1.5–2.0 sec cm^{-1}. However, in Tanzania and Kenya, all measurements with the liquid infiltration technique suggest a reversed diurnal trend in which stomata were wide open in the early morning, closed progressively towards midday, and reopened in the late afternoon (Carr, 1971, 1977a,b; Othieno, 1978b). Analysis of the diurnal course in Malawi showed that aperture and conductance were closely related to irradiance and were unaffected by SD and ψ_{shoot}. On the other hand, the infiltration studies suggest that aperture was unaffected by irradiance but decreased as SD increased. The micrometeorological analysis of Callander and Woodhead (1981b), which showed that canopy resistance was related to *both* irradiance *and* SD, offers a compromise until the inconsistencies are sorted out by more rigorous experimentation. There are several possible reasons for the inconsistencies:

1. One or more techniques may be unreliable: comparison of several techniques on one type of tea at one site is required.

2. Clones or seedling varieties may differ in their response to environmental variables. Carr in Kenya found a varietal difference in the relation between infiltration score and ψ_{shoot}. In Fig. 4a, which is typical of most measurements in Kenya and Tanzania, stomata close as ψ_{shoot} falls; in Fig. 4b, which is similar to the relation observed in Malawi by Squire (1978), stomata are hardly affected by large changes in ψ shoot.

3. Tea stomata may be able to respond to the immediate environment in different ways, depending on their history. For example, measurements in Malawi revealed several different diurnal courses of stomatal aperture. For much of the year the course was roughly symmetrical about midday on cloudless days, but it changed during the dry season. The most pronounced feature of this change was wide opening in the early morning, followed commonly by a transient closing in the middle of the morning. On some days the course was similar to that found by the infiltration measurements in Kenya and Tanzania. This change in pattern at the beginning of the dry season was not caused by a shortage of water in the soil because it occurred on irrigated tea also; furthermore, the change was noticeable as early as 07:00 hr before ψ_{shoot} had fallen below -2 bars.

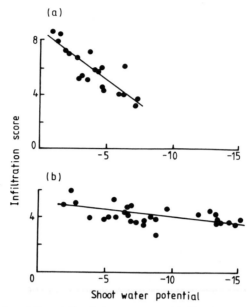

Fig. 4. Examples of the relation between stomatal infiltration score and shoot water potential for two clones (a and b) in Tanzania. Measurements were made on one day; the range of potential was obtained from plants growing at different soil moisture deficits. (From Carr, 1971.)

However, most measurements showed that, whatever the diurnal course, daily mean values of conductance, aperture, or infiltration score remained fairly constant for much of the year and decreased progressively during the dry season, so that in some cases the stomata were probably fully closed for a few hours around midday. Aftereffects of the dry season on stomatal aperture have been reported to last for several weeks or months (Fordham, 1970; Squire, 1978).

The causes of seasonal change in stomatal aperture are not yet understood. The relative importance of soil water and atmospheric humidity will influence the stomatal response shown by tea growing in different soils and climates and will affect the benefits of cultural practices such as irrigation. If tea stomata were very sensitive to SD—a response shown by many other plants—then irrigation applied to soil would have little effect on the heat balance of leaves if the air was very dry. But if stomata did not respond to SD then continued evaporative cooling during drought would depend on the reserves of water in the soil and the distribution of roots. What is known of the importance of soil and atmospheric factors is summarized below, but no definite conclusions can be reached, largely because soil water balances have been constructed without reference to plants, and stomatal behavior has been examined without reference to roots and water.

2. Effects of Soil Moisture Deficit on E/E_o

Examination of the effect of soil moisture deficit cannot proceed without information of soil depth and rooting depth at least. Most tea soils are deep and well-drained (Mann, 1935). Tea, although often reported to be a shallow-rooting crop, has been found to put roots down to 4 m in Tanzania (Carr, 1974), 6 m in Kenya (Kerfoot, 1962; Cooper, 1981), and 6 m in Malawi (Laycock, 1964). Tea is able to extract water over most of its rooting range (Laycock and Wood, 1963b; Cooper, 1981; Eeles, 1981). Willatt (1971, 1973) in Malawi showed that most water was extracted from the top meter when the soil was wetted by irrigation or rain, and that the depth of extraction went lower and lower for unirrigated tea as the dry season progressed. This is demonstrated in Fig. 5 (Cooper, 1981) for ma-

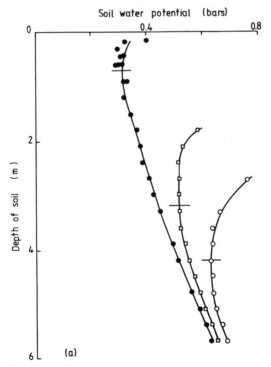

Fig. 5. Profiles of (a) soil moisture potential and (b) soil moisture content, beneath mature tea in Kenya during a period of little rainfall: ●, December 10, 1974; □, January 18, 1975; ○, 26 February 1975. Horizontal lines in (a) show the depth of the zero flux plane. (From Cooper, 1981.)   (Continued)

G. R. Squire and B. A. Callander

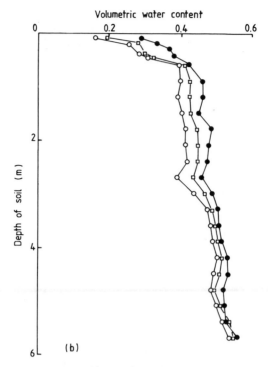

Fig. 5. (*Continued*)

ture tea growing in the 6-m deep latosols of Western Kenya. The sheer depth of many tea soils allows development of soil moisture deficits that seem extraordinarily high by the standards of other crops, yet still leave plenteous water in the soil. Cooper (1981) found that tea roots were able to extract soil moisture against moisture tensions of several bars, but even at the maximum soil moisture deficits he encountered, 350 mm, at least 1 m of soil was permeated by roots where soil tension was less than one bar. In contrast to these deep African soils, tea in parts of Assam and Bangladesh is forced to be shallow-rooting by a high groundwater table and water-logged soil (Dutta, 1971; Van der Laan, 1971). In the "low flats" of the Bangladesh tea areas (Van der Laan, 1971) nearly all the roots are confined to the top 60 cm of soil. It is doubtful whether plants grown in such soils could withstand even moderately large soil moisture deficits. In summary, there is no way of generalizing how and when soil moisture deficits affect the crop factor. This even seems to apply at one particular site. For example, in attempting to relate E/E_o to soil moisture deficit in southern Malawi, Willatt (1971) noted that water use was also affected by factors such as radiation and humidity, so that a complete model based on moisture deficit alone was not possible. That soil moisture deficit must eventually halt transpiration is unquestioned, but whether soil moisture acts as a progressive control or as a final end-stop is not known.

3. Effects of Atmospheric Humidity on E/E_0

At constant irradiance, the stomata of many species close progressively as atmospheric SD rises. The sensitivity of tea stomata to SD is at present contentious (Section III,C,1). Detailed investigations of stomatal behavior of tea have been few, and no general statements can be made. This section examines the possible effects of a stomatal response to SD, based on one series of micrometeorological measurements at one site in Kenya, where canopy resistance (or its reciprocal, conductance) was closely related to only two environmental variables, irradiance and SD.

According to the analysis of Thom and Oliver (1977) the factor E/E_0 for a given crop will depend on the ratio n of the canopy and aerodynamic resistances to vapor exchange. This ratio will vary through the day, but its effective mean value will be weighted heavily towards periods of high evaporation. The relation between E/E_0 and n calculated for tea is shown in Fig. 6. The crop factor exceeds unity at small values of n because the

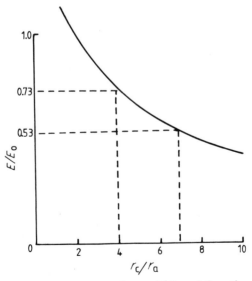

Fig. 6. Predicted relation between crop factor, E/E_0, and the ratio n of the canopy (c) and aerodynamic (a) resistances. For theory see Thom and Oliver (1977); in notation of that paper the values assumed are: $M = 10$; $Q_0 : Q_n : E_{ap} = 4:3:1$; $\Delta = 1.5$ mbar K^{-1}; $\gamma = 0.52$ mbar K^{-1} (2000 m altitude).

crop, being aerodynamically rougher than water, would evaporate faster than a water surface if stomatal control were absent. Mean daytime values of canopy and aerodynamic resistances were derived from the studies of Callander and Woodhead (1981b) and the ratio n shows a seasonal variation from 4 in the wet season to 7 in the dry season. Much of this range could be explained by a seasonal change in minimum canopy resistance from 80 to 150 sec m^{-1}, which was related to a change in maximum SD from 10–15 to 20–25 mbars. Thus, seasonal change in E/E_0 of about 0.2 *could* be attributed to seasonal change in overhead climate.

Thus, during the dry season in Kenya, a combination of lack of intercepted rainfall and a response of canopy conductance to SD could reduce E/E_0 from 0.85 to 0.50. A similar fall of 0.2–0.3 E_0 was found by Willatt (1973) on irrigated tea (with near-zero soil moisture deficit) in Malawi during periods when the atmosphere was hot and dry. Many of the estimates of E/E_0 during the dry season in Kenya in Table I could, therefore, be accounted for without any effect of soil water stress on canopy conductance. In particular, the measurements at Kericho of soil water (Cooper, 1981) and the atmosphere (Callander and Woodhead, 1981) show that even after 90 days with little rain and a soil moisture deficit of 370 mm, the plants were still capable of extracting considerable amounts of water (about 3.5 mm day^{-1}). The difference between these measurements at Kericho and Willatt's (1971, 1973) in Malawi, where E/E_0 was reduced as soil moisture deficit exceeded 200 mm, may reflect differences in either the amount of available water (Section II,D) or the age and rooting depth of the plants examined. Bushes in the Kericho study were 20 yr old and had roots down to 6 m; those in Malawi were 8 yr old with roots down to 4.5 m.

IV. EFFECTS OF DROUGHT ON GROWTH AND YIELD

Most tea plantations are subjected to a dry season, varying in severity and duration in different parts of the world (Section III,A). Dry season may influence the susceptibility of bushes to attack by pests and diseases (Carr, 1974; Shanmuganathan and Rodrigo, 1967), and may disturb the balance of nutrients in growing shoots (Tolhurst, 1971). But the major effect of dry seasons is on the distribution and amount of yield.

A. THE RELATION BETWEEN YIELD AND PHOTOSYNTHESIS

For many years it was assumed that the yield of tea is dependent on the current rate of photosynthesis. Consequently, many studies have emphasized factors such as radiation which affect photosynthesis directly, rather than ones such as temperature and turgor, which affect the growth of shoots directly. Evidence of the independence of yield and photosynthesis was available as early as 1957 (Barua, 1970, p. 321), but only recently (Tanton, 1979) has it been clearly stated that the growth of shoots *in plantations* is rarely limited by shortage of photosynthetic products. The independence of yield and photosynthesis in tea is not yet generally recognized, so the evidence is summarized below.

1. *Harvested material is a small proportion of growth.* The annual growth of tea bushes has rarely been measured because it is difficult to extract all the roots which make up a large part of their weight. Attempts to estimate growth in Malawi (Tanton, 1979) and Kenya (Callander, 1978) have suggested that annual yield might be only 5–15% of annual growth, and a smaller fraction of the total dry weight of the bush. The harvest index is therefore small compared with that for many other crops.

2. *Photosynthesis continues when growth is slow.* Mature leaves on the plucking table are the main source of photosynthetic products. These leaves intercept all photosynthetically active radiation (Hadfield, 1974b), and are active throughout most of the year (Sakai, 1975). Photosynthesis in these leaves proceeds rapidly when shoots are growing very slowly, for example on irrigated tea during the dry season in Malawi (Squire, 1977).

3. *The number of shoots per unit area is the main component of yield.* The components of yield are the rate of growth of shoots, the weight of shoots at harvest, and the number of shoots per unit area of plucking table. Differences in yield between clones or seedling varieties are largely caused by differences in the number of shoots per unit area (Toyao, 1965; Amma, 1971; T. W. Tanton, unpublished). Higher yielding clones have more developing shoots (i.e., sinks) to which carbohydrates can be translocated.

4. *The method of harvesting determines yield.* Usually, young shoots with two leaves and a bud are harvested when they are 10–15 cm long, because longer shoots with more leaves produce tea of poor quality. For most of their life shoots grow very slowly (Fig. 11). About two-thirds of their harvested weight is put on in the final week before harvest (Tanton, 1979). If they were left one more week before harvesting, their weight would approximately double. Long-term experiments in India, Sri Lanka, and Malawi (for references, see Templar, 1976) have shown that yield can be increased by lengthening the plucking round—the time between harvests. The method of harvesting is therefore the main constraint on yield.

B. EFFECTS OF TEMPERATURE ON SHOOT GROWTH

Whereas the number of shoots per unit area is the main discriminant of differences in yield between varieties, the rate of growth of shoots is the main discriminant of the differences in yield from season to season. The rate at which shoots grow depends mainly on the rates of cell division and cell extension and also on whether or not the terminal bud passes through a period of dormancy before reaching harvestable size. The main environmental factor determining the rate of growth is air temperature. The importance of mean temperature to establishment and profitability of plantations has been recognized for many years. References to experimental work in Japan and to statistical correlations between yield and mean temperature (or temperature sums) in India, Africa, and the U.S.S.R. are given in reviews by Carr (1972) and Fordham (1977).

More recently in Malawi, the rate of extension of shoots, measured in glasshouse experiments and in field observations, increased linearly with mean temperature above a base of 12–13°C (Squire, 1979; T. W. Tanton, unpublished). It is not known whether this base temperature is universal, or whether varieties grown at the northernmost limit of the plant's range have adapted to local climate. However, 12–13°C is consistent with the mean temperature at which tea stops growing in many parts of the world (Eden, 1965; Harler, 1964; Carr, 1972). In Malawi, the rate of shoot growth was proportional to mean temperature alone for about 8 months of the year, but the relation shown in Fig. 7 did not hold during the annual dry season. Factors responsible for reducing the rate of growth during dry seasons are discussed in the next section.

C. EFFECTS OF WATER DEFICITS ON SHOOT GROWTH

The annual dry seasons of tea areas in many parts of the world are often characterized by a combination of dry soil, dry air, and high temperature. Without some form of experimentation in controlled environments, it is difficult to isolate the independent effects of these three factors. One of the first attempts to achieve a separation of soil and atmospheric factors was made in Georgia, U.S.S.R. In these studies, the refractive index of sap squeezed from leaves was measured with a hand-held refractometer. The resulting value—the refractometric index of cell-sap concentration (RiCSC)—is closely related to the water content of leaves (Filippov, 1970). Maximum growth of tea shoots occurred only when RiCSC remained below 10% in the hottest part of the day. Increases in RiCSC above 10% were related to the amount of water in the upper 50 cm of soil, but by a mathematical treatment of the data Filippov (1968) showed an

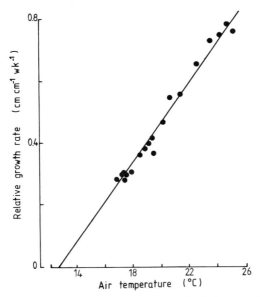

Fig. 7. The relation between mean air temperature and relative growth rate (RGR) of tea shoots in Malawi during periods of frequent rainfall. Each value of RGR was derived from an exponential equation fitted to weekly measurements of shoot length from the start of growth to harvest. Air temperature is the mean screen temperature during the period of growth; data are restricted to periods when the minimum temperature did not fall below the base temperature of 12.5°C. (From unpublished measurements by T. W. Tanton.)

effect of atmospheric humidity on RiCSC which was independent of soil water, such that if the air was very dry, growth of shoots slowed even when the soil was saturated with water.

An effect of atmospheric humidity on the growth of shoots has also been found in parts of Africa. During the dry season in Malawi, shoot extension slowed even when the soil was irrigated, and the linear response to temperature described at the end of the last section broke down. For irrigated tea at a given air temperature, Squire (1979) found a close inverse relation between the weekly rate of shoot extension and mean SD measured at about 14:00 hr at a meteorological site. In these preliminary measurements, the relation was approximately linear over the range of SD from 8 to 32 mbars, but in more extensive (unpublished) measurements by T. W. Tanton, SD had no depressive effect on the rate of shoot growth until the weekly mean SD at 14:00 hr was 22–23 mbars (Fig. 8). Increases in SD above 23 mbars progressively decreased the rate of shoot extension. This relation between meteorological measurements of SD and the complex process of shoot growth may be explained by the relation between SD and the ψ shoot which is described in the following section.

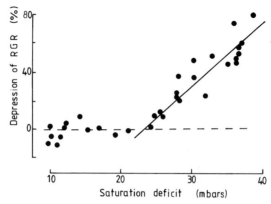

Fig. 8. The relation between mean weekly saturation deficit, measured at 14.00 hr, and the depression of relative growth rate of tea shoots. The dashed line represents the rate predicted from temperature alone by an equation derived from measurements shown in Fig. 7. Measurements were made at different times during the dry season in Malawi. (From unpublished measurements by T. W. Tanton.)

Daily and Seasonal Courses of ψ_{shoot}

The ψ of small branches and shoots at the top of the plucking table has been measured with pressure chambers in Tanzania (Carr, 1971), in Kenya (Carr, 1977a,b; Othieno, 1978b) and in Malawi (Williams, 1971; Green, 1971; Squire, 1976, 1979). Results from all three sites are broadly similar. When the soil was frequently wetted by rain, a condition which may last for 8 to 9 mo., ψ_{shoot} was closely related to atmospheric SD (Williams, 1971; Squire, 1976). SD rarely rose above 20 mbars and ψ_{shoot} rarely fell below -10 bars. Plots of ψ_{shoot} in relation to SD did not usually show hysteresis as SD fell in the afternoon (Fig. 9a).

During the dry season, minimum ψ_{shoot} fell to between -15 and -20 bars (Carr, 1971, 1977a; Othieno, 1978b; Squire, 1979), but by dawn the next day ψ_{shoot} had recovered to between -2 and -3 bars. The fall in ψ_{shoot} in the morning was often arrested, as early as 09.00 hr on some days (Carr, 1971; Squire, 1979), and then remained constant or changed little as SD continued to rise. Plots of potential against SD sometimes showed hysteresis as ψ_{shoot} lagged behind the fall in SD in the late afternoon. The shape of the diurnal course of ψ_{shoot} was similar on irrigated bushes, but minimum values each day were between 4 and 8 bars higher. The relation between ψ_{shoot} and SD during a dry season was, therefore, no longer linear; even on irrigated tea, ψ_{shoot} changed only slightly below -10 bars as SD rose to 30–40 mbars. However, because SD remained high for longer each day during the dry season, and because of hysteresis in the relation between ψ_{shoot} and SD, ψ_{shoot} remained close to the daily minimum for as much as 8 hr; in contrast, during a wet season, ψ_{shoot} rarely remained at its daily minimum for more than one or two hours. After the first shower at the end of the dry season, ψ_{shoot} rapidly returned to a higher value, illustrating that the mat of roots on the surface of the soil had remained potentially active during the drought.

The relative importance of environmental and endogenous factors responsible for the diurnal and seasonal pattern of ψ_{shoot} can only be guessed from the existing information. Because of the effect of SD on ψ_{shoot}, even when soil is saturated, the independent effect of ψ_{soil} on ψ_{shoot} has not been isolated. When leaf and air temperature are similar, atmospheric SD will largely determine the difference in absolute humidity between leaf and atmosphere, and so where the relation between ψ_{shoot} and SD is linear, ψ_{shoot} would appear to fall as the rate of transpiration increases. The importance of stomatal control of transpiration in determining the diurnal pattern of ψ_{shoot} in the dry season, as suggested by Carr

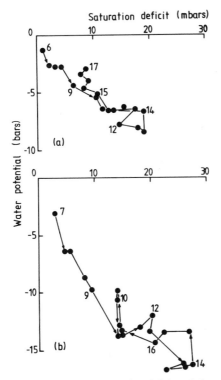

Fig. 9. Relation between atmospheric saturation deficit and shoot water potential during single days in a wet season (a) and a dry season (b). Numbers refer to local time (hr).

(1971), cannot be accurately assessed because there are no measurements relating transpiration from shoots to their ψ.

However, the often close relation between SD and ψ_{shoot} found in Africa permits a tentative explanation of the relation between growth of shoots and meteorological records described in Section IV,C. The rate of extension of shoots was determined by mean temperature until mean SD measured at 14.00 hr rose above 20–25 mbars (Fig. 8). When soil water is not limiting, saturation deficits of 20–25 mbars are equivalent to shoot water potentials of -8 to -10 bars (Williams, 1971; Squire, 1976, 1979). The growth of shoots in Malawi may, therefore, be unaffected by daily changes in ψ_{shoot}, provided that the minimum does not fall below about -8 bars. Similarly, in Tanzania Carr (1971) observed that reduced yields in the dry season coincided with periods when minimum ψ_{shoot} dropped below -7 bars. There is no information on how these low values of ψ act on shoot growth, whether by reducing turgor or by affecting biochemical processes.

D. Effects of Water Deficits on Photosynthesis

The growth of shoots on bushes in plantations is not normally limited by shortage of photosynthetic products (Section IV,A). On irrigated tea, where leaves are photosynthesizing rapidly, shoots extend slowly in the dry season; therefore, the concurrent reductions in photosynthesis and growth of shoots which have been observed on unirrigated tea are probably coincidental and have no causal relation. The importance of the effect of drought on photosynthesis is difficult to assess. After 2 months with little rain in Malawi, photosynthesis was very slow during the middle hours of the day and bushes probably experienced a net loss of carbon. During a long, severe drought, bushes might exhaust their reserves of carbohydrate and the growth of shoots would then be limited by shortage of reserves when the drought ends. However, in areas where tea is currently grown, the effects of dry weather lasting 2 to 4 months seem to be an insufficient drain on the plants' resources to prevent the rapid growth of shoots when the drought is relieved by rain. In Malawi and Tanzania, for example, the largest peak of yield occurs only six weeks after the end of the dry season (Fig. 10).

Plantations in Africa and much of Asia have a LAI of between four and eight (Hadfield, 1974b). Seasonal drought may cause LAI to fall below about three, in which case there is no longer complete interception of light; but there is no information of the levels of drought which bushes must experience before they lose their leaves. Information on the effects of water deficits on photosynthesis of individual leaves is also scanty and inconclusive. Photosynthesis of tea in the field is light-saturated at about

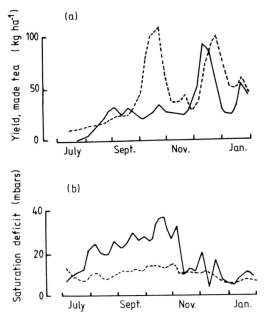

Fig. 10. Weekly yields of irrigated tea (a), and mean weekly saturation deficit at 14:00 hr (b), measured during and after the dry season of 1968 at two sites in East and Central Africa. Irrigation was given until the end of the dry season in November. (From an unpublished report by M. K. V. Carr and M. O. Dale.)

350 W m^{-2} (Sakai, 1975; Squire, 1977), and the whole canopy becomes saturated at 700–800 W m^{-2}—a value equivalent to full sunlight in the "winter" seasons of many tea areas, and about 75% of full sunlight in the "summer." Laboratory studies of photosynthesis (Barua, 1964; Hadfield, 1968) have not concentrated on the effect of water deficits and field studies have failed to separate effects of current water deficits from at least two other phenomena.

The first is a depression in photosynthesis which was observed in Malawi sometimes in the middle of the morning and sometimes in the late afternoon (Squire, 1977). The depression occurred when leaves were light-saturated, when stomata were open, and when ψ_{shoot} was high. Perturbations in the daily course of leaf conductance measured with a diffusion porometer and of stomatal aperture measured with silicone rubber impressions were also observed in the mid-morning (Squire, 1978). The physiological cause of the depression in photosynthesis is not known, but it was recorded only during and just before the dry season when there was little rainfall.

The second is the common association of high leaf temperature with low ψ_{leaf}. In laboratory experiments, Hadfield (1968) found that the rate of photosynthesis dropped rapidly when leaf temperature rose above 35°C, which has since been quoted widely as the optimum temperature for photosynthesis of tea. However, in plantations in many parts of the world, leaf temperatures above 35°C are accompanied by large leaf to air SD and low ψ_{leaf}; it will only be possible to separate the effects of temperature and ψ in the field in an experiment where one is held constant and the other varied, for example in a glasshouse built over established bushes. Knowledge of the relation between ψ and photosynthesis of plantations is confined to measurements made during only 2 yr on one clone part way between the "Assam" and "China" types (Squire, 1977). It would be unwise to generalize from these data, but they suggest that photosynthesis was reduced when ψ_{shoot} fell below -4 bars and stopped when it was around -18 bars.

E. Effects of Water Deficits on the Annual Distribution of Yield

Periods of drought reduce growth of tea shoots and depress yield for perhaps 2 or 3 months in several tea areas. These dry periods are also responsible for cyclical fluctuations of yield with a period of 6 to 8 weeks, which are experienced in Tanzania and Malawi and which occur in the main cropping season following the dry season (Fordham, 1970; Carr, 1974). Because extension of shoots can be very slow in the dry season (for example, small buds on unirrigated tea in Africa may grow only 1–2 mm in a month) many shoots appear to be arrested in the prolonged early phase of growth. Many other shoots develop dormant apical buds. The result is that potential sites for shoot growth accumulate on the plucking table as the dry season progresses. The beginning of the "rains," which immediately raises ψ_{shoot} above -10 bars, synchronizes the rapid extension of these shoots, with the result that most shoots reach harvestable size within 1 or 2 weeks of each other (Fig. 10). The subsequent peaks and troughs of yield are direct consequences of the first peak of yield (Fordham, 1970; Fordham and Palmer-Jones, 1977), and are not the result of current weather or water deficits. The peaks and troughs will not, therefore, be leveled by irrigation, but can be leveled if about half the shoots which make up the first peak are given a "phase-shift" by removing their apical buds after 3 to 4 weeks of growth (Palmer-Jones, 1977).

F. Modeling and Predicting Yields from Weather Data

Methods of management and morphology of bushes vary so much from country to country that it is probably impossible to determine how much differences in yield from country to country are caused by climate. Therefore, nearly all empirical formulas so far derived have attempted to explain the variations in yield from month to month and from year to year in a single plantation. The subject has already been reviewed by Carr (1972). Some formulas include several weather factors such as rainfall, air temperature, humidity, and hours of sunshine (e.g., Sen *et al.,* 1966); others include just one or two factors, such as potential evaporation and rainfall (Laycock, 1964), or calculated soil water deficit in the upper layers of soil (Hanna, 1971). The most striking feature of some of these statistical correlations is their very high level of significance. For example, Devanathan (1975) found a relation, with a correlation coefficient of 0.97, between monthly yield (as a percentage of annual yield) and the product of the previous month's sunshine hours and rainfall.

These formulas are valuable in that they have directed attention to some important meteorological factors, but they may prove to be of limited use in specific areas. They have usually not taken into account possible changes in the components of yield, and some are based on unsound physiological principles. For example, most formulas have assumed that yield is determined by the rate of photosynthesis, but evidence summarized in Section IV,A suggests it is not. Predictions from some formulae have been misleading. Laycock (1964) suggested that yield would be increased during drought if the soil were wetted by irrigation, but irrigation does not always increase yield where there are large SD's.

A more logical, as opposed to empirical, approach to modelling has been adopted by T. W. Tanton (unpublished). His starting points are that (1) assimilates are not normally limiting and (2) the rate of shoot growth depends on mean temperature and can be expressed in day-degrees, provided account is taken of the effect of large saturation deficits. However, the exclusion of a soil moisture variable will preclude use of this model for shallow-rooting tea, particularly under severe drought. Before any model can account for the effects of drought on shoot growth, more information is required on the relations between extension and turgor of shoots and the state of water in the soil and atmosphere.

G. Breeding for Avoidance and Tolerance of Drought

The most usual procedure for selection of tolerance and avoidance of drought is to grow different vegetatively propagated clones or seedling varieties close to each other and observe the ones that grow the best or do not die during a drought. Studies of this type may select the best varieties for a specific soil and climate but do not lead to understanding of the physiology of bushes. Only two aspects of comparative morphology and physiology have been studied in detail: the first is the effect of leaf orientation on transmission of light through the canopy; the second is the relation between ψ_{leaf} and stomatal movements.

China-type bushes with smaller, more upright leaves perform better in heat and drought than Assam-type bushes with larger, more horizontal leaves, but it is not immediately clear why they do. In a detailed study in northeast India, Hadfield (1974b) found that China-type bushes allowed more even distribution of light through the canopy, and suggested that leaves on their plucking table would be cooler because they intercepted less energy per unit area of leaf. Hadfield (1968) also found in a laboratory study that the rate of net photosynthesis of tea leaves declined rapidly if leaf temperature rose above 35°C—a common occurrence in bright sunlight in northeast India. Hadfield then put forward the view that China-type bushes produced greater yields in areas of bright light because their leaves would be just sufficiently cooler than leaves of Assam-types to photosynthesize for longer at the optimum temperature of 35°C. However, the relation between weather, leaf orientation, and yield is not straightforward. The size and angle of leaves affect leaf temperature to some extent, but their actual importance will not be known until other factors in the energy balance (such as stomatal resistance) are known. Although the relation between temperature and rate of photosynthesis can be defined in the laboratory, it is difficult to do so in the field where high temperatures probably occur with low ψ. Because photosynthesis is unlikely to limit yield (Section IV,A), the relation between leaf orientation and photosynthesis may be incidental to the relation between leaf orientation and yield. Assam-type bushes may have lower yields because they have fewer leaves per unit area, and so fewer axils from which shoots can grow.

Carr (1977a,b) and Othieno (1978b) in East Africa have shown that large differences in ψ_{shoot} and liquid filtration score exist between clones and seedling varieties. Carr (1977a) also found that the slope of the linear relation between ψ_{shoot} and infiltration score during a dry period (Fig. 4) might be a guide to the drought tolerance of different cultivars. Those whose stomata appeared to be less sensitive to a decrease in ψ_{shoot} looked as if they had been less affected by the drought. A comparable study was made with a diffusive resistance porometer in Buganda by Renard et al. (1979), who classified 24 clones into three groups distinguished by the relation between ψ_{shoot} and leaf resistance. However, more precise descriptions of drought tolerance (such as the ability of a bush to keep its leaves, or maintain rapid shoot growth) will have to be made before the relation between ψ_{shoot} and leaf resistance or infiltration score can be used for selection. Water potential itself may not be a suitable indicator if osmotic potential (and therefore turgor) also varies among clones and varieties. In fact Carr (1977a) and Squire (1976) obtained evidence in Kenya and Malawi that when the soil is wet, higher-yielding clones may have slightly lower ψ values than lower-yielding clones. Perhaps the greatest problem in selecting for drought tolerance is discussed by Carr (1977b), who found that the desirable qualities of drought tolerance possessed by genetically unique parent bushes were not shown by their cloned offspring. He suggested that qualities such as vigorous rooting, which would favor a superior bush among inferior bushes, would not necessarily be of advantage if all bushes were similar.

V. CULTURAL PRACTICES AND WATER RELATIONS

Several cultural practices affect the water relations of tea plantations. Some, such as irrigation and systems of soil and water conservation, are used to improve the chances of survival or to increase yield during drought. Others, such as pruning and provision of shade or shelter, are used for various other reasons and affect the environment of plantations in many ways.

A. PRUNING

Tea bushes are pruned for several reasons, which include removing diseased branches and maintaining the height of the plucking table (Laycock, 1960; Dutta, 1962; Eden, 1965; Harler, 1964). Pruning temporarily depletes the bushes' stores of carbohydrates (Visser, 1969) and reduces the rate of photosynthesis until new foliage has grown (Sakai, 1975). Pruning also conserves water in the soil. In Malawi, the ratio E/E_o fell to zero immediately after bushes were pruned at the beginning of the dry season, then increased gradually as new leaves grew, until the ratio returned to the original value 10 weeks after pruning (Laycock and Wood, 1963a; Laycock, 1964; Willatt, 1971). In Tanzania, Carr (1971) observed at the end of the dry season that the ψ of shoots was higher for bushes which had been pruned earlier in the dry season than for bushes which had been allowed to keep their foliage throughout the dry season. Some of the benefits of pruning are lost, however, if there is frequent rainfall, because wet soil and a complete canopy evaporate at a similar rate (Dagg, 1970).

B. IRRIGATION

Many tea estates now consider it economical to irrigate during drought, especially before the bushes have achieved complete ground cover. Information on effects of irrigation and conservation of rainwater in different parts of the world is contained in the proceedings of a symposium edited by Carr and Carr (1971). Irrigation during drought increases the chances of survival of young plants, reduces scorching and abscission of leaves, and maintains more open stomata and faster photosynthesis. But irrigation does not always increase the rate of shoot extension and yield.

Most forms of irrigation wet the soil but do not affect the overhead climate. However, when the air is very dry, wet soil alone does not prevent the development of low shoot ψ's, which appear to reduce the rate of shoot extension (Section IV,C). The problem of irrigating in dry atmospheres was examined in the Lenkoran region of Azerbaidzhan (U.S.S.R.) by Lebedev (1962), who compared the effects of irrigation given as a concentrated overhead spray (sprinkler irrigation) every few days and as a fine mist every 15 min. Average relative water content at midday was higher on bushes that received mist irrigation. During four years, 1958–1961, yield of tea between May and October from plots given sprinkler irrigation was 65% of the yield from misted plots; total water applied by sprinkler was 96% of that by misting.

The importance of atmospheric conditions is illustrated in the response of yield to irrigation during the dry season at two sites in Africa (Fig. 10). Sprinkler irrigation (not misting) caused a large increase of yield at one site where SD remained below 20 mbars, but little increase at the other site where SD regularly exceeded 30 mbars. The effects on shoot extension of mist- and sprinkler-irrigation are shown in Fig. 11. The rate of extension was much faster on misted tea and was almost identical to the rate predicted from mean temperature alone according to Tanton's model (Section IV,F).

C. SHADE AND SHELTER

The purpose of shade is to attenuate radiation; the purpose of shelter is to attenuate wind. In practice shade trees provide both shade and shelter. That the effects of shade and shelter are to some extent in opposition has been demonstrated by Ripley (1967). Comparing an open plot with plots separately under artificial shade, artificial shelter, and natural shade (shade trees), he showed that shelter, by reducing advection, increased the daily range of temperature, while shade, by reducing the radiation exchange, decreased the daily range of temperature.

In Tanzania, the effect of 4–5-m-high shelterbelts (windbreaks) of *Hakea saligna* trees on yield and water use of tea at different distances from the windbreak was studied by Carr (1970, 1971). The basis of Carr's

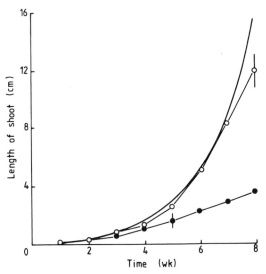

Fig. 11. Growth in length of shoots during the dry season on tea which received mist-irrigation (○), and sprinkler irrigation (●). Growth of shoots was negligible on tea which received no irrigation. The curve shows the rate of growth predicted from mean temperature alone. (From unpublished measurements by T. W. Tanton.)

explanation of the observed complex effects was that wind caused stomatal closure; therefore, sheltered tea had more open stomata. Higher rates of transpiration and growth were recorded on bushes adjacent to the shelter, until soil moisture became limiting when the sheltered tea yielded less than the unsheltered. He concluded that shelterbelts did not improve the water economy of unirrigated tea but that response to irrigation would be greatest for sheltered tea. The shade, as opposed to shelter, afforded by the 4–5-m-high belts was not studied by Carr.

Artificial shade seems to have been used on tea estates primarily for research. Shade trees, on the other hand, have been used for many years in Assam, where they significantly improve yield. Their use in tea gardens has been reviewed by Hadfield (1974a) and Willey (1975). When tea was established in Africa, shade trees were planted as a matter of course, but McCulloch *et al.* (1965) showed that shade trees planted on a 8 m × 14 m grid in Kenya reduced both yield and quality of tea. Also in Kenya, Ripley (1967) demonstrated that shade trees spaced 10 m apart produced little difference in the temperature and humidity regimes compared to an open plot. In Malawi, Laycock and Wood (1963b) showed that yield was regularly depressed by shade trees, although they produced evidence that competition between the tea and the shade trees for soil moisture was *not* the reason for the reduced yields.

The causes of the opposite effects of shade trees in Assam and in Africa are not known. An extensive cover of large, leguminous trees has many effects on the ecology of a plantation, but their main effect is likely to operate through radiation, because in Assam bamboo screening produced the same improvement in yield as did shade trees (Wight, 1958). Photosynthesis of the tea canopy is light-saturated at an irradiance of 700 W m^{-2} (Section IV,D) but under dense shade, radiation reaching bushes may be as low as 200 W m^{-2}, although the level is extremely variable depending on position in relation to surrounding trees (Hadfield, 1974a). Under dense shade, the canopy may change from a condition in which assimilates do not normally limit yield to a condition where they do. Reducing irradiance falling on bushes may also suppress the temperature of leaves during the hottest part of the day. In Assam, air temperatures during the growing season are the highest of all areas where tea is grown, with daily maxima often rising above 32–35°C. In such high temperatures, photosynthesis may be reduced, as suggested by Hadfield (1968) on the strength of laboratory measurements, and the probable association of high leaf temperature with large leaf-to-air saturation deficit may also reduce shoot extension directly. In Africa, where it is cooler in the growing season, the effects of shade on leaf temperature may be insufficient to counter the reduction in photosynthesis.

ACKNOWLEDGMENTS

We are grateful to Drs. T. W. Tanton and M. K. V. Carr for permission to use unpublished information. We are also indebted to the Directors of the following institutes for the provision of reports and other publications: the Tea Research Institute of East Africa, Kericho, Kenya; the National Research Institute of Tea, Japan; the Tea Research Foundation of Central Africa, Malawi; and the United Planters Association of Southern India.

REFERENCES

Amma, S. (1971). Testing young plants for important characters in tea breeding. 2. Multiple regression analysis between yield and shoot characteristics in black tea clones. *Study Tea* 42, 13–19.

Barua, D. N. (1964). Effect of light intensity on assimilation characteristics of detached tea leaves. *J. Agric. Sci.* 63, 265–271.

Barua, D. N. (1970). Light as a factor in the metabolism of the tea plant. *In* "Physiology of Tree Crops" (L. C. Luckwill and C. V. Cutting, eds.), pp. 307–322. Academic Press, New York.

Blackie, J. R. (1972a). Hydrological effects of a change in land use from forest to tea plantation in Kenya. *IASH/UNESCO* Studies and Reports on Hydrology 12, 312–329.

Blackie, J. R. (1972b). Physics report. *East Afr. Agric. For Res. Organ., Annu. Rep.*, p. 41.

Blackie, J. R. (1981). The water balance of the Kericho catchments. *East Afr. Agric. For. J.* **43**, Spec. Issue (in press).

Blackie, J. R., and Edwards, K. A. (1981). The Kericho Research Project. *East Afr. Agric. For. J.* **43**, Spec. Issue (in press).

Busin, P. M. (1970). Moisture requirements and moisture supply to tea crops in the subtropics of the Krasnodar region (in Russian with English summary). *Pochvovedenie* **9**, 34–44.

Callander, B. A. (1978). Eddy correlation measurements of sensible heat flux density. Ph.D. Thesis, University of Strathclyde.

Callander, B. A., and Woodhead, T. (1981a). Eddy correlation measurements of sensible heat flux, and estimation of evaporative heat flux, over growing tea. *East Afr. Agric. For. J.* **43**, Spec. Issue (in press).

Callander, B. A., and Woodhead, T. (1981b). Canopy conductance of estate tea in Kenya. *Agric. Meteorol.* **23** (in press).

Carr, M. K. V. (1970). The role of water in the growth of the tea crop. *In* "Physiology of Tree Crops" (L. C. Luckwill and C. V. Cutting, eds.), pp. 287–305. Academic Press, New York.

Carr, M. K. V. (1971). The internal water status of the tea plant (*Camellia sinensis*) some results illustrating the use of the pressure chamber technique. *Agric. Meteorol.* **9**, 447–460.

Carr, M. K. V. (1972). The climatic requirements of the tea plant: A review. *Exp. Agric.* **8**, 1–14.

Carr, M. K. V. (1974). Irrigating seedling tea in Southern Tanzania: Effects on total yields, distribution of yield and water use. *J. Agric. Sci.* **83**, 363–378.

Carr, M. K. V. (1976). Methods of bringing tea into bearing into relation to water status during dry weather. *Exp. Agric.* **12**, 341–351.

Carr, M. K. V. (1977a). Changes in the water status of tea clones during dry weather in Kenya. *J. Agric. Sci.* **89**, 297–307.

Carr, M. K. V. (1977b). Responses of seedling tea bushes and their clones to water stress. *Exp. Agric.* **13**, 317–324.

Carr, M. K. V., and Carr, S., eds. (1971). "Water and the Tea Plant." Tea Res. Inst. East Africa, Kericho, Kenya.

Cooper, J. D. (1981). Water use of a tea estate from soil moisture measurements. *East Afr. Agric. For. J.* **43**, Spec. Issue (in press).

Dagg, M. (1965). A rational approach to the selection of crops for areas of marginal rainfall in East Africa. *East Afr. Agric. For. J.* **30**, 295–300.

Dagg, M. (1970). A study of the water use of tea in East Africa using an hydraulic lysimeter. *Agric. Meteorol.* **7**, 303–320.

Devanathan, M. A. V. (1975). Weather and the yield of a crop. *Exp. Agric.* **11**, 183–186.

Dutta, S. K. (1962). Effect of method of pruning and different pruning cycles on tea. *Trop. Agric. (Trinidad)* **39**, 83–94.

Dutta, S. K. (1971). Effects of irrigation of tea in Assam. *In* "Water and the Tea Plant" (M. K. V. Carr and S. Carr, eds.), pp. 145–149. Tea Res. Inst. East Africa, Kericho, Kenya.

Eden, T. (1965). "Tea," 2nd ed. Longmans, Green, New York.

Eeles, C. W. D. (1981). Soil moisture deficits under montane rain forest and tea. *East Afr. Agric. For. J.* **43**, Spec. Issue (in press).

Ellis, R. T. (1971). The general background to plant/water relation investigations on tea in Central Africa, and some findings to date. *In* "Water and the Tea Plant" (M. K. V. Carr and S. Carr, eds.), pp. 49–57. Tea Res. Inst. East Africa, Kericho, Kenya.

Filippov, L. A. (1968). A method for assessing the influence of air temperature and humidity on the water regime of the tea plant. *Sov. Plant Physiol. (Engl. Transl.)* **15**, 868–872.

Filippov, L. A. (1969). A physiological method of determining irrigation rates for crop plants. *Sov. Plant Physiol. (Engl. Transl.)* **16**, 785–788.

Filippov, L. A. (1970). Values of cell sap concentrations determined with the aid of a refractometer. *Sov. Plant Physiol. (Engl. Transl.)* **17**, 791–795.

Filippov, L. A., and Busin, P. M. (1969). The relations between soil moisture content, cell sap concentration and the growth of tea shoots. *Sov. Plant Physiol. (Engl. Transl.)* **16**, 50–53.

Food and Agriculture Organization (1979). "FAO Production Yearbook," Vol. 32. FAO/UN, Rome.

Fordham, R. (1970). Factors affecting tea yields in Malawi. *Tea Res. Found. Cent. Afr., Mulanje, Malawi, Annu. Rep.* pp. 71–130.

Fordham, R. (1973). The micrometeorology of an extended area of tea before and after rain. *Agric. Meteorol.* **11**, 99–105.

Fordham, R. (1977). Tea. *In* "Ecophysiology of Tropical Crops" (P. de T. Alvim and T. T. Kozlowski, eds.), pp. 333–349. Academic Press, New York.

Fordham, R., and Palmer-Jones, R. W. (1977). Simulation of intraseasonal yield fluctuations of tea in Malawi. *Exp. Agric.* **13**, 33–42.

Green, R. M. (1971). Weather and the seasonal growth of tea in Malawi. M. Phil. Thesis, University of Nottingham.

Hadfield, W. (1968). Leaf temperature, leaf pose and productivity of the tea bush. *Nature (London)* **219**, 282–284.

Hadfield, W. (1974a). Shade in North-east India tea plantations I. The shade pattern. *J. Appl. Ecol.* **11**, 151–178.

Hadfield, W. (1974b). Shade in North-east Indian tea plantations II. Foliar illumination and canopy characteristics. *J. Appl. Ecol.* **11**, 179–199.

Hanapi, M. M., and Reesinck, J. J. M. (1959). The temperature of leaves in the sun. *Arch. Tea Cult.* **20**, 197–201.

Hanna, L. W. (1971). The effects of water availability on tea yields in Uganda. *J. Appl. Ecol.* **8**, 791–813.

Harler, C. R. (1964). "The Culture and Marketing of Tea," 3rd ed. Oxford Univ. Press, London and New York.

Harler, C. R. (1966). "Tea Growing." Oxford Univ. Press, London and New York.

Kerfoot, O. (1962). Root systems of forest trees, shade trees and tea bushes. *East Afr. Agric. For. J.* **27**, Spec. Issue, 24.

Kingdon-Ward, F. (1950). Does wild tea exist? *Nature (London)* **165**, 297–299.

Laycock, D. H. (1960). A progress report on the first six years of a pruning experiment with Nyasaland tea. *Trop. Agric. (Trinidad)* **37**, 125–134.

Laycock, D. H. (1964). An empirical correlation between weather and yearly tea yields in Malawi. *Trop. Agric. (Trinidad)* **41**, 277–291.

Laycock, D. H., and Wood, R. A. (1963a). Some observations on soil moisture use under tea in Nyasaland. Part 1. The effect of pruning mature tea. *Trop. Agric. (Trinidad)* **40**, 35–42.

Laycock, D. H., and Wood, R. A. (1963b). Some observations on soil moisture use under tea in Nyasaland. Part II. The effect of shade trees. *Trop. Agric. (Trinidad)* **40**, 42–48.

Lebedev, G. V. (1962). New irrigation conditions for agricultural crops. *Sov. Plant Physiol. (Engl. Transl.)* **9**, 400–406.

McCulloch, J. S. G. (1965). Tables for the rapid computation of the Penman estimate of evaporation. *East Afr. Agric. For. J.* **30**, 286–295.

McCulloch, J. S. G., Pereira, H. C., Kerfoot, O., and Goodchild, N. A. (1965). Effect of shade trees on tea yields. *Agric. Meteorol.* **2**, 385–399.

Mann, H. H. (1935). Tea soils. *Commonw. Bur. Soil Sci., Tech. Commun.* No. 32.

Monteith, J. L. (1965). Evaporation and environment. *Symp. Soc. Exp. Biol.* **19**, 205–234.

Monteith, J. L. (1973). "Principles of Environmental Physics." Arnold, London.

Othieno, C. O. (1973). *Tea Res. Inst. East Africa, Kenya, Annu. Rep.*, 38–39.

Othieno, C. O. (1975). Surface run-off and soil erosion in fields of young Tea. *Trop. Agric. (Trinidad)* **52**, 299–308.

Othieno, C. O. (1978a). Supplementary irrigation of young clonal tea in Kenya. I. Survival, growth and yield. *Exp. Agric.* **14**, 229–238.

Othieno, C. O. (1978b). Supplementary irrigation of young clonal tea in Kenya. II. Internal water status. *Exp. Agric.* **14**, 309–316.

Othieno, C. O., and Laycock, D. H. (1977). Factors affecting soil erosion within tea fields. *Trop. Agric. (Trinidad)* **54**, 323–329.

Palmer-Jones, R. W. (1977). The effects of plucking policies on the yield of tea in Malawi. *Exp. Agric.* **13**, 43–49.

Penman, H. L. (1948). Natural evaporation from open water, bare soil and grass. *Proc. R. Soc. London, Ser. A* **194**, 120–145.

Pereira, H. C. (1962). The development of tea estates in tall rain forest. 1A. The research project. *East Afr. Agric. For. J.* **27**, Spec. Issue, 16–18.

Pereira, H. C. (1973). "Land Use and Water Resources in Temperate and Tropical Climates." Cambridge Univ. Press, London and New York.

Renard, C., Flemal, F., and Barampama, D. (1979). Evaluation de la résistance à la sécheresse chez la théier au Burundi. *Cafe, Cacao, The* **23**, 175–182.

Ripley, E. A. (1967). Effects of shade and shelter on the microclimate of tea and their significance. *East Afr. Agric. For. J.* **33**, 67–80.

Sakai, S. (1975). Recent studies and problems of photosynthesis of tea plant. *JARQ* **9**, 101–106.

Sen, A. R., Biswas, A. K., and Sanyal, D. K. (1966). The influence of climatic factors on the yield of tea in the Assam valley. *J. Appl. Meteorol.* **5**, 789–800.

Shanmuganathan, N., and Rodrigo, W. R. F. (1967). Studies in collar and branch canker of young tea (*Phomopsis theae* Petch). II. Influence of soil moisture on the disease. *Tea Q.* **38**, 320–330.

Squire, G. R. (1976). Xylem water potential and yield of tea (*Camellia sinensis*) clones in Malawi. *Exp. Agric.* **12**, 289–297.

Squire, G. R. (1977). Seasonal changes in photosynthesis of tea (*Camellia sinensis*). *J. Appl. Ecol.* **14**, 303–316.

Squire, G. R. (1978). Stomatal behaviour of tea (*Camellia sinensis*) in relation to environment. *J. Appl. Ecol.* **15**, 287–301.

Squire, G. R. (1979). Weather, physiology and seasonality of tea (*Camellia sinensis*) yields in Malawi. *Exp. Agric.* **15**, 321–330.

Szeicz, G., and Long, I. F. (1969). Surface resistance of crop canopies. *Water Resour. Res.* **5**, 622–633.

Tanton, T. W. (1979). Some factors limiting yields of tea (*Camellia sinensis*). *Exp. Agric.* **15**, 187–192.

Templer, J. C. (1976). *Tea Res. Inst. East Africa, Kenya, Annu. Rep.*, 79–82.

Thom A. S., and Oliver, H. R. (1977). On Penman's equation for estimating regional evaporation. *Q. J. R. Meteorol. Soc.* **103**, 345–357.

Tolhurst, J. A. H. (1971). Leaf nutrient content in relation to season and irrigation treatments. *In* "Water and the Tea Plant" (M. K. V. Carr and S. Carr, eds.), pp. 195–212. Tea Res. Inst. East Africa, Kericho, Kenya.

Toyao, T. (1965). Correlation and genetical analyses of several characteristics at the stage of individual selection in the tea plant. 2. On the correlations among yield, number of leaves per plant, and characters of leaf in clonal varieties for green tea (in Japanese). *Study Tea* **30**, 1–4.

UPASI (1973). *Annu. Rep. Tea Sci. Dep., United Planters' Assoc. South. India, Chinchona, Coimbatore, India, 1973.*

Van der Laan, J. C. (1971). Irrigation of tea in East Pakistan. *In* "Water and the Tea Plant" (M. K. V. Carr and S. Carr, eds.), pp. 113–125. Tea Res. Inst. East Africa, Kericho, Kenya.

Visser, T. (1969). The effect of root and shoot damage on the growth of tea plants. *Neth. J. Agric. Sci.* **17**, 234–240.

Wang'ati, F. J., and Blackie, J. R. (1971). A comparison between lysimeter and catchment water balance estimates of evaportranspiration from a tea crop. *In* "Water and the Tea Plant" (M. K. V. Carr and S. Carr, eds.), pp. 9–20. Tea Res. Inst. East Africa, Kericho, Kenya.

Wight, W. (1958). The shade tree tradition in tea gardens of northern India. 1. The value of shade. *Indian Tea Assoc., Sci. Dep., Tocklai Exp. Stn. Proc. Annu. Rep.* pp. 75–97.

Willatt, S. T. (1971). Model of soil water use by tea. *Agric. Meteorol.* **8**, 341–351.

Willatt, S. T. (1973). Moisture use by irrigated tea in Southern Malawi. *Ecol. Stud.* **4**. 331–338.

Willey, R. W. (1975). The use of shade in coffee, cocoa and tea. *Hortic. Abstr.* **45**(12), 791–798.

Williams, E. N. D. (1971). Physiology report. *Tea Res. Found. Cent. Afr., Mulanje, Malawi, Annu. Rep.,* 105–131.

Yanase, Y. (1973). Daily and seasonal variation in evapotranspiration on tea field. *JARQ* **7**, 122–125.

CHAPTER 8

CLOSELY RELATED WOODY PLANTS

Stephen G. Pallardy

SCHOOL OF FORESTRY, FISHERIES AND WILDLIFE, UNIVERSITY OF MISSOURI,
COLUMBIA, MISSOURI

I. INTRODUCTION

In this chapter evidence is presented for genetic variation in water relations among closely related woody plants. Emphasis is placed on intraspecific variation, but some studies utilizing hybrids are included for practical or illustrative purposes. It should be emphasized at the outset that the simple discovery that water relations of certain plants vary is of little benefit unless differences can be interpreted in a useful way. The obvious and perhaps most useful perspective of interpretation is that of relating differences in patterns of plant water relations to relative tolerance to drought. By this approach, basic principles of drought adaptation and useful practical knowledge may be gained.

Because a scarcity of moisture frequently limits plant growth and survival (Kramer, 1969; Fischer and Turner, 1978), selective pressure for adaptation to drought is often high. The morphology and physiology of

Water Deficits and Plant Growth, Vol. VI
Copyright © 1981 by Academic Press, Inc.
All rights of reproduction in any form reserved.
ISBN 0-12-424156-5

terrestrial higher plants reflect this profound influence. Species with ranges that span large geographic areas commonly are subjected to quite distinct moisture regimes and, given sufficient reproductive isolation, specifically drought-adapted lines may emerge. It is also probable that such sufficient genetic isolation and adaptation can occur between populations occupying two adjacent habitats that differ in the frequency of occurrence of drought (e.g., Hermann and Lavender, 1968).

Comparative studies of water relations have been conducted both between and within species of plants. Interspecific studies were concerned with establishing the ecological and physiological basis of plant distribution patterns and superior survival under drought (see Chapter 3 by Hinckley *et al.*), or have investigated the adaptations by which certain agronomic crops may be grown in areas too arid for other species. For example, several studies have been conducted to identify the reasons for the superior capacity of grain sorghum (*Sorghum bicolor*) over *Zea mays* to subsist and produce acceptable yields in regions with low rainfall (Hsiao *et al.*, 1976).

Intraspecific variation in water relations and drought tolerance has been most frequently studied with annual crop plants (Levitt, 1972; Boyer and McPherson, 1975; Begg and Turner, 1976; Taylor and Klepper, 1978) and only rarely in woody plants. Comparative studies of such plants can provide specific identification and information about drought-tolerant genotypes for planting on drought-prone sites, can identify drought-adaptive responses for use in woody plant breeding programs, and can be used in studying adaptations by which plants respond to drought. In many cases, it may be easier to identify adaptive responses to drought by using genotypes of a single species because the confounding influences of large-scale morphological and physiological differences that typically separate species are thereby avoided. However, even within species multiple adaptations to drought may be observed.

II. COMPARATIVE WATER RELATIONS AND DROUGHT TOLERANCE

A. CLASSIFICATIONS AND METHODS OF MEASURING DROUGHT TOLERANCE

Confusion often arises concerning the terminology associated with mechanisms by which plants respond to drought and the plant attribute by which drought tolerance is measured. Therefore, basic concepts that relate to drought response adaptations and measures of drought tolerance as used in this chapter will be briefly discussed.

Drought-avoiding plants (e.g., ephemerals) subsist in areas of inadequate annual precipitation by adjusting their life cycles to avoid physiologically intense phases during periods when sustained drought is likely (Kramer and Kozlowski, 1979). Drought-tolerant plants, as defined by Kramer and Kozlowski (1979), subsist through periods of drought by avoiding or postponing protoplasmic desiccation (desiccation avoidance) or by enduring protoplasmic desiccation in any of several ways (desiccation tolerance). Desiccation-avoiding and desiccation-postponing plants exhibit various morphological and/or physiological characters that tend to conserve or replenish plant water. These may include tissue water storage, heavily cutinized epidermes that resist cuticular transpiration, efficient stomatal control of water loss, leaf abscission, and capacity for extensive rooting. On the other hand, desiccation-tolerant plants are able to endure protoplasmic loss of water or low water potentials without lethal injury. Desiccation-tolerant plants are commonly lower plants such as mosses and bryophytes, but certain higher plants, often called "resurrection plants" (e.g., the angiosperm *Myrothamnus flabellifolia*), can be desiccated to air dryness without loss of viability (Henckel and Pronina, 1969; Bewley, 1979). The majority of drought-tolerant higher plants and most woody plants are primarily desiccation-postponing or desiccation-avoiding rather than desiccation-tolerant (Levitt, 1972).

The plant response by which drought tolerance is measured has varied with the outlook and objectives of investigators in various disciplines. In perhaps the most basic sense, the success of a plant under drought conditions depends on its capacity to survive drought in a viable condition, remaining competitive with its plant associates. A method of integrating quantitative estimates of desiccation-avoidance and desiccation-tolerance capacities of plants in a single measure of drought tolerance has been discussed by Levitt (1972). In practice, however, drought responses are often measured in terms of total dry weight yield or yield of reproductive structures under drought conditions (Boyer, 1976). For example, whereas in forestry maximal production of biomass during drought is a desired objective, in horticulture maximal production of high quality fruit often is the goal. The capacities of a particular plant type to survive, increase in dry weight, and produce fruits and seeds during drought need not be closely related. Therefore, conclusions derived from study of one character under drought conditions do not necessarily hold for others, and extrapolations are often unjustified.

Because the growth environment will affect plant morphology and physiology independent of genetic influences, studies concerned with identifying genetic variation in plants must be conducted under uniform growth conditions. Unless specifically stated otherwise, studies reported in this chapter were conducted accordingly. However, it should be noted that many of these studies utilized seedlings produced from seed collected in habitats or regions characterized by very different environmental conditions. Hence, possible effects of differing environments during reproductive growth on subsequent seedling growth and drought responses may exist. Most investigators have assumed that such effects are so small as to be negligible.

B. QUALITATIVE STUDIES

In several studies, often provenance tests of woody species, variation in rates of survival among different genetic materials have been attributed to presumed drought conditions. For example, genetically related differences in survival under putative drought have been observed in *Quercus rubra* (Deneke, 1974; Kriebel *et al.,* 1976), *Acer saccharum* (Kriebel, 1957), *Populus deltoides* (Farmer, 1970), *Pinus elliottii* (Schultz and Wilhite, 1969), and *Pinus taeda* (Zobel and Goddard, 1955; Goddard and Brown, 1959; Wells and Wakeley, 1966). However, in another study, survival of progeny from wet and dry site populations of *Pinus elliottii* did not differ when planted on wet or dry sites in Florida (Gansel, 1967). Similarly, when planted on upland sites, survival of progeny of seed obtained from upland and lowland sources of *Picea mariana* did not differ (Fowler and Mullin, 1977). Survival of seedlings of *Juglans nigra* obtained from sources of similar latitudes from Kentucky to Kansas in the United States exhibited no apparent relationship to the aridity of the seed source when planted on bottomland or upland sites in southern Illinois (Bey, 1974).

It is obvious that comparisons of survival cannot provide precise comparative information concerning water balance and drought-tolerance adaptations. However, survival tests under water-limited conditions provide a relatively simple technique to preliminarily screen material for potential genetic differences. In this way, time and effort spent in futile study of more arduous aspects of water relations can be minimized. For example, provenances of *Pinus taeda* exhibited significant differences in survival ability under drought in the field (Zobel and Goddard, 1955; Gilmore, 1957; Goddard and Brown, 1959). Subsequent studies, initiated after analysis of field results, examined the morphological and physiological basis for the superior survival under drought conditions of certain sources of *Pinus taeda* (Van Buijtenen *et al.,* 1976). Several species in which simple survival differences under drought have been observed have not been studied further to identify genetic variation in drought-adaptive responses in greater detail (e.g., *Pinus elliottii, Quercus rubra*).

C. Differences in Desiccation Avoidance or Postponement

1. Relative Capacity for Soil Water Absorption

The capacity of a plant to absorb water in relation to potential water loss by the shoot greatly affects its water balance and capability to subsist under conditions of soil and atmospheric drought. The relationship between root absorption capacity and shoot transpiration demand is best measured by comparing transpiring surface with root-absorbing surface. However, accurate measurement of the absorptive surface of roots is a virtually impossible task. Furthermore, the absorptive capacity of the root is extremely variable because of root suberization (Kramer and Kozlowski, 1979) and because water uptake patterns of roots shift with increasing transpiration rate (Brouwer, 1953). As an approximation, the root–shoot ratio is often used to estimate relative absorption-transpiration capacities. Another aspect of root systems that may influence water absorption capacity independent of the root–shoot ratio is the spatial pattern of rooting in the soil profile. Variation in depth, capacity for lateral root extension, and degree of root ramification may significantly increase soil water availability to the plant.

Both root–shoot ratio and root morphological patterns have been shown to vary intraspecifically in woody plants. It has often been possible to relate such differences to the aridity at the seed source. Survival under imposed drought of seedlings of ecotypes from a xeric inland source of *Pseudotsuga menziesii* was greater than that of seedlings of an ecotype from a moist coastal site (Heiner and Lavender, 1972). Survival of the ecotypes was correlated with the capacity of roots of the seedlings to penetrate beyond 30 cm depth in the soil. In another study, a relationship between desiccation-avoidance capacity and root system weight in *Pseudotsuga menziesii* seedlings from interior and maritime seed sources could not be detected (Ferrell and Woodard, 1966). However, seedlings from interior sources that showed greater desiccation-avoidance capacity than maritime seedlings had greater numbers of active root tips, suggesting that ramification by unsuberized roots was more extensive in the interior source seedlings.

A relationship between survival capacity under drought in *Acer saccharum* ecotypes and root morphology has been reported (Kriebel, 1963; Kriebel and Gabriel, 1969). Four-year-old seedlings of a drought-tolerant ecotype from the southeastern United States possessed a root system characterized by a large primary root and a dense mass of fine roots. Drought-susceptible seedlings of ecotypes from the more mesic northeastern United States had diffuse, shallow root systems and lacked a prominent primary root. Although a drought-tolerant ecotype from the more xeric central United States did not have a prominent primary root, it did possess roots which penetrated deeply into the soil.

Seedlings of *Pinus taeda* from two disjunct populations exhibited differences in root distribution patterns. In Texas, isolated stands of this species, located 160 km west of the contiguous range and receiving 40% less annual rainfall, produced seedlings that were more drought-tolerant (in terms of survival under imposed drought), produced deeper roots, greater root proliferation into all higher order lateral roots, and had roots that exhibited steeper angles of downward branching than did seedlings produced from seed obtained from the more mesic contiguous range of *Pinus taeda* (Youngman, 1965; Van Buijtenen *et al.*, 1976; Bilan *et al.*, 1978). Hence, the root systems of the xeric seedlings of disjunct populations were highly modified to gain access to deeper soil water supplies and to exploit intensively available water within the physical limits of the rhizosphere. Differences in root morphology were greatest in early stages of growth, emphasizing the importance of drought adaptations in establishment of young plants. In another study of North Carolina families of *Pinus taeda,* families which produced the greatest 8-year wood volume yields grew fastest under mild water stress as seedlings and had greater root growth rates relative to shoot growth rates as seedlings (Cannell *et al.*, 1978). Relationships between rooting patterns and relative drought tolerance have also been reported for varieties of *Pinus caribaea* (Venator, 1976), ecotypes of *Thuja occidentalis* (Habeck, 1958) and provenances of *Larix leptolepis* (Schönbach *et al.*, 1966).

Experiments conducted with several clones of tea (*Camellia sinensis*) in Kenya illustrate well how differences in drought tolerance are related to rooting characteristics (Carr, 1977; Othieno, 1978a,b). Soil moisture and water relations of two drought-susceptible and two drought-tolerant clones (presumably classified in terms of tea yield under drought) were observed during a dry season under unirrigated conditions. Soil water was depleted to greater depths by drought-tolerant clones 7/3 and 31/11 than by drought susceptible clones 6/8 and 7/9 (Fig. 1). The superior capacity to exploit deep soil water was reflected in significantly higher mean xylem pressure potentials (ψ_x) and hence lower water deficits on days of high atmospheric evaporative demand in the deeper-rooted, drought-tolerant clones (Table I).

Genetically based variation in the root–shoot ratio is frequently correlated with geographic trends in precipitation and potential evaporation. Within a species, seedlings obtained from seed of populations growing in more arid climates commonly have greater root–shoot ratios. For example, when seedlings of *Pseudotsuga menziesii* var. *menziesii* and *P. men-*

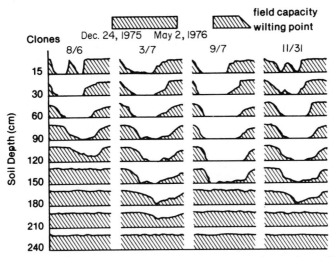

Fig. 1. Weekly soil moisture measurements at several depths for tea clones 6/8, 7/3, 7/9, and 31/11 in Kenya. Values represent data taken from December 24, 1975 to May 2, 1976. (From Othieno, 1978a.)

ziesii var. *glauca* from populations growing in mesic and xeric habitats were grown under several air and soil temperature regimes (Table II), seedlings from xeric habitats had uniformly greater root–shoot ratios under all growth conditions (Lavender and Overton, 1972). In another study using seedlings of Oregon populations of *Pseudotsuga menziesii* var. *menziesii* grown in uniform environments, root–shoot ratios of seedlings obtained from populations on more xeric south-facing aspects were greater than those of seedlings obtained from populations existing on north-facing aspects (Hermann and Lavender, 1968).

Examination of morphological features of 45 provenances of *Pinus sylvestris* grown in the north central United States revealed that provenance root–shoot ratios were related to an "Index of Aridity" combining annual precipitation and a function of mean temperature at the prove-

TABLE I

MEAN ψ_x (BARS) OF TEA CLONES 6/8, 7/3, 7/9, AND 31/11 IN KENYA ON FEBRUARY 6, 1976[a]

	Clone			
	6/8	7/3	7/9	31/11
ψ_x	−14.24[b]	−11.83	−14.38	−11.86

[a]Adapted from Othieno (1978b).
[b]Least significant difference ($p = 0.001$) between two clonal means = 0.57 bar.

TABLE II

Root–Shoot Ratios of Seedlings from Xeric and Mesic Populations of *Pseudotsuga menziesii* vars. *menziesii* and *glauca* in the Northwestern United States[a]

Variety	Seed source	Habitat	Root-shoot ratio[b]
glauca	N. Rocky Mountains	Xeric	0.85
menziesii	S. Oregon	Xeric	0.81
menziesii	Olympic Peninsula	Mesic	0.56
menziesii	Valsetz, Oregon	Mesic	0.58
menziesii	N. W. Oregon	Mesic	0.61
menziesii	Vancouver Island	Mesic	0.60

[a]From Lavender and Overton (1972).

[b]Each value is the average of seveal day/night/soil temperature treatments. Sample size = 96.

nance, root–shoot ratio increasing as the climate at the provenance became more arid (Brown, 1969). Similar genetically based relationships between root–shoot ratio and aridity of habitat or climate have been observed in *Eucalyptus viminalis* (Ladiges, 1974) and *Pinus caribaea* (Venator, 1976).

It appears that in certain species seed size may be related to aridity of the habitat. Seed size of *Pinus sylvestris* provenances grown in Michigan increased with aridity of the locality of origin (Wright and Bull, 1963; Ruby, 1967). Seeds of the more arid Greek, Spanish, and Turkish provenances were larger than those of Scandinavian origin. The authors suggested that larger seeds were an adaptation to soil moisture conditions at the time of germination. Seeds falling on dry soil would benefit from increased food reserves which could be allocated to rapid root growth. Similarly, seeds from drier inland California sources of *Pseudotsuga menziesii* were larger than seeds from wetter coastal sources (Griffin and Ching, 1977). However, no data relating seed size to root growth rate were presented in these two studies. Youngman (1965) did find that seeds of *Pinus taeda* from more arid, disjunct populations in Texas were larger and that seedlings germinated from these seeds exhibited greater rates of root growth than those from more mesic sources. It must be emphasized that the seed collected in the latter two studies developed under very different environmental conditions, and, hence, the variation in seed size may not be entirely attributable to genetic differences. It would be more difficult, but more appropriate, to study seed size of progeny grown in uniform environments (Wright and Bull, 1963), although at the risk of diluting the expression of this trait as a result of uncontrolled outcrossing during reproduction.

Finally, in plants that can be vegetatively propagated by cuttings there is wide variation in the rooting ability of different clones. For example, differences in root numbers and root lengths in hardwood cuttings of several clones of *Populus deltoides* have been reported (Ying and Bagley, 1978). Pallardy and Kozlowski (1979a) noted differences in early root growth among leafy tip cuttings of several hybrid *Populus* clones grown in "minirhizotron" tubes (Bilan, 1964) (Fig. 2). Since fresh leafy cuttings exhibit low water potentials and leaf conductances that recover as rooting proceeds (Loach, 1977; Gay and Loach, 1977), the capacity of certain clones of woody species to root vigorously from cuttings may favor their establishment and early growth.

2. Plant Resistance to Liquid-Phase Water Transport

The potential gradient that is needed to sustain a constant flux of water in the plant increases with resistance to water transport in the roots, stem, and leaves. Water movement from soil into the woody plant root appears to be apoplastic until the Casparian strip of the endodermis is encountered; here a semipermeable membrane must be crossed before entry into xylem elements and transport to the shoot (Hinckley *et al.*, 1978). Because of the properties of cell membranes and coupled water-

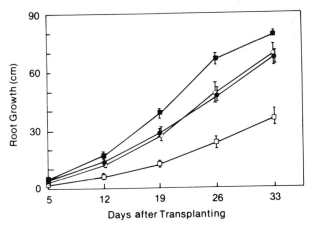

Fig. 2. Root growth of four *Populus* clones in acrylic resin tubes. □ = *Populus deltoides* × *P. caudina;* ○ = *P. nigra* × *P. laurifolia;* ■ = *Populus* sp. (parentage unknown); ● = *P. candicans* × *P. berolinensis.* Bars denote 1 standard error of the mean. (From Pallardy and Kozlowski, 1979a.)

solute uptake by root cells, apparent root resistance may vary inversely with plant transpiration rate (Dalton *et al.,* 1975; Fiscus, 1977). Although species differences in variation in apparent root resistance with transpiration rate have been observed (e.g., Camacho-B. *et al.,* 1974; Kaufmann, 1976), no data are available for comparison at the subspecies level.

Xylem resistance of woody plants and resultant physiological effects can be appreciable, especially in tall forest trees. Most theories of vascular system evolution are based on the view that tracheary elements highly modified to reduce flow resistance are phylogenetically advanced. Carlquist (1975) attempted to interpret available data on xylem evolution from an ecological and physiological perspective. Unfortunately, few intraspecific studies of geographic variation in xylem anatomy of woody plants have been reported. Longer tracheid length has been associated with lower latitudinal origin in *Picea sitchensis* (Dinwoodie, 1963), *Acer negundo* (Winstead, 1978) and *Liquidambar styraciflua* (Randel and Winstead, 1976), but correlations with aridity of the seed source are not apparent. In both angiosperm species most of the water transport would occur not in tracheids, but in vessel elements; hence, the effect of drought-related selective pressure on tracheid dimensions would presumably be minor in the two angiosperms.

It seems possible that species of wide climatic range may be genetically adapted to specific habitats by specialized alterations in xylem architecture. Drawing on concepts presented by Carlquist (1975), the following discussion will suggest likely adaptations based on analysis of the physical aspects of water movement in woody plant stems.

Flow in the lumina of functional axial xylem elements is laminar (Jarvis, 1975) and thus obeys the Hagen-Poiseuille Law for flow in a capillary. Hence, the volume flow q through a length, l, of a xylem element lumen under a hydrostatic pressure difference, ΔP, is

$$q = \frac{\pi r^4}{8 \eta l} \Delta P$$

where r is the radius of the lumen and η is the viscosity of the xylem sap which approximates that of water. The resistance to flow, R, in the element lumen is:

$$R = \Delta P / q$$

In angiosperms that have vessels with simple perforation plates this resistance accounts for much of the flow resistance in the xylem elements. In the tracheids of conifers, an additional resistance associated with the margo of bordered pits must be included (Jarvis, 1975). This resistance can contribute up to 60% of the total resistance to water flow in conifers.

It can be seen that the lumen resistance to sap flow in a single elemtnt (vessel or tracheid) decreases greatly (i.e., with the inverse of the 4th power of the radius) as the radius increases. Therefore, selection for tracheary elements of the largest possible diameter would appear to offer greatest advantage to the plant in reducing flow resistance. However, the selective value of a wide xylem conducting element is positive only when negative xylem tensions are not excessively high (Carlquist, 1975). As the radii of xylem elements increase, the water columns within become increasingly subject to cavitation under negative tension. Thus, low flow resistance, transpirational requirements, and relative immunity to water column cavitation must be balanced. In habitats where xylem tensions will likely be great despite efficient plant regulation of water loss, e.g., areas having very dry soil and/or extreme evaporative demand, one might find populations of a species that possess a genetically based tendency for narrow xylem elements with perhaps thicker cell walls to prevent cell collapse under high negative tension. Since the width and length of xylem elements are highly correlated, it might also be expected that xeric populations would have shorter xylem elements. Total cross sectional area of the elements might be similar to or even greater than that of more mesic populations because of a greater number of xylem elements. In more mesic habitats, characterized by ample soil moisture and moderate evaporative demand, wider xylem elements and low flow resistance might be favored, cavitation being less of a potential problem because of lower xylem tensions.

In *Quercus rubra,* percentage of cross-sectional area of vessels in stems increased and vessel length decreased as annual potential evaporation of the habitat increased (Maeglin, 1976), in agreement with the above suggestions. Unfortunately, vessel diameter was not reported and trees growing in widely differing habitats were compared instead of individuals grown in a common environment. Hence, genetic effects could not be separated from possible environmental effects on xylem anatomy.

A study of xylem anatomy of the west African tree *Terminalia superba* (Longman *et al.,* 1979) has shown that intraspecific variation in vessel diameter occurs in woody plants, although no geographic or environmental correlations were attempted. Similarly, varietal differences in diameters of metaxylem vessels have been observed in wheat (Taylor and Klepper, 1978). Examining intraspecific variation in xylem architecture could provide information on the process of modifying xylem element morphology in response to specific selective pressures.

3. Control of Transpiration

Woody plants commonly restrict transpiration and thereby avoid serious injury from water deficits. The numerous adaptations by which this task is accomplished reflect the complex nature of energy and matter exchange in the plant–atmosphere system. In this section evidence is presented for genetic variation in selected morphological and physiological characters that directly or indirectly reduce transpiration.

a. Variation in Shoot Growth. It has been observed that, in a common, low-stress environment, plants originating from seed collected from more arid portions of a species' range or from xeric habitats have slower rates of shoot growth than those from wetter areas or more mesic habitats. This trend has been reported for *Acer rubrum* (Townsend and Roberts, 1973), *Fraxinus pennsylvanica* (Meuli and Shirley, 1937), *Abies balsamea* (Lester, 1970; Lowe *et al.*, 1977), *Pinus taeda* (Wells and Wakeley, 1966; Woesner, 1972a,b), *Pseudotsuga menziesii* (Griffin and Ching, 1977), *Eucalyptus viminalis* (Ladiges and Ashton, 1974), *Pinus monticola* (Squillace and Bingham, 1958), *Pinus sylvestris* (Wright and Bull, 1963), and *Pinus caribaea* (Venator, 1976). Presumably, slower shoot growth and less leaf surface area exposed to the atmosphere reduce the danger of lethal desiccation by extreme atmospheric and soil drought. The above data support the contention of Grime (1979), who held that a slow rate of growth is a common evolutionary response of plants to habitats where environmental stress is frequent. In experiments cited above, corresponding trends in root system development were not measured but from Section II,C, it can be postulated that the size of the root system is not decreased as much as shoot size. The genetically based tendency of populations from xeric habitats to exhibit restricted shoot growth is of practical significance because it suggests that, for environments prone to water stress, conventional plant breeding and management for maximum shoot growth may not be possible or desirable (Zobel and Kellison, 1978).

b. Variation in Leaf Size and Morphology. Several species of woody plants have been shown to have genetic, intraspecific variations in leaf size and shape. Leaves of *Populus deltoides* clones from mesic eastern United States provenances were larger than those of provenances from the more xeric western portions of the species range (Ying and Bagley, 1976) (Fig. 3). Smaller leaves have also been associated with more xeric origin for *Pinus caribaea* (Venator, 1976), *Pinus sylvestris* (Wright and Bull, 1963; Ruby, 1967), *Betula alleghaniensis* (phenotypically) (Dancik and Barnes, 1975), and *Cercis canadensis* (Donselman, 1976). Seedlings of *Eucalyptus viminalis* obtained from populations growing in xeric habitats produced

Fig. 3. Leaf size of *Populus deltoides* from xeric western (left) to mesic eastern (right) United States provenances. (From Ying and Bagley, 1976.)

narrower leaves than those of seedlings derived from trees of more mesic habitats (Ladiges, 1974). Similarly, Tobiessen (1971) found that the ratio of leaf length to leaf width of the shrub *Isomeris arborea* was greater in seedlings of desert seed sources compared with seedlings grown from seed taken from coastal populations. Phenotypic leaf dissection of populations of *Quercus alba* growing in xeric environments in Iowa, Illinois, and Oklahoma was more pronounced than that of populations growing in mesic environments in Louisiana, Texas, and Florida (Baranski, 1975).

Thus, there appears to be a general reduction in leaf size and an increase in leaf dissection or leaf narrowing in xeric sources of several species. This may be explained by a corresponding effect of leaf size reduction and leaf narrowing and dissection on energy and matter exchange. Small or highly dissected leaves tend to have a thin boundary layer and hence a low boundary layer resistance to sensible heat and water vapor transfer to the atmosphere. The net effect of a low boundary layer resistance on transpiration rate per unit leaf area is difficult to predict. A lower boundary layer resistance reduces the difference between leaf and air temperature by facilitating sensible heat exchange. If temperature of the leaf exceeds air temperature, smaller leaves with a lower boundary layer resistance will be cooler, and hence will have a lower vapor pressure difference between leaf and air, which will tend to lower transpiration. However, the lower boundary layer resistance will also tend to reduce total leaf resistance to water vapor loss, thus tending to increase transpiration.

Whether a specific leaf size or shape reduces transpiration per unit leaf area depends on the leaf environment (i.e., net radiation, leaf and air temperature, vapor pressure deficit, wind speed) and resistance of the leaf to water vapor diffusion (Fig. 4). As environmental factors become decidedly xeric (i.e., high radiation, low relative humidity), transpiration rate decreases with leaf dimensions. At an absorbed radiation level of 1.4 cal cm^{-2} min.$^{-1}$ and 20% relative humidity, a decrease in leaf length from 20 to 5 cm reduces transpiration by 31% with no changes in other leaf conditions. At 20°C, other environmental factors being the same, a 50% reduction in transpiration obtains with a similar decrease in leaf dimension.

Two other advantages of small and highly dissected leaves are worthy of mention. First, because stomatal closure induced by water deficits may frequently restrict transpirational cooling in arid habitats, small and dis-

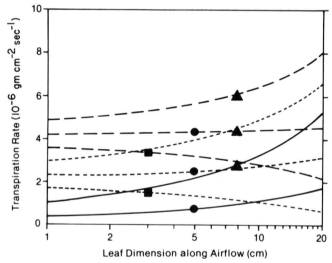

Fig. 4. Variation in transpiration rate with leaf dimension at several values of absorbed radiation and relative humidity. ▲, ●, ■ = 1.4, 1.0, 0.6 cal cm^{-2} min^{-1} absorbed radiation, respectively; — —, -----, ——— = 20%, 60%, 100% relative humidity, respectively. Other environmental and plant factor values: dimension across airflow, 5 cm; leaf diffusion resistance, 10 sec cm^{-1}; air temperature, 40°C; airspeed, 1 m sec^{-1}. (From Gates and Papian, 1971.)

sected leaves aid in avoidance of injuriously high leaf temperatures by being more efficient heat exchangers. Second, the distance that water must travel from main xylem elements to the most remote leaf laminar tissue is reduced in smaller or more dissected leaves.

c. Variation in Leaf Abscission. Although leaf shedding is considered a common desiccation-avoiding adaptation, especially in desert plants (Kozlowski, 1974), there is little information pertaining to variation in leaf abscission responses within a woody species. The only evidence of which the author is aware is that of Tobiessen (1971), who observed that desert sources of *Isomeris arborea* had a slightly greater tendency to abscind older leaves after prolonged drought than did sources from a more mesic coastal environment.

d. Variation in Cuticle. Effects of cuticle on the water balance of plants involve restricting water vapor loss by a hydrophobic barrier and altering spectral properties and energy balance of leaves by the physical structure of the cuticle, especially epicuticular wax structure (Thomas and Barber, 1974). Cuticle-related changes in spectral properties of leaves have indirect effects on water relations, being a result of alterations in leaf temperature. Hence, it is difficult to identify particularly effective light-reflecting wax formations as adaptations to avoid either heat injury or excessive transpiration because both benefits accrue simultaneously.

Differences in cuticular resistance have been observed in detached leaves of different clones of tea (*Camellia sinensis*) (Nagarajah, 1979). Of three clones commonly grown in Sri Lanka, two (2023 and 2025) exhibited similar and more rapid cuticular water loss than did clone DN (Fig. 5). Clone 2023 was susceptible to drought, whereas DN was more drought tolerant. The author attributed at least part of the greater drought tolerance of clone DN to a larger cuticular resistance to water loss. Clone 2025, with a cuticular water loss rate comparable to that of drought-susceptible clone 2023, was more drought tolerant because its deeper root system afforded greater water absorption capacity.

In another series of investigations, it was shown that seedlings from xeric seed sources of *Pinus taeda* had a thicker cutinized epidermis than did seedlings from mesic seed sources (Knauf and Bilan, 1974, 1977) (Table III). Thickly cutinized epidermis was pronounced in cotyledons and primary needles of 2-month-old seedlings, and in needles of 2-yr-old trees but amounts of cutinized epidermis were similar in needles of 16-yr-old trees of both sources. The authors suggested that the survival advantages of a highly cutinized epidermis would be greatest when plants were young and lacked extensive root systems. This would account for

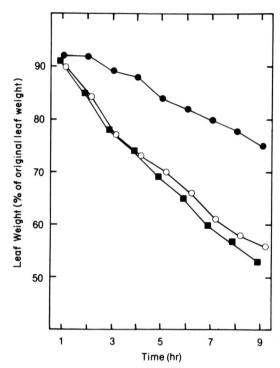

Fig. 5. Relationship between water loss of mature leaves of three clones of tea and time when detached leaves were placed in a forced-draft oven at 33°C. DN, ●; 2025, ○; 2023, ■. (After Nagarajah, 1979.)

the lack of variation in cutinized epidermis in well-established, older trees. It should be cautioned that cuticular wax content and not cuticle thickness determines the resistance of the cuticle to water loss (Schönherr, 1976). If wax contents per unit of cuticle volume were similar in cutinized epidermis of both seed sources, greater cuticular resistance to water loss would presumably be associated with seedlings of the xeric seed source.

Effects of crystalline wax projecting from the leaf surface on boundary layer thickness are small (Reicosky and Hanover, 1976). However, increased leaf reflectance by structured waxes would be an adaptation of great value in arid environments where radiation load and air temperature are high and soil moisture supply is low, thereby decreasing the importance of latent heat exchange by transpiration as a means of dissipating absorbed radiant energy. Variation in leaf surface waxes has been observed among ecotypes of *Eucalyptus urnigera* (Hall *et al.*, 1965), among provenances of *Eucalyptus camaldulensis* (Kaplan, 1974), and among var-

TABLE III

VALUES FOR LEAF MORPHOLOGICAL FEATURES OF 2-MONTH-OLD, 2-YR-OLD, AND 16-YR-OLD *Pinus taeda* PLANTS FROM XERIC AND MESIC SEED SOURCES[a]

Morphological feature	2-month-old seedlings				2-yr-old trees		16-yr-old trees	
	Cotyledons		Primary needles					
	Mesic source	Xeric source	Mesic source	Xeric source	Mesic source	Xeric source	Mesic source	Xeric source
Stomata mm^{-2} leaf surface	30.5	24.9[b]	107.4	101.9	85.5	72.3[b]	113.3	123.5
Stomatal rows (number)								
Abaxial surface	—	—	—	—	8.7	6.8[c]	14.0	11.8
Adaxial surface	—	—	—	—	9.1	6.6[c]	14.5	13.1
Distance between stomatal rows (μm)								
Abaxial surface	—	—	—	—	100.2	118.2	112.3	94.2
Adaxial surface	—	—	—	—	87.6	119.8[c]	84.4	94.0
Depth of stomatal crypt (μm)								
Mean	11.4	11.7	11.9	13.7[c]	—	—	—	—
Abaxial surface	—	—	—	—	10.8	11.0	13.6	12.2
Adaxial surface	—	—	—	—	10.4	11.8[b]	12.4	11.0[c]
Cutinized epidermis thickness (μm)								
Mean	2.3	2.9[c]	2.4	2.8[c]	—	—	—	—
Abaxial surface	—	—	—	—	5.0	5.8[c]	7.2	7.6
Adaxial surface	—	—	—	—	4.8	4.9	7.0	7.6

[a] Adapted from Knauf and Bilan (1974, 1977).
[b] Values of mesic and xeric seed sources significantly different at 5% level.
[c] Values of mesic and xeric seed sources significantly different at 1% level.

ieties of *Picea pungens* (Reicosky and Hanover, 1976). Glaucous types of
these species absorbed less incident radiation than did nonglaucous types
(Thomas and Barber, 1974; Kaplan, 1974; Reicosky and Hanover, 1978).
Lower leaf temperatures have been associated with glaucous-leaved
ecotypes of irrigated and unirrigated *Eucalyptus camaldulensis* (Karschon
and Pinchas, 1969).

e. Variations in Stomatal Anatomy and Control. Because the stomata regulate the bulk of water loss from leaves, it might be expected that stomatal
adaptations within species that affect both the quantity and pattern of
water loss would be fairly common. In fact, this appears to be the situation. Such stomatal adaptations involve reductions in the number and/or
size of stomata, alterations in leaf morphology so that stomata are re-
cessed into crypts so as to increase diffusion resistance to vapor transfer,
and alterations in stomatal response to environment or internal water
stress resulting in water conservation.

   In *Pseudotsuga menziesii* (Zavitkovski and Ferrell, 1968, 1970) and
Acer rubrum (Townsend and Roberts, 1973), seedlings obtained from
mesic sites had higher rates of transpiration than did seedlings from xeric
sites when plants were subjected to environmental conditions favorable
for transpiration. Thus, plants of these two species from xeric sources
appear to have a more restricted potential for gas exchange than do plants
from mesic sources. No data comparing stomatal characteristics of the
different ecotypes were presented, and, hence, differences in transpiration
could not be related to stomatal anatomy.

   In *Pinus taeda,* seedlings from xeric sources possessed fewer stomata
per unit leaf surface area than did seedlings from mesic sources because of
larger distances between rows of stomata (Thames, 1963; Van Buijtenen *et
al.*, 1976; Knauf and Bilan, 1974, 1977) (Table III). Additionally, seedlings
from xeric sources possessed fewer stomata per unit of row length
(Thames, 1963) and were more deeply submerged in stomatal crypts than
were those of seedlings from mesic sources (Van Buijtenen *et al.*, 1976;
Knauf and Bilan, 1977) (Table III). Variation in stomatal dimensions and
stomatal frequency has been observed among *Populus* clones (Siwecki and
Kozlowski, 1973; Ceulemans *et al.*, 1978a; Pallardy and Kozlowski,
1979b) (Table IV), provenances of *Juglans nigra* (Carpenter, 1974) and
among sources of *Pinus palustris* (Snyder *et al.*, 1977). Narrow-sense
heritability for stomatal frequency in *Pinus palustris* was fairly high
($h^2 = 0.56$).

   Jackson *et al.* (1973) reported large differences in transpiration rates
among several pot-grown clones of *Pinus radiata*. Transpiration rates of
two *Pinus radiata* clones differed by about half in a controlled environment

TABLE IV

MEAN ABAXIAL AND ADAXIAL STOMATAL FREQUENCY AND LENGTH OF *Populus* CLONES[a]

Populus species or hybrid	Abaxial frequency (mm^{-2})	Adaxial frequency (mm^{-2})	Abaxial length (μm)	Adaxial length (μm)
4877 *alba*	263.7^{i}*	0^{b}	19.5^{b}	
4878 × *euramericana* "I-214"	159.7^{fgh}	90.4^{j}	24.2^{cde}	24.6^{def}
4879 × *euramericana*	149.6^{efgh}	102.4^{j}	23.3^{cde}	21.3^{b}
5322 × *euramericana* 'Jacometti 788'	165.9^{gh}	89.1^{ij}	23.2^{cd}	23.5^{cd}
5323 × *euramericana* 'Canada Blanc'	140.8^{efg}	53.9^{efg}	23.5^{cde}	21.4^{b}
5326 × *euramericana* 'Eugenii'	125.7^{cde}	65.3^{gh}	21.5^{bc}	21.2^{b}
5328 × *euramericana* 'I-45/51'	88.8^{b}	57.6^{fg}	27.3^{fg}	25.4^{defg}
5377 × *euramericana* 'W-5'	140.5^{defg}	74.9^{hi}	23.9^{cde}	22.4^{bc}
5262 *candicans* × *berolinensis*	176.3^{h}	25.9^{c}	29.9^{gh}	25.3^{defg}
5263 *candicans* × *berolinensis*	167.5^{h}	26.7^{c}	31.3^{h}	27.4^{h}
5331 *betulifolia* × *trichocarpa*	167.2^{h}	30.0^{cd}	32.5^{h}	26.6^{fgh}
5332 *betulifolia* × *trichocarpa*	140.8^{efg}	29.1^{cd}	30.2^{h}	25.2^{defg}
5265 *deltoides* × *trichocarpa*	126.7^{cde}	35.7^{cd}	31.2^{h}	26.3^{efgh}
5266 *deltoides* × *trichocarpa*	123.2^{cde}	57.3^{fg}	30.6^{h}	25.9^{efgh}
5334 *deltoides* × *trichocarpa*	116.0^{bcd}	31.5^{cd}	30.8^{h}	26.5^{efgh}
5264 *deltoides* × *plantierensis*	104.5^{bc}	40.8^{cdef}	29.7^{gh}	26.7^{gh}
5267 *deltoides* × *caudina*	109.1^{bcd}	78.9^{i}	23.6^{cde}	22.1^{bc}
5271 *charkowiensis* × *deltoides*	118.1^{bcd}	46.7^{def}	25.6^{def}	21.8^{bc}
5260 *tristis* × *balsamifera*	130.1^{cdef}	31.2^{cd}	24.7^{def}	24.7^{def}
5258 *Populus* sp. (unknown parentage)	125.1^{cde}	28.5^{c}	26.5^{ef}	24.6^{cde}
5272 *nigra* × *laurifolia*	177.9^{h}	36.8^{cde}	30.0^{gh}	25.6^{efgh}

[a] From Pallardy and Kozlowski (1979b).

* Mean values that do not have the same superscript letter are significantly different at the 5% level.

experiment (Rook and Hobbs, 1976). Additionally, the clone that transpired more rapidly had a much lower survival rate in the field (Bennett and Rook, 1978). Bennett and Rook also studied gas exchange in these two clones, confirmed the trends in transpiration rates noted above, and observed that stomatal resistance in the clone possessing the higher transpiration rate was about half of that of the slowly transpiring clone. No differences in stomatal size, frequency, antechamber depth, or wax occlusion as determined from scanning electron micrographs, or differences in trends of stomatal response to environment were found that could account for the observed variation in stomatal resistance. Possibly, differences in stomatal aperture among clones were similar under all environmental conditions, but this could not be ascertained by the methods used.

More experimentation will be required to establish the exact causes for differing rates of transpiration among plant genotypes. Careful analysis of comparative stomatal anatomy and stomatal movement as they relate to transpiration rates would be a logical starting point; however, examination of mesophyll resistances and cuticular resistance to water loss should also be studied to determine possible causes underlying variation in rates of water loss.

Variation among genotypes in stomatal response to leaf water stress or, indirectly, to soil water stress has been observed in a number of woody species. For example, Nunes (1967) studied drought tolerance in three varieties of cacao (*Theobroma cacao*). The most drought-tolerant varieties (as estimated by resistance to producing drought-induced rolling and yellowing leaves), AV and AA, had transpiration rates that declined much more rapidly as soil moisture was depleted than did the transpiration rate of drought susceptible LA (Fig. 6). Greater transpiration reduction in AV and AA varieties was caused by more rapid and pronounced stomatal closure. *Populus* clones differed in response of leaf resistance to decreasing soil moisture during drying cycles in a constant aerial environment (Ceulemans *et al.*, 1978b). Stomata of a clone of *Populus trichocarpa* closed at much lower soil water potentials than did those of two *Populus trichocarpa* × *P. deltoides* clones and a clone of *P. robusta*. The authors suggested that such effective water loss regulation as shown by the hybrids and *P. robusta* could be related to their superior growth capacity compared with that of *P. trichocarpa*.

In *Pinus taeda*, transpiration rate and stomatal aperture during a drying cycle were reduced in seedlings obtained from xeric, disjunct sources compared with those of seedlings obtained from more mesic sources in east Texas (Bilan *et al.*, 1977) (Table V). Lower transpiration rates at comparable soil moisture contents were also observed in xeric ecotypes compared with those of mesic ecotypes of *Pseudotsuga menziesii* (Fig. 7) (Zavitkovski and Ferrell, 1968, 1970; Unterscheutz *et al.*, 1974).

Genetic variation in stomatal response to atmospheric environmental factors has been reported in certain closely related woody plants. Pallardy (1978) and Pallardy and Kozlowski (1979c) observed distinctive stomatal responses to vapor pressure deficit (VPD) in two *Populus* clones (Fig. 8). At low VPD levels (Fig. 8A), total leaf resistance of clones of *Populus candicans* × *P. berolinensis* and *Populus deltoides* was similar. However, when evaporative demand was high (Fig. 8B), stomatal closure, particularly of adaxial stomata, was much more pronounced in *Populus candicans* × *P. berolinensis*. This accentuated stomatal closure was associated with more rapid recovery from daily minimum ψ_1 in *Populus candicans* × *P. berolinensis,* providing an example of a clear difference in desiccation-avoidance response in these clones. In another experiment, two *Populus* clones exhibited differences in the response of stomatal aperture to two levels of light and VPD (Pallardy and Kozlowski, 1979d)

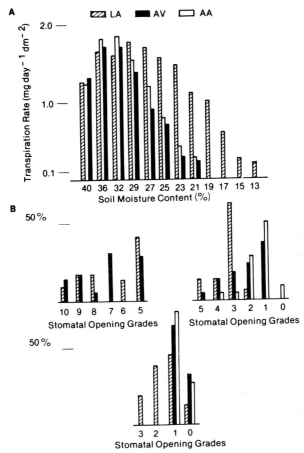

Fig. 6. (A) Comparative transpiration rates of 3 types of *Theobroma cacao* during a soil drying cycle. (B) Relative frequencies of stomatal opening grades at three different soil moisture ranges (left top, 41–30%; right top, 30–24%; bottom, 24–20%). Grade 0 = closed stomata; grade 10 = maximum opening. (From Nunes, 1967.)

(Table VI). Total leaf diffusive resistance of a clone of *Populus candicans × P. berolinensis* was more responsive to changing VPD and less responsive to a change in light intensity than was a clone of *Populus deltoides × P. caudina*. After simple effects of low light and high VPD had been removed, *Populus candicans × P. berolinensis* showed a greater interactive increase in total leaf resistance (r_T) in response to a combination of low light and high VPD than did *Populus deltoides × P. caudina*. *Populus candicans × P. berolinensis*, which exhibited a distinct desiccation-avoidance response in the field and high stomatal sensitivity

TABLE V

PERCENTAGE OF OPEN STOMATA AND TRANSPIRATION RATES OF 1-YR-OLD *Pinus taeda* TREES FROM MESIC AND XERIC SEED SOURCES[a]

Days since watering and seed source	Open stomata (%)	Transpiration rate (μg cm^{-2} $min.^{-1}$)
Day 1		
Mesic source	22.9	10.77
Xeric source	20.7	11.30
Day 3		
Mesic source	16.8[b]	6.78
Xeric source	13.5	7.08
Day 5		
Mesic source	9.4[c]	3.12
Xeric source	5.5	2.92
Day 7		
Mesic source	4.8[c]	1.77[b]
Xeric source	2.3	1.20

[a] From Bilan *et al.* (1977).

[b] Values of mesic and xeric seed sources significantly different at the 5% level.

[c] Values of mesic and xeric seed sources significantly different at the 1% level.

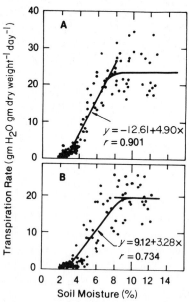

Fig. 7. Effect of decreasing soil moisture on transpiration of *Pseudotsuga menziesii* seedlings from mesic (A) and xeric (B) habitats. Formulas apply to the linear portions of curves between soil moisture contents of 2.94% and 8%. (From Zavitkovski and Ferrell, 1970.)

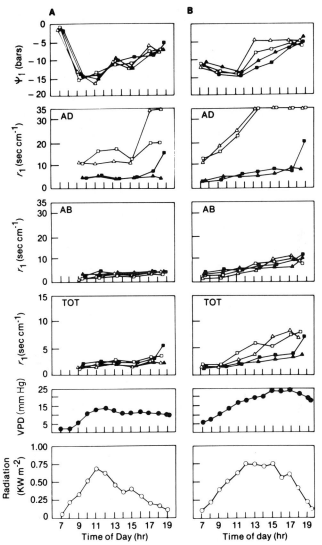

Fig. 8. Diurnal progression of ψ_l and abaxial, adaxial, and total leaf diffusion resistance (r_l) of two plants of a clone of *Populus candicans* × *P. berolinensis* (open symbols) and two plants of a clone of *Populus deltoides* (closed symbols) in northern Wisconsin. (A) Day of relatively low VPD. (B) Day of high VPD. (From Pallardy, 1978.)

to evaporative demand, has a parent that has been described as drought-tolerant (*Populus berolinensis;* Bugala, 1973).

Although data reporting genetic variation in stomatal response to soil moisture and atmospheric factors are fragmentary, they suggest that if

TABLE VI

RESPONSE OF ADAXIAL, ABAXIAL, AND TOTAL LEAF RESISTANCE (r_{AB}, r_{AD}, r_T, sec cm^{-1}) OF TWO *Populus* CLONES TO TWO LEVELS OF PHOTON FLUX DENSITY AND VPD[a]

Photon flux density (μE m^{-2} sec^{-1})	VPD (kPa)	*Populus candicans* × *P. berolinensis*			*Populus deltoides* × *P. caudina*		
		r_{AB}	r_{AD}	r_T	r_{AB}	r_{AD}	r_T
520	0.78	1.68 ± 0.04	3.71 ± 0.10	1.14 ± 0.02	2.27 ± 0.06	3.28 ± 0.13	1.32 ± 0.04
520	3.16	2.81 ± 0.14	10.97 ± 0.52	2.19 ± 0.09	3.31 ± 0.09	5.79 ± 0.25	2.05 ± 0.05
65	0.78	3.15 ± 0.22	8.92 ± 0.89	2.22 ± 0.14	7.20 ± 0.67	17.13 ± 2.41	4.68 ± 0.48
65	3.16	8.04 ± 0.86	51.08 ± 4.09	6.40 ± 0.61	13.02 ± 0.92	29.42 ± 2.88	8.27 ± 0.54

[a]From Pallardy and Kozlowski (1979d). Mean values are shown ± 1 standard error.

plant adaptations for drought tolerance can be correlated with field performance under drought conditions, a groundwork for directed breeding of woody plants for drought tolerance will have been established. Since stomatal response to environmental factors has been shown to be heritable (Roark and Quisenberry, 1977a,b), breeding for certain stomatal response patterns appears feasible.

Genetic differences in the ratio between net photosynthetic carbon gain and water loss by transpiration (i.e., water use efficiency, WUE) have been observed in woody plants. Eickmeier *et al.* (1975) observed that WUE was greater in seedlings of xeric southwest Wisconsin populations of *Tsuga canadensis* compared with WUE of more mesic northeast Wisconsin populations. Additionally, WUE varied among clones of *Hevea brasiliensis* (Samsuddin and Impens, 1978) and among populations of *Populus deltoides* (Drew and Bazzaz, 1978). In a growth situation where water availability limits plant productivity and economic considerations prohibit intensive cultural operations, such as irrigation, selection of planting stock with high WUE may produce acceptable yields with minimal economic investment. Of course WUE measurements under typical field conditions must determine the superiority of certain plant types, and superior WUE must be accompanied by survival ability and growth rates that are acceptable (but not necessarily maximal).

D. DIFFERENCES IN DESSICATION TOLERANCE

The physiological basis of the capacity of certain plants to tolerate water loss sufficient to kill others remains unclear. Two responses conferring desiccation tolerance have been postulated. First is a physical structure where, because of a high cell wall modulus of elasticity, only a moderate decrease in relative water content (RWC) occurs with the development of very low water potentials (Tyree, 1976). It has also been proposed that the presence of large quantities of water-binding hemicelluloses and pectic substances in the cell wall will lower the matric potential of water, thereby resulting in lower ψ_l at a given value of RWC (Ladiges, 1975). The capacity of a plant to exhibit low water potentials at high RWC allows a steep potential gradient between leaf and soil, promoting increased absorption of soil water, presumably without severe effects of low ψ_l on plant metabolism. The advantage in water absorption from the soil by this mechanism decreases as soil texture becomes coarser (Tyree, 1976).

Another type of desiccation-tolerance is protoplasmic in nature. Several desiccation tolerance responses are possible, including osmotic adjustment of the solute potential so that critical turgor loss is avoided and stomatal opening maintained (Kramer and Kozlowski, 1979), and various responses by which cell organelles and other subcellular constituents, particularly membrane systems, are protected or repaired during severe desiccation and subsequent rehydration (Bewley, 1979). Although several studies have reported apparent genetic variation in desiccation tolerance in woody plants, this area of plant water relations is in such a state of flux that the following discussion is necessarily somewhat conjectural.

Seedlings of populations of *Eucalyptus viminalis* from dry Australian habitats had higher RWC at given ψ_l than did seedlings of populations from mesic habitats (Ladiges, 1975). Xeric seedlings also had a higher content of cell wall water, despite similar total amounts of wall material, suggesting that cell walls of the xeric seedlings had higher contents of water-binding macromolecules. Xeric populations also had higher bulk moduli of elasticity. These properties evidently contributed to the observed differences between populations in the ψ_l–RWC relationship. Higher RWC at low water potentials has also been reported in more xeric ecotypes of *Eucalyptus camaldulensis* (Kaplan, 1974).

Inland and coastal sources of *Pseudotsuga menziesii* differed in the moisture content to which needles could be desiccated without being killed (Ferrell and Woodard, 1966; Pharis and Ferrell, 1966). The inland sources, representing xeric environments, generally required lower moisture contents to kill needles than did the mesic coastal sources, although certain sources clearly did not fit into this simple dichotomy. Similarly, moisture contents required to kill needles were slightly lower in xeric sources of *Pinus taeda* than in more mesic sources (Van Buijtenen *et al.*, 1976). Possibly, the observed variations in desiccation tolerance were attributable to protoplasmic differences, although none were identified.

Although osmotic adjustment has been observed in woody plants (e.g., *Quercus robur,* Osonubi and Davies, 1978; *Malus domestica,* Lakso, 1979), few data are available documenting differences in osmoregulation within woody species similar to those seen in crop plants such as wheat (Campbell *et al.*, 1979) and sorghum (Stout and Simpson, 1978). There is some evidence of variation among several *Populus* clones in seasonal patterns of leaf osmotic potential at full hydration and at incipient plasmolysis and of variation in bulk modulus of elasticity (Tyree *et al.*, 1978).

As noted above, the uncertainty as to the basis of physiological desiccation tolerance complicates interpretations of data suggesting differences in protoplasmic response to drought. The stability of chlorophyll under heat treatment of seedlings of *Eucalyptus camaldulensis* from several Australian provenances was positively related to the annual precipitation at the seed source and negatively related to percentage of survival of seedlings grown in a nursery in India (Kaul and Roy, 1967). Kaloyereas

(1958) reported that field observations of drought tolerance in *Pinus taeda* sources were correlated with several plant parameters, including percentage of bound water, chlorophyll stability under heat treatment, and quantity of reduced sulfhydryl groups in leaf extracts (Table VII). Heat stability of chlorophyll correlated most closely with field evaluations of drought tolerance and the relationship was stated to be potentially useful in predicting drought tolerance, although admittedly empirical in nature. There is some support for functional roles of bound water and sulfhydryl groups in protoplasmic desiccation tolerance (Parker, 1968, 1969; Levitt, 1972; Tomati and Galli, 1979).

Progress in analyzing genotypic variation in desiccation tolerance, especially protoplasmic variation, awaits advances in our understanding of the physiological basis of desiccation tolerance itself. The sparse evidence suggests that intraspecific differences in desiccation tolerance may not be uncommon. In future studies, investigators might attempt to correlate desiccation tolerance with comparative response to drought of cell organelle integrity and biochemical processes associated with preservation and reconstitution of membranes (Bewley, 1979). These aspects of desiccation tolerance have been receiving attention recently, and results of preliminary experiments comparing such subcellular aspects of tolerant and nontolerant plants have been encouraging (Bewley, 1979).

TABLE VII

RELATION BETWEEN DROUGHT RESISTANCE, BOUND WATER, EXTRACTABLE-SH GROUPS, AND CHLOROPHYLL STABILITY UNDER HEAT TREATMENT FOR SEVERAL STRAINS OF *Pinus taeda*[a]

Pine strain	Relative drought resistance[b]	Bound water (%)	Extractable –SH groups[c]	Chlorophyll stability index (CSI)[d]
BA-5	1	5.83	93	−16
FA-2	2	6.07	81	− 9
AN-3	3	5.11	79	− 8
BA-2	4	5.49	74	− 4
LA-1	5	5.32	68	+ 2
LA-2	6	4.19	70	+ 4
TY-4	7	4.20	65	+ 8
NC	8	3.97	66	+ 8

[a] From Kaloyereas (1958).

[b] Determined from field observations.

[c] Determined by boiling 2 gm of leaves in 25 ml water for 30 min., and subsequent application of nitro prusside test; values shown are color readings at 520 nm.

[d] Departure from average chlorophyll stability index (CSI), determined by subtracting light absorption of chlorophyll extracts of needles heated at 56°C for 30 min. from that of chlorophyll extracts of unheated needles.

E. Multiple Adaptations

Although drought tolerance adaptations have been discussed individually, it is obvious that with some species variations in drought tolerance are the result of several adaptations. This multiplicity reflects selective pressures affecting certain populations of a species such that a single, individual response in each population results or, perhaps more commonly, a multiple adaptive response of a population to any drought-related selective pressure.

The relatively extensive data on variation in drought-tolerance adaptations of *Pinus taeda* and *Pseudotsuga menziesii* illustrate the variety and complexity of drought responses. In xeric sources of *Pinus taeda* adaptations relating to root growth and root system morphology, shoot growth, and cuticular and stomatal control of transpiration have been related to the arid habitats in which these plants exist. Similarly, variations in desiccation avoidance and desiccation tolerance, transpiration under favorable environmental conditions, stomatal control of transpiration and root–shoot ratio in *Pseudotsuga menziesii* have been related to aridity of habitat. In the latter species, not all the listed responses were detected in every population.

III. APPLICATIONS

A. Planting Recommendations

Although horticulturists and foresters have used empirical drought tolerance evaluations gleaned from field experience, efforts to identify specific adaptive responses conferring drought tolerance to certain varieties of woody plants would be of benefit in making decisions on site suitability. For example, it would be inadvisable to plant a drought-tolerant tree variety in which drought tolerance is known to be associated with capacity to root deeply and absorb water from deep layers of the soil on a site possessing shallow soil or a restrictive soil layer near the surface. In this situation, a variety that would efficiently restrict transpiration and thereby would be able to conserve soil moisture would most likely have a much higher probability of survival. Since planting expenses are often one of the most costly phases of tree crop production, knowledge leading to improved performance of properly selected varieties would justify the research effort required to refine such knowledge of drought tolerance.

B. Breeding Woody Plants for Drought Tolerance and Efficient Water Use

Although directed breeding attempts to improve drought tolerance have been undertaken in only a few crop species and in no woody plants (Wright and Jordan, 1970; Hurd, 1976), numerous investigators have emphasized a need for research on improving drought tolerance by breeding (Boyer and McPherson, 1975; Begg and Turner, 1976; Hurd, 1976; Tyree, 1976; Taylor and Klepper, 1978; Yevjevich *et al.*, 1978). Factors contributing to the current paucity of breeding research include uncertainty as to what constitutes drought tolerance, its measurement, and the nature of physiological mechanisms (and hence traits) by which it functions. For conventional breeding programs, there must be available basic information on desirable traits that are involved in tolerance of drought. Once identified, such traits must be established to be sufficiently heritable to make breeding attempts worthwhile. It is beyond the scope of this chapter to discuss the techniques of plant breeding for drought tolerance. For an extensive discussion of the subject, the reader is referred to Hurd (1976).

Several observations have been offered concerning potential breeding strategies and consequent effects on plant performance. Desiccation avoidance adaptations are more commonly found in higher plants than are desiccation tolerance responses (Levitt, 1972; Kramer and Kozlowski, 1979), and hence would appear to be easier to identify for exploitation in breeding programs. However, desiccation-avoidance adaptations generally, but not invariably, decrease drought effects by restricting gas exchange or by allocating more photosynthate for root system development, total above-ground yield decreasing as a result. In some cases, however (e.g., Hurd, 1975; Cannell *et al.*, 1978), growth of shoots or fruits under drought has correlated in a positive manner with greater root system development.

Tyree (1976) suggested that to increase wood yield, trees might be bred for desiccation-tolerance characteristics such as a high bulk modulus of elasticity in leaves and low osmotic potential at incipient plasmolysis, and for desiccation-avoidance characteristics, i.e., a thick cuticle and efficient stomatal closure (albeit at a very low ψ_l). This nexus of characteristics would theoretically allow relatively high RWC at low ψ_l, promote increased soil water absorption by a steepened water potential gradient, and maintain stomatal opening and high photosynthetic rates until ψ_l values were very low. At a low, critical ψ_l efficient stomatal closure and a thick cuticle would protect the leaves from lethal desiccation. Boyer and McPherson (1975) also supported the acquisition of desiccation tolerance as a breeding objective for crop plants. They suggested that plants be screened for (1) capacity to maintain high rates of cell elongation as seedlings under regimes of limiting moisture, and (2) resistance to leaf senescence in intermediate growth stages. This strategy would produce varieties that would maintain leaf growth under mild water stress and would not be adversely affected by loss of photosynthetic surface as a result of leaf senescence and abscission in response to drought.

Taylor and Klepper (1978) presented several breeding strategies involving root system adaptations. Citing Zobel (1975), they estimated that root system morphology might be easily modified by breeding since, as an example, up to 30% of the genes of *Lycopersicon esculentum* affect root system characters. Breeding for large diameter xylem elements in the vascular system might be a valuable objective since evidence has been obtained that dimensions of vascular elements vary among varieties. Low axial resistance would promote generally higher ψ_l values, a situation that could increase plant productivity. From a more conservative perspective, breeding for high axial resistance might restrict transpiring leaf area development and reduce the likelihood of cavitation of water columns in xylem elements assuring a lower, but more reliable, yield under drought.

Generally, plant breeding programs that have increased yields and efficiency of carbon assimilation have resulted in higher water use efficiencies in new plant varieties (Begg and Turner, 1976). For example, hybrid maize varieties increased WUE by approximately 20% (Pendleton, 1966). Such benefits would also probably accrue to woody plants under similar breeding programs. Other breeding strategies for increasing WUE of crop plants have been directed toward obtaining varieties that mature rapidly or that will tolerate low temperatures such that the plants will not be subjected to periods or seasons of extreme evaporative demand, a situation that increases the amount of water lost for each molecule of CO_2 fixed (Begg and Turner, 1976). Unfortunately, life cycles of woody perennials are rarely amenable to such alterations.

It is possible, however, that woody plants might be bred for stomatal sensitivity to environmental factors, particularly VPD. Such a dynamic response has been shown to increase WUE in the field (Schulze *et al.,* 1975). Breeding for stomatal sensitivity to VPD in woody plants could also be combined with breeding for increased leaf area. Stomatal sensitivity to VPD represents a homeostatic mechanism that could protect a larger leaf area from lethal desiccation without the need of a large investment of photosynthate in root absorptive surface. Although gas exchange would be restricted under high evaporative demand, the larger photosynthetic area performing under most conditions could more than offset periods of restricted photosynthesis caused by stomatal closure induced by high evaporative demand, increased total growth obtaining.

Because of the characteristically lengthy nonflowering juvenile period of woody plants, conventional breeding programs will likely progress more slowly than those for herbaceous annual crops. However, advances in cell culturing and plant regeneration techniques (Cheng, 1978) may allow more rapid progress in developing drought tolerant woody plants. Although culture and regeneration methods for woody plants are at a more primitive stage of development than those for herbaceous plants, the potential utility of these techniques for producing, rapidly and economically, new varieties of woody plants by asexual means warrants their consideration.

Nabors (1976) suggested steps for use in identifying and propagating mutant plants by tissue culture or cell suspension culture: (1) production of callus or suspension cultures; (2) selection of desired mutants; and (3) regeneration of plants from cells or tissues. Successful culturing and regeneration of woody plants have been performed with only a few species (e.g., *Populus* spp.; Winton, 1970, 1971). Suitable selection objectives and procedures are essential for success. Acquisition of a desirable phenotype generally must require alteration of only one allele (expected from natural mutation or induced), and it must be possible to select for the phenotype under culture conditions. For example, with respect to drought tolerance, selection for protoplasmic resistance to desiccation using a medium with a low water potential (imposed with a nonpenetrating osmoticum) would appear possible; however, selection for a large root system or other morphological characters would not appear feasible in tissue culture. Only if higher level morphological mutations manifested innocuous pleiotropic markers at the cellular level could such cell culture selection be accomplished in the latter case. Obstacles to plant improvement by cell and tissue culture methods include, among others, the possibility that the selected plants would be chimeras, the likelihood that many beneficial recessive mutations would be overlooked in the selection process (remedied by culturing haploid tissue) and deleterious pleiotropic mutation effects on the selected cell lines (Durzan and Campbell, 1974; Nabors, 1976).

An even newer array of techniques for producing and genetically manipulating cell protoplasts may offer opportunity for breeding of woody plants (Durzan and Campbell, 1974; Thomas *et al.,* 1979). Protoplasts appear to be genetically malleable because they can be induced to fuse with other protoplasts, to accept foreign genetic material by transduction and transformation, and even to accept extrachromosomal materials such as chloroplasts (Durzan and Campbell, 1974). It is also likely that internal alterations in the protoplast genome are enhanced in protoplast cultures (Thomas *et al.,* 1979). At this time woody plant protoplasts have been isolated, but regeneration has not been obtained (Winton and Huhtinen, 1976). Hence, utilization of such techniques for genetic manipulation of woody plants is contingent on progress in protoplast culture and regeneration methods.

ACKNOWLEDGMENTS

I wish to acknowledge helpful criticisms of the manuscript by Drs. T. M. Hinckley, R. Borchert, and K. D. Lynch. This chapter was written while the author was on the staff of the Department of Forestry, Kansas State University, Manhattan, Kansas and is Kansas Agricultural Experimental Station Contribution 80-165-B.

REFERENCES

Baranski, M. J. (1975). An analysis of variation within white oak. *N. C., Agric. Exp. Stn., Tech. Bull.* **236.**

Begg, J. E., and Turner, N. C. (1976). Crop water deficits. *Adv. Agron.* **28,** 161–217.

Bennett, K. J., and Rook, D. A. (1978). Stomatal and mesophyll resistances in two clones of *Pinus radiata* D. Don. known to differ in transpiration and survival rate. *Aust. J. Plant Physiol.* **5,** 231–238.

Bewley, J. D. (1979). Physiological aspects of desiccation tolerance. *Annu. Rev. Plant Physiol.* **30,** 195–238.

Bey, C. F. (1974). Drought hardiness tests of black walnut seedlings as related to field performance. *Proc. Cent. States For. Tree Improvem. Conf. 9th, 1974* pp. 138–144.

Bilan, M. V. (1964). Acrylic resin tubes for studying root growth in tree seedlings. *For. Sci.* **10,** 461–462.

Bilan, M. V., Hagan, C. T., and Carter, H. B. (1977). Stomatal opening, transpiration, and needle moisture in loblolly pine seedlings from two Texas seed sources. *For. Sci.* **23,** 457–462.

Bilan, M. V., Leach, J. H., and Davies, G. (1978). Root development in loblolly pine (*Pinus taeda* L.) from two Texas seed sources. *In* "Root Form of Planted Trees" (E. van Eerden and J. M. Kinghorn, eds.), Joint Rep. No. 8, pp. 17–22. British Columbia Ministry of Forests/Canadian Forestry Service, Victoria, British Columbia.

Boyer, J. S. (1976). Water deficits and photosynthesis. *In* "Water Deficits and Plant Growth" (T. T. Kozlowski, ed.), Vol. 4, pp. 153–190. Academic Press, New York.

Boyer, J. S., and McPherson, H. G. (1975). Physiology of water deficits in cereal crops. *Adv. Agron.* **27,** 1–23.

Brouwer, R. (1953). Water absorption by the roots of *Vicia faba* plants. I, II. *Proc. K. Ned. Akad. Wet. Ser. C* **56,** 106–115, 129–136.

Brown, J. H. (1969). Variation in roots of greenhouse grown seedlings of different Scotch pine provenances. *Silvae Genet.* **18,** 111–117.

Bugala, W. (1973). Systematics and variability. *In* "The Poplars—*Populus* L." (S. Bialobok, ed.), Nasze Drzewa Lesne Monografie Popularmonaukowe, Vol. 12, pp. 7–108. U. S. Dept. of Commerce, Natl. Tech. Inf. Serv., Springfield, Virginia (transl. from Polish).

Camacho-B., S. E., Hall, A. E., and Kaufmann, M. R. (1974). Efficiency and regulation of water transport in some woody and herbaceous species. *Plant Physiol.* **54,** 169–172.

Campbell, G. S., Papendick, R. T., Rabie, E., and Shajo-Ngowi, A. J. (1979). A comparison of osmotic potential, elastic modulus, and apoplastic water in leaves of dryland winter wheat. *Agron. J.* **71,** 31–36.

Cannell, M. G. R., Bridgewater, F. E., and Greenwood, M. S. (1978). Seedling growth rates, water stress responses and root-shoot relationships related to eight-year volumes among families of *Pinus taeda* L. *Silvae Genet.* **27,** 237–248.

Carlquist, S. (1975). "Ecological Strategies of Xylem Evolution." Univ. of California Press, Berkeley.

Carpenter, S. B. (1974). Variation in leaf morphology in black walnut (*Juglans nigra* L.) and its possible role in photosynthetic efficiency. *Proc. Cent. States For. Tree Improvem. Conf., 8th, 1972* pp. 24–27.

Carr, M. K. V. (1977). Responses of seedling tea bushes and their clones to water stress. *Exp. Agric.* **13,** 317–324.

Ceulemans, I., Impens, I., Lemeur, R., Moermans, R., and Samsuddin, Z. (1978a). Water movement in the soil-poplar-atmosphere system. I. Comparative study of stomatal morphology and anatomy, and the influence of stomatal density and dimensions on the leaf diffusion characteristics in different poplar clones. *Oecol. Plant.* **13,** 1–12.

Ceulemans, I., Impens, I., Lemeur, R., Moermans, R., and Samsuddin, Z. (1978b). Water movement in the soil-poplar-atmosphere system. II. Comparative study of transpiration regulation during water stress situations in four different poplar clones. *Oecol. Plant.* **13,** 139–146.

Cheng, T. Y. (1978). Clonal propagation of woody plant species through tissue culture techniques. *Proc. Int. Plant Propagators' Soc.* **28,** 139–155.

Dalton, F. N., Raats, P. A. C., and Gardner, W. R. (1975). Simultaneous uptake of water and solutes by plant roots. *Agron. J.* **67,** 334–339.

Dancik, B. P., and Barnes, B. V. (1975). Leaf variability in yellow birch (*Betula alleghaniensis*) in relation to environment. *Can. J. For. Res.* **5,** 149–159.

Deneke, F. J. (1974). A red oak provenance trial in Kansas. *Trans. Kans. Acad. Sci.* **77,** 195–199.

Dinwoodie, J. M. (1963). "Variation in Tracheid Length in *Picea sitchensis*," For. Prod. Spec. Rep. 16. Dep. Sci. Ind. Res., HM Stationery Office, London.

Donselman, H. M. (1976). Geographic variation in eastern redbud (*Cercis canadensis* L.). Ph.D. Thesis, Purdue University, West Lafayette, Indiana, *Diss. Abstr. Int. B* **37,** 4852 (1977).

Drew, A. P., and Bazzaz, F. A. (1978). Variation in assimilation among plant parts in three populations of *Populus deltoides. Silvae Genet.* **27,** 189–193.

Durzan, D. J., and Campbell, R. A. (1974). Prospects for the introduction of traits in forest trees by cell and tissue culture. *N. Z. J. For. Sci.* **4,** 261–266.

Eickmeier, W., Adams, M. S., and Lester, D. T. (1975). Two physiological races of *Tsuga canadensis* in Wisconsin. *Can. J. Bot.* **53,** 940–951.

Farmer, R. E. (1970). Variation and inheritance of eastern cottonwood growth and wood properties under two soil moisture regimes. *Silvae Genet.* **19,** 5–8.

Ferrell, W. K., and Woodard, E. S. (1966). Effects of seed origin on drought resistance of Douglas-fir (*Pseudotsuga menziesii* (Mirb.) Franco). *Ecology* **47,** 499–503.

Fischer, R. A., and Turner, N. C. (1978). Plant productivity in arid and semiarid zones. *Annu. Rev. Plant Physiol.* **29,** 277–317.

Fiscus, E. L. (1977). Effects of coupled solute and water flow in plant roots with special reference to Brouwer's experiment. *J. Exp. Bot.* **28,** 71–77.

Fowler, D. P., and Mullin, R. E. (1977). Upland-lowland ecotypes not well developed in black spruce in northern Ontario. *Can. J. For. Res.* **7**, 35–40.

Gansel, C. R. (1967). Do wet- and dry-site ecotypes exist in slash pine? *U.S., For. Serv., Res. Note* **SE-69**, 1–4.

Gates, D. M., and Papian, L. E. (1971). "Atlas of Energy Budgets of Plant Leaves." Academic Press, New York.

Gay, A. P., and Loach, K. (1977). Leaf conductance changes on leafy cuttings of *Cornus* and *Rhododendron* during propagation. *J. Hortic. Sci.* **52**, 509–516.

Gilmore, A. R. (1957). Physical and chemical characteristics of loblolly pine seedlings associated with drought resistance. *Proc. South. Conf. For. Tree Improvem., 4th, 1957*, pp. 34–39.

Goddard, R. E., and Brown, C. L. (1959). "Growth of Drought Resistant Loblolly Pines," Tex. For. Serv. Res. Note 23. Texas For. Service, College Station.

Griffin, A. R., and Ching, K. K. (1977). Geographic variation in Douglas-fir from the coastal ranges of California. *Silvae Genet.* **26**, 149–157.

Grime, J. P. (1979). "Plant Strategies and Vegetation Processes." Wiley, New York.

Habeck, J. R. (1958). White cedar ecotypes in Wisconsin. *Ecology* **39**, 457–463.

Hall, D. M., Matus, A. I., Lamberton, J. A., and Barber, H. N. (1965). Intraspecific variation in wax on leaf surfaces. *Aust. J. Biol. Sci.* **18**, 323–332.

Heiner, T. D., and Lavender, D. P. (1972). Early growth and drought avoidance in Douglas-fir seedlings. *Oreg. State Univ., For. Res. Lab., Res. Pap.* **14**.

Henckel, P. A., and Pronina, N. D. (1969). Anabiosis with desiccation of the poikiloxerophytic flowering plant *Myrothamnus flabellifolia*. *Sov. Plant Physiol. (Engl. Transl.)* **24**, 737–741.

Hermann, R. K., and Lavender, D. P. (1968). Early growth of Douglas-fir from various altitudes and aspects in southern Oregon. *Silvae Genet.* **17**, 143–151.

Hinckley, T. M., Lassoie, J. P., and Running, S. W. (1978). Temporal and spatial variations in the water status of forest trees. *For. Sci. Monogr.* **20**, 1–72.

Hsiao, T. C., Fereres, E., Acevedo, E., and Henderson, D. W. (1976). Water stress and dynamics of growth and yield of crops. *In* "Water and Plant Life: Problems and Modern Approaches" (O. L. Lange, L. Kappen, and E. D. Schulze, eds.), pp. 281–305. Springer-Verlag, Berlin and New York.

Hurd, E. A. (1975). Phenotype and drought tolerance in wheat. *Agric. Meteorol.* **14**, 37–55.

Hurd, E. A. (1976). Plant breeding for drought resistance. *In* "Water Deficits and Plant Growth" (T. T. Kozlowski, ed.), Vol. 4, pp. 317–353. Academic Press, New York.

Jackson, D. S., Gifford, H. H., and Hobbs, I. W. (1973). Daily transpiration rates of radiata pine. *N. Z. J. For. Sci.* **3**, 70–81.

Jarvis, P. G. (1975). Water transfer in plants. *In* "Heat and Mass Transfer in the Biosphere" (D. A. deVries and N. H. Afgan, eds.), Part I, pp. 369–394. Scripta Book Co., Washington, D. C.

Kaloyereas, S. A. (1958). A new method of determining drought resistance. *Plant Physiol.* **33**, 232–233.

Kaplan, J. (1974). The ecology of *Eucalyptus camaldulensis* Dehn. in Israel. *La-Yaaran* **24**, 1–2.

Karschon, R., and Pinchas, L. (1969). Leaf temperatures in ecotypes of *Eucalyptus camaldulensis* Dehn. *Z. Schweiz. Forstver.* **46**, 261–269.

Kaufmann, M. R. (1976). Water transport through plants: current perspectives. *In* "Transport and Transfer Processes in Plants" (I. Wardlaw and J. B. Passioura, eds.), pp. 313–327. Academic Press, New York.

Kaul, R. N., and Roy, R. D. (1967). Chlorophyll stability index, a suitable criterion for rapid screening of tree provenances in arid zones. *Experientia* **23**, 37–38.

Knauf, T. A., and Bilan, M. V. (1974). Needle variation in loblolly pine from mesic and xeric seed sources. *For. Sci.* **20**, 89–90.

Knauf, T. A., and Bilan, M. V. (1977). Cotyledon and primary needle variation in loblolly pine from mesic and xeric seed sources. *For. Sci.* **23**, 33–36.

Kozlowski, T. T. (1974). Extent and significance of shedding of plant parts. *In* "Shedding of Plant Parts" (T. T. Kozlowski, ed.), pp. 1–44. Academic Press, New York.

Kramer, P. J. (1969). "Plant and Soil Water Relationships: A Modern Synthesis." McGraw-Hill, New York.

Kramer, P. J., and Kozlowski, T. T. (1979). "Physiology of Woody Plants." Academic Press, New York.

Kriebel, H. B. (1957). Patterns of variation in sugar maple. *Ohio Agric. Exp. Stn., Res. Bull.* **791**.

Kriebel, H. B. (1963). Selection for drought resistance in sugar maple. *Proc. World Consult. For. Genet. For. Tree Breed., Food Agric. Organ./FORGEN* **63**, Ser. 3/9, No. 2, 1–5.

Kriebel, H. B., and Gabriel, W. J. (1969). Genetics of sugar maple. *U.S. For. Serv., Res. Pap.* **WO-7**, 1–17.

Kriebel, H. B., Bagley, W. T., Deneke, F. J., Funsch, R. W., Roth, P., Jokela, J. J., Merritt, C., Wright, J. W., and Williams, R. D. (1976). Geographic variation in *Quercus rubra* in north central United States plantations. *Silvae Genet.* **25**, 118–122.

Ladiges, P. Y. (1974). Differentiation in some populations of *Eucalyptus viminalis* Labill. in relation to factors affecting seedling establishment. *Aust. J. Bot.* **22**, 471–487.

Ladiges, P. Y. (1975). Some aspects of tissue water relations in three populations of *Eucalyptus viminalis* Labill. *New Phytol.* **75**, 53–62.

Ladiges, P. Y., and Ashton, D. H. (1974). Variation in some central Victorian populations of *Eucalyptus viminalis* Labill. *Aust. J. Bot.* **22**, 81–102.

Lakso, A. N. (1979). Seasonal changes in stomatal response to leaf water potential in apple. *J. Am. Soc. Hortic. Sci.* **104**, 58–60.

Lavender, D. P., and Overton, W. S. (1972). Thermoperiods and soil temperatures as they affect growth and dormancy of Douglas-fir seedlings of different geographic origins. *Oreg. State Univ., For. Res. Lab., Res. Pap.* **13**.

Lester, D. T. (1970). Variation in seedling development of balsam fir associated with seed origin. *Can. J. Bot.* **48**, 1093–1097.

Levitt, J. (1972). "Responses of Plants to Environmental Stresses." Academic Press, New York.

Loach, K. (1977). Leaf water potential and the rooting of cuttings under mist and polythene. *Physiol. Plant.* **40**, 191–197.

Longman, K. A., Leakey, R. B., and Denne, M. P. (1979). Genetic and environmental effects on shoot growth and xylem formation in a tropical tree. *Ann. Bot. (London)* [N.S.] **44**, 377–380.

Lowe, W. J., Hocker, H. W., and McCormack, M. L. (1977). Variation in balsam fir provenances planted in New England. *Can. J. For. Res.* **7**, 63–67.

Maeglin, R. R. (1976). Natural variation of tissue proportions and vessel and fiber length in mature northern red oak. *Silvae Genet.* **25**, 122–126.

Meuli, L. F., and Shirley, H. L. (1937). The effect of seed origin on drought resistance of green ash in the prairie-plains states. *J. For.* **35**, 1060–1062.

Nabors, M. W. (1976). Using spontaneously occurring and induced mutations to obtain agriculturally useful plants. *BioScience* **26**, 761–768.

Nagarajah, S. (1979). Differences in cuticular resistance in relation to transpiration in tea (*Camellia sinensis*). *Physiol. Plant.* **46**, 89–92.

Nunes, M. A. (1967). A comparative study of drought resistance in cacao plants. *Ann. Bot. (London)* [N.S.] **31**, 189–193.

Osonubi, O., and Davies, W. J. (1978). Solute accumulation in leaves and roots of woody plants subjected to water stress. *Oecologia* **32**, 323–332.

Othieno, C. O. (1978a). Supplementary irrigation of young clonal tea in Kenya, I. Survival, growth and yield. *Exp. Agric.* **14**, 229–238.

Othieno, C. O. (1978b). Supplementary irrigation of young clonal tea in Kenya. II. Internal water status. *Exp. Agric.* **14**, 309–316.

Pallardy, S. G. (1978). Water relations of selected *Populus* clones. Ph.D. Thesis, University of Wisconsin, Madison; *Diss. Abstr. Int. B* **40**, 1005–1006 (1979).

Pallardy, S. G., and Kozlowski, T. T. (1979a). Early root and shoot growth of *Populus* clones. *Silvae Genet.* **28**, 153–156.

Pallardy, S. G., and Kozlowski, T. T. (1979b). Frequency and length of stomata of 21 *Populus* clones. *Can J. Bot.* **57**, 2519–2523.

Pallardy, S. G., and Kozlowski, T. T. (1979c). Relationship of leaf diffusion resistance of *Populus* clones to leaf water potential and environment. *Oecologia* **40**, 371–380.

Pallardy, S. G., and Kozlowski, T. T. (1979d). Stomatal response of *Populus* clones to light intensity and vapor pressure deficit. *Plant Physiol.* **64**, 112–114.

Parker, J. (1968). Drought resistance mechanisms. *In* "Water Deficits and Plant Growth" (T. T. Kozlowski, ed.), Vol. 1, pp. 195–234. Academic Press, New York.

Parker, J. (1969). Further studies of drought resistance in woody plants. *Bot. Rev.* **35**, 317–371.

Pendleton, J. W. (1966). Increasing water use efficiency by crop management. *In* "Plant Environment and Efficient Water Use" (W. H. Pierre, D. Kirkham, J. Pisek and R. Shaw, eds.), pp. 236–258. Am. Soc. Agron., Madison, Wisconsin.

Pharis, R. P., and Ferrell, W. K. (1966). Differences in drought resistance between coastal and inland sources of Douglas-fir. *Can. J. Bot.* **44**, 1651–1659.

Randel, W. R., and Winstead, J. E. (1976). Environmental influence on cell and wood characters of *Liquidambar styraciflua*. L. *Bot. Gaz. (Chicago)* **137**, 45–51.

Reicosky, D. A., and Hanover, J. W. (1976). Seasonal changes in leaf surface waxes of *Picea pungens*. *Am. J. Bot.* **63**, 449–456.

Reicosky, D. A., and Hanover, J. W. (1978). Physiological effects of surface waxes. I. Light reflectance for glaucous and nonglaucous *Picea pungens*. *Plant Physiol.* **62**, 101–104.

Roark, B., and Quisenberry, J. E. (1977a). "Evaluation of Cotton Germplasm for Drought Resistance." *In* Proc. of Beltwide Cotton Production Res. Conf., National Cotton Council, Memphis, Tennessee.

Roark, B., and Quisenberry, J. E. (1977b). Environmental and genetic components of stomatal behavior in two genotypes of upland cotton. *Plant Physiol.* **59**, 354–356.

Rook, D. A., and Hobbs, J. F. (1976). Soil temperatures and growth of rooted cuttings of radiata pine. *N. Z. J. For. Sci.* **5**, 296–305.

Ruby, J. L. (1967). The correspondence between genetic, morphological, and climatic variation patterns in Scotch pine. I. Variation in parental characters. *Silvae Genet.* **16**, 50–56.

Samsuddin, Z., and Impens, I. (1978). Water vapor and carbon dioxide diffusion resistances of four *Hevea brasiliensis* clonal seedlings. *Exp. Agric.* **14**, 173–177.

Schönbach, H., Bellmann, E., and Scheumann, W. (1966). Die Jugenwuchslerstung, Dürre- und Frostresistenz verschiedener Provenienzen der japanischen Lärche (*Larix leptolepsis* Gordon). *Silvae Genet.* **15**, 141–147.

Schönherr, J. (1976). Water permeability of isolated cuticular membranes: the effect of cuticular waxes on the diffusion of water. *Planta* **131**, 159–164.

Schultz, R. P., and Wilhite, L. P. (1969). Differential response of slash pine families to drought. *U.S. For. Serv., Res. Note* **SE-104**, 1–3.

Schulze, E. D., Lange, O. L., Evenari, M., Kappen, L., and Buschbom, U. (1975). The role of air humidity and temperature in controlling stromatal resistance of *Prunus armeniaca* L. under desert conditions. III. Effects on water use efficiency. *Oecologia* **19**, 303–314.

Siwecki, R., and Kozlowski, T. T. (1973). Leaf anatomy and water relations of excised leaves of six *Populus* clones. *Arbor. Kornickie* **18**, 84–105.

Snyder, E. B., Dinus, R. J., and Derr, H. J. (1977). Genetics of longleaf pine. *U.S. For. Serv., Res. Pap.* WO-33, 1–24.

Squillace, A. E., and Bingham, R. T. (1958). Localized ecotypic variation in western white pine. *For. Sci.* **4**, 20–34.

Stout, D. G., and Simpson, G. M. (1978). Drought resistance of *Sorghum bicolor*. 1. Drought avoidance mechanisms related to leaf water stress. *Can. J. Plant Sci.* **58**, 213–224.

Taylor, H. M., and Klepper, B. (1978). The role of rooting characteristics in the supply of water to plants. *Adv. Agron.* **30**, 99–128.

Thames, J. L. (1963). Needle variation in loblolly pine from four geographic sources. *Ecology* **44**, 168–169.

Thomas, D. A., and Barber, H. N. (1974). Studies on leaf characteristics of a cline of *Eucalyptus urnigera* from Mount Wellington, Tasmania. II. Reflection, transmission and absorption of radiation. *Aust. J. Bot.* **22**, 701–708.

Thomas, E., King, P. J., and Potrykus, I. (1979). Improvement of crop plants via single cells *in vitro*—an assessment. *Z. Pflanzenzüecht.* **82**, 1–30.

Tobiessen, P. L. (1971). Temperature and drought stress adaptations of desert and coastal populations of *Isomeris arborea*. Ph.D. Thesis, Duke University, Durham, North Carolina; *Diss. Abstr. Int. B* **32**, 1507 (1971).

Tomati, U., and Galli, E. (1979). Water stress and SH-dependent physiological activities in young maize plants. *J. Exp. Bot.* **30**, 557–563.

Townsend, A. M., and Roberts, B. R. (1973). Effect of moisture stress on red maple seedlings from different seed sources. *Can. J. Bot.* **51**, 1989–1995.

Tyree, M. T. (1976). Physical parameters of the soil-plant-atmosphere system: breeding for drought resistance characteristics that might improve wood yield. *In* "Tree Physiology and Yield Improvement" (M. G. R. Cannell and F. T. Last, eds.), pp. 348–359. Academic Press, New York.

Tyree, M. T., Cheung, Y. N. S., MacGregor, M. E., and Talbot, A. J. B. (1978). The characteristics of seasonal and ontogenetic changes in tissue-water relations of *Acer, Populus, Tsuga,* and *Picea. Can. J. Bot.* **56**, 635–647.

Unterschuetz, P., Reutz, W. F., Geppert, R. R., and Ferrell, W. K. (1974). The effect of age, preconditioning and water stress on transpiration rates of Douglas-fir (*Pseudotsuga menziesii*) seedlings of several ecotypes. *Physiol. Plant.* **32**, 214–221.

Van Buijtenen, J. P., Bilan, M. V., and Zimmermann, R. H. (1976). Morpho-physiological characteristics related to drought resistance in *Pinus taeda* L. *In* "Tree Physiology and Yield Improvement" (M. G. R. Cannell and F. T. Last, eds.), pp. 348–359. Academic Press, New York.

Venator, C. R. (1976). Natural selection for drought resistance in *Pinus caribaea* Morelet. *Turrialba* **26**, 381–387.

Wells, C. O., and Wakeley, P. C. (1966). Geographic variation in survival, growth, and fusiform rust infection of planted loblolly pine. *For. Sci. Monogr.* **11**, 40 p.

Winstead, J. E. (1978). Tracheid length as an ecotypic character in *Acer negundo* L. *Am. J. Bot.* **65**, 811–812.

Winton, L. L. (1970). Shoot and tree production from aspen tissue cultures. *Am. J. Bot.* **57**, 904–909.

Winton, L. L. (1971). Tissue culture propagation of European aspen. *For. Sci.* **17**, 348–350.

Wilton, L. L., and Huhtinen, O. (1976). Tissue culture of trees. *In* "Modern Methods in Forest Genetics" (J. P. Miksche, ed.), pp. 243–264. Springer-Verlag, Berlin and New York.

Woesner, R. A. (1972a). Crossing among loblolly pines indigenous to different areas as a means of genetic improvement. *Silvae Genet.* **21**, 35–39.

Woesner, R. A. (1972b). Growth patterns of one-year-old loblolly pine seed sources and inter-provenance crosses under contrasting edaphic conditions. *For. Sci.* **18**, 205–210.

Wright, J. W., and Bull, W. T. (1963). Geographic variation in Scotch pine: results of a 3-year Michigan study. *Silvae Genet.* **12**, 1–25.

Wright, N. L., and Jordan, G. L. (1970). Artificial selection for seedling drought tolerance in Boer lovegrass (*Eragrostis curvala* Nees). *Crop. Sci.* **10**, 99–102.

Yevjevich, V., Hall, W. A., and Salas, J. D. (1978). "Drought Research Needs," Water Resour. Publ., Fort Collins, Colorado.

Ying, C. C., and Bagley, W. T. (1976). Genetic variation of eastern cottonwood in an eastern Nebraska provenance study. *Silvae Genet.* **25**, 67–73.

Ying, C. C., and Bagley, W. T. (1978). Variation in rooting capability of *Populus deltoides*. *Silvae Genet.* **26**, 204–207.

Youngman, A. L. (1965). An ecotypic differentiation approach to the study of isolated populations of *Pinus taeda* in south central Texas. Ph.D. Thesis, University of Texas, Austin; *Diss. Abstr. Int. B* **27**, 3006 (1967).

Zavitkovski, J., and Ferrell, W. K. (1968). Effect of drought upon rates of photosynthesis, respiration, and transpiration of seedlings of two ecotypes of Douglas-fir. *Bot. Gaz. (Chicago)* **129**, 346–350.

Zavitkovski, J., and Ferrell, W. K. (1970). Effect of drought upon the rate of photosynthesis, respiration, and transpiration of seedlings of two ecotypes of Douglas-fir. II. Two-year-old seedlings. *Photosynthetica* **4**, 58–67.

Zobel, B. J., and Goddard, R. E. (1955). "Preliminary Results on Tests of Drought Hardy Strains of Loblolly Pine (*Pinus taeda* L.)," Tex. For. Serv., Res. Note 14. Texas Forest Service, College Station.

Zobel, B. J., and Kellison, R. C. (1978). The rate of growth syndrome. *Silvae Genet.* **27**, 123–124.

Zobel, R. W. (1975). The genetics of root development. *In* "The Development and Function of Roots: Third Cabot Symposium" (J. G. Torrey and D. T. Clarkson, eds.), pp. 261–275. Academic Press, New York.

AUTHOR INDEX

Numbers in italics refer to the pages on which the complete references are listed.

549

SUBJECT INDEX

A

ABA, *see* Abscisic acid
Abscisic acid, 75, 126, 172–176, 186, 187, 350, 448, 453, 459
Abscission, 155, 195
  acorns, 182
  branches, 164, 192
  flower buds, 378
  flowers, 422
  fruits, 187, 352, 378, 396, 422, 453, 458
  leaves, 157, 164, 187, 192, 266, 267, 291, 292, 296, 297, 328, 355, 374, 378, 388, 459, 481, 504, 513, 525, 540
Absorption
  atmospheric moisture, 4–19, 116, 213, 215
  carbon dioxide, 305, 335, 430, *see also* Photosynthesis
  heat, 247
  mineral nutrients, 26, 181, 230, 290, 338, 339, 341, 344, 458
  radiation, 63, 191, 386, 537
  water, 15–19, 24–27, 94, 95, 112, 159, 161, 164–169, 177, 181, 189–191, 215, 223–245, 305, 327, 338, 339, 342, 344, 354, 366, 378, 385–391, 443, 515, 535, 538
Accessory cells, 337
Acclimation, 365, 366, 370, 371, 374, *see also* Adaptation
Acid fuchsin, 84, 117, 118
Activation energy, 123
Adaptation, 19, 448, 458, 459, 512, 518, 520, 525, 528, 535, 538, *see also* Acclimation
Adenosine triphosphate, 184
Advection, 327, 389, 504
Aeration, 20, 24, 27, 165, 168, 226, 239, 242, 260, 341, 345, 376, 377, 402, 439, *see also* Anaerobiosis, Flooding
Aerial roots, 226, 227

Aging, 105, 165, 282, 376
Air blockage, *see* Embolism
Air embolism, *see* Embolism
Air layering, 227
Air pollution, *see* Pollution
Albedo, 386
Alcohols, 329
Alleles, 541
Allometric constant, 289
Aluminum, 169
Anaerobiosis, 169, *see also* Aeration, Flooding
Anatomy
  fruits, 334–337
  leaves, 164, 270–273, 327, 328, 522–525
  roots, 27–30, 344, 345
  stems, 337, 338
  stomata, 528–535
Antitranspirants, 336, 378, 448, 453, 454
Apoplast, 26, 112, 300, 302, 303, 443
Apple orchards, 419–469
Aquifer, 253, 391
Ascent of sap, 125
Aspiration of pits, 77–79, 114, 256
Atacama desert, 8
Atmometer, 260
Atmospheric moisture, 4–19, 116, 213–215, *see also* Dew, Fog
ATP, *see* Adenosine triphosphate
Auxin, 75, 187
Available water, 98–100

B

Bacteria, 76, 82
Basal area, 59, 86–89, 128, 129, 157, 194, 195, 219–221
Benzidene, 83
Biennial bearing, 380
Bitter pit, 454, 455